Lecture Notes in Physics

Volume 834

For further volumes:
http://www.springer.com/series/5304

The Lecture Notes in Physics

The series Lecture Notes in Physics (LNP), founded in 1969, reports new developments in physics research and teaching—quickly and informally, but with a high quality and the explicit aim to summarize and communicate current knowledge in an accessible way. Books published in this series are conceived as bridging material between advanced graduate textbooks and the forefront of research and to serve three purposes:

- to be a compact and modern up-to-date source of reference on a well-defined topic
- to serve as an accessible introduction to the field to postgraduate students and nonspecialist researchers from related areas
- to be a source of advanced teaching material for specialized seminars, courses and schools

Both monographs and multi-author volumes will be considered for publication. Edited volumes should, however, consist of a very limited number of contributions only. Proceedings will not be considered for LNP.

Volumes published in LNP are disseminated both in print and in electronic formats, the electronic archive being available at springerlink.com. The series content is indexed, abstracted and referenced by many abstracting and information services, bibliographic networks, subscription agencies, library networks, and consortia.

Proposals should be sent to a member of the Editorial Board, or directly to the managing editor at Springer:

Christian Caron
Springer Heidelberg
Physics Editorial Department I
Tiergartenstrasse 17
69121 Heidelberg/Germany
christian.caron@springer.com

Diego Dalvit · Peter Milonni · David Roberts
Felipe Rosa

Editors

Casimir Physics

 Springer

Dr. Diego Dalvit
Theoretical Division
Los Alamos National Laboratory
Mail Stop B213
Los Alamos, NM
USA
e-mail: dalvit@lanl.gov

Dr. Peter Milonni
Theoretical Division
Los Alamos National Laboratory
Mail Stop B213
Los Alamos, NM
USA
e-mail: pwm@lanl.gov

Dr. David Roberts
Theoretical Division
Los Alamos National Laboratory
Mail Stop B213
Los Alamos, NM
USA
e-mail: dcroberts1@gmail.com

Dr. Felipe Rosa
Theoretical Division
Los Alamos National Laboratory
Mail Stop B213
Los Alamos, NM
USA
e-mail: siqueira79@gmail.com

ISSN 0075-8450   e-ISSN 1616-6361
ISBN 978-3-642-20287-2   e-ISBN 978-3-642-20288-9
DOI 10.1007/978-3-642-20288-9
Springer Heidelberg Dordrecht London New York

Cover design: eStudio Calamar, Berlin/Figueres

Printed on acid-free paper

Springer is part of Springer Science+Business Media (www.springer.com)

Preface

In 1948 Hendrik Casimir published a paper showing that the existence of electromagnetic zero-point energy implies that there is an attractive force between two uncharged, perfectly conducting, parallel plates. Evgeny Lifshitz in 1955 generalized this theory to the case of dielectric media and finite temperatures. Over the next forty years experiments demonstrated the reality of Casimir forces, while a relatively small number of theoretical papers extended the analyses of Casimir and Lifshitz and explored other aspects of zero-point energies and fluctuations of quantum fields. It was not until the 1990s, however, that these forces were measured unambiguously and found to be in good agreement with predicted values. There followed a rapid growth of interest and research in Casimir physics.

Casimir effects serve as primary examples of directly observable manifestations of the nontrivial properties of quantum fields, and as such are attracting increasing interest from quantum field theorists, particle physicists, and cosmologists. Though very weak except at short distances, Casimir forces are universal in the sense that all material objects are subject to them. They are an increasingly important part of the physics of atom-surface interactions, while in nanotechnology they are being investigated not only as contributors to stiction but also as potential mechanisms for the actuation of micro-electromechanical devices. Analyses of such effects and their potential applications involve theoretical and computational electromagnetism, atomic, molecular and optical physics, and material science, among other specialties.

While the field of Casimir physics is expanding rapidly, it appears to have reached a certain level of maturity in some important respects. This is especially true on the experimental side, where it seems that the main sources of imprecision in force measurements have been identified. Another important achievement has been the development of semi-analytical and numerical methods for the computation of Casimir forces between bodies of practically arbitrary shape. There has also been significant progress in the basic theory of Casimir and related effects, including quantum levitation, quantum friction, and dynamical Casimir effects.

In light of these developments, and with no end yet in sight to the broad-based interest in Casimir physics, we felt that a book consisting of chapters written by internationally recognized leaders in the field would be both timely and of lasting value. The seed idea for this book was a workshop on *New Frontiers in Casimir Force Control* organized by us in Santa Fe, New Mexico, in September 2009. The chapters that follow are approximately evenly divided with regard to theory and experiment and deal mainly, though not exclusively, with surface-surface and atom-surface Casimir effects. Most chapters include a review of a particular aspect of Casimir physics in addition to a detailed presentation of the authors' current research and their perspective on possible future developments in the field. All the chapters include extensive bibliographies.

This volume is not intended to be a unified textbook, but rather a collection of mainly independent chapters written by prominent experts in the field. The ordering of chapters is only by topic and not by degree of depth or specialization. Therefore, the reading order is not at all prescribed by the ordering of the chapters.

We thank all the authors for taking valuable time from their research in order to present detailed and carefully written articles in a style that should appeal to other researchers in the field as well as to a broader audience. Thanks also go to Christian Caron and Gabriele Hakuba at Springer for their editorial support.

Las Alamos, May 2011

D. A. R. Dalvit
P. W. Milonni
D. C. Roberts
F. S. S. Rosa

Contents

Chapter 1
Introduction

Diego A. R. Dalvit, Peter W. Milonni, David C. Roberts
and Felipe S. S. Rosa

1.1 How the Casimir Force was Discovered

Casimir forces are associated with topological constraints on quantum fields. The
most famous such effect was predicted in 1948 by Casimir [1], who found that
there is an attractive force

$$F = -\frac{\pi^2 \hbar c}{240 d^4} \tag{1.1}$$

per unit area between two parallel, uncharged, perfectly conducting plates separated by a distance d. Casimir derived this force as a consequence of the change in
the (infinite) zero-point electromagnetic field energy due to the presence of the
plates. Lifshitz [2], taking a more general approach based on electromagnetic
fluctuations in thermal equilibrium, obtained Casimir's result as the perfect-conductor limit of the force between two dielectric half-spaces separated by a vacuum.
While generally very weak except at very small separations, Casimir effects are of
great interest for both theoretical and practical reasons. Their very existence stands
in contradiction to the prediction of classical electrodynamics that there should
be no force, as is evident from the appearance of $\hbar$ in (1.1). More
practical reasons for the recent interest in Casimir effects are their implications for

D. A. R. Dalvit (✉) · P. W. Milonni · D. C. Roberts · F. S. S. Rosa
Theoretical Division MS B213, Los Alamos National Laboratory,
Los Alamos, NM 87545, USA
e-mail: dalvit@lanl.gov

P. W. Milonni
e-mail: pwm@lanl.gov

D. C. Roberts
e-mail: dcroberts1@gmail.com

F. S. S. Rosa
e-mail: siqueira79@gmail.com

D. Dalvit et al. (eds.), *Casimir Physics*, Lecture Notes in Physics 834,
DOI: 10.1007/978-3-642-20288-9_1, © Springer-Verlag Berlin Heidelberg 2011

micro-electromechanical (MEMS) devices and other systems in which material components are in close proximity.

Casimir was led to his celebrated formula by some rather down-to-earth considerations on the stability of colloids. In this sense the subject can be traced back to the 1873 doctoral thesis of a struggling thirty-six-year-old graduate student, Johannes van der Waals, who many years later (1910) would be awarded the Nobel Prize in Physics. Van der Waals originally suggested a molecular interaction potential of the form $V(r) = -Ar^{-1}e^{-Br}$, where r is the distance between the molecules and A and B are positive constants. It is now well understood, of course, that there are different types of van der Waals forces. The simplest to understand is the orientational (or Keesom) interaction between two molecules with permanent electric dipole moments $p_1, p_2 : V(r) = -p_1^2 p_2^2/3kTr^6$ at temperature T. The force is attractive because attractive orientations are energetically favored over repulsive ones. There is also an induction (or Debye) interaction between two molecules, one of which has a permanent dipole (or quadrupole) moment. Neither of these interactions is sufficiently general to account for the van der Waals equation of state, which requires an attractive interaction even if the molecules have no permanent electric (or magnetic) moments. For this purpose what is required is a third type of van der Waals interaction, first derived by London [3]. For two identical molecules with polarizability $\alpha(\omega)$, this *dispersion force* is

$$V(r) = -\frac{3\hbar}{\pi r^6} \int\limits_0^\infty d\xi \alpha^2(i\xi) \equiv -\frac{C}{r^6}. \tag{1.2}$$

It is called a "dispersion" force because of the appearance of the molecular polarizability, which determines the refractive index $n(\omega)$ via the relation $n^2(\omega) = 1 + 4\pi N\alpha(\omega)$ for a sufficiently dilute medium of N molecules per unit volume. It is important to note that $\alpha(i\xi)$ is a real number, which follows from the Kramers–Kronig relations between the real and imaginary parts of $\alpha(\omega)$.

The dispersion force is, of course, important in many contexts [4], one of them having to do with the stability of certain colloids, such as oil-in-water emulsions, in which interactions between the suspended particles and the molecules of the medium are negligible. Such a colloid is said to be stable if there is no coagulation. An example of a stable colloidal suspension is homogenized milk, in which the suspended particles are fat globules. Homogenization breaks up the globules into pieces sufficiently small that, in addition to other effects, the attractive forces between them are too weak to turn the milk into cream.

Colloidal stability requires that the repulsion between the suspended particles be greater than the attractive dispersion force. Repulsion results from electrostatic surface charges caused by adsorption of electrolyte ions in the liquid medium. In the old but not yet retired Derjaguin-Landau-Verwey-Overbeek (DVLO) theory the repulsive potential energy between two colloidal particles is calculated as the energy required to remove the ions, and the repulsive force is compared to the dispersion force between the particles. If the dispersion force is larger than the repulsive force, the colloid is unstable.

In the DLVO theory it is assumed that the attractive (van der Waals) interaction varies as the inverse sixth power of the distance r between molecules, and that this force is pairwise additive. However, based on their experimental work, Verwey and Overbeek [5] wrote that, "In applying the theory of the attractive forces ... we met with some difficulties ... because the London theory as such is not relativistically invariant, and by working out this idea we found that ... a relativistic correction may well become important." According to them the finite speed of light should cause the intermolecular van der Waals interaction to fall off more rapidly with separation r than r^{-6}. Following the suggestion of Verwey and Overbeek, their colleagues at the Philips Laboratory in Eindhoven, Casimir and Polder [6], revisited the London calculation to include retardation, or in other words to include $\exp(i\mathbf{k} \cdot \mathbf{r})$ in the matrix elements appearing in perturbation theory. Their calculation showed that, for intermolecular separations larger than about 137 Bohr radii typically, the interaction energy between two identical molecules with static (zero-frequency) polarizability α is

$$V(r) = -\frac{23\hbar c}{4\pi r^7}\alpha^2, \qquad (1.3)$$

as opposed to (1.2). It turned out, according to Verwey and Overbeek [5], that this result was "in fair accord" with what they concluded from their experiments.

Another feature of the van der Waals dispersion force is that it is *nonadditive*, as was noted early on by Langmuir [7]. This nonadditivity emerges clearly in the many-body theory summarized below, but it does not appear to have played any significant role in the original DLVO theory.

The simple form of (1.3) led Casimir, after a suggestion by Niels Bohr, to interpret it in terms of zero-point energy: "[Bohr] mumbled something about zero-point energy. That was all, but it put me on a new track ..." [8]. This in turn led him to derive the force (1) as another example of an effect attributable to zero-point energy.

1.2 Zero-Point Energy

The concept of zero-point energy seems to have first appeared in Planck's "second theory" of blackbody radiation, and soon after it played a role in some of Einstein's work [9]. In a paper by Einstein and Stern in 1913 it was noted that, without zero-point energy, the average energy of a harmonic oscillator of frequency ω is

$$E = \frac{\hbar\omega}{e^{\hbar\omega/kT} - 1} \cong k_B T - \frac{1}{2}\hbar\omega \qquad (1.4)$$

if the thermal equilibrium temperature T satisfies $k_B T \gg \hbar\omega$. To satisfy the equipartition theorem to first order in $\hbar$ in this classical limit we must include the zero-point energy $\frac{1}{2}\hbar\omega$. From such considerations Einstein and Stern concluded that "the existence of zero-point energy $[\frac{1}{2}\hbar\omega]$ is probable". A bit later, however, Einstein wrote to Ehrenfest that zero-point energy is "dead as a doornail" [10].

Experimental evidence for zero-point energy was reported by Mulliken in 1924 [11]. Consider an absorptive vibronic transition in a diatomic molecule A in which the vibrational quantum numbers of the upper and lower states of the transition are v' and v'', respectively. The transition frequency is approximately

$$v_A(v',v'') = v_e + \omega'_e\left(v' + \frac{1}{2}\right) - \omega''_e\left(v'' + \frac{1}{2}\right) \tag{1.5}$$

if anharmonic corrections are small. Here v_e is the electronic transition frequency and ω'_e and ω''_e are the vibrational frequencies of the two electronic states. The zero-point vibrational energies of the upper and lower states are then $\frac{1}{2}\hbar\omega'_e$ and $\frac{1}{2}\hbar\omega''_e$, respectively. Now consider a second diatomic molecule B that differs only isotopically from A; for this molecule, similarly, the transition frequency is

$$v_B(v',v'') = v_e + \rho\omega'_e\left(v' + \frac{1}{2}\right) - \rho\omega''_e\left(v'' + \frac{1}{2}\right), \tag{1.6}$$

where $\rho = \sqrt{m_A/m_B}$ and m_A, m_B are the reduced masses of the two molecules. For the 0–0 bands,

$$v_B(0,0) - v_A(0,0) = \frac{1}{2}(\rho - 1)(\omega'_e - \omega''_e), \tag{1.7}$$

which is nonzero only because of the zero-point vibrational energies of the two molecules. Such an isotopic displacement was observed by Mulliken in the 0-0 bands for $B^{10}O^{16}$ and $B^{11}O^{16}$. He concluded that "it is then probable that the minimum vibrational energy of BO (and doubtless other) molecules is 1/2 quantum".

Of course one can cite other evidence for zero-point energy. For example, because of their small masses, He^3 and He^4 do not solidify at small pressures as $T \rightarrow 0$ because their zero-point motion prevents crystallization. Many years ago Debye noted that the zero-point translational energy of the atoms of a crystal lattice causes a reduction in the intensity of radiation in X-ray diffraction even as the temperature approaches absolute zero. In more recent years the zero-point momentum distribution of atoms in Bose-Einstein condensates has been found to have the smallest width consistent with the Heisenberg uncertainty relation [12].

One of the most frequently cited implications of zero-point electromagnetic energy is the Lamb shift, or more specifically the dominant contribution to the Lamb shift in atomic hydrogen. An argument originally due to Feynman goes as follows [13]. Imagine we have a box of volume V containing N identical atoms per unit volume. The frequencies of the allowed field modes in the box are changed from their vacuum values ω to $\omega/n(\omega)$, where $n(\omega) \cong 1 + 2\pi N\alpha(\omega)$ is the refractive index of the (dilute) gas. The change in the zero-point field energy in the box due to the presence of the gas is therefore

$$\Delta E = \sum_{\mathbf{k},\lambda} \left[\frac{1}{2} \frac{\hbar\omega_k}{n(\omega_k)} - \frac{1}{2} \hbar\omega_k \right] \cong - \sum_{\mathbf{k},\lambda} [n(\omega_k) - 1] \frac{1}{2} \hbar\omega_k = -\pi\hbar N \sum_{\mathbf{k},\lambda} \omega_k \alpha(\omega_k),$$

(1.8)

where the $\mathbf{k}$'s and λ's are the mode wave vectors and polarization labels, respectively. For a large box we can replace the discrete summation over modes by an integration: $\sum_{\mathbf{k},\lambda} \to (V/8\pi^3) \sum_\lambda \int d^3 k$. Then, in the limit of a single atom ($NV \to 1$),

$$\Delta E = -\frac{\hbar}{\pi c^3} \int d\omega\omega^3 \alpha(\omega).$$

(1.9)

Subtracting the free-electron energy, and introducing a high-frequency cutoff $mc^2/\hbar$ in this nonrelativistic approach, one obtains Bethe's approximation to the Lamb shift of a one-electron atom in a state with polarizability $\alpha(\omega)$ [14]. For the $2s_{1/2} - 2p_{1/2}$ Lamb shift in hydrogen, Bethe computed 1040 MHz, in good agreement with the measured shift of about 1058 MHz. Thus the largest part of this Lamb shift is attributable to the change in the zero-point field energy due to the mere presence of the atom.

We can think of this a little differently by first recalling that the energy involved in inducing an electric dipole moment $\mathbf{d} = \alpha\mathbf{E}$ in an electric field $\mathbf{E}$ is

$$W = -(1/2)\alpha\mathbf{E}^2.$$

(1.10)

For an atom in a state with polarizability $\alpha(\omega)$ in a field of frequency ω, this energy is just the quadratic ac Stark shift of the state. Now for an atom in vacuum there is a continuum of field frequencies, and we obtain the energy shift due to the zero-point field by integrating over all frequencies. In the integrand in this case we set $(1/4\pi)\mathbf{E}^2$ equal to $\rho_0(\omega)d\omega$, where $\rho_0(\omega)d\omega$ is the zero-point field energy per unit volume in the frequency interval $[\omega, \omega + d\omega]$. Then

$$W = -\frac{1}{2} \int_0^\infty \alpha(\omega) 4\pi\rho_0(\omega)d\omega,$$

(1.11)

which is the same as (1.9), since

$$\rho_0(\omega) = \frac{\omega^2}{\pi^2 c^3} \left(\frac{1}{2}\hbar\omega \right).$$

(1.12)

In other words, the Lamb shift can be regarded as a Stark shift caused by the vacuum electric field.

The same sort of argument can be used to obtain the van der Waals dispersion interaction between two polarizable particles. Thus, for atom A, there is a contribution of the form (1.10) from each frequency ω of the field, and this contribution depends on the polarizability $\alpha_A(\omega)$ of atom A. The field $\mathbf{E}(\mathbf{r}_A, t)$ acting on atom A is the vacuum field $\mathbf{E}_0(\mathbf{r}_A, t)$ at $\mathbf{r}_A$ plus the field $\mathbf{E}_B(\mathbf{r}_A, t)$ at $\mathbf{r}_A$ due to atom B. The latter field is just the field from the electric dipole induced in B by the vacuum field at $\mathbf{r}_B$, and this brings in the polarizability $\alpha_B(\omega)$ of atom B. We thus obtain for atom A (and likewise for atom B) an energy that depends in part on $r = |\mathbf{r}_A - \mathbf{r}_B|$; this r-dependent interaction energy has the general form

$$V(r) \propto -\frac{\hbar}{c^3 r^3} \int_0^\infty d\omega \omega^3 \alpha_A(\omega) \alpha_B(\omega) G(\omega r/c). \tag{1.13}$$

The detailed functional form of $G(\omega r/c)$, which is an oscillatory function of $\omega r/c$, need not concern us, as (1.13) as it stands is sufficient to determine the form of the interaction at large separations: for large r we expect substantial cancellations due to the oscillatory nature of $G(\omega r/c)$, and on this basis we expect to obtain the correct r-dependence of $V(r)$ by cutting off the upper limit of integration at a frequency $\sim c/r$. Then [15]

$$V(r) \propto -\frac{\hbar}{c^3 r^3} \alpha^2 \int_0^{c/r} d\omega \omega^3 G(\omega r/c) = -\frac{\hbar c}{r^7} \alpha^2 \int_0^1 dx x^3 G(x), \tag{1.14}$$

where we have assumed the two atoms to be identical and replaced their polarizability $\alpha(\omega)$ by the static polarizability α under the assumption that c/r is much smaller than the frequency of any transition that makes a significant contribution to $\alpha(\omega)$. Thus we obtain in this way the correct r dependence of the retarded van der Waals interaction.

We can derive the *exact* form of the van der Waals interaction *for all r* by this approach based on zero-point field energy. Neither molecule in the dispersion interaction has a permanent dipole moment; each has a fluctuating dipole moment *induced by the vacuum field* at its position, and from this perspective there is a nonvanishing force because the vacuum field correlation function $\langle E_i(\mathbf{r}_A, t) E_j(\mathbf{r}_B, t) \rangle \neq 0$ [14]. That is, the dipole moments induced by the fluctuating zero-point electric field are correlated over finite differences, leading to a nonvanishing expectation value of the intermolecular interaction energy. In a similar fashion we can obtain, for instance, the van der Waals interaction for magnetically polarizable particles [16].

Another effect of retardation considered by Casimir and Polder [6] is the potential energy of an atom in the vicinity of a perfectly conducting wall. For short distances d between the atom and the wall the potential $V(d)$ may be deduced from the electric dipole-dipole interaction between the atom and its image in the wall, and obviously varies as $1/d^3$. At large distances, however, retardation becomes important and, as in the interatomic van der Waals interaction, the interaction is reduced by a factor 1/(distance), i.e., $V(d) \propto 1/d^4$.

Here again the simple formula (1.10) provides a way to a simple derivation of this result of Casimir and Polder. The Stark shift (1.10) for an atom in a state with polarizability $\alpha(\omega)$ in this example becomes

$$W = -\frac{1}{2} \sum_{\mathbf{k},\lambda} \alpha(\omega) \mathbf{E}^2_{\mathbf{k},\lambda}(\mathbf{x}_A), \tag{1.15}$$

where the summation is over all modes, $\mathbf{k}$ and λ again denoting wave vectors and polarizations, and $\mathbf{E}_{\mathbf{k},\lambda}(\mathbf{x}_A)$ is the zero-point electric field from mode $(\mathbf{k}, \lambda)$ at the position $\mathbf{x}_A$ of the atom. In the half-space in which the atom is located the mode functions are determined, of course, by Maxwell's equations and the boundary conditions on the fields. Consider first a rectangular parallelepiped with sides of length $L_x = L_y = L$ and L_z, its surfaces being assumed to be perfectly conducting. The zero-point electric field $\mathbf{E}$ inside the parallelepiped can be expanded in terms of a complete set of mode functions such that the the the boundary conditions are satisfied and

$$\frac{1}{4\pi} \int d^3r \mathbf{E}^2(\mathbf{r}) = \sum_{\mathbf{k},\lambda} \frac{1}{2} \hbar\omega_k. \tag{1.16}$$

The Cartesian components of $\mathbf{E}$ for each mode $(\mathbf{k}, \mathbf{e}_{\mathbf{k},\lambda})$ are

$$E_x(\mathbf{r}) = \left(\frac{16\pi\hbar\omega}{V}\right)^{1/2} e_x \cos(k_x x) \sin(k_y y) \sin(k_z z),$$

$$E_y(\mathbf{r}) = \left(\frac{16\pi\hbar\omega}{V}\right)^{1/2} e_y \sin(k_x x) \cos(k_y y) \sin(k_z z), \tag{1.17}$$

$$E_z(\mathbf{r}) = \left(\frac{16\pi\hbar\omega}{V}\right)^{1/2} e_z \sin(k_x x) \sin(k_y y) \cos(k_z z),$$

where $e_x^2 + e_y^2 + e_z^2 = 1$, the volume $V = L^2 L_z$, and

$$k_x = \frac{\ell\pi}{L}, \quad k_y = \frac{m\pi}{L}, \quad k_z = \frac{n\pi}{L_z}. \tag{1.18}$$

All positive integers and zero are allowed for ℓ, m, and n. As in the derivation of (1.14) we replace $\alpha(\omega)$ by the static polarizability $\alpha = \alpha(0)$ in (1.15) and write, for an atom at $(L/2, L/2, d)$,

$$
\begin{aligned}
W(d) = -\frac{1}{2}\alpha \sum_{\mathbf{k}} \frac{16\pi\omega_k}{V} \Bigg[&e_x^2 \cos^2\left(\frac{1}{2}k_x L\right) \sin^2\left(\frac{1}{2}k_y L\right) \sin^2(k_z d) \\
&+ e_y^2 \sin^2\left(\frac{1}{2}k_x L\right) \cos^2\left(\frac{1}{2}k_y L\right) \sin^2\left(\frac{1}{2}k_z d\right) \\
&+ e_z^2 \sin^2\left(\frac{1}{2}k_x L\right) \sin^2\left(\frac{1}{2}k_y L\right) \cos^2\left(\frac{1}{2}k_z d\right) \Bigg].
\end{aligned}
\tag{1.19}
$$

The squares of the sines and cosines involving k_x and k_y are rapidly varying and can be replaced by their average value, 1/2, so that (1.19) is replaced by

$$W(d) = -\left(\frac{2\pi\hbar\alpha}{V}\right)\sum_{\mathbf{k}}\omega_k\left[(e_x^2 + e_y^2)\sin^2(k_z d) + e_z^2\cos^2(k_z d)\right]. \qquad (1.20)$$

In the limit $d \to \infty$ in which the atom is infinitely far away from the wall at $z = 0$ we can also replace $\sin^2(k_z d)$ and $\cos^2(k_z d)$ by their averages:

$$W(\infty) = -\left(\frac{2\pi\hbar\alpha}{V}\right)\sum_{\mathbf{k}}\omega_k\left[\frac{1}{2}(e_x^2 + e_y^2 + e_z^2)\right] = -\left(\frac{2\pi\hbar\alpha}{V}\right)\sum_{\mathbf{k}}\frac{1}{2}\omega_k. \qquad (1.21)$$

We define the interaction energy as

$$
\begin{aligned}
V(d) = W(d) - W(\infty) &= -\left(\frac{2\pi\hbar\alpha}{V}\right)\sum_{\mathbf{k}}\omega_k[e_x^2 + e_y^2 - e_z^2]\left[\sin^2(k_z d) - \frac{1}{2}\right]\\
&= \left(\frac{\pi\hbar\alpha}{V}\right)\sum_{\mathbf{k}}\omega_k\cos(2k_z d) \times (2k_z^2/k^2),
\end{aligned}
\qquad (1.22)
$$

where we have used the fact that $\mathbf{e}_{\mathbf{k}\lambda}^2 = 1$ and $\mathbf{k}\cdot\mathbf{e}_{\mathbf{k}\lambda} = 0$ for each mode. $V(d)$ is easily evaluated when we use the fact that there is a continuum of allowed $\mathbf{k}$ vectors and that $\omega_k = kc$:

$$
\begin{aligned}
V(d) &= \left(\frac{2\pi\hbar\alpha}{V}\right)\frac{V}{8\pi^3}\int d^3k\,\frac{k_z^2}{k^2}\cos(2k_z d)\\
&= \left(\frac{\alpha\hbar c}{2\pi}\right)\int_0^\infty dk k^3\int_0^{2\pi}d\theta\sin\theta\cos^2\theta\cos(2kd\cos\theta)\\
&= -\frac{3\alpha\hbar c}{8\pi d^4},
\end{aligned}
\qquad (1.23)
$$

which is exactly the result of Casimir and Polder [6].

We cannot invoke directly such arguments based on (1.10) in the case of the Casimir force (1.1), but, as Casimir showed, we can obtain the force by calculating the change in the zero-point field energy when the plates are separated by a distance d compared to when they are infinitely far apart. Since Casimir's calculation is reproduced in various ways in many other places, we will here present only a simple, heuristic derivation.

Recall that there are $(\omega^2/\pi^2 c^3)d\omega$ modes in the (angular) frequency interval $[\omega, \omega + d\omega]$, and that each mode has a zero-point energy $\hbar\omega/2$. Between the plates the components of the mode vectors perpendicular to the plates are restricted to values $<\pi/d$. There is no such restriction on modes propagating in the two directions parallel to the plates, so we might guess that, to obtain approximately the zero-point field energy in a volume Ad between the plates, we can use the free-space energy density with a lower bound $\beta\pi c/d$, where $\beta \sim 1/3$:

$$E = Ad \int\limits_{\pi c \beta / d}^{\infty} d\omega \left[\frac{1}{2} \hbar\omega \times \frac{\omega^2}{\pi^2 c^3} \right] = Ad \int\limits_{0}^{\infty} d\omega \frac{\hbar\omega^3}{2\pi^2 c^3} - Ad \int\limits_{0}^{\pi c \beta / d} d\omega \frac{\hbar\omega^3}{2\pi^2 c^3}. \quad (1.24)$$

Ignoring the first term in the second equality, which is a bulk-volume contribution and would imply an infinite force, independent of d, we define the potential energy

$$V(d) = -Ad \int\limits_{0}^{\pi c \beta / d} d\omega \frac{\hbar\omega^3}{2\pi^2 c^3} = -Ad \frac{\hbar}{8\pi^2 c^3} \left(\frac{\pi c \beta}{d} \right)^4. \quad (1.25)$$

As noted, β should be around $1/3$; let us take $\beta = 0.325 = (1/90)^{1/4}$. Then

$$V(d) = -A \frac{\pi^2 \hbar c}{720 d^3}, \quad (1.26)$$

or $F(d) = -V'(d)/A = -\pi^2 \hbar c / 240 d^4$, which is Casimir's result (1.1). Of course a more serious derivation should also take account of the effects of zero-point fields in the regions outside the two plates, but in the present example we can ignore them [17].

1.3 The Lifshitz Theory and Its Generalizations

In his generalization of Casimir's theory for perfect conductors at zero temperature, Lifshitz [2] considered two dielectric half-spaces separated by a vacuum region of width d and allowed for finite (equilibrium) temperatures. He calculated the force between the dielectrics using the macroscopic Maxwell equations and the stress tensor in the vacuum region, assuming a noise polarization consistent with the fluctuation-dissipation theorem. Lifshitz's approach employs stochastic fields, but is equivalent to a formulation based on operator-valued fields [18]. The (positive-frequency) Fourier transform $\hat{\mathbf{E}}(\mathbf{r}, \omega)$ of the electric field operator $\hat{\mathbf{E}}(\mathbf{r}, t)$ satisfies

$$-\nabla \times \nabla \times \hat{\mathbf{E}}(\mathbf{r}, \omega) + \frac{\omega^2}{c^2} \varepsilon(\mathbf{r}, \omega) \hat{\mathbf{E}}(\mathbf{r}, \omega) = -\frac{\omega^2}{c^2} \hat{\mathbf{K}}(\mathbf{r}, \omega), \quad (1.27)$$

with the noise polarization operator $\hat{\mathbf{K}}(\mathbf{r}, \omega)$ having the properties

$$\langle \hat{\mathbf{K}}_i^\dagger(\mathbf{r}, \omega) \hat{\mathbf{K}}_j(\mathbf{r}', \omega') \rangle = 4\hbar\varepsilon_I(\omega)\delta_{ij}\delta(\omega - \omega')\delta^3(\mathbf{r} - \mathbf{r}') \frac{1}{e^{\hbar\omega/k_B T} - 1}, \quad (1.28)$$

$$\langle \hat{\mathbf{K}}_i(\mathbf{r}, \omega) \hat{\mathbf{K}}_j^\dagger(\mathbf{r}', \omega') \rangle = 4\hbar\varepsilon_I(\omega)\delta_{ij}\delta(\omega - \omega')\delta^3(\mathbf{r} - \mathbf{r}') \left[\frac{1}{e^{\hbar\omega/k_B T} - 1} + 1 \right], \quad (1.29)$$

where $\varepsilon_I(\omega)$ is the imaginary part of the permittivity $\varepsilon(\omega)$. For simplicity we restrict ourselves here to zero temperature, so that expectation values will refer to the ground state of the matter-field system rather than a finite-temperature thermal equilibrium state.

We first outline a derivation based on a stress tensor that leads to a rather general expression (1.39) from which Lifshitz's formula for the force is obtained. Recall first that in classical electromagnetic theory it follows from the macroscopic Maxwell equations that the force density in a dielectric medium in which there are electric and magnetic fields, but no free charges or currents, has components $f_i = \partial_j T_{ij}$, where the stress tensor

$$T_{ij} = \frac{1}{4\pi}\left[E_i D_j + H_i H_j - \frac{1}{2}(\mathbf{E}\cdot\mathbf{D} + \mathbf{H}\cdot\mathbf{H})\delta_{ij}\right]. \tag{1.30}$$

Then

$$f_i = \frac{1}{8\pi}[(\partial_i E_j)D_j - E_j(\partial_i D_j)] \tag{1.31}$$

when it is assumed that $(\partial/\partial t)(\mathbf{D}\times\mathbf{H})/4\pi c$, the rate of change of the Minkowski expression for the momentum density of the field, can be taken to be zero, as is appropriate under equilibrium conditions. We are also assuming isotropic media, in which case the Minkowski stress tensor (1.30) is symmetric.

When the field is quantized we replace E_j and D_j in (1.31) by operators, symmetrize, and take expectation values:

$$f_i = \frac{1}{8\pi}\text{Re}\left[\langle(\partial_i \hat{E}_j)\hat{D}_j\rangle - \langle\hat{E}_j(\partial_i \hat{D}_j)\rangle\right]. \tag{1.32}$$

In terms of the Fourier transforms of the electric and displacement fields,

$$f_i(\mathbf{r}) = \frac{1}{8\pi}\text{Re}\int\limits_{-\infty}^{\infty} d\omega \int\limits_{-\infty}^{\infty} d\omega'\left\{\langle[\partial_i \hat{E}_j(\mathbf{r},\omega)]\hat{D}_j(\mathbf{r},\omega')\rangle - \langle\hat{E}_j(\mathbf{r},\omega)[\partial_i \hat{D}_j(\mathbf{r},\omega')]\rangle\right\}. \tag{1.33}$$

Since $\hat{\mathbf{D}}(\mathbf{r},\omega) = \epsilon(\mathbf{r},\omega)\hat{\mathbf{E}}(\mathbf{r},\omega) + \hat{\mathbf{K}}(\mathbf{r},\omega)$,

$$\langle\hat{E}_j(\mathbf{r},\omega)\hat{D}_j(\mathbf{r}',\omega')\rangle = \epsilon(\mathbf{r}',\omega')\langle\hat{E}_j(\mathbf{r},\omega)\hat{E}_j(\mathbf{r}',\omega')\rangle + \langle\hat{E}_j(\mathbf{r},\omega)\hat{K}_j(\mathbf{r}',\omega')\rangle. \tag{1.34}$$

In the second term on the right we use

$$\hat{E}_j(\mathbf{r},\omega) = \frac{1}{4\pi}\int d^3r'' G_{ji}(\mathbf{r},\mathbf{r}'',\omega)\hat{K}_i(\mathbf{r}'',\omega), \tag{1.35}$$

where G is the dyadic Green function satisfying

$$-\nabla \times \nabla \times G(\mathbf{r}, \mathbf{r}', \omega) + \frac{\omega^2}{c^2}\varepsilon(\mathbf{r}, \omega)G(\mathbf{r}, \mathbf{r}', \omega) = -4\pi\frac{\omega^2}{c^2}\delta^3(\mathbf{r} - \mathbf{r}'), \quad (1.36)$$

while in the first term [19],

$$\langle \hat{E}_j(\mathbf{r}, \omega)\hat{E}_j(\mathbf{r}', \omega')\rangle = \frac{\hbar}{\pi}\mathrm{Im}G_{jj}(\mathbf{r}, \mathbf{r}', \omega)\delta(\omega - \omega'). \quad (1.37)$$

It follows from the (zero-temperature) fluctuation-dissipation relations (1.28) and (1.29) that

$$\mathrm{Re}\langle \hat{E}_j(\mathbf{r}, \omega)\hat{D}_j(\mathbf{r}', \omega')\rangle = \frac{\hbar}{\pi}\mathrm{Im}[\varepsilon(\mathbf{r}', \omega)G_{jj}(\mathbf{r}, \mathbf{r}', \omega)]\delta(\omega - \omega'), \quad (1.38)$$

and, from (1.33),

$$f_i(\mathbf{r}) = -\frac{\hbar}{8\pi^2}\mathrm{Im}\int_0^\infty d\omega[\partial_i\varepsilon(\mathbf{r}, \omega)]G_{jj}(\mathbf{r}, \mathbf{r}, \omega). \quad (1.39)$$

Equation (1.39) gives the force density due to (non-additive) intermolecular van der Waals dispersion interactions. Its evaluation requires, however, a calculation of the (classical) dyadic Green function, which is generally very hard to do analytically except in a few highly symmetric configurations. In the case of two dielectric half-spaces, for example, (1.39) has been used to derive the Lifshitz expression for the force per unit area between two dielectric half-spaces, and to generalize the original Lifshitz formula to allow for a third dielectric material between the half-spaces [20–23].

Dzyaloshinskii et al. [20–22] obtained (1.39) in a diagrammatic approach leading to a free energy $\mathscr{E}$ having the form of a summation over all n-body van der Waals interactions:

$$\mathscr{E} = -\frac{\hbar}{2\pi}\mathrm{Im}\int_0^\infty d\omega\mathrm{Tr}[\alpha G^0 + \frac{1}{2}\alpha^2 G^0 G^0 + \frac{1}{3}\alpha^3 G^0 G^0 G^0 + \cdots]. \quad (1.40)$$

Here G^0 is the free-space Green function satisfying (1.36) with $\varepsilon(\omega) = 1$ for all frequencies ω, and $\alpha(\omega)$ is again the complex, frequency-dependent polarizability of each particle. The first term in brackets is a single-particle self-energy, while the other terms correspond successively to two-body, three-body, etc. van der Waals interactions. The second term, for example, written out explicitly as

$$\mathrm{Tr}\left[\frac{1}{2}\alpha^2 G^0 G^0\right] = \sum_{m,n=1}^{\mathscr{N}} \alpha^2 G^0_{ij}(\mathbf{r}_n, \mathbf{r}_m, \omega)G^0_{ji}(\mathbf{r}_m, \mathbf{r}_n, \omega), \quad (1.41)$$

gives, excluding terms with $m = n$, the sum of pairwise van der Waals interaction energies of $\mathscr{N}$ atoms for arbitrary interparticle separations $|\mathbf{r}_n - \mathbf{r}_m|$. The third term in brackets in (1.40), similarly, can be shown to yield the three-body van der Waals interaction obtained by Axilrod and Teller [24]. In the model in which the atoms form a continuous medium, equation (1.41), for instance, is replaced by

$$\int d^3r \int d^3r' N(\mathbf{r})N(\mathbf{r}')\alpha^2 G_{ij}^0(\mathbf{r},\mathbf{r}',\omega)G_{ji}^0(\mathbf{r}',\mathbf{r},\omega) \equiv \mathrm{Tr}\left[\frac{\varepsilon-1}{4\pi}\right]^2 [G^0]^2,$$

where $N(\mathbf{r})$ is the number of atoms per unit volume at $\mathbf{r}$ and the formula $\varepsilon(\omega) = 1 + 4\pi N\alpha(\omega)$ has been used to relate the permittivity and the polarizability. Similarly (1.40) is replaced by

$$\mathcal{E} = -\frac{\hbar}{2\pi}\mathrm{Im}\int_0^\infty d\omega\,\mathrm{Tr}\sum_{n=1}^\infty \frac{1}{n}\left[\frac{\varepsilon-1}{4\pi}\right]^n [G^0]^n \tag{1.42}$$

in the continuum approximation.

Since the free-space Green dyadic G^0 is independent of any properties of the medium, a variation $\delta\mathcal{E}$ of $\mathcal{E}$ can depend only on a variation $\delta\varepsilon$ of ε:

$$\delta\mathcal{E} = -\frac{\hbar}{8\pi^2}\mathrm{Im}\int_0^\infty d\omega\,\delta\epsilon\,\mathrm{Tr}\sum_{n=0}^\infty \left[\frac{\varepsilon-1}{4\pi}\right]^n [G^0]^{n+1}. \tag{1.43}$$

From (1.36) we obtain the identity

$$G = G^0 + G^0\left[\frac{\varepsilon-1}{4\pi}\right]G, \tag{1.44}$$

which allows us to rewrite (1.43) as

$$\delta\mathcal{E} = -\frac{\hbar}{8\pi^2}\mathrm{Im}\int_0^\infty d\omega \int d^3r\,\delta\varepsilon(\mathbf{r},\omega)G_{ii}(\mathbf{r},\mathbf{r},\omega). \tag{1.45}$$

Now an infinitesimal local transport $\mathbf{u}(\mathbf{r})$ of the medium implies a variation in ε at $\mathbf{r}$ such that $\varepsilon(\mathbf{r},\omega) + \delta\varepsilon(\mathbf{r},\omega) = \varepsilon(\mathbf{r}-\mathbf{u},\omega)$, or $\delta\varepsilon(\mathbf{r},\omega) = -\nabla\varepsilon\cdot\mathbf{u}$. Therefore

$$\delta\mathcal{E} = -\frac{\hbar}{8\pi^2}\mathrm{Im}\int_0^\infty d\omega[-\nabla\varepsilon\cdot\mathbf{u}]G_{ii}(\mathbf{r},\mathbf{r},\omega) = -\int d^3r\mathbf{f}(\mathbf{r})\cdot\mathbf{u}, \tag{1.46}$$

where the Casimir force density $\mathbf{f}(\mathbf{r})$ is defined by (1.39). The quantum-field-theoretic approach of Dzyaloshinskii et al. [20–22] thus establishes the connection between the non-additive, many-body van der Waals interactions and macroscopic quantum-electrodynamical approaches such as the stress-tensor formulation outlined above.

Another approach to the calculation of Casimir forces between dielectrics, based directly on considerations of zero-point energy, was taken van Kampen et al. [25]

and others in a rederivation of Lifshitz's formula for the force between two dielectric half-spaces. In this approach Maxwell's equations and the appropriate boundary conditions lead to a requirement that the allowed (nontrivial) modes satisfy $F_\beta(\omega, d) = 0$, where the F_β are meromorphic functions of frequency. The total zero-point energy associated with all the allowed modes, i.e., $\hbar/2$ times the sum of the zeros of the functions F_β, is then obtained from a generalization of the argument theorem for meromorphic functions, and from this one can obtain the Lifshitz formula.

A fundamental assumption made by Lifshitz is that the dielectric media are well described by continua characterized by dielectric permittivities. This allows the sum of all the many-body van der Waals forces to be obtained via the *macroscopic* Maxwell equations [26]. This assumption for material media is made, usually implicitly, in practically all theories of Casimir forces, and it appears to be an excellent approximation under practical experimental circumstances. One of the most accurate tests of the Lifshitz formula for the force between dielectric half-spaces was conducted many years ago by Sabisky and Anderson [27]. Using measured data and an analytic fit for the dielectric constants, they compared the predictions of the Lifshitz formula for the variation of the vapor pressure with film thickness of liquid helium, and reported agreement to within a few per cent between theory and experiment.

The formula for the force $F(d)$ per unit area obtained by Lifshitz [2] can be generalized not only to include a material medium between the half-spaces but also to allow all three media to be magnetodielectric:

$$
F(d) = -\frac{\hbar}{2\pi^2} \int_0^\infty dk k \int_0^\infty d\xi K_3 \left(\left[\frac{(\varepsilon_3 K_1 + \varepsilon_1 K_3)(\varepsilon_3 K_2 + \varepsilon_2 K_3)}{(\varepsilon_3 K_1 - \varepsilon_1 K_3)(\varepsilon_3 K_2 - \varepsilon_2 K_3)} e^{2K_3 d} - 1 \right]^{-1} \right.
$$
$$
\left. + \left[\frac{(\mu_3 K_1 + \mu_1 K_3)(\mu_3 K_2 + \mu_2 K_3)}{(\mu_3 K_1 - \mu_1 K_3)(\mu_3 K_2 - \mu_2 K_3)} e^{2K_3 d} - 1 \right]^{-1} \right), \tag{1.47}
$$

where ε_j and μ_j are now respectively the electric permittivity and the magnetic permeability of medium $j(=1, 2, 3)$ evaluated at imaginary frequencies: $\varepsilon_j = \varepsilon_j(i\xi)$, $\mu_j = \mu_j(i\xi)$, and $K_j^2 = k^2 + \varepsilon_j \mu_j \xi^2/c^2$. The terms depending explicitly on ε_j and μ_j in this formula have the form of Fresnel reflection coefficients. In fact in a more computationally useful "scattering" approach, Casimir forces between objects in the case of more general geometries are expressed in terms of the scattering matrices [28, 29]. For example, for media in which only specular reflection occurs, the Casimir force per unit area in the special case of the Lifshitz geometry with vacuum between the half-spaces is given by

$$
F(d) = -\frac{\hbar}{4\pi^3} \int_0^\infty d\xi \int d^2 k K_3 \text{Tr} \frac{\mathbf{R}_1 \cdot \mathbf{R}_2 e^{-2K_3 d}}{1 - \mathbf{R}_1 \cdot \mathbf{R}_2 e^{-2K_3 d}}. \tag{1.48}
$$

$\mathbf{R}_1$ and $\mathbf{R}_2$ are 2×2 reflection matrices for *generally anisotropic* magnetodi-electric media and $K_3^2 = k^2 + \xi^2/c^2$.

1.4 Overview of Experiments

The Casimir force is typically very small, even in the case of perfectly conducting plates. For two 1×1 cm plates separated by 1 μm, for example, equation (1.1) predicts a force of 0.013 dyne. This is comparable to the Coulomb force on the electron in the hydrogen atom, or to the gravitational force between two one-pound weights separated by half an inch, or to about 1/1000 of the weight of a housefly. Not surprisingly, therefore, it took quite a few years before it was unambiguously measured. Here we briefly review some of the earlier experiments. More detailed discussions of some of these experiments, as well as more recent work, can be found in the following chapters.

The earliest experiments measuring van der Waals forces between macroscopic objects were performed by Derjaguin et al. in the 1950s and earlier [30]. The alignment difficulties in experiments with parallel plates led this group to work instead with a spherical lens and a plate. It was shown that the force between a sphere of radius R and a flat surface a distance d away is approximately

$$\mathscr{F}(d) = 2\pi R u(d), \tag{1.49}$$

where $u(d)$ is the interaction energy per unit area between two flat, parallel surfaces separated by d. This approximation, valid when $d \ll R$, has come to be known as the Derjaguin or "proximity force approximation (PFA)", and although it is derived under the assumption of pairwise additive forces between local surface elements, it has been an important and surprisingly accurate tool in the comparison of measured forces with theories for perfectly flat, parallel surfaces [31]. Discussions and references to theoretical analyses of the range of validity of the PFA may be found in several of the following chapters. Measurements of the forces between a flat surface and spheres of different radii have demonstrated that the PFA is fairly accurate for values of d and d / R in the range of many experiments [32].

Experiments of Derjaguin et al. for separations of 0.1–0.4 μm between quartz plates provided the stimulus for Lifshitz's theory [30]. The comparison with the Lifshitz theory was complicated in part by incomplete information about $\varepsilon(i\xi)$ for quartz over all its different absorption regions. On the basis of rough estimates, Lifshitz concluded that "the agreement between the theory and the experimental data is satisfactory" [2], although in his paper he provided few details. Forces measured much later were found to "fit well" with the Lifshitz theory in the retarded regime of large separations, but detailed comparisons of theory and experiments were again hampered by insufficient data for the permittivities required in the Lifshitz formula.

Sparnaay [33] reviewed experimental progress up to 1989, and some of the rapid progress, both theoretical and experimental, in more recent times has been the subject of reviews and special issues of journals that are extensively cited in the following chapters. In Sparnaay's own early experiments [34, 35] the force was inferred from the deflection of a spring attached via an aluminum rod to one of the metal plates. For plate separations of 2–10 μm, Sparnaay measured attractive forces per unit area of magnitude between about 1 and 4×10^{-18} dyn cm$^2/d^4$, compared with the 1.3×10^{-18} dyn cm$^2/d^4$ calculated from Casimir's formula (1.1). Prior to more recent work, Sparnaay's experiments were often cited as the first experimental verification of the Casimir force, but Sparnaay himself concluded only that his observed forces "do not contradict Casimir's theoretical prediction".

Experiments reported by Lamoreaux [36, 37] and Mohideen et al. [38–40] in 1997 and 1998 marked the beginning of a new era of much more precise measurements of Casimir forces. Lamoreaux performed Cavendish-type experiments employing torsion balances, whereas Mohideen's group pioneered the use of an atomic force microscope (AFM) system wherein the Casimir force is determined from the reflection of a laser beam off the AFM cantilever tip and its displacement as measured by photodiodes. The agreement of these experiments with theory appears to be on the order of perhaps several per cent. This great improvement in accuracy compared to earlier experiments stems in part from the availability of much better mechanical, surface characterization, and data acquisition tools. The measurements of Lamoreaux and Mohideen et al. also avoided the problem of stiction associated with previous balance mechanisms.

Another very significant step in the direction of improved accuracy of Casimir force measurements was reported by Chan et al. [41, 42] in 2001, who measured the Casimir force based on the shift in the frequency of a periodically driven MEMS torsional oscillator (MTO) or on the torque exerted on a plate by a metallic sphere. Two years later Decca et al [43] reported the first highly accurate measurement of the Casimir force between two dissimilar metals based on an MTO system.

De Man et al. [44, 45] have described AFM Casimir-force measurements that allow for continuous calibration together with compensation of residual electrostatic effects. Experiments measuring the force in air between a gold-coated sphere and a glass surface coated with either gold or a conductive oxide demonstrated that residual electrostatic effects could be effectively eliminated, while the Casimir force was substantially reduced, as expected, when the gold surface was replaced by the transparent oxide film. These authors have also introduced a "fiber-top cantilever" design that could allow Casimir force measurements under environmental conditions that would preclude measurements employing existing laboratory instrumentation.

Substantial progress has also been made in the measurement of the Casimir-Polder force. A particularly noteworthy experiment was that of Sukenik et al. [46] in which the opacity of a sodium beam passing through a gold-plated channel was measured. Assuming that a sodium atom, hitting a gold surface as a consequence of the Casimir-Polder attractive force, sticks without bouncing, Sukenik et al. found excellent agreement between their data and the atom-surface interaction predicted by Casimir and Polder.

This brief overview of experimental progress is hardly meant to be exhaustive. In the remaining chapters there are $\sim 10^3$ citations to the literature on Casimir effects, a significant portion of it addressing observed Casimir effects and related experimental issues.

1.5 Some Other Aspects of Casimir Forces

The following chapters provide detailed discussions of many of the most interesting theoretical as well as experimental aspects of Casimir effects. As expected from their relation to the ubiquitous van der Waals forces, Casimir effects are involved directly or indirectly in a wide variety of physical phenomena, and it is probably impossible to address all aspects of them in any detail in a single volume. Here we briefly mention a few related topics of current interest that are either not dealt with or only briefly alluded to in this book.

Most of the work on van der Waals–Casimir-Lifshitz forces between material bodies has dealt with perfect conductors or dielectric media. If we take $\varepsilon_1 = \varepsilon_2 \to \infty$ and $\varepsilon_3 = 1$ for all frequencies, we obtain from the Lifshitz formula (1.48) the attractive Casimir force (1.1) for two perfectly conducting parallel plates separated by a vacuum. For identical dielectric half-spaces separated by a vacuum, the force is likewise always attractive. If the dielectric media are different, however, the force can be *repulsive* under certain conditions, e.g., if $\varepsilon_1 > \varepsilon_3 > \varepsilon_2$ over a sufficiently wide range of frequencies. Such a repulsion was predicted by Dzyaloshinksii et al. [20–22] and can be applied, for instance, to explain the tendency of liquid helium to climb the walls of its container: the helium vapor is repelled by the walls, and the helium liquid moves in to fill the space left by the vapor. The repulsive force resulting from regions of different permittivity appears to be well known among colloid scientists [47], and has recently been directly observed by Munday et al. [48], who measured a repulsive force between a gold sphere and a glass surface immersed in bromobenzene.

In recent years it has been recognized that stiction due to attractive Casimir forces should be taken into account in the design of nanomechanical devices. There has consequently been a growing interest in the control of Casimir forces and in particular in the possibility of realizing repulsive Casimir forces between vacuum-separated objects. This in turn has led to the consideration of metamaterials that might be engineered to have permittivities and permeabilities that would allow a degree of control of Casimir forces that would not be possible with naturally occurring materials. An indication of how magnetic properties of materials might be used to control Casimir forces can be seen by taking

$\varepsilon_1 = \mu_2 \to \infty$ and $\varepsilon_3 = 1$ for all frequencies, in which case we obtain from (1.48) a repulsive force with a magnitude of 7/8 times the magnitude of the Casimir force, as first shown in a different way by Boyer [49]. In a similar vein it was shown by Feinberg and Sucher [50] that the van der Waals interaction between two atoms, one electrically polarizable and the other magnetically polarizable, is repulsive.

Computations based on the Lifshitz theory and formulas for permittivities and permeabilities of some existing metamaterials suggest, however, that repulsion will be difficult to realize [51]. In fact Rahi et al. [52] have shown under some rather general conditions of physical interest that Casimir interactions cannot produce a stable equilibrium for any collection of dielectric objects whenever all their permittivities are larger or smaller than the permittivity of the medium in which they are immersed. In the case of metamaterials their arguments indicate that repulsive forces between two objects might only be possible at short distances, in which case a metamaterial surface cannot be assumed to have continuous translational symmetry. Even in this case, however, a repulsion cannot result in a stable equilibrium.

Casimir forces are notoriously difficult to compute for arbitrary geometries. The best-known paradigm here is the Casimir force on a perfectly conducting spherical shell, which Casimir assumed would be attractive. As first shown by Boyer [53] after "a long nightmare of classical special function theory", however, the force in this case is repulsive. In addition to the chapters in this volume addressing numerical approaches to Casimir-force computations, we call attention here to the work of Schaden [54], who has used a semiclassical approach to compute, among other things, the force on a spherical conducting spherical shell to an accuracy within 1% of the exact expression.

In various extensions of the Standard Model of elementary particles there appear non-gravitational, long-range forces between electrically neutral bodies. Experimental constraints on such forces can be determined by Casimir as well as gravitational experiments, and the rapid progress in high-precision Casimir experiments has contributed to an increased interest in this area. We refer the reader to the last chapter of the book by Bordag et al. [38–40] for a discussion of this topic and references to recent work.

Casimir effects are almost always interpreted in terms of zero-point field energy. It is, however, possible to interpret Casimir effects without reference to zero-point field energy. For instance, the Lifshitz formula can be derived from Schwinger's source theory in which "the vacuum is regarded as truly a state with all physical properties equal to zero" [23]. In conventional quantum electrodynamics Casimir forces can also be obtained without explicit reference to vacuum field fluctuations, although of course such fluctuations are implicit in the theory [55]. In a different approach Jaffe [56] has also shown that Casimir effects can be calculated "without reference to zero-point energies", and suggests that approaches based on zero-point energy "won out" because they are much simpler.

Finally we note that there are classical analogs of the Casimir force. A thermodynamic analog occurs in the case of two surfaces in a binary liquid mixture that is close to the critical point [57–59]. The analog of the electromagnetic field fluctuations between plates in the Casimir effect is in this case the fluctuations in the concentrations in the liquid between the two surfaces. The force acting on the surfaces can be attractive or repulsive, depending on whether the adsorptive characteristics of the two surfaces are similar or dissimilar. Both attraction and repulsion in this "critical Casimir effect" have been observed in experiments measuring forces of the same order of magnitude as the forces observed in (quantum) Casimir-force experiments, and the measured forces confirmed theoretical predictions.

1.6 Brief Outline of this Book

The ordering of chapters in this volume is only by topic and not by degree of depth or specialization. The first five chapters are concerned primarily with the theory of surface-surface Casimir effects. Pitaevskii discusses the problems that were encountered in the generalization of the original Lifshitz theory to the case of forces within dielectric media, especially in connection with a general formulation of the stress tensor. He reviews how the desired generalization was finally achieved by many-body diagrammatic methods applicable under conditions of thermal equilibrium. The important role of mechanical equilibrium and pressure gradients in the theory of dielectric bodies separated by a fluid is emphasized, and a physical interpretation is given for the repulsive forces predicted by the generalized Lifshitz theory for dielectrics. The chapter by Milton focuses on the nature of the divergences incurred in calculations of Casimir self-energies, such as that for a conducting spherical shell, based on the model of a massless scalar field and delta-function potentials. He distinguishes between the global divergence associated with the total zero-point energy and the local divergences occurring near boundaries; a unique and finite self-energy is obtained after isolating the global divergence. In the case of parallel plates it is shown that both the finite Casimir self-energy and the divergent self-energies of the plates are consistent with the equivalence principle, and therefore that the divergent self-energies in particular can be absorbed into the masses of the plates. This suggests a general "renormalization" procedure for absorbing divergent self-energies into the properties of boundaries. The following three chapters describe techniques for the computation of Casimir effects for arbitrarily shaped objects. Lambrecht, Canaguier-Durand, Guérout and Reynaud, and Rahi, Emig and Jaffe, focus on the recently developed scattering theory approach and provide valuable introductions to methods they have developed. Lambrecht et al describe computations for several different configurations and make interesting comparisons with the predictions of the PFA, while Rahi et al consider a different set of examples and also address in detail the constraints on stable equilibria presented in Reference [52]. Johnson reviews a wide variety of methods in classical computational electromagnetism that can be applied to the evaluation of various Casimir effects, and offers illuminating perspectives on the strengths and weaknesses of each approach.

   The next four chapters describe various experimental aspects of surface-surface Casimir forces. Lamoreaux reviews some recent experimental progress and then discusses, among other things, some of the implicit approximations that have "wittingly or unwittingly" been adopted in all Casimir force experiments, including his own seminal experiments. These include the PFA and various assumptions about electrostatic calibrations and contact and patch potentials. He cautions against confusing precision and accuracy in Casimir force measurements. Capasso, Munday, and Chan discuss high-precision measurements in MEMS, including applications to nonlinear oscillators for position measurements at the nanoscale, and then describe their experiments demonstrating a reduction in the magnitude of Casimir forces by different effects. They also discuss their experiments on repulsive Casimir forces [41, 42] and possibilities for realizing quantum levitation and Casimir torques between birefringent materials. Decca, Aksyuk, and López discuss advantages of using MEMS for the measurement of Casimir forces; as mentioned earlier, these devices have been used to measure Casimir forces between metallic objects in vacuum to a very high degree of accuracy [43]. Decca et al also discuss electrostatic calibration and other matters involved in Casimir force experiments, and describe potential opto-mechanical experiments that might allow further improvements in the precision of Casimir force measurements. Van Zwol, Svetovoy, and Palasantzas address the requirement, for any comparison of theory and experiment, of having accurate values of optical permittivities; they emphasize that the permittivities of conducting films, which are needed for all frequencies in the Lifshitz theory, are not always reliably given by tabulated data, since they can vary significantly from sample to sample, depending on how the samples are prepared. They describe determinations of the complex permittivities of gold films over a wide range of frequencies. Van Zwol et al. also discuss the characterization of surface roughness by imaging methods, and how such images can be used to characterize surface roughness and to calculate the correction to the surface-surface Casimir force due to it. They discuss the importance of the "distance upon contact" between two rough surfaces and its importance in the determination of their absolute separation and therefore of the Casimir force.

   Atom-surface and dynamical Casimir effects are the subject of the remaining three chapters. Intravaia, Henkel, and Antezza review the theory of the Casimir-Polder interaction and some recent developments relating, among other things, to non-equilibrium systems and experiments with ultracold atoms. They also discuss frictional effects on moving atoms in blackbody fields and near surfaces. De Kieviet, Jentschura, and Lach review the experimental status of the Casimir-Polder force. They discuss other effects that might play a role in future experiments and review their work on the quantum reflection of ground-state atoms from solid surfaces and on the atomic beam spin echo technique. They present new experimental data on the reflectivity of ^{3}He atoms from a gold surface and compare the data with predictions of the Casimir-Polder and non-retarded van der Waals atom-surface interactions. Their approach makes it possible to resolve very detailed features of the atom-surface potential. Finally Dalvit, Maia Neto, and Mazzitelli review the theory of dynamical Casimir effects as well as frictional forces associated with electromagnetic field fluctuations, and discuss possible experimental scenarios for the observation of such effects.

References

1. Casimir, H.B.G.: On the attraction between two perfectly conducting plates. Proc. K. Ned. Akad. Wet. **51**, 793 (1948)
2. Lifshitz, E.M.: The theory of molecular attractive forces between solids. Sov. Phys. JETP **2**, 73 (1956)
3. London, F.: Theory and system of molecular forces. Z. Phys. **63**, 245 (1930)
4. See, for instance, V. A. Parsegian, *Van der Waals Forces: A Handbook for Biologists, Chemists, Engineers, and Physicists* (Cambridge University Press, N. Y., 2006).
5. Verwey, E.J.W., Overbeek, J.T.G.: Theory of the Stability of Lyophobic Colloids. Elsevier, Amsterdam (1948)
6. Casimir, H.B.G., Polder, D.: The influence of retardation on the London-van der Waals forces. Phys. Rev. **73**, 360 (1948)
7. Langmuir, I.: Role of attractive and repulsive forces in formation of tactoids, thixotropic gels, protein crystals and coacervates. J. Chem. Phys. **6**, 873 (1938)
8. Casimir, H.B.G.: Communication to P.W. Milonni, 12 March (1992)
9. Some of Einstein's work related to zero-point energy is reviewed in P.W. Milonni, *The Quantum Vacuum. An Introduction to Quantum Electrodynamics* (Academic, San Diego, 1994).
10. Gearhart, C.A.: 'Astonishing successes' and 'bitter disappointment': The specific heat of hydrogen in quantum theory. Arch. Hist. Exact Sci. **64**, 113 (2010)
11. Mulliken, R.S.: The band spectrum of boron monoxide. Nature **114**, 349 (1924)
12. Stenger, J., Inouye, S., Chikkatur, A.P., Stamper-Kurn, D.M., Pritchard, D.E., Ketterle, W.: Bragg spectroscopy of a Bose-Einstein condensate. Phys. Rev. Lett. **82**, 4569 (1999)
13. See, for instance, Milonni, P.W., Schaden, M., Spruch, L.: The Lamb shift of an atom in a dielectric medium. Phys. Rev. A **59**, 4259 (1999) and references therein
14. See, for example, Reference [9] and references therein
15. Spruch, L.L., Kelsey, E.J.: Vacuum fluctuation and retardation effects in long-range potentials. Phys. Rev. A. **18**, 845 (1978)
16. Feinberg, G., Sucher, J.J., Au, C.K.: The dispersion theory of dispersion forces. Phys. Rep. **180**, 83 (1978)
17. Itzykson, C., Zuber, J.-B.: Quantum Field Theory, pp. 141. McGraw-Hill, N.Y. (1980)
18. Rosa, F.S.S., Dalvit, D.A.R., Milonni, P.W.: Electromagnetic energy, absorption, and Casimir forces: Uniform dielectric media in thermal equilibrium. Phys. Rev. A. **81**, 033812 (2010)
19. Rosa, F.S.S., Dalvit, D.A.R., Milonni, P.W.: to be submitted for publication. Expressions equivalent to (37) but with different choices for the definition of the Green dyadic may be found, for instance, in T. Gruner and D.-G. Welsch, Green-function approach to the radiation-field quantization for homogeneous and inhomogeneous Kramers-Kronig dielectrics. Phys. Rev. A **53**, 1818 (1996) and M.S. Tomas, Casimir force in absorbing monolayers. Phys. Rev. A **66**, 052103 (2002)
20. Dzyaloshinskii, I.E., Pitaevskii, L.P.: Van der Waals forces in an inhomogeneous dielectric. Sov. Phys. JETP **9**, 1282 (1959)
21. Dzyaloshinskii, I.E., Lifshitz, E.M., Pitaevskii, L.P.: The general theory of van der Waals forces. Adv. Phys. **10**, 165 (1961)

22. See also A.A. Abrikosov, L.P. Gorkov, I.E. Dzyaloshinskii, *Methods of Quantum Field Theory in Statistical Physics* (Dover, N.Y., 1975)
23. Schwinger, J., DeRaad, L.L. Jr., Milton, K.A.: Casimir effect in dielectrics. Ann. Phys. (N.Y.) **115**, 1 (1978)
24. Axilrod, B.M., Teller, E.: Interaction of the van der Waals type between three atoms. J.Chem. Phys. **11**, 299 (1943)
25. van Kampen, N.G., Nijboer, B.R.A., Schram, K.: On the macroscopic theory of van der Waals forces. Phys. Lett. **26**, 307 (1968)
26. See, for instance, Reference [9], Sect. 8.3.
27. Sabisky, E.S., Anderson, C.H.: Verification of the Lifshitz theory of the van der Waals potential using liquid-helium films. Phys. Rev. A **7**, 790 (1973)
28. Kats, E.I.: Influence of nonlocality effects on van der Waals interaction. Sov. Phys. JETP **46**, 109 (1977)
29. For a detailed review of the scattering approach see, for instance, the chapters by A. Lambrecht et al and S.J. Rahi et al in this volume
30. Derjaguin, B.V., Rabinovich, Y.I., Churaev, N.V.: Direct measurement of molecular forces. Nature **272**, 313 (1978) and references therein to related work of Derjaguin et al
31. It should be noted that Bressi et al managed to measure with about 15% precision the Casimir force between (nearly) parallel metallic plates: G. Bressi, G. Carugno, R. Onofrio, and G. Ruoso, "Measurement of the Casimir force between parallel metallic surfaces," Phys. Rev. Lett. **88**, 041804 (2002)
32. Krause, D.E., Decca, R.S., López, D., Fischbach, E.: Experimental investigation of the Casimir force beyond the proximity-force approximation. Phys. Rev. Lett. **98**, 050403 (2007)
33. Sparnaay, M.J.: The historical background of the Casimir effect. In: Sarlemijn, A., Sparnaay, M.J. (eds) Physics in the Making, Elsevier, Amsterdam (1989)
34. Sparnaay, M.J.: Attractive forces between flat plates. Nature **180**, 334 (1957)
35. Sparnaay, M.J: Measurements of attractive forces between flat plates. Physica **24**, 751 (1958)
36. Lamoreaux, S.: Demonstration of the Casimir force in the 0.6 to 6 μm range. Phys. Rev. Lett. **78**, 5 (1997)
37. See also S. Lamoreaux, The Casimir force: background, experiments, and applications. Rep. Prog. Phys. **68**, 201 (2005) and "Casimir forces: Still surprising after 60 years," Physics Today (February, 2007), 40-45
38. Mohideen, U., Roy, A.: Precision measurement of the Casimir force from 0.1 to 0.9 μm. Phys. Rev. Lett. **81**, 4549 (1998)
39. See also M. Bordag, U. Mohideen, V.M. Mostepanenko, "New developments in the Casimir effect," Phys. Rep. **353**, 1 (2001)
40. Bordag, M., Klimchitskaya, G.L., Mohideen, U., Mostepanenko, V.M.: Advances in the Casimir Effect. Oxford University Press, N.Y. (2009)
41. Chan, H.B., Aksyuk, V.A., Kleiman, R.N., Bishop, D.J., Capasso, F.: Quantum mechanical actuation of microelectromechanical systems by the Casimir force. Science **291**, 1941 (2001)
42. Chan, H.B., Aksyuk, V.A., Kleiman, R.N., Bishop, D.J., Capasso, F.: Nonlinear micromechanical Casimir oscillator. Phys. Rev. Lett. **87**, 211801 (2001)
43. Decca, R.S., López, D., Fischbach, E., Krause, D.E.: Measurement of the Casimir force between dissimilar metals. Phys. Rev. Lett. **91**, 504021 (2003)
44. de Man, S., HeeckK. Wijngaarden, R.J., Iannuzzi, D.: Halving the Casimir force with conductive oxides. Phys. Rev. Lett. **103**, 040402 (2009)
45. de Man, S., Heeck, K., Smith, K., Wijngaarden, R.J., Iannuzzi, D.: Casimir force measurements in air: two birds with one stone. Int. J. Mod. Phys. A **25**, 2231 (2010)
46. Sukenik, C.I., Boshier, M.G., Cho, D., Sandoghar, V., Hinds, E.A.: Measurement of the Casimir-Polder force. Phys. Rev. Lett. **70**, 560 (1993)
47. See for instance A.A. Feiler, L. Bergstrom, M.W. Rutland, "Superlubricity using repulsive van der Waals forces," Langmuir **24**, 2274 (2008), and references therein.
48. Munday, J.N., Capasso, F., Parsegian, V.A.: Measured long-range Casimir-Lifshitz forces. Nature **457**, 170 (2009)

49. Boyer, T.H.: Van der Waals forces and zero-point energy for dielectric and permeable materials. Phys. Rev. A **9**, 2078 (1974)
50. Feinberg, G., Sucher, J.: General theory of the van der Waals interaction: A model-independent approach. Phys. Rev. A **2**, 2395 (1970)
51. Rosa, F.S.S., Dalvit, D.A.R., Milonni, P.W.: Casimir-Lifshitz theory and metamaterials. Phys. Rev. Lett. **100**, 183602 (2008); "Casimir interactions for anisotropic magnetodielectric metamaterials", Phys. Rev. A **78**, 032117 (2008)
52. Rahi, S.J., Kardar, M., Emig, T.: Constraints on stable equilibria with fluctuation-induced (Casimir) forces. Phys. Rev. Lett. **105**, 070404 (2010)
53. Boyer, T.H.: Quantum zero-point energy and long-range forces. Ann. Phys. (N.Y.) **56**, 474 (1970)
54. Schaden, M.: Semiclassical estimates of electromagnetic Casimir self-energies of spherical and cylindrical metallic shells. Phys. Rev. A **82**, 022113 (2010)
55. See, for instance, Reference [9] and references therein.
56. Jaffe, R.L.: Casimir effect and the quantum vacuum. Phys. Rev. D. **72**, 021301(R) (2005)
57. Fisher, M.E., de Gennes, P.-G.: Wall phenomena in a critical binary mixture. C.R. Acad. Sci. Paris B. **287**, 209 (1978)
58. Garcia, R., Chan, M.H.W.: Critical Casimir effect near the ^{3}He-^{4}He tricritical point. Phys. Rev. Lett. **88**, 086101 (1978)
59. Hertlein, C., Helden, L., Gambassi, A., Dietrich, S., Bechinger, C.: Direct measurement of critical Casimir forces. Nature **451**, 172 (2008)

Chapter 2
On the Problem of van der Waals Forces in Dielectric Media

Lev P. Pitaevskii

Abstract A short review of the problems which arise in the generalization of the Lifshitz theory of van der Waals force in the case of forces inside dielectric media is presented, together with some historical remarks. General properties of the stress tensor of equilibrium electromagnetic field in media are discussed, and the importance of the conditions of mechanical equilibrium is stressed. The physical meaning of the repulsive van der Waals interaction between bodies immersed in a liquid is discussed.

2.1 Introduction

The quantum theory of the long range van der Waals interaction between neutral objects has a long and instructive history [1]. The existence of *attractive* forces between atoms was established by van der Waals on the basis of analysis of experimental data on equations of state of real gases.

It was F. London who understood the electric nature of these forces and built in 1930 a seminal quantum-mechanical theory of the forces at large distances, deriving the famous $1/R^6$ law for the energy of interaction [2].

The next important step was performed by Casimir and Polder. Using quantum electrodynamics, they derived a more general expression for the energy of the atom-atom interaction and showed that, due to retardation effects, London's law at

L. P. Pitaevskii (✉)
INO-CNR BEC Center and Dipartimento di Fisica, Università di Trento, Trento, 38123, Povo, Italy
e-mail: lev@science.unitn.it

L. P. Pitaevskii
Kapitza Institute for Physical Problems, Russian Academy of Sciences, ul. Kosygina 2, 119334, Moscow, Russia

D. Dalvit et al. (eds.), *Casimir Physics*, Lecture Notes in Physics 834,
DOI: 10.1007/978-3-642-20288-9_2, © Springer-Verlag Berlin Heidelberg 2011

large distances is changed to a $1/R^7$ law [3]. In the same paper the authors considered for the first time the problem of including a macroscopic body. They calculated a force between an atom and perfect metal plate. The interaction between perfect metal plates was calculated by Casimir [4].

The most general theory of the van der Waals interaction between any condensed bodies was developed by E. Lifshitz in 1954–1955 [5, 6]. His theory is applicable to a body with arbitrary dielectric properties at any temperature. It also automatically takes into account relativistic retardation effects. To calculate forces one must know the dielectric and magnetic permeabilities of the bodies and solve the Maxwell equations for their given configuration.

It was assumed in the Lifshitz theory (as well as in previous theories) that the space between the bodies was a vacuum. A generalization of the theory where the gap between bodies is filled with some medium was a natural next step. However, to make this step one must overcome some essential difficulties. To clarify the nature of these difficulties, let us recall the main points of the Lifshitz approach. This approach is based on averaging of the Maxwell stress tensor in vacuum,

$$\sigma_{ik}^{(vac)} = \frac{E_i E_k + B_i B_k}{4\pi} - \frac{E^2 + B^2}{8\pi} \delta_{ik}: \tag{2.1}$$

with respect to electromagnetic fluctuations in thermodynamic equilibrium.

Because the theory of equilibrium electromagnetic fluctuations in arbitrary media was already developed by Rytov [7] and Landau and Lifshitz [8], the tensor could be averaged for equilibrium conditions and the forces calculated.

The Rytov theory has a semi-phenomenological character. Rytov considered the fluctuations as created by the Langevin-like sources, namely fluctuating electric and magnetic polarizations. It was assumed that these polarizations at two points r_1 and r_2 of a medium are correlated only when the two points coincide, i.e., their correlation functions are proportional to $\delta(r_1 - r_2)$. The coefficients of proportionality were chosen to obtain the correct density of black-body radiation from the bodies. During preparation of the book [8], Landau and Lifshitz derived equations of the Rytov theory using the exact fluctuation-dissipation theorem, established by Callen and Welton in 1951.

It was natural to think that the generalization for the case of bodies separated by a medium could be obtained if one could find a general expression for the electromagnetic stress tensor for arbitrary time-dependent fields in a medium. Because the Rytov theory describes electromagnetic fluctuations also inside a medium, it would then be possible to perform the average of the tensor. Thus the first step was to calculate the tensor. Note that this problem was formulated in the book [8], and Landau believed that it could be solved. One can read in the end of Sect. 61:

"Considerable interest attaches to the determination of the (time) average stress tensor giving the forces on matter in a variable electromagnetic field. The problem is meaningful for both absorbing and non absorbing media, whereas that concerning the internal energy can be proposed only if absorption is neglected. The corresponding formulae, however, have not yet been derived."

I had an opportunity to read proofs of the book when I joined the Landau department in the Institute for Physical Problems in Moscow as a Ph.D. student in 1955. After reading this paragraph I decided to derive "the corresponding formulae" for the tensor. It was, of course, a quite ambitious goal. After approximately three months of work I met Igor Dzyaloshinskii. Because the authors asked both him and me to help in the proof-reading of Ref. [8], we naturally discussed the topics of the book and I discovered that Dzyaloshinskii had also been working on the tensor problem even longer than I. We decided to join our efforts.

Our attempts were based on the use of the second-order quantum mechanical perturbation theory to calculate the quadratic contribution of the fields into the tensor. One can obtain formal equations; however, for a medium with dissipation we could not express them in terms of dielectric and magnetic permeabilities. Oddly enough, one could easily obtain an equation for the *total* force acting on a body in vacuum. However, trying to obtain from this equation the *force density* inside the body, one necessarily violated the condition of the symmetry of the stress tensor, or other conditions implied for the tensor. We also tried to develop a thermodynamic approach. However, the entropy increase due to the dissipation makes this approach meaningless.

We worked long enough and finally decided that the problem was hopeless. (In any case it was my opinion.)

However, after a period of disappointment, I suddenly recognized that the problem of the van der Waals forces for the case of a liquid film can be solved without involving the general tensor problem. Because, as we will see below, these considerations actually were not employed and never were published, I would like to present them here.

Let us consider a liquid film of thickness d in vacuum. Its free energy per unit of area can be presented in the form

$$\mathscr{F}(T, d) = \varphi_0(T)d + \mathscr{F}^{(elm)}(T, d) \qquad (2.2)$$

where $\varphi_0(T)$ is the density of free energy for a bulk body and $\mathscr{F}^{(elm)}(T, d)$ is the contribution of the van der Waals forces, which can be normalized in such a way that $\mathscr{F}^{(elm)} \to 0$ at $d \to \infty$. The goal of the theory in this case is to calculate $\mathscr{F}^{(elm)}(T, d)$. However, this quantity can be calculated on the basis of the Lifshitz theory. Let us consider our film at the distance l from a half-space *of the same liquid* (see Fig. 2.1). The theory permits the calculation of the force $F(d,l)$ between the film and the half-space and the change of the free energy when the film moves from the surface of the bulk liquid to infinity:

$$\Delta\mathscr{F}(T,d) = \int\limits_0^\infty F dl. \qquad (2.3)$$

This quantity is just $\mathscr{F}^{(elm)}(T, d)$. Indeed, if the film is near the surface of the half space, the configuration corresponds to a bulk body. Actually, the integral

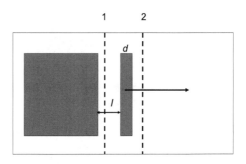

Fig. 2.1 Scheme for calculation of the chemical potential of a film in terms of the stress tensor in vacuum

(2.3) is divergent at small l. However, this infinite contribution does not depend on d and can be omitted altogether with a d-independent constant.

We discussed the idea with Landau and he agreed with the argumentation. We immediately began calculations of the force F for the three-boundary configuration of Fig. 2.1. E. M. Lifshitz, who was interested in the film problem very much, joined us. However, Landau cooled our ardor. He said that it is meaningless to solve a three boundary problem to find an answer for a two-boundary one and that the integration (2.3) must be performed in a general form, "in some symbolic way" as he said.

As a result, Dzyaloshinskii and I began to look for a different approach, being sure this time that the answer exists. It was a lucky coincidence that just in this period Dzyaloshinskii, in collaboration with Abrikosov and Gorkov, worked on developing the Matsubara diagram technique for the solution of *equilibrium* problems in the quantum many-body theory. We decided to try this approach. The first attempt was very successful. We immediately recognized which diagrams are important. As an intermediate result we obtained an equation where the van der Waals contribution to the free energy was expressed in terms of an integration with respect to the charge of the electron. This equation was correct, but not very useful because the dielectric permeability can, of course, be measured only for the actual value of the charge. Finally, however, we discovered that in the Matsubara technique *the variation* of the free energy with respect to the dielectric permeability can be expressed in terms of the Matsubara Green's function of imaginary frequency. This permitted us to calculate the tensor explicitly [9].

The reason why our previous attempts were in vain became clear now. One can obtain the tensor of the *equilibrium* electromagnetic field in a medium, that is, the tensor of van der Waals forces, but not the tensor of arbitrary electromagnetic fields. The possibility to derive the tensor of the equilibrium fluctuation electromagnetic field in an absorbing medium does not mean, of course, the possibility to determine the tensor for an arbitrary variable field. Even if this quantity has physical meaning, according to the Landau conjecture quoted above, it does not mean that it can be expressed in terms of the electric and magnetic permeabilities $\varepsilon(\omega)$ and $\mu(\omega)$. I believe that this is impossible. In any case, the tensor, obtained by direct calculation for a plasma, where the problem can be explicitly solved using the Boltzmann kinetic equation, cannot be expressed in such a way [10].

The difficulty, of course, is the energy dissipation. In a transparent medium, where dissipation is absent, the tensor can be obtained for arbitrary non-equilibrium fields [11].

Before discussing concrete results, let us discuss general properties of the stress tensor in thermal equilibrium. It can be presented in the form

$$\sigma_{ik} = -P_0(T, \rho)\delta_{ik} + \sigma_{ik}^{(elm)}, \qquad (2.4)$$

where $P_0(T, \rho)$ can be defined as the pressure of a uniform infinite liquid at given density ρ and temperature T and $\sigma_{ik}^{(elm)}$ is the contribution from the electromagnetic fluctuations, i.e., the van der Waals interaction. This contribution must satisfy several important conditions:

1. The tensor must be symmetric: $\sigma_{ik}^{(elm)} = \sigma_{ki}^{(elm)}$.
2. The van der Waals part of the force, acting on the liquid, must be derivable from a potential:

$$F_i^{(elm)} = \partial_k \sigma_{ik}^{(elm)} = -\rho \partial_i \zeta^{(elm)}. \qquad (2.5)$$

Actually $\zeta^{(elm)}$ is just the contribution of the van der Waals interaction to the chemical potential of the fluid. The first condition is a direct consequence of the symmetry of the microscopic energy-momentum tensor. The tensor $(-\sigma_{ik})$ is its averaged spatial part. Condition 2 ensures the possibility of mechanical equilibrium of the fluid in the presence of the van der Waals interaction. Indeed, the condition for such an equilibrium is $F_i = \partial_k \sigma_{ik} = 0$. Taking into account that $dP_0 = \rho d\zeta_0(\rho, t)$, we can rewrite this equation as

$$\partial_i \zeta_0 - F_i^{(elm)}/\rho = 0. \qquad (2.6)$$

which implies (2.4) for arbitrary configurations of interactive bodies. Thus violation of the condition 2 would result in permanent flow of the liquid in equilibrium and actually would permit us to build the notorious *Perpetual Motion* machine.

For an analogous reason, on the boundary between a fluid and a solid the tangential components of the tensor must be continuous. This condition is satisfied automatically by virtue of the boundary conditions for the fluctuating fields. If the normal to the surface is directed along z, it must be the case that $\sigma_{\alpha z}^{elm(1)} = \sigma_{\alpha z}^{elm(2)}$, $\alpha = x, y$. Violation of this condition would result in the existence of permanent flow of the liquid near a solid boundary.

Equation (2.6) can be written as a condition of constancy of the total chemical potential ζ:

$$\zeta(\rho, T) = \zeta_0(\rho, t) + \zeta^{(elm)}(\rho, T) = const. \qquad (2.7)$$

Notice that, neglecting the change in density of the liquid under the influence of the van der Waals forces, one can express this condition also as

Fig. 2.2 Diagrammatic
representation of corrections
to the free energy

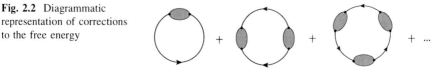

$$P_0(\rho, T)/\rho + \zeta^{(elm)} = const. \tag{2.8}$$

For calculations of the forces in the state of mechanical equilibrium one can omit this part of the tensor and exclude the pressure, i.e., use instead of (2.4) the tensor

$$\sigma'_{ik} = \sigma^{(elm)}_{ik} + \rho\zeta^{(elm)}(T, \rho)\delta_{ik}. \tag{2.9}$$

Notice that the tensor σ'_{ik} by *definition* satisfies the equation

$$\partial_k \sigma'_{ik} = 0. \tag{2.10}$$

This obvious property results sometimes in misunderstandings (Ref. [12]).

2.2 Free Energy of the Equilibrium Electromagnetic Field in an Absorbing Medium

I present now a simplified version of our deviation of the force tensor in a liquid obtained for the first time in [9]. It is sufficient to take into account only the electromagnetic interaction in the system. Nuclear forces are obviously irrelevant to our problem. Then the correction due to interaction to the free energy can be presented in the Matsubara technique as a set of "ring" diagrams (see Fig. 2.2) where the dashed lines represent the Matsubara Green's functions of the electromagnetic field $\mathscr{D}_0$ without interaction [13]. Every "bead" is the polarization operator Π, which includes all diagrams which cannot be separated into parts, connected by one dashed line. It is important that the diagrams of Fig. 2.2 cannot be summed up into the exact Green's function, because of the extra factor $1/n$ in each term, where n is the number of the "beads".

Let us calculate now the variation of the free energy with respect to a small change $\delta\Pi$ of the density of the liquid. The crucial point is that the factor $1/n$ will be canceled as a result of the variation and the parts of the diagrams with the non-variated beads will be summed up to the exact Green's function. The result is

$$\delta\mathscr{F} = \delta\mathscr{F}_0 - \sum_{s=0}^{\infty}{}' \int \mathscr{D}_{ik}(\mathbf{r}, \mathbf{r}'; \xi_s)\delta\Pi_{ik}(\mathbf{r}, \mathbf{r}'; \xi_s)d\mathbf{r}d\mathbf{r}', \tag{2.11}$$

where $\xi_s = 2s\pi T$ and the term with $s = 0$ is taken with a factor $(1/2)$ [14]. This is *an exact* equation of quantum electrodynamics. It is valid also in a vacuum, where

Π describes the radiation correction to $\mathscr{D}$. However, it is practically worthless, because explicit expressions for $\mathscr{D}$ and Π cannot be obtained. It also contains ultraviolet divergences. However, these divergences are due to contributions from the short wave-length fluctuations, whereas we are interested in effects due to the inhomogeneity of the medium. i.e., the presence of boundaries of bodies etc. This permits us to produce a renormalization of this equation. To do this, let us write the $\mathscr{D}$-function as

$$\mathscr{D}_{ik}(\mathbf{r},\mathbf{r}';\xi_s) = [\mathscr{D}_{ik}(\mathbf{r},\mathbf{r}';\xi_s) - \bar{\mathscr{D}}_{ik}(\mathbf{r},\mathbf{r}';\xi_s)] + \bar{\mathscr{D}}_{ik}(\mathbf{r},\mathbf{r}';\xi_s), \qquad (2.12)$$

where $\bar{\mathscr{D}}_{ik}(\mathbf{r},\mathbf{r}';\xi_s)$ is the Green's function of an auxiliary homogeneous infinite medium whose permeabilities are the same as that of the actual medium at the point $\mathbf{r}'$. After substitution into (2.11) the third term can be absorbed in the term δF_0. This term acquires the meaning of the variation of the free energy of this uniform medium.

$$\delta\mathscr{F} = \delta\mathscr{F}_0 - \sum_{s=0}^{\infty}{}' \int [\mathscr{D}_{ik}(\mathbf{r},\mathbf{r}';\xi_s) - \bar{\mathscr{D}}_{ik}(\mathbf{r},\mathbf{r}';\xi_s)]\delta\Pi_{ik}(\mathbf{r},\mathbf{r}';\xi_s)d\mathbf{r}d\mathbf{r}'. \quad (2.13)$$

In (2.13) the important fluctuations are those whose wavelengths are of the same order of magnitude as the inhomogeneities of the system (e.g., the thickness of films and separations of bodies). These lengths are assumed to be large compared to interatomic dimensions. However, these long-wavelength fluctuations can be described by macroscopic electrodynamics. According to the general theory, the Matsubara Green's function $\mathscr{D}_{ik}(\mathbf{r},\mathbf{r}';\xi_s) = D_{ik}^R(\mathbf{r},\mathbf{r}';i\xi_s)$, where D^R is the "usual" retarded Green's function for the vector-potential of the electromagnetic field. Accordingly, $\mathscr{D}$ satisfies in the macroscopic limit the explicit equation

$$[\partial_i\partial_l - \delta_{il}\Delta + (\xi_s^2/c_2)\varepsilon(i|\xi_s|,\mathbf{r})\delta_{il}]\mathscr{D}_{ik}(\mathbf{r},\mathbf{r}';|\xi_s|) = -4\pi\hbar\delta_{ik}\delta(\mathbf{r}-\mathbf{r}'). \quad (2.14)$$

The equation for $\bar{\mathscr{D}}$ can be obtained by changing in (2.14) the permeability $\varepsilon(i|\xi_s|,\mathbf{r})$ to $\varepsilon(i|\xi_s|,\mathbf{r}')$. However, in the majority of practical problems the excluding of the divergences can be achieved simply by omitting terms which do not depend on the spatial parameters, e.g., on the distances between bodies. Taking this into account, and to avoid complications of the equations, we will denote in the future the difference (2.12) as $\mathscr{D}$. It is worth noting that the left-hand side of (2.14) coincides with the Maxwell equations for the vector potential $\mathbf{A}$ of the electromagnetic field with the frequency $i\xi_s$ in the gauge where the scalar potential $\phi = 0$. The final results of the theory do not depend, of course, on the gauge. The Green's function $\mathscr{D}_0$ satisfies the same (2.14) with $\varepsilon = 1$. Let us write (2.14) symbolically as $\hat{\mathscr{D}}^{-1}\mathscr{D} = \delta_{il}\delta(\mathbf{r}-\mathbf{r}')$ and the equation for $\mathscr{D}_0$ as $\hat{\mathscr{D}}_0^{-1}\mathscr{D}_0 = \delta_{il}\delta(\mathbf{r}-\mathbf{r}')$. Then from the definition $\Pi_{ik} = [\mathscr{D}_{0ik}^{-1} - \mathscr{D}_{ik}^{-1}]$ we find an equation for Π:

$$\Pi_{kl}(\xi_s;\mathbf{r}_1,\mathbf{r}_2) = \frac{\xi_s^2}{4\pi}\delta_{kl}\delta(\mathbf{r}-\mathbf{r}')[\varepsilon(i|\xi_s|;\mathbf{r}_1) - 1]. \quad (2.15)$$

This permits us to express the variation of the free energy in terms of ε:

$$\delta\mathscr{F} = \delta\mathscr{F}_0 - \frac{T}{4\pi}\sum_{s=0}^{\infty}{}' \int \left[\xi_s^2 \mathscr{D}_{ll}(\mathbf{r},\mathbf{r};\xi_s)\right]\delta\varepsilon(i\xi_s;\mathbf{r})d\mathbf{r}. \qquad (2.16)$$

It is convenient for further equations to introduce a new function

$$\mathscr{D}_{ik}^E(\mathbf{r},\mathbf{r}';\xi_s) = -\xi_s^2\mathscr{D}_{ik}(\mathbf{r},\mathbf{r}';\xi_s). \qquad (2.17)$$

Function $\mathscr{D}$ describes the quadratic fluctuations of the vector potential in the Matsubara technique. Accordingly, the function $\mathscr{D}^E$ describes fluctuations of the electric field. We will also use the function

$$\mathscr{D}_{ik}^H(\mathbf{r},\mathbf{r}';\xi_s) = \text{curl}_{il}\text{curl}'_{km}\mathscr{D}_{lm}(\mathbf{r},\mathbf{r}';\xi_s) \qquad (2.18)$$

describing fluctuations of the magnetic field. Now we can rewrite (2.16) as

$$\delta\mathscr{F} = \delta\mathscr{F}_0 + \frac{T}{4\pi}\sum_{s=0}^{\infty}{}' \int \mathscr{D}_{ll}^E(\mathbf{r},\mathbf{r};\xi_s)\delta\varepsilon(i\xi_s;\mathbf{r})d\mathbf{r}. \qquad (2.19)$$

2.3 Stress Tensor of the van der Waals Interaction Inside an Absorbing Medium

We can use (2.19) to calculate the tensor of van der Waals forces in a fluid. It is instructive, however, as a first step to compare the equation for a free energy variation for given sources of field in a dielectric in the absence of dispersion (Ref. [8], (15.19)):

$$\delta\mathscr{F} = \delta\mathscr{F}_0 - \int \frac{E^2}{8\pi}\delta\varepsilon d\mathbf{r} \qquad (2.20)$$

This equation permits the calculation of the force $\mathbf{f}$ and finally to find the stress tensor (see Ref. [8], (15.9) and (35.2), in the presence of both electric and magnetic fields we must take the sum of these equations):

$$\sigma_{ik}^A = -P_0\delta_{ik} + \frac{\varepsilon E_i E_k + H_i H_k}{4\pi} - \frac{E^2}{8\pi}\left[\varepsilon - \rho\left(\frac{\partial\varepsilon}{\partial\rho}\right)_T\right]\delta_{ik} - \frac{H^2}{8\pi}\delta_{ik}. \qquad (2.21)$$

This equation was derived by M. Abraham around 1909 and is one of the most important results of the electrodynamics of continuous media.

Now we can write the tensor of the van der Waals forces by direct analogy with (2.21). Indeed, the functions $\mathscr{D}_{ik}^E(\mathbf{r},\mathbf{r}';\xi_s)$ and $\mathscr{D}_{ik}^H(\mathbf{r},\mathbf{r}';\xi_s)$ satisfy equations which are similar to the products $E_i(\mathbf{r})E_k(\mathbf{r}')$ and $H_i(\mathbf{r})H_k(\mathbf{r}')$. The presence of the δ-function term on the right-hand side of (2.14) is not important because this term in

any case will be eliminated in the course of the renormalization. Thus, the general
expression for the stress tensor for a fluid with $\mu = 1$ is [9]:

$$\sigma_{ik} = -P_0\delta_{ik} - \frac{\hbar T}{2\pi} \left\{ \sum_{s=0}^{\infty}{}' \left(\varepsilon D_{ik}^E + D_{ik}^H - \frac{1}{2} D_{ii}^E \left[\varepsilon - \rho \left(\frac{\partial \varepsilon}{\partial \rho} \right)_T \right] \delta_{ik} - \frac{1}{2} D_{ii}^H \delta_{ik} \right) \right\},$$

(2.22)

where D_{ik}^E and D_{ik}^H were defined above, $\varepsilon = \varepsilon(\rho, T, i\zeta_n)$, $\zeta_n = 2nT/\hbar$, and $P_0(\rho, T)$
is the pressure as a function of density and temperature in the absence of an
electric field.

Equation (2.22) assumes the system to be in thermal, but still not in *mechanical*
equilibrium. As was said before, the last condition can be formulated as a condition of the constancy of the chemical potential ζ, which can be defined by the
equation $\delta F = \int \zeta \delta \rho d\mathbf{r}$. The variation must be taken at fixed boundaries of the
bodies. One has from (2.19)

$$\zeta(\rho, T) = \zeta_0(\rho, T) + \frac{\hbar T}{4\pi} \sum_{s=0}^{\infty}{}' \mathscr{D}_{ll}^E(\mathbf{r}, \mathbf{r}; \xi_s) \frac{\partial \varepsilon(i\xi_s; \mathbf{r})}{\partial \rho}.$$

(2.23)

The condition of mechanical equilibrium means that $\zeta(\rho, T) = const.$ Let the
fluid have uniform density in the absence of the van der Waals forces. Taking into
account that $dP_0 = \rho d\zeta_0$ and neglecting in the second term any change of ρ due
the van der Waals interaction, we can rewrite the condition of equilibrium as:

$$P_0 + \frac{\hbar T}{4\pi} \sum_s{}' D_{ii}^E \rho \left(\frac{\partial \varepsilon}{\partial \rho} \right)_T = const.$$

(2.24)

This equation can be used to calculate the perturbation $\delta\rho$ of the density of the
liquid due to the van der Waals forces. Expanding the first term with respect to $\delta\rho$,
we easily find

$$\delta\rho = -\frac{\hbar T}{4\pi} \sum_s{}' D_{ii}^E \rho \left(\frac{\partial \varepsilon}{\partial P} \right)_T.$$

(2.25)

Equation (2.24) implies that a part of the stress tensor (2.22) is constant through
the fluid, being a uniform compressing or expanding pressure. This part can be
omitted in many problems, for example in the calculation of the full force acting
on a body embedded in the fluid. Subtracting the constant tensor
$\left[-P_0 - \frac{\hbar T}{4\pi} \sum_n D_{ii}^E \rho \left(\frac{\partial \varepsilon}{\partial \rho} \right)_T \right] \delta_{ik}$ from (2.22) one arrives to the "contracted" tensor,
which was obtained for the first time in [15] (see also Ref. [16]):

$$\sigma_{ik}' = -\frac{\hbar T}{2\pi} \left\{ \sum_s{}' \left(\varepsilon \mathscr{D}_{ik}^E + \mathscr{D}_{ik}^H - \frac{1}{2} \varepsilon \mathscr{D}_{ii}^E \delta_{ik} - \frac{1}{2} \mathscr{D}_{ii}^E \delta_{ik} \right) \right\}.$$

(2.26)

I would like to stress that the "P_0" term in the tensor (2.22) plays an important role. Ignoring this term would lead to wrong results. In this connection it is appropriate to quote Landau and Lifshitz's remark (see Ref. [8], Sect. 15): "The problem of calculating the forces (called *pondermotive* forces) which act on a dielectric in an arbitrary non-uniform electric field is fairly complicated..."

Notice that $\partial \sigma'_{ik}/\partial x_k = 0$ and hence $\oint \sigma'_{ik} dS_k = 0$ for integration over any closed surface, surrounding a volume of a uniform fluid, just due to the fact that in mechanical equilibrium electromagnetic forces are compensated by a pressure gradient. Analogous integration over any surface surrounding a solid body gives the total force acting on the body.

Equivalent theories of the force between bodies separated by a liquid were developed by Barash and Ginzburg [17] and Schwinger, DeRead and Milton [18]. The method of [17] is based on a very interesting and new physical idea. I cannot discuss it here. Notice only, that the method permits us to calculate forces on the basis of the solution of the imaginary frequencies "dispersion relation" $\mathscr{D}^{-1}(i\xi_s) = 0$, without an actual calculation of $\mathscr{D}$. This results in further simplification of calculations. The authors of [18] performed the free energy variation assuming actually the condition of the mechanical equilibrium from the very beginning and obtained directly (2.26).

2.4 Van der Waals Forces Between Bodies Separated by a Liquid

Now we can calculate the force acting on bodies separated by a dielectric liquid. It is worth noticing, however, that even for bodies in vacuum the method, based on using the imaginary-frequencies Green's functions, involves simpler calculations than the original Lifshitz method, because the solution of the equation for the Green's functions is simpler than the procedure of averaging of the stress tensor.

It was shown in [15] that if the problem has been solved for bodies in a vacuum, the answer for bodies in liquid can be found by a simple scaling transformation. Let us denote the dielectric permeability of the liquid as ε. If we perform a coordinate transformation $\mathbf{r} = \tilde{\mathbf{r}}/\varepsilon^{1/2}$ and introduce the new functions $\mathscr{D}_{ik} = \tilde{\mathscr{D}}_{ik}/\varepsilon^{1/2}$ and $\mathscr{D}^E_{ik} = \tilde{\mathscr{D}}^E_{ik}\varepsilon^{1/2}$, $\mathscr{D}^H_{ik} = \tilde{\mathscr{D}}^H_{ik}\varepsilon^{3/2}$, then

$$\sigma'_{ik} = -\frac{\hbar T}{2\pi}\left\{\sum_{s=0}^{\infty}{}' \varepsilon^{3/2}\left(\tilde{\mathscr{D}}^E_{ik} + \tilde{\mathscr{D}}^H_{ik} - \frac{1}{2}\tilde{\mathscr{D}}^E_{ii}\delta_{ik} - \frac{1}{2}\tilde{\mathscr{D}}^E_{ii}\delta_{ik}\right)\right\}. \tag{2.27}$$

One can see easily that the new functions $\tilde{\mathscr{D}}_{ik}$ satisfy in the new coordinates $\tilde{\mathbf{r}}$ equations of the same form (2.14) for bodies in vacuum, while the permeabilities ε_α of the bodies were changed to $\varepsilon_\alpha/\varepsilon$.

One can usually neglect the influence of the temperature on the forces between bodies in a liquid. Then one can change $T\sum_{s=0}^{\infty}{}'\dots \rightarrow \frac{\hbar}{2\pi}\int_0^\infty \dots d\xi_s$. We will consider below only this case. We also will consider only the small "London" distances where the characteristic distance between bodies $l \ll \lambda$, where λ is the characteristic wavelength of the absorption spectra of the media. In this case one can neglect the magnetic Green's function $\mathscr{D}_{ik}^{H}$ and the electric function can be presented as $\mathscr{D}_{ik}^{E} = \hbar \partial_i \partial_k' \phi$, where the "electrostatic" Green's function ϕ satisfies [19, 20]

$$\partial_i[\varepsilon(i\xi; \mathbf{r})\partial_i\phi(\xi; \mathbf{r}, \mathbf{r}')] = -4\pi\delta(\mathbf{r} - \mathbf{r}'). \tag{2.28}$$

Thus ϕ is just the potential of a unit charge placed at point $\mathbf{r}'$. We will present here results for two important problems.

2.4.1 Interaction of a Small Sphere with a Plane Body

As a first example we consider a dielectric sphere in the vicinity of a plane surface of a bulk body. Let the radius R of the sphere be small in comparison with the distance l between the sphere and the surface. We consider first the problem in vacuum. Then the energy of interaction can be obtained directly from (2.19), taking into account that the change of the dielectric permeability due to the presence of the sphere at point $\mathbf{r}_0$ is $\delta\varepsilon(\omega) = 4\pi\alpha(\omega)\delta(\mathbf{r} - \mathbf{r}_0)$, where $\alpha(\omega)$ is the polarizability of the sphere. In the zero-temperature London regime we get

$$V(l) = \frac{\hbar}{2\pi} \int\limits_0^\infty \alpha(i\xi) \left[\mathscr{D}_{ll}^{E}(\xi; \mathbf{r}, \mathbf{r}') \right]_{\mathbf{r} \to \mathbf{r} \to \mathbf{r}_0} d\xi. \tag{2.29}$$

The "potential" ϕ can be taken from [8], Sect. 7, Problem 1. A simple calculation then gives

$$\mathscr{D}_{ll}^{E}(\xi; \mathbf{r}_0, \mathbf{r}_0) = -\frac{\hbar}{2l^3} \frac{\varepsilon_1(i\xi) - 1}{\varepsilon_1(i\xi) + 1}. \tag{2.30}$$

Taking into account that

$$\alpha(i\xi) = R^3 \frac{\varepsilon_2(i\xi) - 1}{\varepsilon_2(i\xi) + 2}, \tag{2.31}$$

we find

$$V(l) = -\frac{\hbar R^3}{4\pi l^3} \int\limits_0^\infty \frac{(\varepsilon_2(i\xi) - 1)(\varepsilon_1(i\xi) - 1)}{(\varepsilon_2(i\xi) + 2)(\varepsilon_1(i\xi) + 1)} d\xi. \tag{2.32}$$

The force acting on the sphere is

$$F(l) = -\frac{dV}{dl} = -\frac{3\hbar R^3}{4\pi l^4} \int\limits_0^\infty \frac{(\varepsilon_2(i\xi) - 1)(\varepsilon_1(i\xi) - 1)}{(\varepsilon_2(i\xi) + 2)(\varepsilon_1(i\xi) + 1)} d\xi. \tag{2.33}$$

One must be careful when rewriting this equation for a case of bodies separated by liquid. The transformation was formulated for the tensor σ'. Taking into account that $F = \int \sigma'_{zz} dx dy$, we conclude that it is enough to change $\varepsilon_1 \to \varepsilon_1/\varepsilon$ and $\varepsilon_2 \to \varepsilon_2/\varepsilon$:

$$F(l) = \frac{3\hbar R^3}{4\pi l^4} \int\limits_0^\infty \frac{(\varepsilon_2(i\xi) - \varepsilon(i\xi))(\varepsilon_1(i\xi) - \varepsilon(i\xi))}{(\varepsilon_2(i\xi) + \varepsilon(i\xi))(\varepsilon_1(i\xi) + \varepsilon(i\xi))} d\xi. \tag{2.34}$$

2.4.2 Interaction Between Two Parallel Plates

Let us consider now the force between solid bodies 1 and 2 separated by very small distances. It should be noted that, for a rigorous statement of the problem, it is necessary to consider at least one of the bodies as being of finite size and surrounded by the liquid. Then $F_i = \oint \sigma'_{ik} dS_k$ is the total force acting on the body. However, since the van der Waals forces decrease very quickly with distance, the integrand is actually different from zero only inside the gap and the force can be calculated as $F = F_z = \int \sigma'_{zz} dx dy$. Notice that, due to (2.10), the quantity σ'_{zz} does not depend on z. Finally the force per unit area can be expressed as [15]

$$F = \frac{\hbar}{16\pi l^3} \int\limits_0^\infty \int\limits_0^\infty x^2 \left[\frac{(\varepsilon_1 + \varepsilon)(\varepsilon_2 + \varepsilon)}{(\varepsilon_1 - \varepsilon)(\varepsilon_1 - \varepsilon)} e^x - 1 \right] dx d\xi \tag{2.35}$$

where the dielectric permeabilities must be taken as functions of the imaginary frequency $i\xi$.

2.5 Remarks about Repulsive Interactions

It is well known that forces between bodies in vacuum are attractive. In the cases considered in the previous section it is obvious, because for any body $\varepsilon(i\xi) > 1$ for $\xi > 0$. It also follows from (2.34) and (2.35) that forces are attractive for bodies of the same media ($\varepsilon_1 = \varepsilon_2$).

If, however, the bodies are different, the force can be either attractive or repulsive. It is clear from (2.34) and (2.35) that if the differences $\varepsilon_1 - \varepsilon$ and $\varepsilon_2 - \varepsilon$ have different signs in the essential region of values ξ, we have $F < 0$, that is, the bodies repel one another.

To understand better the physical meaning of this repulsion, let us consider the problem of the body-sphere interaction and assume that the materials of both the sphere and the liquid (but not the body) are optically rarefied, i.e., that $\varepsilon_2 - 1 \ll 1$ and $\varepsilon - 1 \ll 1$. Then (2.34) can be simplified as

$$F(l) \approx \frac{3\hbar R^3}{8\pi l^4} \int\limits_0^\infty \frac{(\varepsilon_2(i\xi) - 1)}{(\varepsilon_2(i\xi) + 1)} (\varepsilon_1(i\xi) - \varepsilon(i\xi)) d\xi. \tag{2.36}$$

The force is now expressed as a difference of two terms, with clear physical meaning. The first term is the force that acts on the sphere in vacuum. The second term is the force that would act in vacuum on an identical sphere, but with optical properties of the liquid. This second term is an exact analogy of the Archimedes' buoyant force, which acts on a body embedded in a liquid in a gravitational field. This remark again stresses the importance of the condition of the mechanical equilibrium in the liquid. Of course, such a simple interpretation is possible only in the limit of rarefied media.

The existence of the repulsive van der Waals forces, predicted in [15] was confirmed in several experiments (see [21, 22] and references therein). See also the Chap. 8 by Capasso et al. in this volume. Corresponding experiments are, how-ever, quite difficult. Forces at large distances are quite small, while at small distances the atomic structure of the media becomes essential.

2.6 Liquid Films

The van der Waals forces play an important part in the physics of surface phe-nomena, and in the properties of thin films in particular. A fundamental problem here is the dependence of the chemical potential ζ on the thickness d of a film. For example, the thickness of a film on a solid surface in equilibrium with the vapour at pressure P is given by the equation

$$\zeta(P, d) = \zeta_0(P) + \frac{T}{m} \ln \frac{P}{P_{sat}}. \tag{2.37}$$

If the thickness d of the film is large compared to interatomic distances, this dependence is defined mainly by the van der Waals forces. Actually the contri-bution of these forces to the chemical potential is given by the general (2.23). However, this equation cannot be used directly, because it gives the chemical potential ζ in terms of the density ρ, while (2.37) requires ζ as a function of the pressure P.

According to the conditions of mechanical equilibrium, the normal component σ_{zz} of the stress tensor must be continuous at the surface of the film: $-P_0(\rho, T) + \sigma_{zz}^{(elm)} = -P$. Then $\rho(P_0) = \rho(P + \sigma_{zz}^{(elm)}) \approx \rho(P) + (\partial \rho)/(\partial P)\sigma_{zz}^{(elm)}$ and $\zeta_0(\rho) \approx \zeta_0(P) + (\partial \zeta/\partial \rho)(\partial \rho/\partial P)\sigma_{zz}^{(elm)} = \zeta_0(P) + \sigma_{zz}^{(elm)}/\rho$, where we took into account that $(\partial \zeta/\partial \rho) = 1/\rho$. Thus we have

$$\zeta(P, d) = \zeta_0(P) + \sigma_{zz}^{(elm)}/\rho + \frac{\hbar T}{4\pi} \sum_{s=0}^\infty {}' \mathscr{D}_{ll}^E(\mathbf{r}, \mathbf{r}; \xi_s) \frac{\partial \varepsilon(i\xi_s; \mathbf{r})}{\partial \rho}$$

$$= \zeta_0(P) + \sigma'_{zz}/\rho = \zeta_0(P) + F(d)/\rho, \tag{2.38}$$

where $F(d)$ is the force, which in the London regime is given by (2.35) with $\varepsilon_2 = 1, l \to d$. (As far as electromagnetic properties of the vapour are concerned, we can treat it as a vacuum.) One can now rewrite (2.37) in the form

$$F(d) = -\frac{T}{m} \ln \frac{P}{P_{sat}}. \tag{2.39}$$

If the film is placed on a solid wall situated vertically in the gravitational field, the dependence of the film thickness on the altitude is given by the equation

$$F(d) = \rho g x, \tag{2.40}$$

where x is the height.

In conclusion, let us consider a "free" film in vacuum. Then the chemical potential can be written as

$$\zeta(P, T, d) = \zeta_0(P, T) + F(d)/\rho \tag{2.41}$$

where F can be obtained from (2.35) with $\varepsilon_1 = \varepsilon_2 = 1, l \to d$:

$$F = \frac{\hbar}{16\pi d^3} \int\limits_0^\infty \int\limits_0^\infty x^2 \left[\frac{(1+\varepsilon)^2}{(1-\varepsilon)^2} e^x - 1 \right] dx d\xi. \tag{2.42}$$

Note that this is just the quantity which can be calculated by integration of the force in the three-boundary geometry of Fig. 2.1 However, these calculations have never been performed, and the correctness of the corresponding considerations has not been proved [23].

Acknowledgment I thank R.Scott for critical reading of this paper and useful suggestions.

References

1. I use the generic term "van der Waals forces" for long-range forces between neutral objects in any conditions. Thus I do not distinguish between the London, Casimir, Casimir-Polder and Lifshitz forces
2. London, F.: Theory and system of molecular forces. Z. Phys. **63**, 245 (1930)
3. Casimir, H.B., Polder, D.: The influence of retardation on the London-van der Waals forces. Phys. Rev. **73**, 360 (1948)
4. Casimir, H.B.: On the attraction between two perfectly conducting plates. Proc. K. Ned. Akad. Wet. **51**, 793 (1948)
5. Lifshitz, E.M.: Theory of molecular attraction forces between condensed bodies. Doklady Akademii Nauk SSSR **97**, part 4, 643 (1954); Influence of temperature on molecular attraction forces between condensed bodies. **100**, part 5, 879 (1955)
6. Lifshitz, E.M.: The theory of molecular attractive forces between solids. Sov. Phys. JETP **2**, 73 (1956)

7. Rytov, S.M.: Theory of the Electric Fluctuations and Thermal Radiation [in Russian] Publication of Acad. of Sciences of USSR, Moscow (1953), English translation: Air Force Cambridge Research Center, Bedford, MA (1959)
8. Landau, L.D., Lifshitz, E.M.: Electrodynamics of Continuous Media, Pergamon Press, Oxford, 1960. Russian edition was published in 1957
9. Dzyaloshinskii, I.E., Pitaevskii, L.P.: Van der Waals forces in an inhomogeneous dielectric. Sov. Phys. JETP. **9**, 1282 (1959)
10. Perel, V.I., Pinskii Ya, M.: Stress tensor for a plasma in a high frequency electromagnetic field with account of collisions. Sov. Phys. JETP. **27**, 1014 (1968)
11. Pitaevskii, L.P.: Electric forces in a transparent dispersive medium. Sov. Phys. JETP. **12**, 1008 (1961)
12. Pitaevskii, L.P.: Comment on "Casimir force acting on magnetodielectric bodies embedded in media". Phys. Rev. A. **73**, 047801 (2006)
13. Abrikosov, A.A., Gorkov, L.P., Dzyaloshinskii I.E. (1963) Methods of Quantum Field Theory in Statistical Physics, Prentice-Hall, Englewood Cliffs
14. Here and below I put $k_B = 1$. Likewise I put $\hbar = 1$ in intermediate equations. I use the CGSE system of electromagnetic units and for simplicity neglect the magnetic properties of media, i.e., put $\mu = 1$
15. Dzyaloshinskii, I.E., Lifshitz, E.M., Pitaevskii, L.P.: Van der Waals forces in liquid films. Sov. Phys. JETP. **10**, 161 (1960)
16. Dzyaloshinskii, I.E., Lifshitz, E.M., Pitaevskii, L.P.: The general theory of van der Waals forces. Adv. Phys. **10**, 165 (1961)
17. Barash Yu, S., Ginzburg, V.L.: Electromagnetic fluctuations in matter and molecular (van der Waals) forces between them. Sov. Phys. Uspekhi. **18**, 305 (1975)
18. Schwinger, J., DeRaad, L.L., Milton, K.A.: Casimir effect in dielectrics. Ann. Phys. (N.Y.) **115**, 1 (1978)
19. Volokitin, A.I., Persson, B.N.P.: Radiative heat transfer between nanostructures. Phys. Rev. B. **63**, 205404 (2001)
20. Pitaevskii, L.P.: Thermal Lifshitz force between an atom and a conductor with small density of carriers. Phys. Rev. Lett. **101**, 163202 (2008)
21. Munday, J.N., Capasso, F., Parsegian, A.V.: Measured long-range repulsive Casimir-Lifshitz forces. Nature **457**, 170 (2009)
22. Munday, J.N., Capasso, F.: Repulsive Casimir and van der Waals forces: from measurements to future technologies. In: Milton, K.A., Bordag, M. (eds) Quantum Field Theory under the Influence of External Conditions., pp. 127. Word Scientific, New Jersey (2010)
23. Note added in proofs: After this article was submitted, a preprint by Zheng and Narayanaswamy [24] appeared, where the authors independently developed a method based on the "three-boundary geometry." Their results coincide with our Green's functions approach.
24. Zheng, Y., Narayanaswamy, A.: Phys. Rev. A. **83**, 042504 (2011); e-print arXiv: 1011.5433

Chapter 3
Local and Global Casimir Energies: Divergences, Renormalization, and the Coupling to Gravity

Kimball A. Milton

Abstract From the beginning of the subject, calculations of quantum vacuum energies or Casimir energies have been plagued with two types of divergences: The total energy, which may be thought of as some sort of regularization of the zero-point energy, $\sum \frac{1}{2}\hbar\omega$, seems manifestly divergent. And local energy densities, obtained from the vacuum expectation value of the energy-momentum tensor, $\langle T_{00}\rangle$, typically diverge near boundaries. These two types of divergences have little to do with each other. The energy of interaction between distinct rigid bodies of whatever type is finite, corresponding to observable forces and torques between the bodies, which can be unambiguously calculated. The divergent local energy densities near surfaces do not change when the relative position of the rigid bodies is altered. The self-energy of a body is less well-defined, and suffers divergences which may or may not be removable. Some examples where a unique total self-stress may be evaluated include the perfectly conducting spherical shell first considered by Boyer, a perfectly conducting cylindrical shell, and dilute dielectric balls and cylinders. In these cases the finite part is unique, yet there are divergent contributions which may be subsumed in some sort of renormalization of physical parameters. The finiteness of self-energies is separate from the issue of the physical observability of the effect. The divergences that occur in the local energy-momentum tensor near surfaces are distinct from the divergences in the total energy, which are often associated with energy located exactly on the surfaces. However, the local energy-momentum tensor couples to gravity, so what is the significance of infinite quantities here? For the classic situation of parallel plates there are indications that the divergences in the local energy density are consistent with divergences in Einstein's equations; correspondingly, it has been shown that divergences in the total Casimir energy serve to precisely renormalize the masses of the plates, in accordance with the equivalence principle. This should be a

K. A. Milton (✉)
Homer L. Dodge Department of Physics and Astronomy,
University of Oklahoma, Norman, OK 73019, USA
e-mail: milton@nhn.ou.edu

D. Dalvit et al. (eds.), *Casimir Physics*, Lecture Notes in Physics 834,
DOI: 10.1007/978-3-642-20288-9_3, © Springer-Verlag Berlin Heidelberg 2011

general property, but has not yet been established, for example, for the Boyer sphere. It is known that such local divergences can have no effect on macroscopic causality.

3.1 Introduction

For more than 60 years it has been appreciated that quantum fluctuations can give rise to macroscopic forces between bodies [1]. These can be thought of as the sum, in general nonlinear, of the van der Waals forces between the constituents of the bodies, which, in the 1930s had been shown by London [2] to arise from dipole-dipole interactions in the nonretarded regime, and in 1947 to arise from the same interactions in the retarded regime, giving rise to so-called Casimir-Polder forces [3]. Bohr [4] apparently provided the incentive to Casimir to rederive the macroscopic force between a molecule and a surface, and then derive the force between two conducting surfaces, directly in terms of zero-point fluctuations of the electromagnetic fields in which the bodies are immersed. But these two points of view—action at a distance and local action—are essentially equivalent, and one implies the other, not withstanding some objections to the latter [5].

The quantum-vacuum-fluctuation force between two parallel surfaces—be they conductors or dielectrics [6–8] —was the first situation considered, and still the only one accessible experimentally. (For a current review of the experimental situation, see the chapters by Lamoreaux, Capasso et al., Decca et al., Van Zwol et al., and De Kieviet et al. in this volume, and also [9, 10]) Actually, most experiments measure the force between a spherical surface and a plane, but the surfaces are so close together that the force may be obtained from the parallel plate case by a geometrical transformation, the so-called proximity force approximation (PFA) [11–13]. However, it is not possible to find an extension to the PFA beyond the first approximation of the separation distance being smaller than all other scales in the problem. In the last few years, advances in technique have allowed quasi-analytical and numerical calculations to be carried out between bodies of essentially any shape, at least at medium to large separation, so the limitations of the PFA may be largely transcended. (See also the chapters by Rahi et al., by Johnson and by Lambrecht et al. in this volume for additional discussions about advances in numerical and analytical calculations. For earlier references, see, for example [14].) These advances have shifted calculational attention away from what used to be the central challenge in Casimir theory, how to define and calculate Casimir energies and self-stresses of single bodies.

There are, of course, sound reasons for this. Forces between distinct bodies are necessarily physically finite, and can, and have, been observed by experiment. Self-energies or self-stresses typically involve divergent quantities which are difficult to remove, and have obscure physical meaning. For example, the self-stress on a perfectly conducting spherical shell of negligible thickness was calculated by Boyer in 1968 [15], who found a repulsive self-stress that has subsequently been confirmed by a variety of techniques. Yet it remains unclear what physical significance this energy has. If the sphere is bisected and the two halves pulled apart, there will be an attraction (due to the closest parts of the hemispheres) not a repulsion. The same remarks, although exacerbated, apply to the self-stress on a rectangular box [16–19].The situation in that case is worse because (3.1) the sharp corners give rise to additional divergences not present in the case of a smooth boundary (it has been proven that the self-energy of a smooth closed infinitesimally thin conducting surface is finite [20, 21]), and (3.2) the exterior contributions cannot be computed because the vector Helmholtz equation cannot be separated. But calculational challenges aside, the physical significance of self-energy remains elusive.

The exception to this objection is provided by gravity. Gravity couples to the local energy-momentum or stress tensor, and, in the leading quantum approximation, it is the vacuum expectation value of the stress tensor that provides the source term in Einstein's equations. Self energies should therefore in principle be observable. This is largely uncharted territory, except in the instance of the classic situation of parallel plates. There, after a bit of initial confusion, it has now been established that the divergent self-energies of each plate in a two-plate apparatus, as well as the mutual Casimir energy due to both plates, gravitates according to the equivalence principle, so that indeed it is consistent to absorb the divergent self-energies of each plate into the gravitational and inertial mass of each [22, 23]. This should be a universal feature.

In this paper, for pedagogical reasons, we will concentrate attention on the Casimir effect due to massless scalar field fluctuations, where the potentials are described by δ-function potentials, so-called semitransparent boundaries. In the limit as the coupling to these potentials becomes infinitely strong, this imposes Dirichlet boundary conditions. At least in some cases, Neumann boundary conditions can be achieved by the strong coupling limit of the derivative of δ-function potentials. So we can, for planes, spheres, and circular cylinders, recover in this way the results for electromagnetic field fluctuations imposed by perfectly conducting boundaries. Since the mutual interaction between distinct semitransparent bodies have been described in detail elsewhere [24–26], we will, as implied above, concentrate on the self-interaction issues.

A summary of what is known for spheres and circular cylinders is given in Table 3.1.

3.2 Casimir Effect Between Parallel Plates: A δ-Potential Derivation

In this section, we will rederive the classic Casimir result for the force between parallel conducting plates [1]. Since the usual Green's function derivation may be found in monographs [38], and was for example reviewed in connection with current controversies over finiteness of Casimir energies [36], we will here present

Table 3.1 Casimir energy (E) for a sphere and Casimir energy per unit length ($\mathscr{E}$) for a cylinder, both of radius a

Type	$E_{\text{Sphere}}a$	$\mathscr{E}_{\text{Cylinder}}a^2$	References
EM	$+0.04618$	-0.01356	[15, 27]
D	$+0.002817$	$+0.0006148$	[28, 29]
$(\varepsilon - 1)^2$	$+0.004767 = \frac{23}{1536\pi}$	0	[30, 31]
ξ^2	$+0.04974 = \frac{5}{32\pi}$	0	[32, 33]
δe^2	± 0.0009	0	[34, 35]
λ^2/a^2	$+0.009947 = \frac{1}{32\pi}$	0	[36, 37]

Here the different boundary conditions are perfectly conducting for electromagnetic fields (EM), Dirichlet for scalar fields (D), dilute dielectric for electromagnetic fields [coefficient of $(\varepsilon - 1)^2$], dilute dielectric for electromagnetic fields with media having the same speed of light (coefficient of $\xi^2 = [(\varepsilon - 1)/(\varepsilon + 1)]^2$), perfectly conducting surface with eccentricity δe (coefficient of δe^2), and weak coupling for scalar field with δ-function boundary given by (3.60), (coefficient of λ^2/a^2). The references given are, to the author's knowledge, the first paper in which the results in the various cases were found

a different approach, based on δ-function potentials, which in the limit of strong coupling reduce to the appropriate Dirichlet or Robin boundary conditions of a perfectly conducting surface, as appropriate to TE and TM modes, respectively. Such potentials were first considered by the Leipzig group [39, 40], but more recently have been the focus of the program of the MIT group [41–44]. The discussion here is based on a paper by the author [45]. (See also [46].) (A multiple scattering approach to this problem has also been given in [25].)

We consider a massive scalar field (mass μ) interacting with two δ-function potentials, one at $x = 0$ and one at $x = a$, which has an interaction Lagrange density

$$\mathscr{L}_{\text{int}} = -\frac{1}{2}\lambda\delta(x)\phi^2(x) - \frac{1}{2}\lambda'\delta(x - a)\phi^2(x), \tag{3.1}$$

where the positive coupling constants λ and λ' have dimensions of mass. In the limit as both couplings become infinite, these potentials enforce Dirichlet boundary conditions at the two points:

$$\lambda, \lambda' \to \infty: \qquad \phi(0), \phi(a) \to 0. \tag{3.2}$$

The Casimir energy for this situation may be computed in terms of the Green's function G,

$$G(x, x') = i\langle T\phi(x)\phi(x')\rangle, \tag{3.3}$$

which has a time Fourier transform,

$$G(x, x') = \int \frac{d\omega}{2\pi} e^{-i\omega(t-t')} \mathscr{G}(x, x'; \omega). \tag{3.4}$$

Actually, this is a somewhat symbolic expression, for the Feynman Green's function (3.3) implies that the frequency contour of integration here must pass below the singularities in ω on the negative real axis, and above those on the positive real axis [47, 48]. Because we have translational invariance in the two directions parallel to the plates, we have a Fourier transform in those directions as well:

$$\mathscr{G}(x, x'; \omega) = \int \frac{(d\mathbf{k})}{(2\pi)^2} e^{i\mathbf{k}\cdot(\mathbf{r}-\mathbf{r}')_\perp} g(x, x'; \kappa), \tag{3.5}$$

where $\kappa^2 = \mu^2 + k^2 - \omega^2$.

The reduced Green's function in (3.5) in turn satisfies

$$\left[-\frac{\partial^2}{\partial x^2} + \kappa^2 + \lambda\delta(x) + \lambda'\delta(x-a) \right] g(x, x') = \delta(x - x'). \tag{3.6}$$

This equation is easily solved, with the result

$$g(x, x') = \frac{1}{2\kappa} e^{-\kappa|x-x'|} + \frac{1}{2\kappa\Delta} \left[\frac{\lambda\lambda'}{(2\kappa)^2} 2\cosh\kappa|x - x'| \right.$$
$$\left. - \frac{\lambda}{2\kappa}\left(1 + \frac{\lambda'}{2\kappa}\right) e^{2\kappa a} e^{-\kappa(x+x')} - \frac{\lambda'}{2\kappa}\left(1 + \frac{\lambda}{2\kappa}\right) e^{\kappa(x+x')} \right] \tag{3.7a}$$

for both fields inside, $0 < x, x' < a$, while if both field points are outside, $a < x, x'$,

$$g(x, x') = \frac{1}{2\kappa} e^{-\kappa|x-x'|} + \frac{1}{2\kappa\Delta} e^{-\kappa(x+x'-2a)}$$
$$\times \left[-\frac{\lambda}{2\kappa}\left(1 - \frac{\lambda'}{2\kappa}\right) - \frac{\lambda'}{2\kappa}\left(1 + \frac{\lambda}{2\kappa}\right) e^{2\kappa a} \right]. \tag{3.7b}$$

For $x, x' < 0$,

$$g(x, x') = \frac{1}{2\kappa} e^{-\kappa|x-x'|} + \frac{1}{2\kappa\Delta} e^{\kappa(x+x')}$$
$$\times \left[-\frac{\lambda'}{2\kappa}\left(1 - \frac{\lambda}{2\kappa}\right) - \frac{\lambda}{2\kappa}\left(1 + \frac{\lambda'}{2\kappa}\right) e^{2\kappa a} \right]. \tag{3.7c}$$

Here, the denominator is

$$\Delta = \left(1 + \frac{\lambda}{2\kappa}\right)\left(1 + \frac{\lambda'}{2\kappa}\right) e^{2\kappa a} - \frac{\lambda\lambda'}{(2\kappa)^2}. \tag{3.8}$$

Note that in the strong coupling limit we recover the familiar results, for example, inside

$$\lambda, \lambda' \to \infty: \quad g(x, x') \to -\frac{\sinh \kappa x_< \sinh \kappa (x_> - a)}{\kappa \sinh \kappa a} \tag{3.9}$$

Here $x_>, x_<$ denote the greater, lesser, of x, x'. Evidently, this Green's function vanishes at $x = 0$ and at $x = a$.

Let us henceforward consider $\mu = 0$, since otherwise there are no long-range forces. (There is no nonrelativistic Casimir effect—for example, see [38], p. 30.) We can now calculate the force on one of the δ-function plates by calculating the discontinuity of the stress tensor, obtained from the Green's function (3.3) by

$$\langle T^{\mu\nu} \rangle = \left(\partial^\mu \partial^{\nu'} - \frac{1}{2} g^{\mu\nu} \partial^\lambda \partial'_\lambda \right) \frac{1}{i} G(x, x') \bigg|_{x=x'}. \tag{3.10}$$

Writing a reduced stress tensor by

$$\langle T^{\mu\nu} \rangle = \int \frac{d\omega}{2\pi} \int \frac{(d\mathbf{k})}{(2\pi)^2} t^{\mu\nu}, \tag{3.11}$$

we find inside, just to the left of the plate at $x = a$,

$$t_{xx} \big|_{x=a-} = \frac{1}{2i} (-\kappa^2 + \partial_x \partial_{x'}) g(x, x') \bigg|_{x=x'=a-} \tag{3.12a}$$

$$= -\frac{\kappa}{2i} \left\{ 1 + 2 \frac{\lambda \lambda'}{(2\kappa)^2} \frac{1}{\Delta} \right\}. \tag{3.12b}$$

From this we must subtract the stress just to the right of the plate at $x = a$, obtained from (3.7b), which turns out to be in the massless limit

$$t_{xx} \big|_{x=a+} = -\frac{\kappa}{2i}, \tag{3.13}$$

which just cancels the 1 in braces in (3.12b). Thus the pressure on the plate at $x = a$ due to the quantum fluctuations in the scalar field is given by the simple, finite expression

$$P = \langle T_{xx} \rangle \big|_{x=a-} - \langle T_{xx} \rangle \big|_{x=a+}$$

$$= -\frac{1}{32\pi^2 a^4} \int_0^\infty dy\, y^2 \frac{1}{(y/(\lambda a) + 1)(y/(\lambda' a) + 1) e^y - 1}, \tag{3.14}$$

which coincides with the result given in [44, 49]. The leading behavior for small $\lambda = \lambda'$ is

$$P^{\mathrm{TE}} \sim -\frac{\lambda^2}{32\pi^2 a^2}, \quad \lambda \ll 1, \tag{3.15a}$$

while for large λ it approaches half of Casimir's result [1] for perfectly conducting parallel plates,

$$P^{\text{TE}} \sim -\frac{\pi^2}{480a^4}, \qquad \lambda \gg 1. \tag{3.15b}$$

We can also compute the energy density. Integrating the energy density over all space should give rise to the total energy. Indeed, the above result may be easily derived from the following expression for the total energy,

$$E = \int (\mathbf{dr})\langle T^{00}\rangle = \frac{1}{2i} \int (\mathbf{dr})(\partial^0\partial'^0 - \nabla^2)G(x,x')\Big|_{x=x'}$$
$$= \frac{1}{2i} \int (\mathbf{dr}) \int \frac{d\omega}{2\pi} 2\omega^2 \mathcal{G}(\mathbf{r},\mathbf{r}), \tag{3.16}$$

if we integrate by parts and omit the surface term. Integrating over the Green's functions in the three regions, given by (3.7a–c), we obtain for $\lambda = \lambda'$,

$$\mathcal{E} = \frac{1}{48\pi^2 a^3} \int_0^\infty dy\, y^2 \frac{1}{1+y/(\lambda a)} - \frac{1}{96\pi^2 a^3} \int_0^\infty dy\, y^3 \frac{1+2/(y+\lambda a)}{(y/(\lambda a)+1)^2 e^y - 1}, \tag{3.17}$$

where the first term is regarded as an irrelevant constant (λ is constant so the a can be scaled out), and the second term coincides with the massless limit of the energy first found by Bordag et al. [39], and given in [44, 49]. When differentiated with respect to a, (3.17), with λ fixed, yields the pressure (3.14). (We will see below that the divergent constant describe the self-energies of the two plates.)

If, however, we integrate the interior and exterior energy density directly, one gets a different result. The origin of this discrepancy with the naive energy is the existence of a surface contribution to the energy. To see this, we must include the potential in the stress tensor,

$$T^{\mu\nu} = \partial^\mu\phi\partial^\nu\phi - \frac{1}{2}g^{\mu\nu}(\partial^\lambda\phi\partial_\lambda\phi + V\phi^2), \tag{3.18}$$

and then, using the equation of motion, it is immediate to see that the energy density is

$$T^{00} = \frac{1}{2}\partial^0\phi\partial^0\phi - \frac{1}{2}\phi(\partial^0)^2\phi + \frac{1}{2}\nabla\cdot(\phi\nabla\phi), \tag{3.19}$$

so, because the first two terms here yield the last form in (3.16), we conclude that there is an additional contribution to the energy,

$$\hat{E} = -\frac{1}{2i} \int d\mathbf{S}\cdot\nabla G(x,x')\Big|_{x'=x} \tag{3.20a}$$

$$= -\frac{1}{2i} \int_{-\infty}^{\infty} \frac{d\omega}{2\pi} \int \frac{(dk)}{(2\pi)^2} \sum \frac{d}{dx} g(x,x') \Big|_{x'=x}, \tag{3.20b}$$

where the derivative is taken at the boundaries (here $x = 0, a$) in the sense of the outward normal from the region in question. When this surface term is taken into account the extra terms incorporated in (3.17) are supplied. The integrated formula (3.16) automatically builds in this surface contribution, as the implicit surface term in the integration by parts. That is,

$$E = \int (dr)\langle T^{00}\rangle + \hat{E}. \tag{3.21}$$

(These terms are slightly unfamiliar because they do not arise in cases of Neumann or Dirichlet boundary conditions.) See Fulling [50] for further discussion. That the surface energy of an interface arises from the volume energy of a smoothed interface is demonstrated in [45], and elaborated in Sect. 3.2.2

In the limit of strong coupling, we obtain

$$\lim_{\lambda \to \infty} \mathscr{E} = -\frac{\pi^2}{1440a^3}, \tag{3.22}$$

which is exactly one-half the energy found by Casimir for perfectly conducting plates [1]. Evidently, in this case, the TE modes (calculated here) and the TM modes (calculated in the following subsection) give equal contributions.

3.2.1 TM Modes

To verify this last claim, we solve a similar problem with boundary conditions that the derivative of g is continuous at $x = 0$ and a,

$$\frac{\partial}{\partial x} g(x,x') \Big|_{x=0,a} \quad \text{is continuous}, \tag{3.23a}$$

but the function itself is discontinuous,

$$g(x,x') \Big|_{x=a-}^{x=a+} = \lambda \frac{\partial}{\partial x} g(x,x') \Big|_{x=a}, \tag{3.23b}$$

and similarly at $x = 0$. (Here the coupling λ has dimensions of length.) These boundary conditions reduce, in the limit of strong coupling, to Neumann boundary conditions on the planes, appropriate to electromagnetic TM modes:

$$\lambda \to \infty: \qquad \frac{\partial}{\partial x} g(x,x') \Big|_{x=0,a} = 0. \tag{3.23c}$$

It is completely straightforward to work out the reduced Green's function in this case. When both points are between the planes, $0 < x, x' < a$,

$$g(x, x') = \frac{1}{2\kappa} e^{-\kappa|x-x'|} + \frac{1}{2\kappa\tilde{\Delta}} \left\{ \left(\frac{\lambda\kappa}{2}\right)^2 2\cosh\kappa(x - x')\right.$$
$$\left. + \frac{\lambda\kappa}{2}\left(1 + \frac{\lambda\kappa}{2}\right)\left[e^{\kappa(x+x')} + e^{-\kappa(x+x'-2a)}\right]\right\}, \tag{3.24a}$$

while if both points are outside the planes, $a < x, x'$,

$$g(x, x') = \frac{1}{2\kappa} e^{-\kappa|x-x'|}$$
$$+ \frac{1}{2\kappa\tilde{\Delta}}\frac{\lambda\kappa}{2} e^{-\kappa(x+x'-2a)}\left[\left(1 - \frac{\lambda\kappa}{2}\right) + \left(1 + \frac{\lambda\kappa}{2}\right)e^{2\kappa a}\right], \tag{3.24b}$$

where the denominator is

$$\tilde{\Delta} = \left(1 + \frac{\lambda\kappa}{2}\right)^2 e^{2\kappa a} - \left(\frac{\lambda\kappa}{2}\right)^2. \tag{3.25}$$

It is easy to check that in the strong-coupling limit, the appropriate Neumann boundary condition (3.23c) is recovered. For example, in the interior region, $0 < x, x' < a$,

$$\lim_{\lambda\to\infty} g(x, x') = \frac{\cosh\kappa x_< \cosh\kappa(x_> - a)}{\kappa\sinh\kappa a}. \tag{3.26}$$

Now we can compute the pressure on the plane by computing the xx component of the stress tensor, which is given by (3.12a), so we find

$$t_{xx}\big|_{x=a-} = \frac{1}{2i}\left[-\kappa - \frac{2\kappa}{\tilde{\Delta}}\left(\frac{\lambda\kappa}{2}\right)^2\right], \tag{3.27a}$$

$$t_{xx}\big|_{x=a+} = -\frac{1}{2i}\kappa, \tag{3.27b}$$

and the flux of momentum deposited in the plane $x = a$ is

$$t_{xx}\big|_{x=a-} - t_{xx}\big|_{x=a+} = \frac{i\kappa}{\left(\frac{2}{\lambda\kappa} + 1\right)^2 e^{2\kappa a} - 1}, \tag{3.28}$$

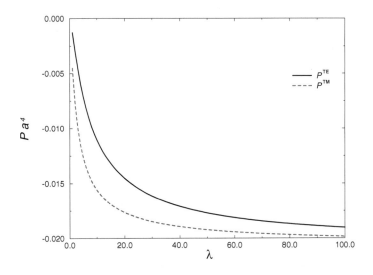

Fig. 3.1 TE and TM Casimir pressures between δ-function planes having strength λ and separated by a distance a. In each case, the pressure is plotted as a function of the dimensionless coupling, λa or λ/a, respectively, for TE and TM contributions

and then by integrating over frequency and transverse momentum we obtain the pressure:

$$P^{\mathrm{TM}} = -\frac{1}{32\pi^2 a^4} \int_0^\infty dy\, y^3 \frac{1}{\left(\frac{4a}{\lambda y}+1\right)^2 e^y - 1}. \tag{3.29}$$

In the limit of weak coupling, this behaves as follows:

$$P^{\mathrm{TM}} \sim -\frac{15}{64\pi^2 a^6}\lambda^2, \tag{3.30}$$

which is to be compared with (3.15a). In strong coupling, on the other hand, it has precisely the same limit as the TE contribution, (3.15b), which confirms the expectation given at the end of the previous subsection. Graphs of the two functions are given in Fig. 3.1.

For calibration purposes we give the Casimir pressure in practical units between ideal perfectly conducting parallel plates at zero temperature:

$$P = -\frac{\pi^2}{240a^4}\hbar c = -\frac{1.30\,\mathrm{mPa}}{(a/1\,\mu\mathrm{m})^4}. \tag{3.31}$$

3.2.2 Self-energy of Boundary Layer

Here we show that the divergent self-energy of a single plate, half the divergent term in (3.17), can be interpreted as the energy associated with the boundary layer. We do this in a simple context by considering a scalar field interacting with the background

$$\mathcal{L}_{\text{int}} = -\frac{\lambda}{2}\phi^2\sigma, \tag{3.32}$$

where the background field σ expands the meaning of the δ function,

$$\sigma(x) = \begin{cases} h, & -\frac{\delta}{2}<x<\frac{\delta}{2}, \\ 0, & \text{otherwise,} \end{cases} \tag{3.33}$$

with the property that $h\delta = 1$. The reduced Green's function satisfies

$$\left[-\frac{\partial^2}{\partial x^2} + \kappa^2 + \lambda\sigma(x)\right]g(x,x') = \delta(x-x'). \tag{3.34}$$

This may be easily solved in the region of the slab, $-\frac{\delta}{2}<x<\frac{\delta}{2}$,

$$g(x,x') = \frac{1}{2\kappa'}\left\{e^{-\kappa'|x-x'|} + \frac{1}{\Delta}\left[\lambda h\cosh\kappa'(x+x')\right.\right.$$
$$\left.\left. + (\kappa'-\kappa)^2 e^{-\kappa'\delta}\cosh\kappa'(x-x')\right]\right\}. \tag{3.35}$$

Here $\kappa' = \sqrt{\kappa^2 + \lambda h}$, and

$$\hat{\Delta} = 2\kappa\kappa'\cosh\kappa'\delta + (\kappa^2 + \kappa'^2)\sinh\kappa'\delta. \tag{3.36}$$

This result may also easily be derived from the multiple reflection formulas given in [46], and agrees with that given by Graham and Olum [51].

Let us proceed here with more generality, and consider the stress tensor with an arbitrary conformal term [52],

$$T^{\mu\nu} = \partial^\mu\phi\partial^\nu\phi - \frac{1}{2}g^{\mu\nu}(\partial_\lambda\phi\partial^\lambda\phi + \lambda h\phi^2) - \xi(\partial^\mu\partial^\nu - g^{\mu\nu}\partial^2)\phi^2, \tag{3.37}$$

in $d+2$ dimensions, d being the number of transverse dimensions, and ξ is an arbitrary parameter, sometimes called the conformal parameter. Applying the corresponding differential operator to the Green's function (3.35), introducing polar coordinates in the (ζ, k) plane, with $\zeta = \kappa\cos\theta$, $k = \kappa\sin\theta$, and

$$\langle\sin^2\theta\rangle = \frac{d}{d+1}, \tag{3.38}$$

we get the following form for the energy density within the slab.

$$\langle T^{00}\rangle = \frac{2^{-d-2}\pi^{-(d+1)/2}}{\Gamma((d+3)/2)} \int\limits_0^\infty \frac{d\kappa\kappa^d}{\kappa'\hat\Delta} \left\{ \lambda h \left[(1-4\xi)(1+d)\kappa'^2 - \kappa^2\right] \cosh 2\kappa'x \right.$$

$$\left. - (\kappa'-\kappa)^2 e^{-\kappa'\delta}\kappa^2 \right\}, \quad -\delta/2 < x < \delta/2. \tag{3.39}$$

We can also calculate the energy density on the other side of the boundary, from the Green's function for $x, x' < -\delta/2$,

$$g(x,x') = \frac{1}{2\kappa}\left[e^{-\kappa|x-x'|} - e^{\kappa(x+x'+\delta)}\lambda h \frac{\sinh\kappa'\delta}{\hat\Delta} \right], \tag{3.40}$$

and the corresponding energy density is given by

$$\langle T^{00}\rangle = -\frac{d(1-4\xi(d+1)/d)}{2^{d+2}\pi^{(d+1)/2}\Gamma((d+3)/2)} \int\limits_0^\infty d\kappa\kappa^{d+1}\frac{1}{\hat\Delta}\lambda h e^{2\kappa(x+\delta/2)}\sinh\kappa'\delta, \tag{3.41}$$

which vanishes if the conformal value of ξ is used. An identical contribution comes from the region $x > \delta/2$.

Integrating $\langle T^{00}\rangle$ over all space gives the vacuum energy of the slab

$$E_{\text{slab}} = -\frac{1}{2^{d+2}\pi^{(d+1)/2}\Gamma((d+3)/2)} \int\limits_0^\infty d\kappa\kappa^d \frac{1}{\kappa'\hat\Delta}\left[(\kappa'-\kappa)^2\kappa^2 e^{-\kappa'\delta}\delta \right.$$

$$\left. + (\lambda h)^2\frac{\sinh\kappa'\delta}{\kappa'} \right]. \tag{3.42}$$

Note that the conformal term does not contribute to the total energy. If we now take the limit $\delta \to 0$ and $h \to \infty$ so that $h\delta = 1$, we immediately obtain the self-energy of a single δ-function plate:

$$E_\delta = \lim_{h\to\infty} E_{\text{slab}} = \frac{1}{2^{d+2}\pi^{(d+1)/2}\Gamma((d+3)/2)} \int\limits_0^\infty d\kappa\kappa^d \frac{\lambda}{\lambda+2\kappa}. \tag{3.43}$$

which for $d=2$ precisely coincides with one-half the constant term in (3.17). There is no surface term in the total Casimir energy as long as the slab is of finite width, because we may easily check that $\frac{d}{dx}g\big|_{x=x'}$ is continuous at the boundaries $\pm\frac{\delta}{2}$. However, if we only consider the energy internal to the slab we encounter not only the integrated energy density but a surface term from the integration by parts—see (3.21). It is the complement of this boundary term that gives rise to E_δ, (3.43), in this way of proceeding. That is, as $\delta \to 0$,

$$-\int\limits_{\text{slab}} (d\mathbf{r}) \int d\zeta\, \zeta^2 \mathscr{G}(\mathbf{r},\mathbf{r}) = 0, \tag{3.44}$$

so

$$E_\delta = \hat{E}\big|_{x=-\delta/2} + \hat{E}\big|_{x=\delta/2}, \tag{3.45}$$

with the normal defining the surface energies pointing into the slab. This means that in this limit, the slab and surface energies coincide.

Further insight is provided by examining the local energy density. In this we follow the work of Graham and Olum [51, 53]. From (3.39) we can calculate the behavior of the energy density as the boundary is approached from the inside:

$$\langle T^{00} \rangle \sim \frac{\Gamma(d+1)\lambda h}{2^{d+4}\pi^{(d+1)/2}\Gamma((d+3)/2)} \frac{1 - 4\xi(d+1)/d}{(\delta - 2|x|)^d}, \quad |x| \to \delta/2. \tag{3.46}$$

For $d = 2$ for example, this agrees with the result found in [51] for $\xi = 0$:

$$\langle T^{00} \rangle \sim \frac{\lambda h}{96\pi^2} \frac{(1 - 6\xi)}{(\delta/2 - |x|)^2}, \quad |x| \to \frac{\delta}{2}. \tag{3.47}$$

Note that, as we expect, this surface divergence vanishes for the conformal stress tensor [52], where $\xi = d/4(d+1)$. (There will be subleading divergences if $d > 2$.) The divergent term in the local energy density from the outside, (3.41), as $x \to -\delta/2$, is just the negative of that found in (3.46). This is why, when the total energy is computed by integrating the energy density, it is finite for $d < 2$, and independent of ξ. The divergence encountered for $d = 2$ may be handled by renormalization of the interaction potential [51].

Note, further, that for a thin slab, close to the exterior but such that the slab still appears thin, $x \gg \delta$, the sum of the exterior and interior energy density divergences combine to give the energy density outside a δ-function potential:

$$u_\delta = -\frac{\lambda}{96\pi^2}(1 - 6\xi)\left[\frac{h}{(x - \delta/2)^2} - \frac{h}{(x + \delta/2)^2}\right] = -\frac{\lambda}{48\pi^2}\frac{1 - 6\xi}{x^3}, \tag{3.48}$$

for small x. Although this limit might be criticized as illegitimate, this result is correct for a δ-function potential, and we will see that this divergence structure occurs also in spherical and cylindrical geometries, so that it is a universal surface divergence without physical significance, barring gravity.

For further discussion on surface divergences, see Sect. 3.3

3.3 Surface and Volume Divergences

It is well known as we have just seen that in general the Casimir energy density diverges in the neighborhood of a surface. For flat surfaces and conformal theories (such as the conformal scalar theory considered above [36], or electromagnetism)

those divergences are not present.[1] In particular, Brown and Maclay [58] calculated the local stress tensor for two ideal plates separated by a distance a along the
z axis, with the result for a conformal scalar

$$\langle T^{\mu\nu} \rangle = -\frac{\pi^2}{1440 a^4} [4\hat{z}^\mu \hat{z}^\nu - g^{\mu\nu}]. \tag{3.49}$$

This result was given more recent rederivations in [59, 36]. Dowker and
Kennedy [60] and Deutsch and Candelas [61] considered the local stress tensor
between planes inclined at an angle α, with the result, in cylindrical coordinates
(t, r, θ, z),

$$\langle T^{\mu\nu} \rangle = -\frac{f(\alpha)}{720\pi^2 r^4} \begin{pmatrix} 1 & 0 & 0 & 0 \\ 0 & -1 & 0 & 0 \\ 0 & 0 & 3 & 0 \\ 0 & 0 & 0 & -1 \end{pmatrix}, \tag{3.50}$$

where for a conformal scalar, with Dirichlet boundary conditions,

$$f(\alpha) = \frac{\pi^2}{2\alpha^2} \left(\frac{\pi^2}{\alpha^2} - \frac{\alpha^2}{\pi^2} \right), \tag{3.51}$$

and for electromagnetism, with perfect conductor boundary conditions,

$$f(\alpha) = \left(\frac{\pi^2}{\alpha^2} + 11 \right) \left(\frac{\pi^2}{\alpha^2} - 1 \right). \tag{3.52}$$

For $\alpha \to 0$ we recover the pressures and energies for parallel plates, (3.15b) and
(3.31). (These results were later discussed in [62].)

Although for perfectly conducting flat surfaces, the energy density is finite, for
electromagnetism the individual electric and magnetic fields have divergent RMS
values,

$$\langle E^2 \rangle \sim -\langle B^2 \rangle \sim \frac{1}{\epsilon^4}, \qquad \epsilon \to 0, \tag{3.53}$$

a distance ϵ above a conducting surface. However, if the surface is a dielectric,
characterized by a plasma dispersion relation, these divergences are softened

$$\langle E^2 \rangle \sim \frac{1}{\epsilon^3}, \qquad -\langle B^2 \rangle \sim \frac{1}{\epsilon^2}, \qquad \epsilon \to 0, \tag{3.54}$$

so that the energy density also diverges [63, 64]

---

[1] In general, this need not be the case. For example, Romeo and Saharian [54] show that with
mixed boundary conditions the surface divergences need not vanish for parallel plates. For
additional work on local effects with mixed (Robin) boundary conditions, applied to spheres and
cylinders, and corresponding global effects, see [55–57, 50]. See also Sect. 3.2.2 and [51, 53].

$$\langle T^{00} \rangle \sim \frac{1}{\epsilon^3}, \quad \epsilon \to 0. \tag{3.55}$$

The null energy condition $(n_\mu n^\mu = 0)$

$$T^{\mu\nu} n_\mu n_\nu \geq 0 \tag{3.56}$$

is satisfied, so that gravity still focuses light.

Graham [65, 66] examined the general relativistic energy conditions required by causality. In the neighborhood of a smooth domain wall, given by a hyperbolic tangent, the energy density is always negative at large enough distances. Thus the weak energy condition is violated, as is the null energy condition (3.56). However, when (3.56) is integrated over a complete geodesic, positivity is satisfied. It is not clear if this last condition, the Averaged Null Energy Condition, is always obeyed in flat space. Certainly it is violated in curved space, but the effects always seem small, so that exotic effects such as time travel are prohibited.

However, as Deutsch and Candelas [61] showed many years ago, in the neighborhood of a curved surface for conformally invariant theories, $\langle T_{\mu\nu} \rangle$ diverges as ϵ^{-3}, where ϵ is the distance from the surface, with a coefficient proportional to the sum of the principal curvatures of the surface. In particular they obtain the result, in the vicinity of the surface,

$$\langle T_{\mu\nu} \rangle \sim \epsilon^{-3} T^{(3)}_{\mu\nu} + \epsilon^{-2} T^{(2)}_{\mu\nu} + \epsilon^{-1} T^{(1)}_{\mu\nu}, \tag{3.57}$$

and obtain explicit expressions for the coefficient tensors $T^{(3)}_{\mu\nu}$ and $T^{(2)}_{\mu\nu}$ in terms of the extrinsic curvature of the boundary.

For example, for the case of a sphere, the leading surface divergence has the form, for conformal fields, for $r = a + \epsilon$, $\epsilon \to 0$

$$\langle T_{\mu\nu} \rangle = \frac{A}{\epsilon^3} \begin{pmatrix} 2/a & 0 & 0 & 0 \\ 0 & 0 & 0 & 0 \\ 0 & 0 & a & 0 \\ 0 & 0 & 0 & a\sin^2\theta \end{pmatrix}, \tag{3.58}$$

in spherical polar coordinates, where the constant is $A = 1/720\pi^2$ for a scalar field satisfying Dirichlet boundary conditions, or $A = 1/60\pi^2$ for the electromagnetic field satisfying perfect conductor boundary conditions. Note that (3.58) is properly traceless. The cubic divergence in the energy density near the surface translates into the quadratic divergence in the energy found for a conducting ball [67]. The corresponding quadratic divergence in the stress corresponds to the absence of the cubic divergence in $\langle T_{rr} \rangle$.

This is all completely sensible. However, in their paper Deutsch and Candelas [61] expressed a certain skepticism about the validity of the result of [68] for the spherical shell case (described in part in Sect. 3.4.2) where the divergences cancel. That skepticism was reinforced in a later paper by Candelas [31], who criticized the authors of [68] for omitting δ function terms, and constants in the energy.

These objections seem utterly without merit. In a later critical paper by the same author [70], it was asserted that errors were made, rather than a conscious removal of unphysical divergences.

Of course, surface curvature divergences are present. As Candelas noted [69, 70], they have the form

$$E = E^S \int dS + E^C \int dS(\kappa_1 + \kappa_2) + E_I^C \int dS(\kappa_1 - \kappa_2)^2 + E_{II}^C \int dS\kappa_1\kappa_2 + \cdots,$$

(3.59)

where κ_1 and κ_2 are the principal curvatures of the surface. The question is to what extent are they observable. After all, as has been shown in [38, 36] and in Sect. 3.2.2, we can drastically change the local structure of the vacuum expectation value of the energy-momentum tensor in the neighborhood of flat plates by merely exploiting the ambiguity in the definition of that tensor, yet each yields the same finite, observable (and observed!) energy of interaction between the plates. For curved boundaries, much the same is true. A priori, we do not know which energy-momentum tensor to employ, and the local vacuum-fluctuation energy density is to a large extent meaningless. It is the global energy, or the force between distinct bodies, that has an unambiguous value. It is the belief of the author that divergences in the energy which go like a power of the cutoff are probably unobservable, being subsumed in the properties of matter. Moreover, the coefficients of the divergent terms depend on the regularization scheme. Logarithmic divergences, of course, are of another class [40]. Dramatic cancellations of these curvature terms can occur. It might be thought that the reason a finite result was found for the Casimir energy of a perfectly conducting spherical shell [15, 20, 68] is that the term involving the squared difference of curvatures in (3.59) is zero only in that case. However, it has been shown that at least for the case of electromagnetism the corresponding term is not present (or has a vanishing coefficient) for an arbitrary smooth cavity [21], and so the Casimir energy for a perfectly conducting ellipsoid of revolution, for example, is finite.[2] This finiteness of the Casimir energy (usually referred to as the vanishing of the second heat-kernel coefficient [71]) for an ideal smooth closed surface was anticipated already in [20], but contradicted by [61]. More specifically, although odd curvature terms cancel inside and outside for any thin shell, it would be anticipated that the squared-curvature term, which is present as a surface divergence in the energy density, would be reflected as an unremovable divergence in the energy. For a closed surface the last term in (3.59) is a topological invariant, so gives an irrelevant constant, while no term of the type of the penultimate term can appear due to the structure of the traced cylinder expansion [50].

---

[2] The first steps have been made for calculating the Casimir energy for an ellipsoidal boundary [34, 35], but only for scalar fields since the vector Helmholtz equation is not separable in the exterior region.

3.4 Casimir Forces on Spheres via δ-Function Potentials

This section is an adaptation and an extension of calculations presented in [45, 46]. This investigation was carried out in response to the program of the MIT group [41–44, 49]. They first rediscovered irremovable divergences in the Casimir energy for a circle in $2 + 1$ dimensions first discovered by Sen [72, 73], but then found divergences in the case of a spherical surface, thereby casting doubt on the validity of the Boyer calculation [15]. Some of their results, as we shall see, are spurious, and the rest are well known [40]. However, their work has been valuable in sparking new investigations of the problems of surface energies and divergences.

We now carry out the calculation we presented in Sect. 3.2 in three spatial dimensions, with a radially symmetric background

$$\mathcal{L}_{\text{int}} = -\frac{1}{2}\frac{\lambda}{a^2}\delta(r-a)\phi^2(x), \tag{3.60}$$

which would correspond to a Dirichlet shell in the limit $\lambda \to \infty$. The scaling of the coupling, which here has dimensions of length, is demanded by the requirement that the spatial integral of the potential be independent of a. The time-Fourier transformed Green's function satisfies the equation ($\kappa^2 = -\omega^2$)

$$\left[-\nabla^2 + \kappa^2 + \frac{\lambda}{a^2}\delta(r-a)\right]\mathcal{G}(\mathbf{r},\mathbf{r}') = \delta(\mathbf{r}-\mathbf{r}'). \tag{3.61}$$

We write $\mathcal{G}$ in terms of a reduced Green's function

$$\mathcal{G}(\mathbf{r},\mathbf{r}') = \sum_{lm} g_l(r,r')Y_{lm}(\Omega)Y_{lm}^*(\Omega'), \tag{3.62}$$

where g_l satisfies

$$\left[-\frac{1}{r^2}\frac{d}{dr}r^2\frac{d}{dr} + \frac{l(l+1)}{r^2} + \kappa^2 + \frac{\lambda}{a^2}\delta(r-a)\right]g_l(r,r') = \frac{1}{r^2}\delta(r-r'). \tag{3.63}$$

We solve this in terms of modified Bessel functions, $I_\nu(x)$, $K_\nu(x)$, where $\nu = l + 1/2$, which satisfy the Wronskian condition

$$I_\nu'(x)K_\nu(x) - K_\nu'(x)I_\nu(x) = \frac{1}{x}. \tag{3.64}$$

The solution to (3.63) is obtained by requiring continuity of g_l at each singularity, at r' and a, and the appropriate discontinuity of the derivative. Inside the sphere we then find ($0 < r, r' < a$)

$$g_l(r,r') = \frac{1}{\kappa r r'}\left[e_l(\kappa r_>)s_l(\kappa r_<) - \frac{\lambda}{\kappa a^2}s_l(\kappa r)s_l(\kappa r')\frac{e_l^2(\kappa a)}{1 + \frac{\lambda}{\kappa a^2}s_l(\kappa a)e_l(\kappa a)}\right]. \tag{3.65}$$

Here we have introduced the modified Riccati-Bessel functions,

$$s_l(x) = \sqrt{\frac{\pi x}{2}} I_{l+1/2}(x), \quad e_l(x) = \sqrt{\frac{2x}{\pi}} K_{l+1/2}(x). \tag{3.66}$$

Note that (3.65) reduces to the expected Dirichlet result, vanishing as $r \to a$, in the limit of strong coupling:

$$\lim_{\lambda \to \infty} g_l(r, r') = \frac{1}{\kappa r r'} \left[e_l(\kappa r_>)s_l(\kappa r_<) - \frac{e_l(\kappa a)}{s_l(\kappa a)} s_l(\kappa r)s_l(\kappa r') \right]. \tag{3.67}$$

When both points are outside the sphere, $r, r' > a$, we obtain a similar result:

$$g_l(r, r') = \frac{1}{\kappa r r'} \left[e_l(\kappa r_>)s_l(\kappa r_<) - \frac{\lambda}{\kappa a^2} e_l(\kappa r)e_l(\kappa r') \frac{s_l^2(\kappa a)}{1 + \frac{\lambda}{\kappa a^2} s_l(\kappa a)e_l(\kappa a)} \right]. \tag{3.68}$$

which similarly reduces to the expected result as $\lambda \to \infty$.

Now we want to get the radial–radial component of the stress tensor to extract the pressure on the sphere, which is obtained by applying the operator

$$\partial_r \partial_{r'} - \frac{1}{2}(-\partial^0 \partial'^0 + \nabla \cdot \nabla') \to \frac{1}{2} \left[\partial_r \partial_{r'} - \kappa^2 - \frac{l(l+1)}{r^2} \right] \tag{3.69}$$

to the Green's function, where in the last term we have averaged over the surface of the sphere. Alternatively, we could notice that [74]

$$\nabla \cdot \nabla' P_l(\cos \gamma) \Big|_{\gamma \to 0} = \frac{l(l+1)}{r^2}, \tag{3.70}$$

where γ is the angle between the two directions. In this way we find, from the discontinuity of $\langle T_{rr} \rangle$ across the $r = a$ surface, the net stress

$$\mathscr{S} = -\frac{\lambda}{2\pi a^3} \sum_{l=0}^{\infty} (2l+1) \int_0^{\infty} dx \frac{(e_l(x)s_l(x))' - \frac{2e_l(x)s_l(x)}{x}}{1 + \frac{\lambda a e_l(x)s_l(x)}{x}}. \tag{3.71}$$

(Notice that there was an error in the sign of the stress, and of the scaling of the coupling, in [45, 46].)

The same result can be deduced by computing the total energy (3.16). The free Green's function, the first term in (3.65) or (3.68), evidently makes no significant contribution to the energy, for it gives a term independent of the radius of the sphere, a, so we omit it. The remaining radial integrals are simply

$$\int_0^x dy \, s_l^2(y) = \frac{1}{2x} \left[(x^2 + l(l+1))s_l^2(x) + x s_l(x)s_l'(x) - x^2 s_l'^2(x) \right], \tag{3.72a}$$

$$\int_{x}^{\infty} dy\, e_l^2(y) = -\frac{1}{2x}\left[(x^2 + l(l+1))e_l^2(x) + xe_l(x)e_l'(x) - x^2 e_l'^2(x)\right]. \quad (3.72b)$$

Then using the Wronskian (3.64), we find that the Casimir energy is

$$E = -\frac{1}{2\pi a}\sum_{l=0}^{\infty}(2l+1)\int_{0}^{\infty} dx\, x\frac{d}{dx}\ln\left[1 + \frac{\lambda}{a}I_\nu(x)K_\nu(x)\right]. \quad (3.73)$$

If we differentiate with respect to a we immediately recover the force (3.71). This expression, upon integration by parts, coincides with that given by Barton [75], and was first analyzed in detail by Scandurra [76]. This result has also been rederived using the multiple-scattering formalism [25]. For strong coupling, it reduces to the well-known expression for the Casimir energy of a massless scalar field inside and outside a sphere upon which Dirichlet boundary conditions are imposed, that is, that the field must vanish at $r = a$:

$$\lim_{\lambda\to\infty} E = -\frac{1}{2\pi a}\sum_{l=0}^{\infty}(2l+1)\int_{0}^{\infty} dx\, x\frac{d}{dx}\ln[I_\nu(x)K_\nu(x)], \quad (3.74)$$

because multiplying the argument of the logarithm by a power of x is without effect, corresponding to a contact term. Details of the evaluation of (3.74) are given in [36], and will be considered in Sect. 3.4.2 below. (See also [77–79].)

The opposite limit is of interest here. The expansion of the logarithm is immediate for small λ. The first term, of order λ, is evidently divergent, but irrelevant, since that may be removed by renormalization of the tadpole graph. In contradistinction to the claim of [42–44, 49], the order λ^2 term is finite, as established in [36]. That term is

$$E^{(\lambda^2)} = \frac{\lambda^2}{4\pi a^3}\sum_{l=0}^{\infty}(2l+1)\int_{0}^{\infty} dx\, x\frac{d}{dx}[I_{l+1/2}(x)K_{l+1/2}(x)]^2. \quad (3.75)$$

The sum on l can be carried out using a trick due to Klich [80]: The sum rule

$$\sum_{l=0}^{\infty}(2l+1)e_l(x)s_l(y)P_l(\cos\theta) = \frac{xy}{\rho}e^{-\rho}, \quad (3.76)$$

where $\rho = \sqrt{x^2 + y^2 - 2xy\cos\theta}$, is squared, and then integrated over θ, according to

$$\int_{-1}^{1} d(\cos\theta)P_l(\cos\theta)P_{l'}(\cos\theta) = \delta_{ll'}\frac{2}{2l+1}. \quad (3.77)$$

In this way we learn that

$$\sum_{l=0}^{\infty}(2l+1)e_l^2(x)s_l^2(x) = \frac{x^2}{2}\int_0^{4x}\frac{dw}{w}e^{-w}. \tag{3.78}$$

Although this integral is divergent, because we did not integrate by parts in (3.75), that divergence does not contribute:

$$E^{(\lambda^2)} = \frac{\lambda^2}{4\pi a^3}\int_0^{\infty}dx\frac{1}{2}x\frac{d}{dx}\int_0^{4x}\frac{dw}{w}e^{-w} = \frac{\lambda^2}{32\pi a^3}, \tag{3.79}$$

which is exactly the result (4.25) of [36].

However, before we are too euphoric, we recognize that the order λ^3 term appears logarithmically divergent, just as [44, 49] claim. This does not signal a breakdown in perturbation theory. Suppose we subtract off and add back in the two leading terms,

$$E = -\frac{1}{2\pi a}\sum_{l=0}^{\infty}(2l+1)\int_0^{\infty}dx\,x\frac{d}{dx}\left[\ln\left(1+\frac{\lambda}{a}I_vK_v\right)-\frac{\lambda}{a}aI_vK_v+\frac{\lambda^2}{2a^2}(I_vK_v)^2\right]+\frac{\lambda^2}{32\pi a^3}.$$

$$\tag{3.80}$$

To study the behavior of the sum for large values of l, we can use the uniform asymptotic expansion (Debye expansion), for $v\to\infty$,

$$\begin{aligned}I_v(x) &\sim \sqrt{\frac{t}{2\pi v}}e^{v\eta}\left(1+\sum_k\frac{u_k(t)}{v^k}\right), \\ K_v(x) &\sim \sqrt{\frac{\pi t}{2v}}e^{-v\eta}\left(1+\sum_k(-1)^k\frac{u_k(t)}{v^k}\right),\end{aligned} \tag{3.81}$$

where

$$x=mz, \quad t=1/\sqrt{1+z^2}, \quad \eta(z)=\sqrt{1+z^2}+\ln\left[\frac{z}{1+\sqrt{1+z^2}}\right], \quad \frac{d\eta}{dz}=\frac{1}{zt}. \tag{3.82}$$

The polynomials in t appearing in (3.81) are generated by

$$u_0(t)=1, \quad u_k(t)=\frac{1}{2}t^2(1-t^2)u_{k-1}'(t)+\frac{1}{8}\int_0^t ds(1-5s^2)u_{k-1}(s). \tag{3.83}$$

We now insert these expansions into (3.80) and expand not in λ but in v; the leading term is

$$E^{(\lambda^3)} \sim \frac{\lambda^3}{24\pi a^4} \sum_{l=0}^{\infty} \frac{1}{\nu} \int_0^{\infty} \frac{dz}{(1+z^2)^{3/2}} = \frac{\lambda^3}{24\pi a^4} \zeta(1). \tag{3.84}$$

Although the frequency integral is finite, the angular momentum sum is divergent. The appearance here of the divergent $\zeta(1)$ seems to signal an insuperable barrier to extraction of a finite Casimir energy for finite λ. The situation is different in the limit $\lambda \to \infty$ —See Sect. 3.4.2.

This divergence has been known for many years, and was first calculated explicitly in 1998 by Bordag et al. [40], where the second heat kernel coefficient gave an equivalent result,

$$E \sim \frac{\lambda^3}{48\pi a^4} \frac{1}{s}, \quad s \to 0. \tag{3.85}$$

A possible way of dealing with this divergence was advocated in [76]. More recently, Bordag and Vassilevich [81] have reanalyzed such problems from the heat kernel approach. They show that this $O(\lambda^3)$ divergence corresponds to a surface tension counterterm, an idea proposed by me in 1980 [82, 83] in connection with the zero-point energy contribution to the bag model. Such a surface term corresponds to λ fixed, which then necessarily implies a divergence of order λ^3. Bordag argues that it is perfectly appropriate to insert a surface tension counterterm so that this divergence may be rendered finite by renormalization.

3.4.1 TM Spherical Potential

Of course, the scalar model considered in the previous subsection is merely a toy model, and something analogous to electrodynamics is of far more physical relevance. There are good reasons for believing that cancellations occur in general between TE (Dirichlet) and TM (Robin) modes. Certainly they do occur in the classic Boyer energy of a perfectly conducting spherical shell [15, 20, 68], and the indications are that such cancellations occur even with imperfect boundary conditions [75]. Following the latter reference, let us consider the potential

$$\mathcal{L}_{\text{int}} = \frac{1}{2} \lambda \frac{1}{r} \frac{\partial}{\partial r} \delta(r-a) \phi^2(x). \tag{3.86}$$

Here λ again has dimensions of length. In the limit $\lambda \to \infty$ this corresponds to TM boundary conditions. The reduced Green's function is thus taken to satisfy

$$\left[-\frac{1}{r^2} \frac{\partial}{\partial r} r^2 \frac{\partial}{\partial r} + \frac{l(l+1)}{r^2} + \kappa^2 - \frac{\lambda}{r} \frac{\partial}{\partial r} \delta(r-a) \right] g_l(r,r') = \frac{1}{r^2} \delta(r-r'). \tag{3.87}$$

At $r = r'$ we have the usual boundary conditions, that g_l be continuous, but that its derivative be discontinuous,

$$r^2 \frac{\partial}{\partial r} g_l \Bigg|_{r=r'-}^{r=r'+} = -1, \tag{3.88}$$

while at the surface of the sphere the derivative is continuous,

$$\frac{\partial}{\partial r} r g_l \Bigg|_{r=a-}^{r=a+} = 0, \tag{3.89a}$$

while the function is discontinuous,

$$g_l \Bigg|_{r=a-}^{r=a+} = -\frac{\lambda}{a} \frac{\partial}{\partial r} r g_l \Bigg|_{r=a}. \tag{3.89b}$$

Equations (3.89a) and (3.89b) are the analogues of the boundary conditions (3.23a, b) treated in Sect. 3.2.1.

It is then easy to find the Green's function. When both points are inside the sphere,

$$r, r' < a: \quad g_l(r, r') = \frac{1}{\kappa r r'} \left[s_l(\kappa r_<) e_l(\kappa r_>) - \frac{\lambda \kappa [e_l'(\kappa a)]^2 s_l(\kappa r) s_l(\kappa r')}{1 + \lambda \kappa e_l'(\kappa a) s_l'(\kappa a)} \right], \tag{3.90a}$$

and when both points are outside the sphere,

$$r, r' > a: \quad g_l(r, r') = \frac{1}{\kappa r r'} \left[s_l(\kappa r_<) e_l(\kappa r_>) - \frac{\lambda \kappa [s_l'(\kappa a)]^2 e_l(\kappa r) e_l(\kappa r')}{1 + \lambda \kappa e_l'(\kappa a) s_l'(\kappa a)} \right]. \tag{3.90b}$$

It is immediate that these supply the appropriate Robin boundary conditions in the $\lambda \to \infty$ limit:

$$\lim_{\lambda \to 0} \frac{\partial}{\partial r} r g_l \Bigg|_{r=a} = 0. \tag{3.91}$$

The Casimir energy may be readily obtained from (3.16), and we find, using the integrals (3.72a, b)

$$E = -\frac{1}{2\pi a} \sum_{l=0}^{\infty} (2l+1) \int_0^{\infty} dx \, x \frac{d}{dx} \ln \left[1 + \frac{\lambda}{a} x e_l'(x) s_l'(x) \right]. \tag{3.92}$$

The stress may be obtained from this by applying $-\partial/\partial a$, and regarding λ as constant, or directly, from the Green's function by applying the operator,

$$t_{rr} = \frac{1}{2i} \left[\nabla_r \nabla_{r'} - \kappa^2 - \frac{l(l+1)}{r^2} \right] g_l \Bigg|_{r'=r}, \tag{3.93}$$

which is the same as that in (3.69), except that

$$\nabla_r = \frac{1}{r}\partial_r r, \tag{3.94}$$

appropriate to TM boundary conditions (see [84], for example). Either way, the total stress on the sphere is

$$\mathcal{S} = -\frac{\lambda}{2\pi a^3}\sum_{l=0}^{\infty}(2l+1)\int_0^{\infty}dx\, x^2 \frac{\left[e_l'(x)s_l'(x)\right]'}{1+\frac{\lambda}{a}xe_l'(x)s_l'(x)}. \tag{3.95}$$

The result for the energy (3.92) is similar, but not identical, to that given by Barton [75].

Suppose we now combine the TE and TM Casimir energies, (3.73) and (3.92):

$$E^{\mathrm{TE}} + E^{\mathrm{TM}} = -\frac{1}{2\pi a}\sum_{l=0}^{\infty}(2l+1)\int_0^{\infty}dx\, x\frac{d}{dx}\ln\left[\left(1+\frac{\lambda}{a}\frac{e_l s_l}{x}\right)\left(1+\frac{\lambda}{a}xe_l's_l'\right)\right]. \tag{3.96}$$

In the limit $\lambda \to \infty$ this reduces to the familiar expression for the perfectly conducting spherical shell [68]:

$$\lim_{\lambda \to \infty} E = -\frac{1}{2\pi a}\sum_{l=1}^{\infty}(2l+1)\int_0^{\infty}dx\, x\left(\frac{e_l'}{e_l}+\frac{e_l''}{e_l'}+\frac{s_l'}{s_l}+\frac{s_l''}{s_l'}\right). \tag{3.97}$$

Here we have, as appropriate to the electrodynamic situation, omitted the $l = 0$ mode. This expression yields a finite Casimir energy, as we will see in Sect. 3.4.2. What about finite λ? In general, it appears that there is no chance that the divergence found in the previous section in order λ^3 can be cancelled. But suppose the coupling for the TE and TM modes are different. If $\lambda^{\mathrm{TE}}\lambda^{\mathrm{TM}} = 4a^2$, a cancellation appears possible, as discussed in [46].

3.4.2 Evaluation of Casimir Energy for a Dirichlet Spherical Shell

In this section we will evaluate the above expression (3.74) for the Casimir energy for a massless scalar in three space dimensions, with a spherical boundary on which the field vanishes. This corresponds to the TE modes for the electrodynamic situation first solved by Boyer [15, 20, 68]. The purpose of this section (adapted from [36, 46]) is to emphasize anew that, contrary to the implication of [42–44, 49], the corresponding Casimir energy is also finite for this configuration.

The general calculation in D spatial dimensions was given in [77]; the pressure is given by the formula

$$P = -\sum_{l=0}^{\infty} \frac{(2l+D-2)\Gamma(l+D-2)}{l!2^D\pi^{(D+1)/2}\Gamma(\frac{D-1}{2})a^{D+1}} \int_0^{\infty} dx\, x\frac{d}{dx}\ln\left[I_\nu(x)K_\nu(x)x^{2-D}\right]. \quad (3.98)$$

Here $\nu = l - 1 + D/2$. For $D = 3$ this expression reduces to

$$P = -\frac{1}{8\pi^2 a^4}\sum_{l=0}^{\infty}(2l+1)\int_0^{\infty} dx\, x\frac{d}{dx}\ln\left[I_{l+1/2}(x)K_{l+1/2}(x)/x\right]. \quad (3.99)$$

This precisely corresponds to the strong limit $\lambda \to \infty$ given in (3.74), if we recall the comment made about contact terms there. In [77] we evaluated expression (3.98) by continuing in D from a region where both the sum and integrals existed. In that way, a completely finite result was found for all positive D not equal to an even integer.

Here we will adopt a perhaps more physical approach, that of allowing the time-coordinates in the underlying Green's function to approach each other, temporal point-splitting, as described in [68]. That is, we recognize that the x integration above is actually a (dimensionless) imaginary frequency integral, and therefore we should replace

$$\int_0^{\infty} dx\, f(x) = \frac{1}{2}\int_{-\infty}^{\infty} dy\, e^{iy\delta}f(|y|), \quad (3.100)$$

where at the end we are to take $\delta \to 0$. Immediately, we can replace the x^{-1} inside the logarithm in (3.99) by x, which makes the integrals converge, because the difference is proportional to a δ function in the time separation, a contact term without physical significance.

To proceed, we use the uniform asymptotic expansions for the modified Bessel functions, (3.81). This is an expansion in inverse powers of $\nu = l + 1/2$, low terms in which turn out to be remarkably accurate even for modest l. The leading terms in this expansion are, using (3.81),

$$\ln\left[xI_{l+1/2}(x)K_{l+1/2}(x)\right] \sim \ln\frac{zt}{2} + \frac{1}{\nu^2}g(t) + \frac{1}{\nu^4}h(t) + \cdots, \quad (3.101)$$

$$g(t) = \frac{1}{8}(t^2 - 6t^4 + 5t^6), \quad (3.102a)$$

$$h(t) = \frac{1}{64}(13t^4 - 284t^6 + 1062t^8 - 1356t^{10} + 565t^{12}). \quad (3.102b)$$

The leading term in the pressure is therefore

$$P_0 = -\frac{1}{8\pi^2 a^4}\sum_{l=0}^{\infty}(2l+1)\nu\int_0^{\infty} dz\, t^2 = -\frac{1}{8\pi a^4}\sum_{l=0}^{\infty}\nu^2 = \frac{3}{32\pi a^4}\zeta(-2) = 0, \quad (3.103)$$

where in the last step we have used the formal zeta function evaluation[3]

$$\sum_{l=0}^{\infty} v^{-s} = (2^s - 1)\zeta(s).$$ (3.104)

Here the rigorous way to argue is to recall the presence of the point-splitting factor $e^{ivz\delta}$ and to carry out the sum on l using

$$\sum_{l=0}^{\infty} e^{ivz\delta} = -\frac{1}{2i}\frac{1}{\sin z\delta/2}$$ (3.105)

so

$$\sum_{l=0}^{\infty} v^2 e^{ivz\delta} = -\frac{d^2}{d(z\delta)^2}\frac{i}{2\sin z\delta/2} = \frac{i}{8}\left(-\frac{2}{\sin^3 z\delta/2} + \frac{1}{\sin z\delta/2}\right).$$ (3.106)

Then P_0 is given by the divergent expression

$$P_0 = \frac{i}{4\pi^2 a^4 \delta^3} \int_{-\infty}^{\infty} \frac{dz}{z^3}\frac{1}{1+z^2},$$ (3.107)

which we argue is zero because the integrand is odd, as justified by averaging over contours passing above and below the pole at $z = 0$.

The next term in the uniform asymptotic expansion (3.101), that involving g, likewise gives zero pressure, as intimated by (3.104), which vanishes at $s = 0$. The same conclusion follows from point splitting, using (3.105) and arguing that the resulting integrand $\sim z^2 t^3 g'(t)/z\delta$ is odd in z. Again, this cancellation does not occur in the electromagnetic case because there the sum starts at $l = 1$.

So here the leading term which survives is that of order v^{-4} in (3.101), namely

$$P_2 = \frac{1}{4\pi^2 a^4}\sum_{l=0}^{\infty}\frac{1}{v^2}\int_0^{\infty} dz\, h(t),$$ (3.108)

where we have now dropped the point-splitting factor because this expression is completely convergent. The integral over z is

$$\int_0^{\infty} dz\, h(t) = \frac{35\pi}{32768}$$ (3.109)

and the sum over l is $3\zeta(2) = \pi^2/2$, so the leading contribution to the stress on the sphere is

$$\mathscr{S}_2 = 4\pi a^2 P_2 = \frac{35\pi^2}{65536 a^2} = \frac{0.00527094}{a^2}.$$ (3.110)

---

[3] Note that the corresponding TE contribution the electromagnetic Casimir pressure would not be zero, for there the sum starts from $l = 1$.

Numerically this is a terrible approximation.

What we must do now is return to the full expression and add and subtract the leading asymptotic terms. This gives

$$\mathscr{S} = \mathscr{S}_2 - \frac{1}{2\pi a^2} \sum_{l=0}^{\infty} (2l+1) R_l, \tag{3.111}$$

where

$$R_l = Q_l + \int_0^{\infty} dx \left[\ln zt + \frac{1}{v^2} g(t) + \frac{1}{v^4} h(t) \right], \tag{3.112}$$

where the integral

$$Q_l = - \int_0^{\infty} dx \ln[2xI_v(x)K_v(x)] \tag{3.113}$$

was given the asymptotic form in [77, 38] ($l \gg 1$):

$$Q_l \sim \frac{v\pi}{2} + \frac{\pi}{128v} - \frac{35\pi}{32768v^3} + \frac{565\pi}{1048577v^5} - \frac{1208767\pi}{2147483648v^7}$$
$$+ \frac{138008357\pi}{137438953472v^9} + \cdots. \tag{3.114}$$

The first two terms in (3.114) cancel the second and third terms in (3.112), of course. The third term in (3.114) corresponds to $h(t)$, so the last three terms displayed in (3.114) give the asymptotic behavior of the remainder, which we call $w(v)$. Then we have, approximately,

$$\mathscr{S} \approx \mathscr{S}_2 - \frac{1}{\pi a^2} \sum_{l=0}^{n} v R_l - \frac{1}{\pi a^2} \sum_{l=n+1}^{\infty} v w(v). \tag{3.115}$$

For $n = 1$ this gives $\mathscr{S} \approx 0.00285278/a^2$, and for larger n this rapidly approaches the value first given in [77], and rederived in [78, 79, 85]

$$\mathscr{S}^{\mathrm{TE}} = 0.002817/a^2, \tag{3.116}$$

a value much smaller than the famous electromagnetic result [15, 86, 68, 20],

$$\mathscr{S}^{\mathrm{EM}} = \frac{0.04618}{a^2}, \tag{3.117}$$

because of the cancellation of the leading terms noted above. Indeed, the TM contribution was calculated separately in [84], with the result

$$\mathscr{S}^{\mathrm{TM}} = -0.02204 \frac{1}{a^2}, \tag{3.118}$$

and then subtracting the $l = 0$ modes from both contributions we obtain (3.117),

$$\mathscr{E}^{\mathrm{EM}} = \mathscr{E}^{\mathrm{TE}} + \mathscr{E}^{\mathrm{TM}} + \frac{\pi}{48a^2} = \frac{0.0462}{a^2}. \tag{3.119}$$

3.4.3 Surface Divergences in the Energy Density

The following discussion is based on [74]. Using (3.70), we immediately find the following expression for the energy density inside or outside the sphere:

$$\langle T^{00} \rangle = \int_0^\infty \frac{d\kappa}{2\pi} \sum_{l=0}^\infty \frac{2l+1}{4\pi} \left\{ \left[-\kappa^2 + \partial_r \partial_{r'} + \frac{l(l+1)}{r^2} \right] g_l(r,r') \Big|_{r'=r} \right.$$
$$\left. - 2\xi \frac{1}{r^2} \frac{\partial}{\partial r} r^2 \frac{\partial}{\partial r} g_l(r,r) \right\}, \tag{3.120}$$

where ξ is the conformal parameter as seen in (3.37). To find the energy density in either region we insert the appropriate Green's functions (3.65) or (3.68), but delete the free part,

$$g_l^0 = \frac{1}{\kappa r r'} s_l(\kappa r_<) e_l(\kappa r_>), \tag{3.121}$$

which corresponds to the *bulk energy* which would be present if either medium filled all of space, leaving us with for $r > a$

$$u(r) = -(1 - 4\xi) \int_0^\infty \frac{d\kappa}{2\pi} \sum_{l=0}^\infty \frac{2l+1}{4\pi} \frac{\frac{\lambda}{\kappa a^2} s_l^2(\kappa a)}{1 + \frac{\lambda}{\kappa a^2} e_l(\kappa a) s_l(\kappa a)} \left\{ \frac{e_l^2(\kappa r)}{\kappa r^2} \left[-\kappa^2 \frac{1+4\xi}{1-4\xi} \right. \right.$$
$$\left. \left. + \frac{l(l+1)}{r^2} + \frac{1}{r^2} \right] - \frac{2}{r^3} e_l(\kappa r) e_l'(\kappa r) + \frac{\kappa}{r^2} e_l'^2(\kappa r) \right\}. \tag{3.122}$$

Inside the shell, $r < a$, the energy is given by a similar expression obtained from (3.122) by interchanging e_l and s_l.

We want to examine the singularity structure as $r \to a$ from the outside. For this purpose we use the leading uniform asymptotic expansion, $l \to \infty$, obtained from (3.81)

$$e_l(x) \sim \sqrt{zt} \, e^{-\nu\eta}, \quad s_l(x) \sim \frac{1}{2} \sqrt{zt} \, e^{\nu\eta},$$
$$e_l'(x) \sim -\frac{1}{\sqrt{zt}} e^{-\nu\eta}, \quad s_l'(x) \sim \frac{1}{2} \frac{1}{\sqrt{zt}} e^{\nu\eta}, \tag{3.123}$$

where $\nu = l + 1/2$, and z, t, and η are given in (3.82). The coefficient of $e_l(\kappa r) e_l(\kappa r')$ occurring in the δ-function potential Green's function (3.68), in strong and weak coupling, becomes

$$\frac{\lambda}{a} \to \infty : \to \frac{s_l(\kappa a)}{e_l(\kappa a)} \tag{3.124a}$$

$$\frac{\lambda}{a} \to 0 : \to \frac{\lambda}{\kappa a^2} s_l^2(\kappa a). \tag{3.124b}$$

In either case, we carry out the asymptotic sum over angular momentum using (3.123) and the analytic continuation of (3.105)

$$\sum_{l=0}^{\infty} e^{-\nu \chi} = \frac{1}{2 \sinh \frac{\chi}{2}} \tag{3.125}$$

Here ($r \approx a$)

$$\chi = 2\left[\eta(z) - \eta\left(z\frac{a}{r}\right)\right] \approx 2z\frac{d\eta}{dz}(z)\frac{r-a}{r} = \frac{2}{t}\frac{r-a}{r}. \tag{3.126}$$

The remaining integrals over z are elementary, and in this way we find that the leading divergences in the energy density are as $r \to a+$,

$$\frac{\lambda}{a} \to \infty : \quad u \sim -\frac{1}{16\pi^2} \frac{1 - 6\xi}{(r-a)^4} \tag{3.127a}$$

$$\frac{\lambda}{a} \to 0 : \quad u^{(n)} \sim \left(-\frac{\lambda}{a}\right)^n \frac{\Gamma(4-n)}{96\pi^2 a^4}(1 - 6\xi)\left(\frac{a}{r-a}\right)^{4-n}, \quad n < 4, \tag{3.127b}$$

where the latter is the leading divergence in order n. These results clearly seem to demonstrate the virtue of the conformal value of $\xi = 1/6$; but see below. (The value for the Dirichlet sphere (127a) first appeared in [61]; it more recently was rederived in [87], where, however, the subdominant term, the leading term if $\xi = 1/6$, namely (3.130), was not calculated. Of course, this result is the same as the surface divergence encountered for parallel Dirichlet plates [36, 38].) The perturbative divergence for $n = 1$ in (3.127b) is exactly that found for a plate—see (3.48).

Thus, for $\xi = 1/6$ we must keep subleading terms. This includes keeping the subdominant term in χ,[4]

$$\chi \approx \frac{2}{t}\frac{r-a}{r} + t\left(\frac{r-a}{r}\right)^2, \tag{3.128}$$

the distinction between $t(z)$ and $\tilde{t} = t(\tilde{z} = za/r)$,

$$\tilde{z}\tilde{t} \approx zt - t^2 z\frac{r-a}{r}, \tag{3.129}$$

---

[4] Note there is a sign error in (4.8) of [74].

as well as the next term in the uniform asymptotic expansion of the Bessel functions (3.81). Including all this, it is straightforward to recover the well-known result (3.58) [61] for strong coupling (Dirichlet boundary conditions):

$$\frac{\lambda}{a} \to \infty: \quad u \sim \frac{1}{360\pi^2} \frac{1}{a(r-a)^3}, \tag{3.130}$$

Following the same process for weak coupling, we find that the leading divergence in order n, $1 \leq n < 3$, is $(r \to a\pm)$

$$\lambda \to 0: \quad u^{(n)} \sim \left(\frac{\lambda}{a^2}\right)^n \frac{1}{1440\pi^2} \frac{1}{a(a-r)^{3-n}} (n-1)(n+2)\Gamma(3-n). \tag{3.131}$$

Note that the subleading $O(\lambda)$ term again vanishes. Both (3.130, 3.131) apply for the conformal value $\xi = 1/6$.

3.4.4 Total Energy and Renormalization

As discussed in [74] we may consider the potential, in the spirit of (3.32),

$$\mathscr{L}_{\text{int}} = -\frac{\lambda}{2a^2}\phi^2\sigma(r), \tag{3.132a}$$

where

$$\sigma(r) = \begin{cases} 0, & r < a_-, \\ h, & a_- < r < a_+, \\ 0, & a_+ < r. \end{cases} \tag{3.132b}$$

Here $a_\pm = a \pm \delta/2$, and we set $h\delta = 1$. That is, we have expanded the δ-function shell so that it has finite thickness.

In particular, the integrated local energy density inside, outside, and within the shell is E_{in}, E_{out}, and E_{sh}, respectively. The total energy of a given region is the sum of the integrated local energy and the surface energy (3.20a) bounding that region ($\xi = 1/6$):

$$\tilde{E}_{\text{in}} = E_{\text{in}} + \hat{E}_-, \tag{3.133a}$$

$$\tilde{E}_{\text{out}} = E_{\text{out}} + \hat{E}_+, \tag{3.133b}$$

$$\tilde{E}_{\text{sh}} = E_{\text{sh}} + \hat{E}'_+ + \hat{E}'_-, \tag{3.133c}$$

where $\hat{E}_\pm$ is the outside (inside) surface energy on the surface at $r = a_\pm$, while $\hat{E}'_\pm$ is the inside (outside) surface energy on the same surfaces. E_{in}, E_{out}, and E_{sh}

represent $\int (d\mathbf{r}) \langle T^{00} \rangle$ in each region. Because for a nonsingular potential the surface energies cancel across each boundary,

$$\hat{E}_+ + \hat{E}'_+ = 0, \quad \hat{E}_- + \hat{E}'_- = 0, \tag{3.134}$$

the total energy is

$$E = \tilde{E}_{in} + \tilde{E}_{out} + \tilde{E}_{sh} = E_{in} + E_{out} + E_{sh}. \tag{3.135}$$

In the singular thin shell limit, the integrated local shell energy is the total surface energy of a thin Dirichlet shell:

$$E_{sh} = \hat{E}_+ + \hat{E}_- \neq 0. \tag{3.136}$$

See the remark at the end of Sect. 3.2.2. This shell energy, for the conformally coupled theory, is finite in second order in the coupling (in at least two plausible regularization schemes), but diverges in third order. We showed in [74] that the latter precisely corresponds to the known divergence of the total energy in this order. Thus we have established the suspected correspondence between surface divergences and divergences in the total energy, which has nothing to do with divergences in the local energy density as the surface is approached. This precise correspondence should enable us to absorb such global divergences in a renormalization of the surface energy, and should lead to further advances of our understanding of quantum vacuum effects. We will elaborate on this point in the following.

3.5 Semitransparent Cylinder

This section is based on [37]. We consider a massless scalar field ϕ in a δ-cylinder background,

$$\mathscr{L}_{int} = -\frac{\lambda}{2a} \delta(r - a)\phi^2, \tag{3.137}$$

a being the radius of the "semitransparent" cylinder. The massive case was earlier considered by Scandurra [88]. We will continue to assume that the dimensionless coupling $\lambda > 0$ to avoid the appearance of negative eigenfrequencies. The time-Fourier transform of the Green's function satisfies

$$\left[-\nabla^2 - \omega^2 + \lambda\delta(r - a) \right] \mathscr{G}(\mathbf{r}, \mathbf{r}') = \delta(\mathbf{r} - \mathbf{r}'). \tag{3.138}$$

Adopting cylindrical coordinates, we write

$$\mathscr{G}(\mathbf{r}, \mathbf{r}') = \int \frac{dk}{2\pi} e^{ik(z-z')} \sum_{m=-\infty}^{\infty} \frac{1}{2\pi} e^{im(\varphi-\varphi')} g_m(r, r'; k), \tag{3.139}$$

where the reduced Green's function satisfies

$$\left[-\frac{1}{r}\frac{\partial}{\partial r}r\frac{\partial}{\partial r} + \kappa^2 + \frac{m^2}{r^2} + \frac{\lambda}{a}\delta(r-a)\right]g_m(r,r';k) = \frac{1}{r}\delta(r-r'), \qquad (3.140)$$

where $\kappa^2 = k^2 - \omega^2$. Let us immediately make a Euclidean rotation,

$$\omega \to i\zeta, \qquad (3.141)$$

where ζ is real, so κ is likewise always real. Apart from the δ functions, this is the modified Bessel equation.

Because of the Wronskian (3.64) satisfied by the modified Bessel functions, we have the general solution to (3.140) as long as $r \ne a$ to be

$$g_m(r,r';k) = I_m(\kappa r_<)K_m(\kappa r_>) + A(r')I_m(\kappa r) + B(r')K_m(\kappa r), \qquad (3.142)$$

where A and B are arbitrary functions of r'. Now we incorporate the effect of the δ function at $r = a$ in (3.140). It implies that g_m must be continuous at $r = a$, while it has a discontinuous derivative,

$$\frac{\partial}{\partial r}g_m(r,r';k)\bigg|_{r=a-}^{r=a+} = \frac{\lambda}{a}g_m(a,r';k), \qquad (3.143)$$

from which we rather immediately deduce the form of the Green's function inside and outside the cylinder:

$$r,r' < a: \quad g_m(r,r';k) = I_m(\kappa r_<)K_m(\kappa r_>)$$
$$- \frac{\lambda K_m^2(\kappa a)}{1 + \lambda I_m(\kappa a)K_m(\kappa a)}I_m(\kappa r)I_m(\kappa r'), \qquad (3.144a)$$

$$r,r' > a: \quad g_m(r,r';k) = I_m(\kappa r_<)K_m(\kappa r_>)$$
$$- \frac{\lambda I_m^2(\kappa a)}{1 + \lambda I_m(\kappa a)K_m(\kappa a)}K_m(\kappa r)K_m(\kappa r'). \qquad (3.144b)$$

Notice that in the limit $\lambda \to \infty$ we recover the Dirichlet cylinder result, that is, that g_m vanishes at $r = a$.

3.5.1 Cylinder Pressure and Energy

The easiest way to calculate the total energy is to compute the pressure on the cylindrical walls due to the quantum fluctuations in the field. This may be computed, at the one-loop level, from the vacuum expectation value of the stress tensor,

$$\langle T^{\mu\nu}\rangle = \left(\partial^\mu\partial^{\prime\nu} - \frac{1}{2}g^{\mu\nu}\partial^\lambda\partial'_\lambda\right)\frac{1}{i}G(x,x')\Big|_{x=x'} - \xi(\partial^\mu\partial^\nu - g^{\mu\nu}\partial^2)\frac{1}{i}G(x,x), \quad (3.145)$$

which we have written in a Cartesian coordinate system. Here we have again included the conformal parameter ξ, which is equal to 1/6 for the stress tensor that makes conformal invariance manifest. The conformal term does not contribute to the radial-radial component of the stress tensor, however, because then only transverse and time derivatives act on $G(x,x)$, which depends only on r. The discontinuity of the expectation value of the radial-radial component of the stress tensor is the pressure of the cylindrical wall:

$$P = \langle T_{rr}\rangle_{\text{in}} - \langle T_{rr}\rangle_{\text{out}}$$

$$= -\frac{1}{16\pi^3}\sum_{m=-\infty}^{\infty}\int_{-\infty}^{\infty}dk\int_{-\infty}^{\infty}d\zeta\frac{\lambda\kappa^2}{1+\lambda I_m(\kappa a)K_m(\kappa a)}$$

$$\times\left[K_m^2(\kappa a)I_m'^2(\kappa a) - I_m^2(\kappa a)K_m'^2(\kappa a)\right]$$

$$= -\frac{1}{16\pi^3}\sum_{m=-\infty}^{\infty}\int_{-\infty}^{\infty}dk\int_{-\infty}^{\infty}d\zeta\frac{\kappa}{a}\frac{d}{d\kappa a}\ln[1+\lambda I_m(\kappa a)K_m(\kappa a)], \quad (3.146)$$

where we have again used the Wronskian (3.64) . Regarding ka and ζa as the two Cartesian components of a two-dimensional vector, with magnitude $x \equiv \kappa a = \sqrt{k^2a^2 + \zeta^2a^2}$, we get the stress on the cylinder per unit length to be

$$\mathscr{S} = 2\pi a P = -\frac{1}{4\pi a^3}\int_0^\infty dx\, x^2\sum_{m=-\infty}^{\infty}\frac{d}{dx}\ln[1+\lambda I_m(x)K_m(x)], \quad (3.147)$$

which possesses the expected Dirichlet limit as $\lambda \to \infty$. The corresponding expression for the total Casimir energy per unit length follows by integrating

$$\mathscr{S} = -\frac{\partial}{\partial a}\mathscr{E}, \quad (3.148)$$

that is,

$$\mathscr{E} = -\frac{1}{8\pi a^2}\int_0^\infty dx\, x^2\sum_{m=-\infty}^{\infty}\frac{d}{dx}\ln[1+\lambda I_m(x)K_m(x)]. \quad (3.149)$$

This expression, the analog of (3.73) for the spherical case, is, of course, completely formal, and will be regulated in various ways, for example, with an analytic or exponential regulator as we will see in the following, or by using zeta-function regularization [37].

Alternatively, we may compute the energy directly from the general formula (3.16). To evaluate (3.16) in this case, we use the standard indefinite integrals over squared Bessel functions. When we insert the above construction of the Green's function (3.144a, b), and perform the integrals over the regions interior and exterior to the cylinder we obtain (3.149) immediately.

3.5.2 Weak-coupling Evaluation

Suppose we regard λ as a small parameter, so let us expand (3.149) in powers of λ. The first term is

$$\mathscr{E}^{(1)} = -\frac{\lambda}{8\pi a^2} \sum_{m=-\infty}^{\infty} \int_0^\infty dx\, x^2 \frac{d}{dx} K_m(x) I_m(x). \qquad (3.150)$$

The addition theorem for the modified Bessel functions is

$$K_0(kP) = \sum_{m=-\infty}^{\infty} e^{im(\phi-\phi')} K_m(k\rho) I_m(k\rho'), \quad \rho > \rho', \qquad (3.151)$$

where $P = \sqrt{\rho^2 + \rho'^2 - 2\rho\rho' \cos(\phi - \phi')}$. If this is extrapolated to the limit $\rho' = \rho$ we conclude that the sum of the Bessel functions appearing in (3.150) is $K_0(0)$, that is, a constant, so there is no first-order contribution to the energy. For a rigorous derivation of this result, see [37].

We can proceed the same way to evaluate the second-order contribution,

$$\mathscr{E}^{(2)} = \frac{\lambda^2}{16\pi a^2} \int_0^\infty dx\, x^2 \frac{d}{dx} \sum_{m=-\infty}^{\infty} I_m^2(x) K_m^2(x). \qquad (3.152)$$

By squaring the sum rule (3.151), and taking the limit $\rho' \to \rho$, we evaluate the sum over Bessel functions appearing here as

$$\sum_{m=-\infty}^{\infty} I_m^2(x) K_m^2(x) = \int_0^{2\pi} \frac{d\varphi}{2\pi} K_0^2(2x \sin \varphi/2). \qquad (3.153)$$

Then changing the order of integration we find that the second-order energy can be written as

$$\mathscr{E}^{(2)} = -\frac{\lambda^2}{64\pi^2 a^2} \int_0^{2\pi} \frac{d\varphi}{\sin^2 \varphi/2} \int_0^\infty dz\, z\, K_0^2(z), \qquad (3.154)$$

where the Bessel-function integral has the value 1/2. However, the integral over φ is divergent. We interpret this integral by adopting an analytic regularization based on the integral [31]

$$\int_0^{2\pi} d\varphi \left(\sin \frac{\varphi}{2}\right)^s = \frac{2\sqrt{\pi} \Gamma\left(\frac{1+s}{2}\right)}{\Gamma\left(1 + \frac{s}{2}\right)}, \qquad (3.155)$$

which holds for $\mathrm{Re}\, s > -1$. Taking the right-side of this equation to define the φ integral for all s, we conclude that the φ integral in (3.154), and hence the second-order energy $\mathscr{E}^{(2)}$, is zero.

3.5.2.1 Numerical Evaluation

Given that the above argument evidently formally omits divergent terms, it may be more satisfactory, as in [31], to offer a numerical evaluation of $\mathscr{E}^{(2)}$. (The corresponding argument for $\mathscr{E}^{(1)}$ is given in [37].) We can very efficiently do so using the uniform asymptotic expansions (3.81). Thus the asymptotic behavior of the product of Bessel functions appearing in (3.152) is

$$I_m^2(x)K_m^2(x) \sim \frac{t^2}{4m^2}\left(1 + \sum_{k=1}^{\infty} \frac{r_k(t)}{m^{2k}}\right). \tag{3.156}$$

The first three polynomials occurring here are

$$r_1(t) = \frac{t^2}{4}(1 - 6t^2 + 5t^4), \tag{3.157a}$$

$$r_2(t) = \frac{t^4}{16}(7 - 148t^2 + 554t^4 - 708t^6 + 295t^8), \tag{3.157b}$$

$$\begin{aligned} r_3(t) = \frac{t^6}{16}(36 &- 1666t^2 + 13775t^4 - 44272t^6 \\ &+ 67162t^8 - 48510t^{10} + 13475t^{12}). \end{aligned} \tag{3.157c}$$

We now write the second-order energy (3.152) as

$$\begin{aligned} \mathscr{E}^{(2)} = -\frac{\lambda^2}{8\pi a^2}\Bigg\{ & \int_0^{\infty} dx\, x\left[I_0^2(x)K_0^2(x) - \frac{1}{4(1+x^2)}\right] \\ & - \frac{1}{4}\lim_{s\to 0}\left(\frac{1}{2} + \sum_{m=1}^{\infty} m^{-s}\right)\int_0^{\infty} dz\, z^{2-s}\frac{d}{dz}\frac{1}{1+z^2} \\ & + 2\int_0^{\infty} dz\, z\frac{t^2}{4}\sum_{m=1}^{\infty}\sum_{k=1}^{3}\frac{r_k(t)}{m^{2k}} \\ & + 2\sum_{m=1}^{\infty}\int_0^{\infty} dx\, x\left[I_m^2(x)K_m^2(x) - \frac{t^2}{4m^2}\left(1 + \sum_{k=1}^{3}\frac{r_k(t)}{m^{2k}}\right)\right]\Bigg\}. \end{aligned} \tag{3.158}$$

In the final integral $z = x/m$. The successive terms are evaluated as

$$\mathscr{E}^{(2)} \approx -\frac{\lambda^2}{8\pi a^2}\left[\frac{1}{4}(\gamma + \ln 4) - \frac{1}{4}\ln 2\pi - \frac{\zeta(2)}{48} + \frac{7\zeta(4)}{1920} - \frac{31\zeta(6)}{16128}\right.$$

$$\left. + 0.000864 + 0.000006\right] = -\frac{\lambda^2}{8\pi a^2}(0.000000), \tag{3.159}$$

where in the last term in (3.158) only the $m = 1$ and 2 terms are significant. Therefore, we have demonstrated numerically that the energy in order λ^2 is zero to an accuracy of better than 10^{-6}.

The astute reader will note that we used a standard, but possibly questionable, analytic regularization in defining the second term in (3.158), where the initial sum and integral are only defined for $1 < s < 2$, and then the result is continued to $s = 0$. Alternatively, we could follow [31] and insert there an exponential regulator in each integral of $e^{-x\delta}$, with δ to be taken to zero at the end of the calculation. For $m \neq 0$ x becomes mz, and then the sum on m becomes

$$\sum_{m=1}^{\infty} e^{-mz\delta} = \frac{1}{e^{z\delta} - 1}. \tag{3.160}$$

Then when we carry out the integral over z we obtain for that term

$$\frac{\pi}{8\delta} - \frac{1}{4}\ln 2\pi. \tag{3.161}$$

Thus we obtain the same finite part as above, but in addition an explicitly divergent term

$$\mathscr{E}^{(2)}_{\text{div}} = -\frac{\lambda^2}{64a^2\delta}. \tag{3.162}$$

If we think of the cutoff in terms of a vanishing proper time τ, $\delta = \tau/a$, this divergent term is proportional to $1/a$, so the divergence in the energy goes like L/a, if L is the (very large) length of the cylinder. This is of the form of the shape divergence encountered in [31].

3.5.2.2 Divergences in the Total Energy

In this subsection we are going to use heat-kernel knowledge to determine the divergence structure in the total energy. We consider a general cylinder of the type $\mathscr{C} = \mathbb{R} \times Y$, where Y is an arbitrary smooth two dimensional region rather than merely being the disc. As a metric we have $ds^2 = dz^2 + dY^2$ from which we obtain that the zeta function (density) associated with the Laplacian on $\mathscr{C}$ is ($\text{Re } s > 3/2$)

$$\zeta(s) = \frac{1}{2\pi} \int\limits_{-\infty}^{\infty} dk \sum_{\lambda_Y} (k^2 + \lambda_Y)^{-s} = \frac{1}{2\pi} \frac{\sqrt{\pi}\Gamma\left(s - \frac{1}{2}\right)}{\Gamma(s)} \sum_{\lambda_Y} \lambda_Y^{1/2-s}$$

$$= \frac{1}{2\pi} \frac{\sqrt{\pi}\Gamma\left(s - \frac{1}{2}\right)}{\Gamma(s)} \zeta_Y\left(s - \frac{1}{2}\right) \tag{3.163}$$

Here λ_Y are the eigenvalues of the Laplacian on Y, and $\zeta_Y(s)$ is the zeta function associated with these eigenvalues. In the zeta-function scheme the Casimir energy is defined as

$$E_{\text{Cas}} = \frac{1}{2} \mu^{2s} \zeta\left(s - \frac{1}{2}\right)\bigg|_{s=0}, \tag{3.164}$$

which, in the present setting, turns into

$$E_{\text{Cas}} = \frac{1}{2} \mu^{2s} \frac{\Gamma(s-1)}{2\sqrt{\pi}\Gamma\left(s - \frac{1}{2}\right)} \zeta_Y(s-1)\bigg|_{s=0}. \tag{3.165}$$

Expanding this expression about $s = 0$, one obtains

$$E_{\text{Cas}} = \frac{1}{8\pi s} \zeta_Y(-1) + \frac{1}{8\pi} \left(\zeta_Y(-1)[2\ln(2\mu) - 1] + \zeta_Y'(-1)\right) + \mathcal{O}(s). \tag{3.166}$$

The contribution associated with $\zeta_Y(-1)$ can be determined solely from the heat-kernel coefficient knowledge, namely

$$\zeta_Y(-1) = -a_4, \tag{3.167}$$

in terms of the standard 4th heat-kernel coefficient. The contribution coming from $\zeta_Y'(-1)$ can in general not be determined. But as we see, at least the divergent term can be determined entirely by the heat-kernel coefficient.

The situation considered in the Casimir energy calculation is a δ-function shell along some smooth line Σ in the plane (here, a circle of radius a). The manifolds considered are the cylinder created by the region inside of the line, and the region outside of the line; from the results the contribution from free Minkowski space has to be subtracted to avoid trivial volume divergences (the representation in terms of the Bessel functions already has Minkowski space contributions subtracted). The δ-function shell generates a jump in the normal derivative of the eigenfunctions; call the jump U (here, $U = \lambda/a$). The leading heat-kernel coefficients for this situation, namely for functions which are continuous across the boundary but which have a jump of the first normal derivative at the boundary, have been determined in [89]; the relevant a_4 coefficient is given in Theorem 7.1, p. 139 of that reference. The results there are very general; for our purpose there is exactly one term that survives, namely

$$a_4 = -\frac{1}{24\pi} \int\limits_{\Sigma} dl\, U^3, \tag{3.168}$$

which shows that

$$E_{\text{Cas}}^{\text{div}} = \frac{1}{192\pi^2 s} \int_\Sigma dl\, U^3. \tag{3.169}$$

So no matter along which line the δ-function shell is concentrated, the first two orders in a weak-coupling expansion do not contribute any divergences in the total energy. But the third order does, and the divergence is given above.

For the example considered, as mentioned, $U = \lambda/a$ is constant, and the integration leads to the length of the line which is $2\pi a$. Thus we get for this particular example

$$\mathscr{E}_{\text{Cas}}^{\text{div}} = \frac{1}{96\pi s} \frac{\lambda^3}{a^2}. \tag{3.170}$$

[Compare this with the corresponding divergence for a sphere, (3.85).] This can be easily checked from the explicit representation we have for the energy. We have already seen that the first two orders in λ identically vanish, while the part of the third order that potentially contributes a divergent piece is

$$\mathscr{E}^{(3)} = -\frac{1}{8\pi a^2} \sum_{m=-\infty}^{\infty} \int_0^\infty dx\, x^{2-2s} \frac{d}{dx} \frac{1}{3} \lambda^3 K_m^3(x) I_m^3(x). \tag{3.171}$$

The $m = 0$ contribution is well behaved about $s = 0$; while for the remaining sum using

$$K_m^3(mz) I_m^3(mz) \sim \frac{1}{8m^3} \frac{1}{(1+z^2)^{3/2}}, \tag{3.172}$$

we see that the leading contribution is

$$\begin{aligned}
\mathscr{E}^{(3)} &\sim -\frac{\lambda^3}{12\pi a^2} \sum_{m=1}^{\infty} m^{2-2s} \int_0^\infty dz\, z^{2-2s} \frac{d}{dz} \frac{1}{8m^3} \frac{1}{(1+z^2)^{3/2}} \\
&= -\frac{\lambda^3}{96\pi a^2} \zeta_R(1+2s) \int_0^\infty dz\, z^{2-2s} \frac{d}{dz} \frac{1}{(1+z^2)^{3/2}} \\
&= \frac{\lambda^3}{96\pi a^2} \zeta_R(1+2s) \frac{\Gamma(2-s)\Gamma\left(s+\frac{1}{2}\right)}{\Gamma(3/2)} = \frac{\lambda^3}{96\pi a^2 s} + \mathcal{O}(s^0),
\end{aligned} \tag{3.173}$$

in perfect agreement with the heat-kernel prediction (3.170).

3.5.3 Strong Coupling

The strong-coupling limit of the energy (3.149), that is, the Casimir energy of a
Dirichlet cylinder,

$$\mathscr{E}^D = -\frac{1}{8\pi a^2} \sum_{m=-\infty}^{\infty} \int_0^{\infty} dx \, x^2 \frac{d}{dx} \ln I_m(x) K_m(x), \tag{3.174}$$

was worked out to high accuracy by Gosdzinsky and Romeo [29],

$$\mathscr{E}^D = \frac{0.000614794033}{a^2}. \tag{3.175}$$

It was later redone with less accuracy by Nesterenko and Pirozhenko [90].

For completeness, let us sketch the evaluation here. We carry out a numerical
calculation (very similar to that of [90]) in the spirit of Sect. 3.5.2.1. We add and
subtract the leading uniform asymptotic expansion (for $m = 0$ the asymptotic
behavior) as follows:

$$
\begin{aligned}
\mathscr{E}^D = -\frac{1}{8\pi a^2} \Bigg\{ &-2 \int_0^{\infty} dx \, x \left[\ln(2xI_0(x)K_0(x)) - \frac{1}{8}\frac{1}{1+x^2} \right] \\
&+ 2 \sum_{m=1}^{\infty} \int_0^{\infty} dx \, x^2 \frac{d}{dx} \left[\ln(2xI_m(x)K_m(x)) - \ln\left(\frac{xt}{m}\right) - \frac{1}{2}\frac{r_1(t)}{m^2} \right] \\
&- 2\left(\frac{1}{2} + \sum_{m=1}^{\infty}\right) \int_0^{\infty} dx \, x^2 \frac{d}{dx} \ln 2x + 2 \sum_{m=1}^{\infty} \int_0^{\infty} dx \, x^2 \frac{d}{dx} \ln xt \\
&+ \sum_{m=1}^{\infty} \int_0^{\infty} dx \, x^2 \frac{d}{dx} \left[\frac{r_1(t)}{m^2} - \frac{1}{4}\frac{1}{1+x^2} \right] \\
&+ \frac{1}{4}\left(\frac{1}{2} + \sum_{m=1}^{\infty}\right) \int_0^{\infty} dx \, x^2 \frac{d}{dx}\frac{1}{1+x^2} \Bigg\}. \tag{3.176}
\end{aligned}
$$

In the first two terms we have subtracted the leading asymptotic behavior so the
resulting integrals are convergent. Those terms are restored in the fourth, fifth, and
sixth terms. The most divergent part of the Bessel functions are removed by
the insertion of $2x$ in the corresponding integral, and its removal in the third term.
(As we've seen above, such terms have been referred to as "contact terms,"
because if a time-splitting regulator, $e^{i\zeta\tau}$, is inserted into the frequency integral, a
term proportional to $\delta(\tau)$ appears, which is zero as long as $\tau \neq 0$.) The terms
involving Bessel functions are evaluated numerically, where it is observed that the
asymptotic value of the summand (for large m) in the second term is $1/32m^2$. The
fourth term is evaluated by writing it as

$$2 \lim_{s \to 0} \sum_{m=1}^{\infty} m^{2-s} \int_0^{\infty} dz \frac{z^{1-s}}{1+z^2} = 2\zeta'(-2) = -\frac{\zeta(3)}{2\pi^2}, \tag{3.177}$$

while the same argument, as anticipated, shows that the third "contact" term is zero,[5] while the sixth term is

$$-\frac{1}{2} \lim_{s \to 0} \left[\zeta(s) + \frac{1}{2} \right] \frac{1}{s} = \frac{1}{4} \ln 2\pi. \tag{3.178}$$

The fifth term is elementary. The result then is

$$\mathcal{E}^D = \frac{1}{4\pi a^2} (0.010963 - 0.0227032 + 0 + 0.0304485 + 0.21875 - 0.229735)$$
$$= \frac{1}{4\pi a^2} (0.007724) = \frac{0.0006146}{a^2}, \tag{3.179}$$

which agrees with (3.175) to the fourth significant figure.

3.5.3.1 Exponential Regulator

As in Sect. 3.5.2.1, it may seem more satisfactory to insert an exponential regulator rather than use analytic regularization. Now it is the third, fourth, and sixth terms in (3.176) that must be treated. The latter is just the negative of (3.161). We can combine the third and fourth terms to give using (3.160)

$$-\frac{1}{\delta^2} - \frac{2}{\delta^2} \int_0^{\infty} \frac{dz\, z^3}{z^2 + \delta^2} \frac{d^2}{dz^2} \frac{1}{e^z - 1}. \tag{3.180}$$

The latter integral may be evaluated by writing it as an integral along the entire z axis, and closing the contour in the upper half plane, thereby encircling the poles at $i\delta$ and at $2in\pi$, where n is a positive integer. The residue theorem then gives for that integral

$$-\frac{2\pi}{\delta^3} - \frac{\zeta(3)}{2\pi^2}, \tag{3.181}$$

so once again we obtain the same finite part as in (3.177). In this way of proceeding, then, in addition to the finite part in (3.179), we obtain divergent terms

---

[5] This argument is a bit suspect, since the analytic continuation that defines the integrals has no common region of existence. Thus the argument in the following subsection may be preferable. However, since that term is properly a contact term, it should in any event be spurious.

$$\mathcal{E}^D_{\text{div}} = \frac{1}{64a^2\delta} + \frac{1}{8\pi a^2\delta^2} + \frac{1}{4a^2\delta^3}, \tag{3.182}$$

which, with the previous interpretation for δ, implies divergent terms in the energy proportional to L/a (shape), L (length), and aL (area), respectively. Such terms presumably are to be subsumed in a renormalization of parameters in the model. Had a logarithmic divergence occurred [as does occur in weak coupling in $\mathcal{O}(\lambda^3)$] such a renormalization would apparently be impossible— however, see [37].

3.5.4 Local Energy Density

We compute the energy density from the stress tensor (3.145), or

$$\langle T^{00} \rangle = \frac{1}{2i} \left(\partial^0 \partial^{0\prime} + \nabla \cdot \nabla^\prime \right) G(x,x^\prime) \Big|_{x^\prime=x} - \frac{\xi}{i} \nabla^2 G(x,x)$$

$$= \frac{1}{16\pi^3 i} \int_{-\infty}^{\infty} dk \int_{-\infty}^{\infty} d\omega \sum_{m=-\infty}^{\infty} \left[\left(\omega^2 + k^2 + \frac{m^2}{r^2} + \partial_r \partial_{r^\prime} \right) g(r,r^\prime) \Big|_{r^\prime=r} \right.$$

$$\left. - 2\xi \frac{1}{r} \partial_r r \partial_r g(r,r) \right]. \tag{3.183}$$

We omit the free part of the Green's function, since that corresponds to the energy that would be present in the vacuum in the absence of the cylinder. When we insert the remainder of the Green's function (3.144b), we obtain the following expression for the energy density outside the cylindrical shell:

$$u(r) = \langle T^{00} - T^{00}_{(0)} \rangle = -\frac{\lambda}{16\pi^3} \int_{-\infty}^{\infty} d\zeta \int_{-\infty}^{\infty} dk \sum_{m=-\infty}^{\infty} \frac{I_m^2(\kappa a)}{1 + \lambda I_m(\kappa a) K_m(\kappa a)}$$

$$\times \left[\left(2\omega^2 + \kappa^2 + \frac{m^2}{r^2} \right) K_m^2(\kappa r) + \kappa^2 K_m^{\prime 2}(\kappa r) - 2\xi \frac{1}{r} \frac{\partial}{\partial r} r \frac{\partial}{\partial r} K_m^2(\kappa r) \right],$$

$$r > a. \tag{3.184}$$

The factor in square brackets can be easily seen to be, from the modified Bessel equation,

$$2\omega^2 K_m^2(\kappa r) + \frac{1 - 4\xi}{2} \frac{1}{r} \frac{\partial}{\partial r} r \frac{\partial}{\partial r} K_m^2(\kappa r). \tag{3.185}$$

For the interior region, $r < a$, we have the corresponding expression for the energy density with $I_m \leftrightarrow K_m$.

3.5.5 Total and Surface Energy

We first need to verify that we recover the expression for the energy found in Sect. 3.5.1. So let us integrate expression (3.184) over the region exterior of the cylinder, and the corresponding interior expression over the inside region. The second term in (3.185) is a total derivative, while the first is exactly the one evaluated in Sect. 3.5.1. The result is

$$
2\pi \int_0^\infty dr\, r\, u(r) = -\frac{1}{8\pi a^2} \sum_{m=-\infty}^\infty \int_0^\infty dx\, x^2 \frac{d}{dx} \ln[1 + \lambda I_m(x) K_m(x)]
$$

$$
- (1 - 4\xi) \frac{\lambda}{4\pi a^2} \int_0^\infty dx\, x \sum_{m=-\infty}^\infty \frac{I_m(x) K_m(x)}{1 + \lambda I_m(x) K_m(x)}. \tag{3.186}
$$

The first term is the total energy (3.149), but what do we make of the second term? In strong coupling, it would represent a constant that should have no physical significance (a contact term—it is independent of a if we revert to the physical variable κ as the integration variable). In general, however, there is another contribution to the total energy, residing precisely on the singular surface. This surface energy is given in general by [60, 91, 92, 55, 50, 45]

$$
\hat{E} = -\frac{1 - 4\xi}{2i} \oint_S d\mathbf{S} \cdot \nabla G(x, x') \Big|_{x'=x}, \tag{3.187}
$$

as given for $\xi = 0$ in (3.20a), where the normal to the surface is out of the region in question. In this case it is easy to see that $\hat{E}$ exactly equals the negative of the second term in (3.186). This is an example of the general theorem (3.21)

$$
\int (d\mathbf{r}) u(\mathbf{r}) + \hat{E} = E, \tag{3.188}
$$

that is, the total energy E is the sum of the integrated local energy density and the surface energy. The generalization of this theorem, (3.187, 3.188), to curved space is given in [57]. A consequence of this theorem is that the total energy, unlike the local energy density, is independent of the conformal parameter ξ. (Note that this surface energy vanishes when $\xi = 1/4$ as Fulling has stressed [93].)

3.5.6 Surface Divergences

We now turn to an examination of the behavior of the local energy density (3.184) as r approaches a from outside the cylinder. To do this we use the uniform asymptotic expansion (3.81). Let us begin by considering the strong-coupling limit, a Dirichlet cylinder. If we stop with only the leading asymptotic behavior, we obtain the expression

$$u(r) \sim -\frac{1}{8\pi^3} \int\limits_0^\infty d\kappa \kappa 2 \sum_{m=1}^\infty e^{-m\chi} \left\{ \left[-\kappa^2 + (1-4\xi)\left(\kappa^2 + \frac{m^2}{r^2}\right) \right] \frac{\pi t}{2m} \right.$$

$$\left. + (1-4\xi)\kappa^2 \frac{\pi}{2mt} \frac{1}{z^2} \right\}, \qquad (\lambda \to \infty), \tag{3.189}$$

where

$$\chi = -2\left[\eta(z) - \eta\left(z\frac{a}{r}\right)\right], \tag{3.190}$$

and we have replaced the integral over k and ζ by one over the polar variable κ as before. Here we ignore the difference between r and a except in the exponent, and we now replace κ by mz/a. Close to the surface,

$$\chi \sim \frac{2}{t} \frac{r-a}{r}, \quad r-a \ll r, \tag{3.191}$$

and we carry out the sum over m according to

$$2 \sum_{m=1}^\infty m^3 e^{-m\chi} \sim -2\frac{d^3}{d\chi^3} \frac{1}{\chi} = \frac{12}{\chi^4} \sim \frac{3}{4} \frac{t^4 r^4}{(r-a)^4}. \tag{3.192}$$

Then the energy density behaves, as $r \to a+$,

$$u(r) \sim -\frac{3}{64\pi^2} \frac{1}{(r-a)^4} \int\limits_0^\infty dz\, z[t^5 + t^3(1-8\xi)]$$

$$= -\frac{1}{16\pi^2} \frac{1}{(r-a)^4} (1-6\xi). \tag{3.193}$$

This is the universal surface divergence first discovered by Deutsch and Candelas [61] and seen for the sphere in (3.127a) [74]. It therefore occurs, with precisely the same numerical coefficient, near a Dirichlet plate [36]. Unless gravity is considered, it is utterly without physical significance, and may be eliminated with the conformal choice for the parameter ξ, $\xi = 1/6$.

We will henceforth make this conformal choice. Then the leading divergence depends upon the curvature. This was also worked out by Deutsch and Candelas [61]; for the case of a cylinder, that result is

$$u(r) \sim \frac{1}{720\pi^2} \frac{1}{r(r-a)^3}, \quad r \to a+, \tag{3.194}$$

exactly 1/2 that for a Dirichlet sphere of radius a (3.130) [74], as anticipated from the general analysis summarized in (3.59). Here, this result may be straightforwardly derived by keeping the $1/m$ corrections in the uniform asymptotic expansion (3.81), as well as the next term in the expansion of χ, (3.128).

3.5.6.1 Weak Coupling

Let us now expand the energy density (3.184) for small coupling,

$$
u(r) = -\frac{\lambda}{16\pi^3} \int\limits_{-\infty}^{\infty} d\zeta \int\limits_{-\infty}^{\infty} dk \sum_{m=-\infty}^{\infty} I_m^2(\kappa a) \sum_{n=0}^{\infty} (-\lambda)^n I_m^n(\kappa a) K_m^n(\kappa a)
$$
$$
\times \left\{ \left[-\kappa^2 + (1-4\xi)\left(\kappa^2 + \frac{m^2}{r^2}\right) \right] K_m^2(\kappa r) + (1-4\xi)\kappa^2 K_m'^2(\kappa r) \right\}.
$$
(3.195)

If we again use the leading uniform asymptotic expansions for the Bessel functions, we obtain the expression for the leading behavior of the term of order λ^n,

$$
u^{(n)}(r) \sim \frac{1}{8\pi^2 r^4} \left(-\frac{\lambda}{2}\right)^n \int\limits_0^{\infty} dz\, z \sum_{m=1}^{\infty} m^{3-n} e^{-m\chi} t^{n-1}(t^2 + 1 - 8\xi).
$$
(3.196)

The sum on m is asymptotic to

$$
\sum_{m=1}^{\infty} m^{3-n} e^{-m\chi} \sim (3-n)! \left(\frac{tr}{2(r-a)}\right)^{4-n}, \quad r \to a+,
$$
(3.197)

so the most singular behavior of the order λ^n term is, as $r \to a+$,

$$
u^{(n)}(r) \sim (-\lambda)^n \frac{(3-n)!(1-6\xi)}{96\pi^2 r^n (r-a)^{4-n}}.
$$
(3.198)

This is exactly the result found for the weak-coupling limit for a δ-sphere (3.127b) [74] and for a δ-plane (3.48) [45], so this is also a universal result, without physical significance. It may be made to vanish by choosing the conformal value $\xi = 1/6$.

With this conformal choice, once again we must expand to higher order. We use the corrections noted above, in (3.81) and (3.128, 3.129). Then again a quite simple calculation gives

$$
u^{(n)} \sim (-\lambda)^n \frac{(n-1)(n+2)\Gamma(3-n)}{2880\pi^2 r^{n+1}(r-a)^{3-n}}, \quad r \to a+,
$$
(3.199)

which is analytically continued from the region $1 \leq \mathrm{Re}\,n < 3$. Remarkably, this is exactly one-half the result found in the same weak-coupling expansion for the leading conformal divergence outside a sphere (3.131) [74]. Therefore, like the strong-coupling result (3.194), this limit is universal, depending on the sum of the principal curvatures of the interface.

In [37] we considered a annular shell of finite thickness, which as the thickness δ tended to zero gave a finite residual energy in the annulus, in terms of the energy density u in the annulus,

$$\mathscr{E}_{\text{ann}} = 2\pi\delta au \sim (1 - 4\xi)\frac{\lambda}{4\pi a^2} \sum_{m=-\infty}^{\infty} \int_0^{\infty} d\kappa a \, \kappa a \frac{I_m(\kappa a)K_m(\kappa a)}{1 + \lambda I_m(\kappa a)K_m(\kappa a)} = \hat{\mathscr{E}}, \quad (3.200)$$

which is exactly the form of the surface energy given by the negative of the second term in (3.186). In particular, note that the term in $\hat{\mathscr{E}}$ of order λ^3 is, for the conformal value $\xi = 1/6$, exactly equal to that term in the total energy $\mathscr{E}$ (3.149): [see (3.171)]

$$\hat{\mathscr{E}}^{(3)} = \mathscr{E}^{(3)}. \quad (3.201)$$

This means that the divergence encountered in the global energy (3.170) is exactly accounted for by the divergence in the surface energy, which would seem to provide strong evidence in favor of the renormizablity of that divergence.

3.6 Gravitational Acceleration of Casimir Energy

We will here show that a body undergoing uniform acceleration (hyperbolic motion) imparts the same acceleration to the quantum vacuum energy associated with this body. This is consistent with the equivalence principle that states that all forms of energy should gravitate equally. A general variational argument, which, however, did not deal with the divergent parts of the energy, was given in [22]. This section is based on [23].

3.6.1 Green's Functions in Rindler Coordinates

Relativistically, uniform acceleration is described by hyperbolic motion,

$$z = \xi \cosh \tau \quad \text{and} \quad t = \xi \sinh \tau. \quad (3.202)$$

Here the proper acceleration of the particle described by these equations is ξ^{-1}, and we have chosen coordinates so that at time $t = 0$, $z(0) = \xi$. Here we are going to consider the corresponding metric

$$ds^2 = -dt^2 + dz^2 + dx^2 + dy^2 = -\xi^2 d\tau^2 + d\xi^2 + dx^2 + dy^2. \quad (3.203)$$

In these coordinates, the d'Alembertian operator takes on cylindrical form

$$-\left(\frac{\partial}{\partial t}\right)^2 + \left(\frac{\partial}{\partial z}\right)^2 + \nabla_\perp^2 = -\frac{1}{\xi^2}\left(\frac{\partial}{\partial \tau}\right)^2 + \frac{1}{\xi}\frac{\partial}{\partial \xi}\left(\xi\frac{\partial}{\partial \xi}\right) + \nabla_\perp^2, \quad (3.204)$$

where $\perp$ refers to the x-y plane.

3.6.1.1 Green's Function for One Plate

For a scalar field in these coordinates, subject to a potential $V(x)$, the action is

$$W = \int d^4x \sqrt{-g(x)} \mathcal{L}(\phi(x)), \tag{3.205}$$

where $x \equiv (\tau, x, y, \xi)$ represents the coordinates, $d^4x = d\tau \, d\xi \, dx \, dy$ is the coordinate volume element, $g_{\mu\nu}(x) = \text{diag}(-\xi^2, +1, +1, +1)$ defines the metric, $g(x) = \det g_{\mu\nu}(x) = -\xi^2$ is the determinant of the metric, and the Lagrangian density is

$$\mathcal{L}(\phi(x)) = -\frac{1}{2} g_{\mu\nu}(x) \partial^\mu \phi(x) \partial^\nu \phi(x) - \frac{1}{2} V(x) \phi(x)^2, \tag{3.206}$$

where for a single semitransparent plate located at ξ_1

$$V(x) = \lambda \delta(\xi - \xi_1), \tag{3.207}$$

and $\lambda > 0$ is the coupling constant having dimensions of mass. More explicitly we have

$$W = \int d^4x \frac{\xi}{2} \left[\frac{1}{\xi^2} \left(\frac{\partial \phi}{\partial \tau} \right)^2 - \left(\frac{\partial \phi}{\partial \xi} \right)^2 - (\nabla_\perp \phi)^2 - V(x) \phi^2 \right]. \tag{3.208}$$

Stationarity of the action under an arbitrary variation in the field leads to the equation of motion

$$\left[-\frac{1}{\xi^2} \frac{\partial^2}{\partial \tau^2} + \frac{1}{\xi} \frac{\partial}{\partial \xi} \xi \frac{\partial}{\partial \xi} + \nabla_\perp^2 - V(x) \right] \phi(x) = 0. \tag{3.209}$$

The corresponding Green's function satisfies the differential equation

$$-\left[-\frac{1}{\xi^2} \frac{\partial^2}{\partial \tau^2} + \frac{1}{\xi} \frac{\partial}{\partial \xi} \xi \frac{\partial}{\partial \xi} + \nabla_\perp^2 - V(x) \right] G(x, x') = \frac{\delta(\xi - \xi')}{\xi} \delta(\tau - \tau') \delta(\mathbf{x}_\perp - \mathbf{x}'_\perp). \tag{3.210}$$

Since in our case $V(x)$ has only ξ dependence we can write this in terms of the reduced Green's function $g(\xi, \xi')$,

$$G(x, x') = \int_{-\infty}^{\infty} \frac{d\omega}{2\pi} \int \frac{d^2k_\perp}{(2\pi)^2} e^{-i\omega(\tau - \tau')} e^{i\mathbf{k}_\perp \cdot (\mathbf{x} - \mathbf{x}')_\perp} g(\xi, \xi'), \tag{3.211}$$

where $g(\xi, \xi')$ satisfies

$$-\left[\frac{1}{\xi} \frac{\partial}{\partial \xi} \xi \frac{\partial}{\partial \xi} + \frac{\omega^2}{\xi^2} - k_\perp^2 - V(x) \right] g(\xi, \xi') = \frac{\delta(\xi - \xi')}{\xi}. \tag{3.212}$$

We recognize this equation as defining the semitransparent cylinder problem discussed in Sect. 3.5 [37], with the replacements

$$a \to \xi_1, \quad m \to \zeta = -i\omega, \quad \kappa \to k = k_\perp, \quad \lambda \to \lambda\xi_1, \tag{3.213}$$

so that from (3.144a, b) we may immediately write down the solution in terms of modified Bessel functions,

$$g(\xi, \xi') = I_\zeta(k\xi_<)K_\zeta(k\xi_>) - \frac{\lambda\xi_1 K_\zeta^2(k\xi_1)I_\zeta(k\xi)I_\zeta(k\xi')}{1 + \lambda\xi_1 I_\zeta(k\xi_1)K_\zeta(k\xi_1)}, \quad \xi, \xi' < \xi_1, \tag{3.214a}$$

$$= I_\zeta(k\xi_<)K_\zeta(k\xi_>) - \frac{\lambda\xi_1 I_\zeta^2(k\xi_1)K_\zeta(k\xi)K_\zeta(k\xi')}{1 + \lambda\xi_1 I_\zeta(k\xi_1)K_\zeta(k\xi_1)}, \quad \xi, \xi' > \xi_1. \tag{3.214b}$$

Note that in the strong-coupling limit, $\lambda \to \infty$, this reduces to the Green's function satisfying Dirichlet boundary conditions at $\xi = \xi_1$.

3.6.1.2 Minkowski-space Limit

To recover the Minkowski-space Green's function for the semitransparent plate, we use the uniform asymptotic expansion (Debye expansion), based on the limit

$$\xi \to \infty, \quad \xi_1 \to \infty, \quad \xi - \xi_1 \text{ finite}, \quad \zeta \to \infty, \quad \zeta/\xi_1 \text{ finite}. \tag{3.215}$$

For large ζ we use (3.81) with $x = \zeta z = k\xi$, for example. Expanding the above expressions (3.214a, b) around some arbitrary point ξ_0, chosen such that the differences $\xi - \xi_0$, $\xi' - \xi_0$, and $\xi_1 - \xi_0$ are finite, we find for the leading term, for example,

$$\sqrt{\xi\xi'}I_\zeta(k\xi)K_\zeta(k\xi') \sim \frac{1}{2\kappa}e^{\kappa(\xi-\xi')}, \tag{3.216}$$

where $\kappa^2 = k^2 + \hat{\zeta}^2$, $\hat{\zeta} = \zeta/\xi_0$. In this way, taking for simplicity $\xi_0 = \xi_1$, we find the Green's function for a single plate in Minkowski space,

$$\xi_1 g(\xi, \xi') \to g^{(0)}(\xi, \xi') = \frac{1}{2\kappa}e^{-\kappa|\xi-\xi'|} - \frac{\lambda}{\lambda + 2\kappa}\frac{1}{2\kappa}e^{-\kappa|\xi-\xi_1|}e^{-\kappa|\xi'-\xi_1|}. \tag{3.217}$$

3.6.1.3 Green's Function for Two Parallel Plates

For two semitransparent plates perpendicular to the ξ-axis and located at ξ_1, ξ_2, with couplings λ_1 and λ_2, respectively, we find the following form for the Green's function:

$$g(\xi, \xi') = I_< K_> - \frac{\lambda_1\xi_1 K_1^2 + \lambda_2\xi_2 K_2^2 - \lambda_1\lambda_2\xi_1\xi_2 K_1 K_2(K_2 I_1 - K_1 I_2)}{\Delta}II', \quad \xi, \xi' < \xi_1,$$

$$\tag{3.218a}$$

$$= I_< K_> - \frac{\lambda_1 \xi_1 I_1^2 + \lambda_2 \xi_2 I_2^2 + \lambda_1 \lambda_2 \xi_1 \xi_2 I_1 I_2 (I_2 K_1 - I_1 K_2)}{\Delta} K K_\prime, \quad \xi, \xi' > \xi_2,$$

(3.218b)

$$= I_< K_> - \frac{\lambda_2 \xi_2 K_2^2 (1 + \lambda_1 \xi_1 K_1 I_1)}{\Delta} I I_\prime - \frac{\lambda_1 \xi_1 I_1^2 (1 + \lambda_2 \xi_2 K_2 I_2)}{\Delta} K K_\prime$$
$$+ \frac{\lambda_1 \lambda_2 \xi_1 \xi_2 I_1^2 K_2^2}{\Delta} (I K_\prime + K I_\prime), \quad \xi_1 < \xi, \xi' < \xi_2,$$

(3.218c)

where

$$\Delta = (1 + \lambda_1 \xi_1 K_1 I_1)(1 + \lambda_2 \xi_2 K_2 I_2) - \lambda_1 \lambda_2 \xi_1 \xi_2 I_1^2 K_2^2, \qquad (3.219)$$

and we have used the abbreviations $I_1 = I_\zeta(k\xi_1)$, $I = I_\zeta(k\xi)$, $I_\prime = I_\zeta(k\xi')$, etc.

Again we can check that these formulas reduce to the well-known Minkowski-space limits. In the $\xi_0 \to \infty$ limit, the uniform asymptotic expansion (3.81) gives, for $\xi_1 < \xi, \xi' < \xi_2$

$$\xi_0 g(\xi, \xi') \to g^{(0)}(\xi, \xi') = \frac{1}{2\kappa} e^{-\kappa|\xi - \xi'|} + \frac{1}{2\kappa\tilde{\Delta}} \left[\frac{\lambda_1 \lambda_2}{4\kappa^2} 2 \cosh \kappa(\xi - \xi') \right.$$
$$\left. - \frac{\lambda_1}{2\kappa} \left(1 + \frac{\lambda_2}{2\kappa} \right) e^{-\kappa(\xi + \xi' - 2\xi_2)} - \frac{\lambda_2}{2\kappa} \left(1 + \frac{\lambda_1}{2\kappa} \right) e^{\kappa(\xi + \xi' - 2\xi_1)} \right],$$

(3.220)

where $(a = \xi_2 - \xi_1)$

$$\tilde{\Delta} = \left(1 + \frac{\lambda_1}{2\kappa} \right) \left(1 + \frac{\lambda_2}{2\kappa} \right) e^{2\kappa a} - \frac{\lambda_1 \lambda_2}{4\kappa^2}, \qquad (3.221)$$

which is exactly the expected result (3.7a, 3.8). The correct limit is also obtained in the other two regions.

3.6.2 Gravitational Acceleration of Casimir Apparatus

We next consider the situation when the plates are forced to "move rigidly" [94] in such a way that the proper distance between the plates is preserved. This is achieved if the two plates move with different but constant proper accelerations.

The canonical energy-momentum or stress tensor derived from the action (3.205) is

$$T_{\alpha\beta}(x) = \partial_\alpha \phi(x) \partial_\beta \phi(x) + g_{\alpha\beta}(x) \mathcal{L}(\phi(x)), \qquad (3.222)$$

where the Lagrange density includes the δ-function potential. The components referring to the pressure and the energy density are

$$T_{33}(x) = \frac{1}{2} \frac{1}{\xi^2} \left(\frac{\partial \phi}{\partial \tau} \right)^2 + \frac{1}{2} \left(\frac{\partial \phi}{\partial \xi} \right)^2 - \frac{1}{2} (\nabla_\perp \phi)^2 - \frac{1}{2} V(x) \phi^2, \qquad (3.223a)$$

$$\frac{1}{\xi^2}T_{00}(x) = \frac{1}{2}\frac{1}{\xi^2}\left(\frac{\partial\phi}{\partial\tau}\right)^2 + \frac{1}{2}\left(\frac{\partial\phi}{\partial\xi}\right)^2 + \frac{1}{2}(\nabla_\perp\phi)^2 + \frac{1}{2}V(x)\phi^2. \qquad (3.223b)$$

The latter may be written in an alternative convenient form using the equations of motion (3.209):

$$T_{00} = \frac{1}{2}\left(\frac{\partial\phi}{\partial\tau}\right)^2 - \frac{1}{2}\phi\frac{\partial^2}{\partial\tau^2}\phi + \frac{\xi}{2}\frac{\partial}{\partial\xi}\left(\phi\xi\frac{\partial}{\partial\xi}\phi\right) + \frac{\xi^2}{2}\nabla_\perp\cdot(\phi\nabla_\perp\phi), \qquad (3.224)$$

which is the appropriate version of (3.19) here. The force density is given by [95] $-\nabla_\nu T^\nu{}_\lambda$, or

$$f_\lambda = -\frac{1}{\sqrt{-g}}\partial_\nu(\sqrt{-g}T^\nu{}_\lambda) + \frac{1}{2}T^{\mu\nu}\partial_\lambda g_{\mu\nu}, \qquad (3.225)$$

or in Rindler coordinates

$$f_\xi = -\frac{1}{\xi}\partial_\xi(\xi T^{\xi\xi}) - \xi T^{00}. \qquad (3.226)$$

When we integrate over all space to get the force, the first term is a surface term which does not contribute[6]:

$$\mathscr{F} = \int d\xi\ \xi f_\xi = -\int \frac{d\xi}{\xi^2}T_{00}. \qquad (3.227)$$

This could be termed the Rindler coordinate force per area, defined as the change in momentum per unit Rindler coordinate time τ per unit cross-sectional area. If we multiply $\mathscr{F}$ by the gravitational acceleration g we obtain the gravitational force per area on the Casimir energy. This result (3.227) seems entirely consistent with the equivalence principle, since $\xi^{-2}T_{00}$ is the energy density. Using the expression (3.224) for the energy density, taking the vacuum expectation value, and rescaling $\zeta = \hat{\zeta}\xi$, we see that the gravitational force per cross sectional area is merely

---

[6] Note that in previous works, such as [45, 46], the surface term was included, because the integration was carried out only over the interior and exterior regions. Here we integrate over the surface as well, so the additional so-called surface energy is automatically included. This is described in the argument leading to (3.20a). Note, however, if (3.226) is integrated over a small interval enclosing the δ-function potential,

$$\int\limits_{\xi_1-\epsilon}^{\xi_1+\epsilon} d\varsigma\xi f_\xi = -\xi_1\Delta T^{\xi\xi},$$

where $\Delta T^{\xi\xi}$ is the discontinuity in the normal-normal component of the stress density. Dividing this expression by ξ_1 gives the usual expression for the force on the plate.

$$\mathcal{F} = \int d\xi \xi \int \frac{d\hat{\zeta} d^2 \mathbf{k}}{(2\pi)^3} \hat{\zeta}^2 g(\xi, \xi). \tag{3.228}$$

This result for the energy contained in the force equation (3.228) is an immediate consequence of the general formula for the Casimir energy (3.16) [38].

Alternatively, we can start from the following formula for the force density for a single semitransparent plate, following directly from the equations of motion (3.209),

$$f_\xi = \frac{1}{2} \phi^2 \partial_\xi \lambda \delta(\xi - \xi_1). \tag{3.229}$$

The vacuum expectation value of this yields the force in terms of the Green's function,

$$\mathcal{F} = -\lambda \frac{1}{2} \int \frac{d\zeta\, d^2 \mathbf{k}}{(2\pi)^3} \partial_\xi [\xi g(\xi, \xi)] \Big|_{\xi=\xi_1}. \tag{3.230}$$

3.6.2.1 Gravitational Force on a Single Plate

For example, the force on a single plate at ξ_1 is given by

$$\mathcal{F} = -\partial_{\xi_1} \frac{1}{2} \int \frac{d\zeta\, d^2 \mathbf{k}}{(2\pi)^3} \ln[1 + \lambda \xi_1 I_\zeta(k\xi_1) K_\zeta(k\xi_1)], \tag{3.231}$$

Expanding this about some arbitrary point ξ_0, with $\zeta = \hat{\zeta}\xi_0$, using the uniform asymptotic expansion (3.81), we get ($\kappa^2 = k^2 + \hat{\zeta}^2$)

$$\xi_1 I_\zeta(k\xi_1) K_\zeta(k\xi_1) \sim \frac{\xi_1}{2\zeta} \frac{1}{\sqrt{1 + (k\xi_1/\zeta)^2}} \approx \frac{\xi_1}{2\kappa\xi_0} \left(1 - \frac{k^2}{\kappa^2} \frac{\xi_1 - \xi_0}{\xi_0}\right). \tag{3.232}$$

From this, if we introduce polar coordinates for the $\mathbf{k}$-$\hat{\zeta}$ integration, the coordinate force is

$$\begin{aligned}
\mathcal{F} &= -\frac{1}{2} \partial_{\xi_1} \frac{\xi_0}{2\pi^2} \int_0^\infty d\kappa\, \kappa^2 \frac{\lambda}{2\kappa + \lambda} \left(1 + \frac{\xi_1 - \xi_0}{\xi_0}\right) \left(1 - \frac{\langle k^2 \rangle}{\kappa^2} \frac{\xi_1 - \xi_0}{\xi_0}\right) \\
&= -\frac{\lambda}{4\pi^2} \partial_{\xi_1} (\xi_1 - \xi_0) \int_0^\infty \frac{d\kappa}{2\kappa + \lambda} \langle \hat{\zeta}^2 \rangle \\
&= -\frac{1}{96\pi^2 a^3} \int_0^\infty \frac{dy\, y^2}{1 + y/\lambda a},
\end{aligned} \tag{3.233}$$

where for example

$$\langle \hat{\zeta}^2 \rangle = \frac{1}{2} \int_{-1}^{1} d\cos\theta \cos^2\theta \kappa^2 = \frac{1}{3}\kappa^2. \tag{3.234}$$

The divergent expression (3.233) is just the negative of the quantum vacuum energy of a single plate, seen in (3.17) and (3.43).

3.6.2.2 Parallel Plates Falling in a Constant Gravitational Field

In general, we have two alternative forms for the gravitational force on the two-plate system:

$$\mathcal{F} = -(\partial_{\xi_1} + \partial_{\xi_2})\frac{1}{2} \int \frac{d\zeta \, d^2\mathbf{k}}{(2\pi)^3} \ln\Delta, \tag{3.235}$$

Δ given in (3.219), which is equivalent to (3.228). (In the latter, however, bulk energy, present if no plates are present, must be omitted.) From either of the above two methods, we find the coordinate force [as defined below (3.227)] is given by

$$\mathcal{F} = -\frac{1}{4\pi^2} \int_0^{\infty} d\kappa\kappa^2 \ln\Delta_0, \tag{3.236}$$

where $\Delta_0 = e^{-2\kappa a}\tilde{\Delta}$, $\tilde{\Delta}$ given in (3.221). The integral may be easily shown to be

$$\mathcal{F} = \frac{1}{96\pi^2 a^3} \int_0^{\infty} dy y^3 \frac{1 + \frac{1}{y+\lambda_1 a} + \frac{1}{y+\lambda_2 a}}{\left(\frac{y}{\lambda_1 a} + 1\right)\left(\frac{y}{\lambda_2 a} + 1\right)e^y - 1}$$

$$- \frac{1}{96\pi^2 a^3} \int_0^{\infty} dy y^2 \left[\frac{1}{\frac{y}{\lambda_1 a} + 1} + \frac{1}{\frac{y}{\lambda_2 a} + 1}\right] \tag{3.237a}$$

$$= -(\mathcal{E}_c + \mathcal{E}_{d1} + \mathcal{E}_{d2}), \tag{3.237b}$$

which is just the negative of the Casimir energy of the two semitransparent plates including the divergent pieces—See (3.17) [45, 46]. Note that $\mathcal{E}_{di}$, $i = 1, 2$, are simply the divergent energies (3.233) associated with a single plate.

3.6.2.3 Renormalization

The divergent terms in (3.237b) simply renormalize the masses (per unit area) of each plate:

$$E_{\text{total}} = m_1 + m_2 + \mathscr{E}_{d1} + \mathscr{E}_{d2} + \mathscr{E}_c$$
$$= M_1 + M_2 + \mathscr{E}_c, \tag{3.238}$$

where m_i is the bare mass of each plate, and the renormalized mass is $M_i = m_i + \mathscr{E}_{di}$. Thus the gravitational force on the entire apparatus obeys the equivalence principle

$$g\mathscr{F} = -g(M_1 + M_2 + \mathscr{E}_c). \tag{3.239}$$

The minus sign reflects the downward acceleration of gravity on the surface of the earth. Note here that the Casimir interaction energy $\mathscr{E}_c$ is negative, so it reduces the gravitational attraction of the system.

3.6.3 Summary

We have found, in conformation with the result given in [22], an extremely simple answer to the question of how Casimir energy accelerates in a weak gravitational field: Just like any other form of energy, the gravitational force F divided by the area of the plates is

$$\frac{F}{A} = -g\mathscr{E}_c. \tag{3.240}$$

This is the result expected by the equivalence principle, but is in contradiction to some earlier disparate claims in the literature [95–99]. Bimonte et al. [100] now agree completely with our conclusions. This result perfectly agrees with that found by Saharian et al. [101] for Dirichlet, Neumann, and perfectly conducting plates for the finite Casimir interaction energy. The acceleration of Dirichlet plates follows from our result when the strong coupling limit $\lambda \to \infty$ is taken. What makes our conclusion particularly interesting is that it refers not only to the finite part of the Casimir interaction energy between semitransparent plates, but to the divergent parts as well, which are seen to simply renormalize the gravitational mass of each plate, as they would the inertial mass. The reader may object that by equating gravitational force with uniform acceleration we have built in the equivalence principle, and so does any procedure based on Einstein's equations; but the real nontriviality here is that quantum fluctuations obey the same universal law. The reader is also referred to the important work on this subject by Jaekel and Reynaud [102], and extensive references therein.

3.7 Conclusions

In this review, I have illustrated the issues involved in calculating self-energies in the simple context of massless scalar fields interacting with δ-function potentials, so-called semitransparent boundaries. This is not as unrealistic as it might sound, since in the strong coupling limit this yields Dirichlet boundary conditions, and by using derivative of δ-function boundaries, we can recover Neumann boundary conditions. Thus, where the boundaries admit the separation into TE and TM modes, we can recover perfect-conductor boundaries imposed on electromagnetic fields.

We have examined both divergences occurring in the total energy, and divergences which appear in the local energy density as boundaries are approached. The latter divergences often have little to do with the former, because the local divergences may cancel across the boundaries, and they typically depend on the form (canonical or conformal, for example) of the local stress-energy tensor. The global divergences apparently can always be uniquely isolated, leaving a unique finite self-energy; in some cases at least the divergent parts can be absorbed into a renormalization of properties of the boundaries, such as their masses. It is expected that if the ideal boundaries were represented as a solitonic structure arising from a background field, this "renormalization" idea could be put on a more rigorous footing.

Evidence for the consistency of this view occurs in the parallel plate configuration, where we show that the finite interaction energy and the divergent self-energies of each plate exhibit the same inertial and gravitational properties, that is, are each consistent with the equivalence principle. Thus it is indeed consistent to absorb the self-energies into the masses of each plate. We hope to prove in the future that this renormalization consistency is a general feature.

In spite of the length of this review, we have barely scratched the surface. In particular, we have not discussed how the divergent contributions of the local stress tensor are consistent with Einstein's equations [103]. We have also only discussed simple separable geometries, where the equations for the Green's functions can be solved on both the inside and the outside of the boundaries. This excludes the extensive work on rectangular cavities, where only the sum over interior eigenvalues can be carried out [16–19, 104]. There are some numerical coincidences, for example between the energy for a sphere and a cube, but since divergences have been simply omitted by zeta-function regularization, the significance of the latter results remains unclear. There are a few other examples where the interior Casimir contribution can be computed exactly, while the exterior problem cannot be solved, an example being a cylinder with cross section of an equilateral triangle. Such results seem more problematic than those we have discussed here.

We also have not discussed semiclassical and numerical techniques. For example, there is the extremely interesting work of Schaden [105], who computes a very accurate approximation for the Casimir energy of a spherical shell using optical path techniques. The same technique gives zero for the cylindrical shell, not the attractive value found in [27], which is not surprising. Not unrelated to this technique is the exact worldline method of Gies and collaborators [106–108], which is able to capture edge effects. The optical path work of Scardicchio and Jaffe [109–111] should also be cited, although it is largely restricted to examining the forces between distinct bodies. This review also does not refer to the remarkable progress in numerical techniques, some of which are related to the multiple scattering approach—for some recent references see [112, 113], (See also the chapters by Rahi et al., by Johnson and by Lambrecht et al. in this volume for additional discussions about the multiple scattering approach)—, which however, have not yet been turned to examining self-interactions.

The central issue is the meaning of Casimir self-energy, and how, in principle, it might be observed. Probably the right direction to address such issues is in terms of quantum corrections to solitons—for example, see [114–116]. The issues being considered go to the very heart of renormalized quantum field theory, and likely to the meaning and origin of mass, a subject about which we in fact know very little.

Acknowledgements I thank the US Department of Energy and the US National Science Foundation for partial support of this work. I thank my many collaborators, including Carl Bender, Iver Brevik, Inés Cavero-Peláez, Lester DeRaad, Steve Fulling, Ron Kantowski, Klaus Kirsten, Vladimir Nesterenko, Prachi Parashar, August Romeo, K.V. Shajesh, and Jef Wagner, for their contributions to the work described here.

References

1. Casimir, H.B.G.: On the attraction between two perfectly conducting plates. Proc. Kon. Ned. Akad. Wetensch. **51**, 793 (1948)
2. London, F.: Theory and system of molecular forces. Z. Physik **63**, 245 (1930)
3. Casimir, H.B.G., Polder, D.: The influence of retardation on the London-Van Der Waals forces. Phys. Rev. **73**, 360 (1948)
4. Casimir, H.B.G.: In: Bordag, M. (ed.) The Casimir Effect 50 Years Later: The Proceedings of the Fourth Workshop on Quantum Field Theory Under the Influence of External Conditions, World Scientific, Singapore, p. 3, (1999)
5. Jaffe, R.L.: Unnatural acts: Unphysical consequences of imposing boundary conditions on quantum fields. AIP Conf. Proc. **687**, p. 3 (2003). arXiv:hep-th/0307014
6. Lifshitz, E.M.: Zh. Eksp. Teor. Fiz. **29**, 94 (1956), [English translation: The theory of molecular attractive forces between solids. Soviet Phys. JETP **2**,73 (1956)]
7. Dzyaloshinskii, I.D., Lifshitz, E. M., Pitaevskii, L.P.: Zh. Eksp. Teor. Fiz. **37**, 229 (1959), [English translation: Van der Waals forces in liquid films. Soviet Phys. JETP **10**, 161 (1960)]
8. Dzyaloshinskii, I.D., Lifshitz, E.M., Pitaevskii, L.P., Usp. Fiz. Nauk **73**, 381(1961), [English translation: General theory of van der Waals forces. Soviet Phys. Usp. **4**, 153 (1961)]
9. Bordag, M., Klimchitskaya, G.L., Mohideen, U., Mostepanenko, V.M.: *Advances in the Casimir Effect.* Int. Ser. Monogr. Phys. **145**, 1 (2009). (Oxford University Press, Oxford, 2009)
10. Klimchitskaya, G.L., Mohideen, U., Mostepanenko, V.M.: The Casimir force between real materials: experiment and theory. Rev. Mod. Phys. **81**, 1827 (2009). arXiv:0902.4022[cond-mat.other]
11. Deryagin(Derjaguin), B.V.: Analysis of friction and adhesion IV: The theory of the adhesion of small particles. Kolloid Z. **69**, 155 (1934)
12. Deryagin(Derjaguin), B.V. et al.: Effect of contact deformations on the adhesion of particles. J. Colloid. Interface Sci. **53**, 314 (1975)
13. Blocki, J., Randrup, J., Świątecki, W. J., Tsang, C.F.: Proximity forces. Ann. Phys. (N.Y.) **105**, 427 (1977)
14. Milton, K.A.: Recent developments in the Casimir effect. J. Phys. Conf. Ser. **161**, 012001 (2009). [hep-th]]

15. Boyer, T.H.: Quantum electromagnetic zero point energy of a conducting spherical shell and the Casimir model for a charged particle. Phys. Rev. **174**, 1764 (1968)
16. Lukosz, W.: Electromagnetic zero-point energy and radiation pressure for a rectangular cavity. Physica **56**, 109 (1971)
17. Lukosz, W.: Electromagnetic zero-point energy shift induced by conducting closed surfaces. Z. Phys. **258**, 99 (1973)
18. Lukosz, W.: Electromagnetic zero-point energy shift induced by conducting surfaces. II. The infinite wedge and the rectangular cavity. Z. Phys. **262**, 327 (1973)
19. Ambjørn, J., Wolfram, S.: Properties of the vacuum. I. Mechanical and thermodynamic. Ann. Phys. (N.Y.) **147**, 1 (1983)
20. Balian, R., Duplantier, B.: Electromagnetic waves near perfect conductors. II. Casimir effect. Ann. Phys. (N.Y.) **112**, 165 (1978)
21. Bernasconi, F., Graf, G.M., Hasler, D.: The heat kernel expansion for the electromagnetic field in a cavity. Ann. Henri Poincaré **4**, 1001 (2003). arXiv:math-ph/0302035
22. Fulling, S.A., Milton, K.A., Parashar, P., Romeo, A., Shajesh, K.V., Wagner, J.: How does Casimir energy fall?. Phys. Rev. D **76**, 025004 (2007). arXiv:hep-th/0702091
23. Milton, K.A., Parashar, P., Shajesh, K.V., Wagner, J.: How does Casimir energy fall? II. Gravitational acceleration of quantum vacuum energy. J. Phys. A **40**, 10935 (2007). [hep-th]]
24. Milton, K.A., Wagner, J.: Exact Casimir Interaction Between Semitransparent Spheres and Cylinders. Phys. Rev. D **77**, 045005 (2008). [arXiv:0711.0774 [hep-th]]
25. Milton, K.A., Wagner, J.: Multiple Scattering Methods in Casimir Calculations. J. Phys. A **41**, 155402 (2008). [hep-th]]
26. Wagner, J., Milton, K.A., Parashar, P.: Weak Coupling Casimir Energies for Finite Plate Configurations. J. Phys. Conf. Ser. **161**, 012022 (2009). [arXiv:0811.2442 [hep-th]]
27. DeRaad, L.L. Jr., Milton, K.A.: Casimir Selfstress On A Perfectly Conducting Cylindrical Shell. Ann. Phys. (N.Y.) **136**, 229 (1981)
28. Bender, C.M., Milton, K.A.: Casimir effect for a D-dimensional sphere. Phys. Rev. D **50**, 6547 (1994). arXiv:hep-th/9406048
29. Gosdzinsky, P., Romeo, A.: Energy of the vacuum with a perfectly conducting and infinite cylindrical surface. Phys. Lett. B **441**, 265 (1998). arXiv:hep-th/9809199
30. Brevik, I., Marachevsky, V.N., Milton, K.A.: Identity of the van der Waals force and the Casimir effect and the irrelevance of these phenomena to sonoluminescence. Phys. Rev. Lett. **82**, 3948 (1999). arXiv:hep-th/9810062
31. Cavero-Peláez, I., Milton, K.A.: Casimir energy for a dielectric cylinder. Ann. Phys. (N.Y.) **320**, 108 (2005). arXiv:hep-th/0412135
32. Klich, I.: Casimir's energy of a conducting sphere and of a dilute dielectric ball. Phys. Rev. D **61**, 025004 (2000). arXiv:hep-th/9908101
33. Milton, K.A., Nesterenko, A.V., Nesterenko, V.V.: Mode-by-mode summation for the zero point electromagnetic energy of an infinite cylinder. Phys. Rev. D **59**, 105009 (1999)
34. Kitson, A.R., Signal, A.I.: Zero-point energy in spheroidal geometries. J. Phys. A **39**, 6473 (2006). arXiv:hep-th/0511048
35. Kitson, A.R., Romeo, A.: Perturbative zero-point energy for a cylinder of elliptical section. Phys. Rev. D **74**, 085024 (2006). arXiv:hep-th/0607206
36. Milton, K.A.: Calculating Casimir energies in renormalizable quantum field theory. Phys. Rev. D **68**, 065020 (2003). arXiv:hep-th/0210081.
37. Cavero-Peláez, I., Milton, K.A., Kirsten, K.: Local and global Casimir energies for a semitransparent cylindrical shell. J. Phys. A **40**, 3607 (2007). arXiv:hep-th/0607154
38. Milton, K.A.: The Casimir Effect: Physical Manifestations of Zero-Point Energy. World Scientific, Singapore (2001)
39. Bordag, M., Hennig, D., Robaschik, D.: Vacuum energy in quantum field theory with external potentials concentrated on planes. J. Phys. A **25**, 4483 (1992)
40. Bordag, M., Kirsten, K., Vassilevich, D.: Ground state energy for a penetrable sphere and for a dielectric ball. Phys. Rev. D **59**, 085011 (1999). arXiv:hep-th/9811015

41. Graham, N., Jaffe, R.L., Weigel, H.: Casimir effects in renormalizable quantum field theories. Int. J. Mod. Phys. A **17**, 846 (2002). arXiv:hep-th/0201148

42. Graham, N., Jaffe, R.L., Khemani, V., Quandt, M., Scandurra, M., Weigel, H.: Calculating vacuum energies in renormalizable quantum field theories: a new approach to the Casimir problem. Nucl. Phys. B **645**, 49 (2002). arXiv:hep-th/0207120

43. Graham, N., Jaffe, R.L., Khemani, V., Quandt, M., Scandurra, M., Weigel, H.: Casimir energies in light of quantum field theory. Phys. Lett. B **572**, 196 (2003). arXiv:hep-th/0207205

44. Graham, N., Jaffe, R.L., Khemani, V., Quandt, M., Scandurra, M., Weigel, H.: The Dirichlet Casimir problem. Nucl. Phys. B **677**, 379 (2004). arXiv:hep-th/0309130

45. Milton, K.A.: Casimir energies and pressures for delta-function potentials. J. Phys. A **37**, 6391 (2004). arXiv:hep-th/0401090

46. Milton, K.A.: The Casimir effect: Recent controversies and progress. J. Phys. A **37**, R209 (2004). arXiv:hep-th/0406024

47. Kantowski, R., Milton, K.A.: Scalar Casimir energies in $M^4 \times S^N$. Phys. Rev. D **35**, 549 (1987)

48. Brevik, I., Jensen, B., Milton, K.A.: Comment on "Casimir energy for spherical boundaries". Phys. Rev. D **64**, 088701 (2001). arXiv:hep-th/0004041

49. Weigel H.: Dirichlet spheres in continuum quantum field theory. In: Milton, K.A. (ed.) Proceedings of the 6th Workshop on Quantum Field Theory Under the Influence of External Conditions, p. 195, (Rinton Press, Princeton, N.J., 2004) arXiv:hep-th/0310301

50. Fulling, S.A.: Systematics of the relationship between vacuum energy calculations and heat kernel coefficients. J. Phys. A **36**, 6857 (2003)

51. Graham, N., Olum, K.D.: Negative energy densities in quantum field theory with a background potential. Phys. Rev. D **67**, 085014 (2003). arXiv:quant-ph/0302117

52. Callan, C.G. Jr., Coleman, S., Jackiw, R.: A new improved energy-momentum tensor. Ann. Phys. (N.Y.) **59**, 42 (1970)

53. Olum, K.D., Graham, N.: Static negative energies near a domain wall. Phys. Lett. B **554**, 175 (2003). arXiv:gr-qc/0205134

54. Romeo, A., Saharian, A.A.: Casimir effect for scalar fields under Robin boundary conditions on plates. J. Phys. A **35**, 1297 (2002). arXiv:hep-th/0007242

55. Romeo, A., Saharian, A.A.: Vacuum densities and zero-point energy for fields obeying Robin conditions on cylindrical surfaces. Phys. Rev. D **63**, 105019 (2001). arXiv:hepth/0101155

56. Saharian, A.A.: Scalar Casimir effect for D-dimensional spherically symmetric Robin boundaries. Phys. Rev. D **6**, 125007 (2001). arXiv:hep-th/0012185

57. Saharian, A.A.: On the energy-momentum tensor for a scalar field on manifolds with boundaries. Phys. Rev. D **69**, 085005 (2004). arXiv:hep-th/0308108

58. Brown, L.S., Maclay, G.J.: Vacuum stress between conducting plates: An Image solution. Phys. Rev. **184**, 1272 (1969)

59. Actor, A.A., Bender, I.: Boundaries immersed in a scalar quantum field. Fortsch. Phys. **44**, 281 (1996)

60. Dowker, J.S., Kennedy, G.: Finite temperature and boundary effects in static space-times. J. Phys. A **11**, 895 (1978)

61. Deutsch, D., Candelas, P.: Boundary effects in quantum field theory. Phys. Rev. D **20**, 3063 (1979)

62. Brevik, I., Lygren, M.: Casimir effect for a perfectly conducting wedge. Ann. Phys. (N.Y.) **251**, 157 (1996)

63. Sopova, V., Ford, L.H.: The electromagnetic field stress tensor near dielectric half-spaces. In: Milton, K.A. (ed.) Proceedings of the 6th Workshop on Quantum Field Theory Under the Influence of External Conditions, p.140. Rinton Press, Princeton, NJ, (2004)

64. Sopova, V., Ford, L.H.: The Electromagnetic Field Stress Tensor between Dielectric Half-Spaces. Phys. Rev. D **72**, 033001 (2005). arXiv:quant-ph/0504143

65. Graham, N.: Do casimir energies obey general relativity energy conditions?. In: Milton, K.A. (ed.) Proceedings of the 6th Workshop on Quantum Field Theory Under the Influence of External Conditions, Rinton Press, Princeton, NJ (2004)
66. Graham, N., Olum, K.D.: Plate with a hole obeys the averaged null energy condition. Phys. Rev. D **72**, 025013 (2005). arXiv:hep-th/0506136
67. Milton, K.A.: Semiclassical electron models: Casimir self-stress in dielectric and conducting balls. Ann. Phys. (N.Y.) **127**, 49 (1980)
68. Milton, K.A., DeRaad, L.L. Jr., Schwinger, J.: Casimir self-stress on a perfectly conducting spherical shell. Ann. Phys. (N.Y.) **115**, 388 (1978)
69. Candelas, P.: Vacuum energy in the presence of dielectric and conducting surfaces. Ann. Phys. (N.Y.) **143**, 241 (1982)
70. Candelas, P.: Vacuum energy in the bag model. Ann. Phys. (N.Y.) **167**, 257 (1986)
71. Bordag, M., Mohideen, U., Mostepanenko, V.M.: New developments in the Casimir effect. Phys. Rept. **353**, 1 (2001). arXiv:quant-ph/0106045
72. Sen, S.: Geometrical determination of the sign of the Casimir force in two spatial dimensions. Phys. Rev. D **24**, 869 (1981)
73. Sen, S.: A calculation of the Casimir force on a circular boundary. J. Math. Phys. **22**, 2968 (1981)
74. Cavero-Peláez, I., Milton, K.A., Wagner, J.: Local casimir energies for a thin spherical shell. Phys. Rev. D **73**, 085004 (2006). arXiv:hep-th/0508001
75. Barton, G.: Casimir energies of spherical plasma shells. J. Phys. A **37**, 1011 (2004)
76. Scandurra, M.: The ground state energy of a massive scalar field in the background of a semi-transparent spherical shell. J. Phys. A **32**, 5679 (1999). arXiv:hep-th/9811164
77. Bender, C.M., Milton, K.A.: Scalar Casimir effect for a D-dimensional sphere. Phys. Rev. D **50**, 6547 (1994). arXiv:hep-th/9406048
78. Leseduarte, S., Romeo, A.: Complete zeta-function approach to the electromagnetic Casimir effect for a sphere. Europhys. Lett. **34**, 79 (1996)
79. Leseduarte, S., Romeo, A.: Complete zeta-function approach to the electromagnetic Casimir effect for spheres and circles. Ann. Phys. (N.Y.) **250**, 448 (1996). arXiv:hepth/9605022
80. Klich, I.: Casimir energy of a conducting sphere and of a dilute dielectric ball. Phys. Rev. D **61**, 025004 (2000). arXiv:hep-th/9908101
81. Bordag, M., Vassilevich, D.V.: Nonsmooth backgrounds in quantum field theory. Phys. Rev. D **70**, 045003 (2004). arXiv:hep-th/0404069
82. Milton, K.A.: Zero-point energy in bag models. Phys. Rev. D **22**, 1441 (1980)
83. Milton, K.A.: Zero-point energy of confined fermions. Phys. Rev. D **22**, 1444 (1980)
84. Milton, K.A.: Vector Casimir effect for a D-dimensional sphere. Phys. Rev. D **55**, 4940 (1997). arXiv:hep-th/9611078
85. Leseduarte, S., Romeo, A.: Influence of a magnetic fluxon on the vacuum energy of quantum fields confined by a bag. Commun. Math. Phys. **193**, 317 (1998). arXiv:hep-th/9612116
86. Davies, B.: Quantum electromagnetic zero-point energy of a conducting spherical shell. J. Math. Phys. **13**, 1324 (1972)
87. Schwartz-Perlov, D., Olum, K.D.: Energy conditions for a generally coupled scalar field outside a reflecting sphere. Phys. Rev. D **72**, 065013 (2005). arXiv:hep-th/0507013
88. Scandurra, M.: Vacuum energy of a massive scalar field in the presence of a semi-transparent cylinder. J. Phys. A **33**, 5707 (2000). arXiv:hep-th/0004051
89. Gilkey, P.B., Kirsten, K., Vassilevich, D.V.: Heat trace asymptotics with transmittal boundary conditions and quantum brane-world scenario. Nucl. Phys. B **601**, 125 (2001)
90. Nesterenko, V.V., Pirozhenko, I.G.: Spectral zeta functions for a cyllinder and a circle. J. Math. Phys. **41**, 4521 (2000)
91. Kennedy, G., Critchley, R., Dowker, J.S.: Finite temperature field theory with boundaries: stress tensor and surface action renormalization. Ann. Phys. (N.Y.) **125**, 346 (1980)
92. Romeo, A., Saharian, A.A.: Casimir effect for scalar fields under Robin boundary conditions on plates. J. Phys. A **35**, 1297 (2002). arXiv:hep-th/0007242

93. Fulling, S.A., Kaplan, L., Kirsten, K., Liu, Z.H., Milton, K.A.: Vacuum stress and closed paths in rectangles, pistons, and pistols. J. Phys. A **42**, 155402 (2009). arXiv:0806.2468[hep-th]

94. Born, M.: The theory of the rigid electron in the kinematics of the relativity principle. Ann. Phys. (Leipzig) **30**, 1 (1909)

95. Calloni, E., Di Fiore, L., Esposito, G., Milano, L., Rosa, L.: Vacuum fluctuation force on a rigid Casimir cavity in a gravitational field. Phys. Lett. A **297**, 328 (2002)

96. Karim, M., Bokhari, A.H., Ahmedov, B.J.: The Casimir force in the Schwarzchild metric. Class. Quant. Grav. **17**, 2459 (2000)

97. Caldwell, R.R.: Gravitation of the Casimir effect and the cosmological non-constant. arXiv:astro-ph/0209312

98. Sorge, F.: Casimir effect in a weak gravitational field. Class. Quant. Grav. **22**, 5109 (2005)

99. Bimonte, G., Calloni, E., Esposito, G., Rosa, L.: Energy-momentum tensor for a Casimir apparatus in a weak gravitational field. Phys. Rev. D **74**, 085011 (2006)

100. Bimonte, G., Esposito, G., Rosa, L.: From Rindler space to the electromagnetic energy-momentum tensor of a Casimir apparatus in a weak gravitational field. Phys. Rev. D **78**, 024010 (2008). arXiv:0804.2839 [hep-th]

101. Saharian, A.A., Davtyan, R.S., Yeranyan, A.H.: Casimir energy in the Fulling-Rindler vacuum. Phys. Rev. D **69**, 085002 (2004). arXiv:hep-th/0307163

102. Jaekel. M.T., Reynaud, S.: Mass, inertia and gravitation. arXiv:0812.3936 [gr-qc]

103. Estrada, R., Fulling, S.A., Liu, Z., Kaplan, L., Kirsten, K., Milton, K.A.: Vacuum stress-energy density and its gravitational implications. J. Phys. A **41**, 164055 (2008)

104. Actor, A.A.: Scalar quantum fields confined by rectangular boundaries. Fortsch. Phys. **43**, 141 (1995)

105. Schaden, M.: Semiclassical electromagnetic Casimir self-energies. arXiv:hep-th/0604119

106. Gies, H., Klingmuller, K.: Casimir edge effects. Phys. Rev. Lett. **97**, 220405 (2006). arXiv:quant-ph/0606235

107. Gies, H., Klingmuller, K.: Worldline algorithms for Casimir configurations Phys. Rev. D **74**, 045002 (2006). arXiv:quant-ph/0605141

108. Gies, H., Klingmuller, K.: Casimir effect for curved geometries: PFA validity limits. Phys. Rev. Lett. **96**, 220401 (2006). arXiv:quant-ph/0601094

109. Jaffe, R.L., Scardicchio, A.: The casimir effect and geometric optics. Phys. Rev. Lett. **92**, 070402 (2004). arXiv:quant-ph/0310194

110. Scardicchio, A., Jaffe, R.L.: Casimir effects: an optical approach I. foundations and examples. Nucl. Phys. B **704**, 552 (2005). arXiv:quant-ph/0406041

111. Schroeder, O., Scardicchio, A., Jaffe, R.L.: The Casimir energy for a hyperboloid facing a plate in the optical approximation. Phys. Rev. A **72**, 012105 (2005). arXiv:hep-th/0412263

112. Graham, N., Shpunt, A., Emig, T., Rahi, S.J., Jaffe, R.L., Kardar, M.: Casimir force at a knife's edge. Phys. Rev. D **81**, 061701 (2010). arXiv:0910.4649 [quant-ph]

113. Rahi, S.J., Rodriguez, A.W., Emig, T., Jaffe, R.L., Johnson, S.G., Kardar, M.: Nonmonotonic effects of parallel sidewalls on Casimir forces between cylinders. Phys. Rev. A **77**, 030101 (2008). arXiv:0711.1987 [cond-mat.stat-mech]

114. Farhi, E., Graham, N., Haagensen, P., Jaffe, R.L.: Finite quantum fluctuations about static field configurations. Phys. Lett. B **427**, 334 (1998). arXiv:hep-th/9802015

115. Graham, N., Jaffe, R.L.: Energy, central charge, and the BPS bound for 1+1 dimensional supersymmetric solitons. Nucl. Phys. B **544**, 432 (1999). arXiv:hep-th/9808140

116. Cavero-Peláez, I., Guilarte, J.M.: Local analysis of the sine-Gordon kink quantum fluctuations. to appear In: Milton, K. A., Bordag, M. (eds.) Proceedings of the 9th Conference on Quantum Field Theory Under the Influence of External Conditions, World Scientific, Singapore (2010). arXiv:0911.4450 [hep-th]

Chapter 4
Casimir Effect in the Scattering Approach: Correlations Between Material Properties, Temperature and Geometry

**Astrid Lambrecht, Antoine Canaguier-Durand,
Romain Guérout and Serge Reynaud**

Abstract We present calculations of the quantum and thermal Casimir interaction between real mirrors in electromagnetic fields using the scattering approach. We begin with a pedagogical introduction of this approach in simple cases where the scattering is specular. We then discuss the more general case of stationary arbitrarily shaped mirrors and present in particular applications to two geometries of interest for experiments, that is corrugated plates and the plane-sphere geometry. The results nicely illustrate the rich correlations existing between material properties, temperature and geometry in the Casimir effect.

4.1 Introduction

The Casimir effect [1] is an observable effect of vacuum fluctuations in the mesoscopic world, to be tested with the greatest care as a crucial prediction of quantum field theory [2–8]. It also constitutes a fascinating interface between quantum field theory and other important aspects of fundamental physics, for example through its connection with the problem of vacuum energy [9–11].

A. Lambrecht (✉) · A. Canaguier-Durand · R. Guérout · S. Reynaud
Laboratoire Kastler Brossel, CNRS, ENS, Université Pierre et Marie Curie case 74,
Campus Jussieu, 75252 Paris Cedex 05, France
e-mail: astrid.lambrecht@upmc.fr

A. Canaguier-Durand
e-mail: antoine.canaguier@upmc.fr

R. Guérout
e-mail: romain.guerout@upmc.fr

S. Reynaud
e-mail: serge.reynaud@upmc.fr

D. Dalvit et al. (eds.), *Casimir Physics*, Lecture Notes in Physics 834,
DOI: 10.1007/978-3-642-20288-9_4, © Springer-Verlag Berlin Heidelberg 2011

Casimir physics plays an important role in the tests of gravity at sub-millimeter ranges [12, 13]. Strong constraints have been obtained in short range Cavendish-like experiments [14]. A hypothetical new force of Yukawa-like form could not exceed the gravitational force in the range above 56 μm. For ranges of the order of the micrometer, similar tests are performed by comparing the results of Casimir force measurements with theoretical predictions [15–17]. At even shorter scales, those tests can be performed using atomic [18] or nuclear [19] force measurements. In any of these short-range gravity tests, a new hypothetical force would appear as a difference between the experimental result F_{exp} and the theoretical prediction F_{th}. This implies that F_{th} and F_{exp} have to be assessed independently from each other and necessarily forbids use of the theory-experiment comparison for proving (or disproving) some specific experimental result or theoretical model.

Finally, the Casimir force and the closely related Van der Waals force are dominant at micron or sub-micron distances, entailing their strong connections with various important domains, such as atomic and molecular physics, condensed matter and surface physics, chemical and biological physics, micro- and nano-technology [20].

4.2 Comparison of Casimir Force Measurements with Theory

Casimir calculated the force between a pair of perfectly smooth, flat and parallel plates in the limit of zero temperature and perfect reflection which led him to the universal expressions for the force F_{Cas} and energy E_{Cas}

$$F_{\mathrm{Cas}} = -\frac{\hbar c \pi^2 A}{240 L^4}, \quad E_{\mathrm{Cas}} = -\frac{\hbar c \pi^2 A}{720 L^3}. \tag{4.1}$$

with L the mirrors' separation, A their surface, c the speed of light and $\hbar$ the Planck constant. The universality of these ideal Casimir formulas is explained by the saturation of the optical response of perfect mirrors which exactly reflect 100% of the incoming fields. This idealization does not correspond to any real mirror. In fact, the effect of imperfect reflection is large in most experiments, and a precise knowledge of its frequency dependence is essential for obtaining reliable theoretical predictions to be compared with Casimir force measurements [21–35]. See also the chapter of van Zwol et al. in this volume for additional discussions of characterization of optical properties in Casimir force experiments.

4.2.1 The Description of Metallic Mirrors

The most precise experiments are performed with metallic mirrors which are good reflectors at frequencies smaller than their plasma frequency ω_{P}. Their optical response at a frequency ω is described by a reduced dielectric function written as

$$\varepsilon[\omega] = \bar{\varepsilon}[\omega] + \frac{\sigma[\omega]}{-i\omega}, \quad \sigma[\omega] = \frac{\omega_P^2}{\gamma - i\omega}. \tag{4.2}$$

The function $\bar{\varepsilon}[\omega]$ represents the contribution of interband transitions and it is regular at the limit $\omega \rightarrow 0$. Meanwhile $\sigma[\omega]$ is the reduced conductivity, measured as a frequency (the SI conductivity is $\epsilon_0\sigma$,) which describes the contribution of the conduction electrons.

A simplified description corresponds to the lossless limit $\gamma \rightarrow 0$ often called the plasma model. As γ is much smaller than ω_P for good conductors, this simple model captures the main effect of imperfect reflection. However it cannot be considered as an accurate description since a much better fit of tabulated optical data is obtained with a non null value of γ [36, 37]. Furthermore, the Drude model, with $\gamma \neq 0$, meets the important property of ordinary metals which have a finite static conductivity $\sigma_0 = \frac{\omega_P^2}{\gamma}$, in contrast to the lossless limit which corresponds to an infinite value for σ_0.

Another correction to the Casimir expressions is associated with the effect of thermal fluctuations [38–41]. Boström and Sernelius have remarked that the small non zero value of γ had a significant effect on the force evaluation at ambient temperature [42]. This significant difference is attributed to the vanishing contribution of TE modes at zero frequency for dissipative mirrors entailing that for the Casimir force, contrary to the dielectric function, there is no continuity from the Drude to the plasma model at the limit of a vanishing relaxation. The ratio between the predictions evaluated at $\gamma = 0$ and $\gamma \neq 0$ even reaches a factor of 2 at the limit of large temperatures or large distances. Unfortunately it has not yet been possible to test this striking prediction since the current experiments do not explore this domain.

The current status of Casimir experiments appears to favor theoretical predictions obtained with the lossless plasma model $\gamma = 0$ rather than those corresponding to the Drude model with $\gamma \neq 0$ as one might have expected (see Fig. 4.1 in [29]). See the chapter by Decca et al. in this volume for additional discussions of this observation. We thus have to face a discrepancy between theory and experiment. This discrepancy may have various origins, in particular artifacts in the experiments or inaccuracies in the calculations. They may also come from yet unmastered differences between the situations studied in theory and the experimental realizations.

These remarks have led to a blossoming of papers devoted to the thermal effect on the Casimir force, for reviews see e.g. [43–47]. It is worth emphasizing that microscopic descriptions of the Casimir interaction between two metallic bulks lead to predictions agreeing with the lossy Drude model rather than the lossless plasma model at the limit of large temperatures or large distances [48–50].

It is also important to note that the Drude model leads to a negative contribution of the Casimir interaction to entropy, in contrast to the plasma model [51]. There is no principle inconsistency with the laws of thermodynamics at this point since the negative contribution is nothing but a difference of entropies (see for example [52]).

Fig. 4.1 Two plane parallel
plates at distance L facing
each other constitute the
Casimir cavity

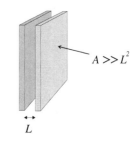

$$A \gg L^2$$

L

4.2.2 The Role of Geometry

The geometry plays an important role in the context of theory/experiment comparison for Casimir forces. Precise experiments are indeed performed between a plane and a sphere whereas most exact calculations are devoted to the geometry of two parallel plates. The estimation of the force in the plane-sphere geometry thus involves the so-called *Proximity Force Approximation* (PFA) [53] which amounts to averaging the force calculated in the parallel-plates geometry over the distribution of local inter-plate distances, the force being deduced from the Lifshitz formula [54, 55], the meaning of which will be discussed below.

This trivial treatment of geometry cannot reproduce the rich interconnection expected to take place between the Casimir effect and geometry [56–59]. In the plane-sphere geometry in particular, the PFA can only be valid when the radius R is much larger than the separation L [60–62]. But even if this limit is met in experiments, the PFA gives no information about its accuracy for a given ratio of L/R and how this accuracy depends on the properties of the mirror, on the distance or temperature.

Answers to these questions can only be obtained by pushing the theory beyond the PFA, which has been done in the past few years [63–67]. A multipolar expansion of the Casimir effect between perfect mirrors in electromagnetic vacuum was proposed in [68, 69]. These calculations have now been performed for plane and spherical metallic surfaces coupled to electromagnetic vacuum, at zero [70] or non zero temperature [71, 72], which has opened the way to a comparison with theory of the only experimental study devoted to a test of PFA in the plane-sphere geometry [73]. As we will see at the end of this article, the features of the thermal Casimir force mentioned in Sect. 4.2 are considerably altered when the geometry is properly taken into account. The factor of 2 between the force values within Drude and plasma model is reduced to a factor of 3/2, decreasing even more below this value when small spheres are considered. Negative entropies are not only found for the Drude model but also for perfect reflector and plasma models, which means that negative contributions of the Casimir interaction to entropy can be found even in the absence of dissipation.

Another specific geometry of great interest, that we will present in the following, is that of surfaces with periodic corrugations. As lateral translation symmetry is broken, the Casimir force contains a lateral component which is smaller than the normal one, but has nevertheless been measured in dedicated experiments [74, 75]. Calculations beyond the PFA were first performed with the simplifying assumptions of perfect reflection [76] or shallow corrugations [77–79]. As expected, the PFA was found to be accurate only in the limit of large corrugation wavelengths. Very recently, experiments have been able to probe the beyond-PFA regime [80, 81] and exact calculations of the forces between real mirrors with deep corrugations [82, 83] have been performed. More discussions on these topics will be presented below.

4.3 The Scattering Approach

In the following, we will focus our attention on the scattering approach, which is an efficient and elegant method for addressing the aforementioned questions.

This method has been used for years for describing the optical properties of non-perfectly reflecting mirrors in terms of scattering amplitudes [84, 85]. These scattering amplitudes are often deduced from Fresnel reflection amplitudes calculated for mirrors described by local dielectric response functions, in which case the expression of the Casimir force is reduced to the Lifshitz expression [54, 55]. However the scattering approach is much more general than the Lifshitz one since real mirrors are always described by some scattering amplitudes but not necessarily by local dielectric response functions. This point will be discussed in more detail below.

The interest in the scattering approach has considerably increased since it has become clear that it is also an extremely efficient method for calculating the Casimir effect in non-trivial geometries. This was realized by several groups employing different theoretical techniques and using different notations (see [86] for an historical overview). Besides the already quoted papers, one may cite the following references which used different versions of the scattering approach [64, 87–92] or alternative methods [93–98]. This topic has seen recently an impressive number of new applications proposed, among which one may cite [99–106]. See also the chapter of Rahi et al. in this volume for additional discussions of the scattering approach in Casimir physics.

An early application of the scattering approach to non-trivial geometries and non-perfect reflectors was developed in [107, 108] to calculate the roughness correction to the Casimir force between two planes, in a perturbative expansion with respect to the roughness amplitude. The same perturbative formalism was also applied to compute the lateral Casimir force [77–79] and the Casimir torque [109] between two corrugated surfaces made of real material, and then to derive the Casimir-Polder potential for an atom near a corrugated surface [111, 112].

Let us recall that results applicable to the non-retarded case have been available [113, 114] before those corresponding to the full retarded theory, and also that the scattering theory has been used for a long time for studying the Casimir-Polder force between atoms and molecules [115, 116].

We begin the review of the scattering approach by an introduction considering the two simple cases of the Casimir force between two scatterers on a 1-dimensional line and between two parallel plates coupled through specular scattering to 3-dimensional electromagnetic fields [84]. We then address the general case of non-specular scatterers in 3-dimensional electromagnetic fields [8].

4.3.1 Mirrors on a 1-Dimensional Line

The first case corresponds to the quantum field theory in 2-dimensional spacetime (1-d space plus time). In this simple case, we have to consider only two scalar fields counter-propagating along opposite directions. The results summarized below are drawn from a series of papers devoted to the study of static or dynamic Casimir force between mirrors coupled to these scalar fields [9, 84, 117–125]. For example, it was established in [118] that the Casimir energy does contribute to the inertia of the cavity as it should according to the principles of relativity.

In this simple model, a mirror M_1 at rest at position q_1 is described by a 2×2 scattering matrix S_1 containing reflection and transmission amplitudes r_1 and t_1

$$S_1 = \begin{bmatrix} t_1 & r_1 e^{-2i\omega q_1/c} \\ r_1 e^{2i\omega q_1/c} & t_1 \end{bmatrix}. \tag{4.3}$$

Two mirrors M_1 and M_2 at rest at positions q_1 and q_2 form a Fabry–Perot cavity described by a global scattering matrix S_{12} which can be deduced from the elementary matrices S_1 and S_2 associated with the two mirrors.

$$S_{12} = \frac{1}{d} \begin{bmatrix} t_1 t_2 & dr_2 e^{-i\omega L/c} + t_2^2 r_1 e^{i\omega L/c} \\ dr_1 e^{-i\omega L/c} + t_1^2 r_2 e^{i\omega L/c} & t_1 t_2 \end{bmatrix}. \tag{4.4}$$

The denominator d is given by

$$d = 1 - r_1 r_2 e^{2i\omega L/c}, \quad L \equiv q_2 - q_1, \tag{4.5}$$

and its zeros (the poles of S_{12}) represent the resonances of the cavity. It turns out that the forthcoming discussion of the Casimir effect depend only on the expression of d and not on all the other details in the form of S_{12}. The reason explaining this property is the following relation between the determinants of the S–matrices (all supposed to be unitary in the simple model):

$$\det S_{12} = (\det S_1)(\det S_2)\left(\frac{d^*}{d}\right). \tag{4.6}$$

From this relation, it is easy to derive the Casimir free energy as a variation of field energy (vacuum energy at $T = 0$, vacuum plus thermal energy otherwise). The presence of a scatterer indeed shifts the field modes and thus induces a variation of the global field energy. The Casimir free energy is then obtained as the variation of field energy in presence of the cavity corrected by the effects of each mirror taken separately [84]

$$\mathcal{F} \equiv \delta \mathcal{F}_{\text{field},12} - \delta \mathcal{F}_{\text{field},1} - \delta \mathcal{F}_{\text{field},2} = -\int_0^\infty \frac{d\omega}{2\pi} N\hbar\Delta. \tag{4.7}$$

Δ is a function of the frequency ω representing the phase-shift produced by the Fabry–Perot cavity, again corrected by the effects of each mirror taken separately

$$\Delta(\omega) = \frac{\ln \det S_{12} - \ln \det S_1 - \ln \det S_2}{i} = \frac{1}{i} \ln\left(\frac{d^*}{d}\right). \tag{4.8}$$

N is the mean number of thermal photons per mode, given by the Planck law, augmented by the term $\frac{1}{2}$ which represents the contribution of the vacuum

$$N(\omega) = \frac{1}{2} + \frac{1}{\exp\frac{\hbar\omega}{k_B T} - 1} = \frac{1}{2 \tanh\frac{\hbar\omega}{2k_B T}}. \tag{4.9}$$

This phase-shift formula can be given alternative interpretations [84]. In particular, when the Casimir force F is derived from the free energy

$$F = -\frac{\partial \mathcal{F}(L,T)}{\partial L} = \int_0^\infty \frac{d\omega}{\pi} \frac{N\hbar\omega}{c}(f + f^*) = \int_0^\infty \frac{d\omega}{\pi} \frac{N\hbar\omega}{c}(g - 1),$$

$$f \equiv \frac{re^{2i\omega L/c}}{1 - re^{2i\omega L/c}}, \quad g \equiv \frac{1 - |re^{2i\omega L/c}|^2}{|1 - re^{2i\omega L/c}|^2}, \tag{4.10}$$

it is seen as resulting from the difference of radiation pressures exerted onto the inner and outer sides of the mirrors by the field fluctuations. For each field mode at frequency ω, $\frac{N\hbar\omega}{c}$ represents the field momentum while g is the ratio of fluctuation energies inside and outside the Fabry–Perot cavity.

Using the analytic properties of the causal function $\ln d$, the Casimir free energy can also be written as an integral over imaginary frequencies $\omega = i\xi$ (Wick rotation)

$$\mathcal{F} = \hbar \int \frac{d\xi}{2\pi} \cot\left(\frac{\hbar\xi}{2k_B T}\right) \ln d(i\xi). \tag{4.11}$$

Using the pole decomposition of the cotangent function and the analytic properties of $\ln d$, this expression can finally be written as a sum over Matsubara frequencies

$$\mathcal{F} = k_B T \sum_m{}' \ln d(i\xi_m), \quad \xi_m \equiv \frac{2\pi m k_B T}{\hbar}. \tag{4.12}$$

The Matsubara sum $\sum_m'$ is the sum over positive integers m with $m = 0$ counted with a weight $\frac{1}{2}$.

For completeness, let us recall also that the contribution to entropy of the Casimir interaction is simply written as

$$S \equiv -\frac{\partial \mathcal{F}(L, T)}{\partial T}.$$

(4.13)

Hence, it is defined as a difference of entropies just as the free energy $\mathcal{F}$ has been defined in (4.7) above as a difference of free energies.

4.3.2 Specular Reflection in 3-d Space

The same lines of reasoning can be followed when studying the case of two specularly reflecting mirrors coupled to electromagnetic fields in 3-dimensional space. The geometry is sketched in Fig. 4.1 with two plane parallel mirrors aligned along the transverse directions x and y (longitudinal direction denoted by z).

Due to the symmetry of this configuration, the frequency ω, the transverse vector $\mathbf{k} \equiv (k_x, k_y)$ and the polarization $p = \mathrm{TE}, \mathrm{TM}$ are preserved by all scattering processes. The mirrors are described by reflection and transmission amplitudes which depend on these scattering parameters. We assume thermal equilibrium for the whole "cavity + fields" system, and proceed with the derivation as in the simpler case of a 1-dimensional space. Some elements have to be treated with greater care now [85, 8]. First there is a contribution of evanescent waves besides that of ordinary modes freely propagating outside and inside the cavity and it has to be taken carefully into account. The properties of the evanescent waves are described through an analytical continuation of those of ordinary ones, using the well defined analytic behavior of the scattering amplitudes. Then dissipation inside the mirrors may also play a role which implies considering the additional fluctuation lines coming along with dissipation [8, 85].

At the end of this derivation the free energy may still be written as a Matsubara sum

$$\mathcal{F} = k_\mathrm{B} T \sum_{\mathbf{k}} \sum_{p} \sum_{m}{}' \ln d(i\xi_m, \mathbf{k}, p), \quad \xi_m \equiv \frac{2\pi m k_\mathrm{B} T}{\hbar},$$

$$\sum_{\mathbf{k}} \equiv A \int \frac{d^2\mathbf{k}}{4\pi^2} \equiv A \int \frac{dk_x dk_y}{4\pi^2}.$$

(4.14)

$\sum_{\mathbf{k}}$ is the sum over transverse wavevectors with A the area of the plates, $\sum_p$ the sum over polarizations and $\sum_m{}'$ the same Matsubara sum as in the 1-d case. The denominator is now written in terms of the result κ of Wick rotation on the longitudinal wavevector k_z

$$d(i\xi, \mathbf{k}, p) = 1 - r_1(i\xi, \mathbf{k}, p) r_2(i\xi, \mathbf{k}, p) \exp^{-2\kappa L},$$

$$\kappa \equiv \sqrt{\mathbf{k}^2 + \frac{\xi^2}{c^2}}. \tag{4.15}$$

This expression reproduces the ideal Casimir formula (4.1) in the limits of perfect reflection $r_1 r_2 \to 1$ and zero temperature $T \to 0$. It is valid and regular at thermal equilibrium at any temperature and for any optical model of mirrors obeying causality and high frequency transparency properties [8, 84, 85]. It can thus be used for calculating the Casimir force between arbitrary mirrors, as soon as the reflection amplitudes are specified. These amplitudes are commonly deduced from models of mirrors, the simplest of which is the well known Lifshitz model [54, 55] which corresponds to semi-infinite bulk mirrors characterized by a local dielectric response function $\varepsilon(\omega)$ and reflection amplitudes deduced from the Fresnel law

$$r_{\mathrm{TE}}(\mathbf{k}, \xi) = \frac{\kappa - \kappa_t}{\kappa + \kappa_t}, \quad r_{\mathrm{TM}}(\mathbf{k}, \xi) = \frac{\varepsilon\kappa - \kappa_t}{\varepsilon\kappa + \kappa_t}, \tag{4.16}$$

$$\kappa_t \equiv \sqrt{\mathbf{k}^2 + \varepsilon\frac{\xi^2}{c^2}}. \tag{4.17}$$

ε is the dielectric function (4.2) and κ_t denotes the result of Wick rotation of the longitudinal wavevector inside the medium.

In the most general case, the optical response of the mirrors cannot be described by a local dielectric response function. The expression (4.14) of the free energy is still valid in this case with the reflection amplitudes to be determined from microscopic models of mirrors. Recent attempts in this direction can be found for example in [126–133].

At this stage, several remarks can be addressed to the readers interested in historical details:

- The Lifshitz expression was not written in terms of reflection amplitudes until Kats noticed that this formulation was natural [134]. To our best knowledge, the first appearance of an expression of the Casimir effect in terms of reflection amplitudes corresponding to an arbitrary microscopic model (not necessarily a dielectric response function) is in [84].
- The fact that the expression (4.14) of the free energy is valid for lossy as well as lossless mirrors is far from obvious. In the lossy case, one has indeed to take into account the contributions of fluctuations coming from the additional modes associated with dissipation. This property has been demonstrated with an increasing range of validity in [84, 85] and [8] (see also [135] for a theorem playing a crucial role in this demonstration).
- The question had been asked in [43] whether the regularity conditions needed to write the Matsubara sum were met for the Drude model which shows discontinuities at $\xi \to 0$. This question has been answered positively in [52].

4.3.3 The Non-specular Scattering Formula

We now present a more general scattering formula allowing one to calculate the Casimir force between stationary objects with arbitrary shapes. We restrict our attention to the case of disjoint objects, exterior to each other, which corresponds to the configuration initially considered by Casimir (for interior configurations, which may be treated with similar techniques, see for example [136–139]).

The main generalization with respect to the already discussed specular cases is that the scattering matrix $\mathbb{S}$ has now to account for non-specular reflection. It is therefore a much larger matrix which mixes different wavevectors and polarizations while preserving frequency as long as the scatterers are stationary [8]. Of course, the non-specular scattering formula is the generic one while specular reflection can only be an idealization.

As previously, the Casimir free energy can be written as the sum of all the phase-shifts contained in the scattering matrix

$$\mathcal{F} = i\hbar \int_0^\infty \frac{d\omega}{2\pi} N(\omega) \ln \det \mathbb{S} = i\hbar \int_0^\infty \frac{d\omega}{2\pi} N(\omega) \mathrm{Tr} \ln \mathbb{S}. \tag{4.18}$$

The symbols det and Tr refer to determinant and trace over the modes of the scattering matrix at a given frequency ω. After a Wick rotation the formula can still be written as a Matsubara sum

$$\mathcal{F} = k_B T \sum_m {}' \mathrm{Tr} \ln D(i\xi_m), \quad \xi_m \equiv \frac{2\pi m k_B T}{\hbar}. \tag{4.19}$$

The matrix D (here written at Matsubara frequencies $\omega_m = i\xi_m$) is the denominator of the scattering matrix. It describes the resonance properties of the cavity formed by the two objects 1 and 2 and may be written as

$$D = 1 - R_1 \exp^{-KL} R_2 \exp^{-KL}. \tag{4.20}$$

The matrices R_1 and R_2 represent reflection on the two objects 1 and 2 respectively while $\exp^{-KL}$ describes propagation in between reflections on the two objects. Note that the matrices D, R_1 and R_2, which were diagonal in the plane wave basis for specular scattering, are no longer diagonal in the general case of non-specular scattering. The propagation factors remain diagonal in this basis with their eigenvalues κ written as in (4.14). Clearly the expression (4.19) does not depend on the choice of a specific basis. We remark also that (4.19) takes a simpler form in the limit of zero temperature

$$F = -\frac{dE}{dL}, \quad E = \hbar \int_0^\infty \frac{d\xi}{2\pi} \ln \det D(i\xi). \tag{4.21}$$

Applications to be presented in the next sections will also involve the Casimir force gradient G which is often measured in experiments and defined as

$$G = -\frac{dF}{dL}. \tag{4.22}$$

A number of the following applications will be discussed within the zero temperature limit, with a change of notation from the free energy $\mathcal{F}$ to the ordinary energy E at zero temperature.

4.4 Applications to Non-trivial Geometries

Formula (4.21) has been used to evaluate the effect of roughness or corrugation of the surfaces on the value of the Casimir force [107, 77–79, 31] in a perturbative manner with respect to the roughness or corrugation amplitudes. It has also allowed one to study a Bose–Einstein condensate used as a local probe of vacuum above a nano-grooved plate [111, 112]. The scattering approach has clearly a larger domain of applicability, not limited to the perturbative regime, as soon as techniques are available for computing the large matrices involved in its evaluation [82, 83, 140].

Another important application, which we will summarize also in the present section, corresponds to the plane-sphere geometry used in most Casimir force experiments and for which explicit "exact calculations" (see a discussion of the meaning of this expression below) have recently become available [68–72].

4.4.1 Perturbative Treatment of Shallow Corrugations

As already stated, the lateral Casimir force between corrugated plates is a topic of particular interest. It could in particular allow for a new test of quantum electrodynamics through the dependence of the lateral force on the corrugation wavevector [77–79].

Here, we consider a geometry with two plane mirrors, M_1 and M_2, having corrugated surfaces described by uniaxial sinusoidal profiles such as shown in Fig. 4.2. We denote h_1 and h_2 the local heights with respect to mean planes $z_1 = 0$ and $z_2 = L$

$$h_1 = a_1 \cos(k_C x), \quad h_2 = a_2 \cos(k_C(x - b)), \quad k_C = 2\pi/\lambda_C. \tag{4.23}$$

h_1 and h_2 have null spatial averages and L is the mean distance between the two surfaces; h_1 and h_2 are both counted as positive when they correspond to a decrease in the separation; λ_C is the corrugation wavelength, k_C the corresponding wavevector, and b the spatial mismatch between the corrugation crests. At lowest

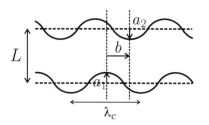

Fig. 4.2 Parallel corrugated surfaces, with L representing the mean separation distance, a_1 and a_2 the corrugation amplitudes and b the lateral mismatch between the crests. When the corrugations are supposed to be the smallest length scales, the effect of the corrugations can be studied in the perturbative expansion. This approximation will be dropped later on

order in the corrugation amplitudes, when $a_1, a_2 \ll \lambda_C, \lambda_P, L$ (with λ_P the plasma wavelength describing the properties of the metallic mirror), the Casimir energy may be obtained by expanding up to second order the general formula (4.21). This perturbative approximation will be dropped in the next subsection.

The part of the Casimir energy able to produce a lateral force is then found to be

$$F^{\text{lat}} = -\frac{\partial \delta E^{\text{corrug}}}{\partial b},$$

$$\delta E^{\text{corrug}} = -\hbar \int_0^\infty \frac{d\xi}{2\pi} \text{Tr}\left(\delta R_1 \frac{\exp^{-KL}}{D_{\text{plane}}} \delta R_2 \frac{\exp^{-KL}}{D_{\text{plane}}} \right). \tag{4.24}$$

δR_1 and δR_2 are the first-order variation of the reflection matrices R_1 and R_2 induced by the corrugations; D_{plane} is the matrix D evaluated at zeroth order in the corrugations; it is diagonal on the basis of plane waves and commutes with K.

Explicit calculations of (4.24) have been performed for the simplest case of experimental interest, with two corrugated metallic plates described by the plasma dielectric function. These calculations have led to the following expression of the lateral part of the Casimir energy

$$\delta E^{\text{corrug}} = \frac{A}{2} G_C(k_C) a_1 a_2 \cos(k_C b). \tag{4.25}$$

The spectral sensitivity function $G_C(k_C)$ has been given and discussed in [79]. Using its expression, it is possible to prove a properly defined "Proximity Force Theorem" which states that the PFA is recovered at the limit of long corrugation wavelengths $k_C \rightarrow 0$. Obviously, this theorem does not imply that the PFA is always valid or, in other words, that $G_C(k_C)$ may be replaced by $G_C(0)$.

To assess the validity of the PFA for the lateral Casimir force description, we now introduce the dimensionless quantity

$$\rho(k_C) = \frac{G_C(k_C)}{G_C(0)}. \tag{4.26}$$

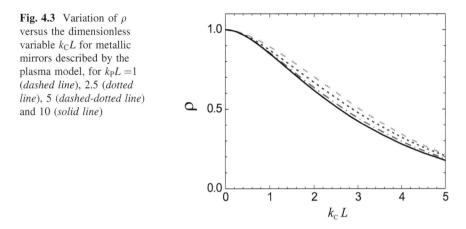

Fig. 4.3 Variation of ρ versus the dimensionless variable $k_C L$ for metallic mirrors described by the plasma model, for $k_P L = 1$ (*dashed line*), 2.5 (*dotted line*), 5 (*dashed-dotted line*) and 10 (*solid line*)

The variation of this ratio ρ with the parameters of interest is shown in Fig. 4.3 for gold covered plates with $\lambda_P = 137\,\text{nm}$. The ratio ρ is smaller than unity as soon as k_C significantly deviates from 0 which means that the PFA overestimates the lateral Casimir force. For large values of k_C, it even decays exponentially to zero, leading to an extreme deviation of the real lateral force from the PFA prediction.

Another situation of interest arises when the corrugation plates are rotated with respect to each other. Assuming as previously corrugations of sinusoidal shape with corrugation wavevectors $\mathbf{k}_j$ having the same modulus $k = 2\pi/\lambda_C$ on both plates, it is possible to derive the second-order correction δE^{torque} to the Casimir energy which depends on the angle θ between the corrugations and thus has the ability to induce a Casimir torque [77, 78, 109]. Only crossed terms, proportional to the corrugation amplitudes on both plates, contribute to this expression, as the square terms are independent of the angle θ. The expression δE^{torque} contains as the special case $\theta = 0$ the pure lateral energy discussed above. Note that the dependence on the material properties and corrugation wavevector are captured by the same response function G_C already calculated.

For quantitative estimations, we assume that the corrugations are restricted to a rectangular section of area $L_x L_y$ with transverse dimensions L_x and L_y much larger than the plate separation L and neglect diffraction at the borders of the plates. In Fig. 4.4, we plot δE^{torque} obtained in this manner, in arbitrary units, as a function of b and θ. The Casimir energy is found to be minimal at $\theta = 0$ and $b = 0, \lambda_C, 2\lambda_C, \ldots$, which corresponds to a situation where corrugations are aligned. Starting from $\theta = b = 0$ and rotating plate 2 around its center, one follows the line $b = 0$ in Fig. 4.4. Clearly, for small angles the plate is attracted back to $\theta = b = 0$ without sliding laterally.

The Casimir torque is then deduced by deriving the energy with respect to the angle θ

$$\tau = -\frac{\partial}{\partial \theta} \delta E^{\text{torque}}. \tag{4.27}$$

Fig. 4.4 Casimir energy as a function of the rotation angle θ and the lateral displacement b

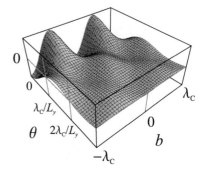

Its maximum is at $\theta = 0.66\lambda_C/L_y$ where it is given by

$$\frac{\tau}{L_x L_y} = 0.109 a_1 a_2 k G_C(k) L_y. \tag{4.28}$$

The maximum torque per unit area is proportional to the length L_y of the corrugation lines, which plays the role of the moment arm.

In contrast with the similar torque appearing between misaligned birefringent plates [141], the torque is here coupled to the lateral force. This could induce complicated behaviors in an experiment and would probably have to be controlled. This can be clearly seen on Fig. 4.4: if the plate is released after a rotation of $\theta > \lambda_C/L_y$ it will move in a combination of rotation and lateral displacement. The energy correction vanishes at $\theta = \lambda_C/L_y$, defining the range of stability of the configuration $b = \theta = 0$. Rotation is favored over lateral displacements only for $\theta < \lambda_C/L_y$.

However, the advantage of the configuration with corrugated plates is that the torque has a larger magnitude. Fig. 4.5 shows the maximum torque as a function of mean separation between the two corrugated gold plates with a plasma wavelength $\lambda_P = 137$ nm. At a plate separation of about 100 nm the torque per unit area can be as high as 10^{-7} N m^{-1} These results on lateral forces and Casimir torques suggest that non-trivial effects of geometry, i.e. effects beyond the PFA, can be observed with dedicated experiments. It is however difficult to achieve this goal with corrugation amplitudes a_1, a_2 meeting the conditions of validity of the perturbative expansion. This approximation is dropped in the next subsection.

4.4.2 Non-perturbative Calculations with Deep Gratings

As already stated, recent experiments have been able to probe the beyond-PFA regime with deep corrugations [80, 81] and it has also become possible to calculate exact expressions of the forces between nanostructures without using the perturbative assumption. This necessarily involves the non-specular scattering formula

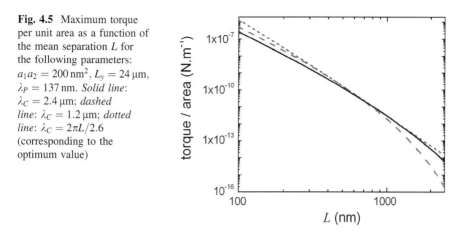

Fig. 4.5 Maximum torque per unit area as a function of the mean separation L for the following parameters: $a_1 a_2 = 200\,\text{nm}^2$, $L_y = 24\,\mu\text{m}$, $\lambda_P = 137\,\text{nm}$. *Solid line*: $\lambda_C = 2.4\,\mu\text{m}$; *dashed line*: $\lambda_C = 1.2\,\mu\text{m}$; *dotted line*: $\lambda_C = 2\pi L/2.6$ (corresponding to the optimum value)

(4.19) and the evaluation of scattering properties mixing different wavevectors and polarizations.

In the following we briefly discuss the Casimir interaction energy in a typical device made of two nanostructured surfaces of intrinsic silicon, such as shown in Fig. 4.6.

To model the material properties of intrinsic silicon, we use a Drude–Lorentz model for which the dielectric function is well approximated by [142]

$$\varepsilon(i\xi) = \varepsilon_\infty + \frac{(\varepsilon_0 - \varepsilon_\infty)\xi_0^2}{\xi^2 + \xi_0^2}, \qquad (4.29)$$

with $\varepsilon_0 \approx 11.87$ the value of the dielectric function at zero frequency, $\varepsilon_\infty \approx 1.035$ the high frequency limit of the dielectric function and $\omega_0 = i\xi_0 \approx 4.34\,\text{eV}$. Calculated with the proximity force approximation, the Casimir force between the two gratings is given by the geometric sum of two contributions corresponding to the Casimir force between two plates F_{PP} at distances L and $L - 2h$, which is independent of the corrugation period d.

To assess quantitatively the validity of the PFA, we plot as before the dimensionless quantity

$$\rho = \frac{F}{F_{\text{PFA}}}. \qquad (4.30)$$

Fig. 4.7 shows this ratio for two silicon gratings, separated by $L = 250$ nm, of height $h = 100$ nm as a function of the corrugation period d with $d_1 = d/2$ [82]. Clearly, the PFA is not a valid approximation except for two limiting cases, that is a vanishing corrugation period $d \to 0$ and a very large corrugation period $d \to \infty$, meaning in either case that the structured surfaces become flat. In between, the exact result for the Casimir force is always smaller than the PFA prediction, meaning that the PFA overestimates the force. This has to be contrasted with calculations for perfect conductors where the PFA always underestimates the real force.

Fig. 4.6 Two surfaces with
rectangular gratings of depth
h, gap width d and trench
width $d - d_1$

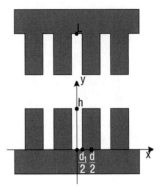

Fig. 4.7 Casimir force
normalized by its PFA value
for two gratings of intrinsic
silicon with amplitude
$h = 100$ nm and $d_1 = \frac{d}{2}$ as a
function of d at a fixed
distance $L = 250$ nm

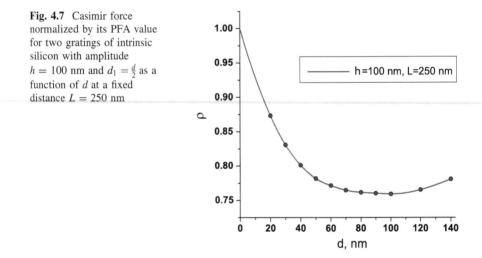

One important parameter to keep in mind is the number of diffraction orders
that has to be retained in the calculation in order for the Casimir energy to converge in the numerical calculation. This is illustrated in Fig. 4.8 for two silicon
gratings. For the sake of convenience, we plot the Casimir energy normalized by
the energy for perfectly reflecting plane mirrors, i.e. the energy reduction factor.
The blue curve corresponds to the situation of two gratings of period 400 nm
separated by a distance $L = 50$ nm. Clearly around five orders of diffraction are
sufficient for the calculation of the Casimir energy in this case. The number of
necessary diffraction orders decreases with increasing distance between the gratings. This is illustrated by the red curve where the two aforementioned gratings are
now separated by a distance $L = 400$ nm and where the Casimir energy has
basically converged to its final value with only one order of diffraction retained.
The fast convergence is here due to that fact that oblique diffraction orders are

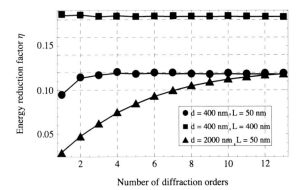

Fig. 4.8 Convergence of the calculated Casimir energy between two gratings as a function of the number of diffraction orders retained in the calculation. Gratings with different periods are plotted as circles and squares (400 nm) and triangles (2 μm). The convergence of the calculations becomes slower when increasing the grating period d or decreasing the separation L

exponentially suppressed with increasing distance [112]. Finally, the greater the period of the grating the more orders of diffractions are needed as shows the curve with triangles where the period of the two gratings is now 2 μm. In this case, the Casimir energy has not yet fully converged to its final value even with as much as 13 orders of diffraction. This can be understood because the momentum transferred by the grating $q = \frac{2\pi}{d}$ is now small so that different orders of diffraction are nearly collinear with the specular one and therefore greatly contribute to the final energy.

If attention is paid to the issue of convergence this calculation method is essentially exact and allows for direct comparisons with experimental results. In a recent experiment, Chan et al. have measured the Casimir force gradient between a gold sphere and a grating of doped silicon [80]. Two samples of silicon gratings have been used. Both have a corrugation depth of 1μm, but different periods of 400 nm and 1μm respectively. The experimental data points of the ratio between the force gradient and its PFA approximation for both samples have been kindly provided by Ho Bun Chan and are plotted in Fig. 4.9. Experimentally the trench arrays are created with duty cycle close to but not exactly equal to 50%. This results in a filling factor p which gives the top part of the grating with respect to the period. See also the chapter by Capasso et al. in this volume for further details of this experiment.

Concerning the calculation we model the optical properties of silicon by the dielectric function (4.29). We have also taken into account the doping of the silicon by adding a Drude part to this dielectric function, but this has not led to noticeable changes for the Casimir interaction in the distance range up to 500 nm which has been explored in the experiment. To model the optical properties of gold we have used available optical data, extrapolated at low frequencies by a Drude model $\varepsilon(i\xi) = \frac{\omega_p^2}{\xi(\xi+\gamma)}$ with $\omega_p = 9\,\mathrm{eV}$ and $\gamma = 35\,\mathrm{meV}$. The method is described in detail in [36]. The calculations were run up to $N = 3$ diffraction orders, after which

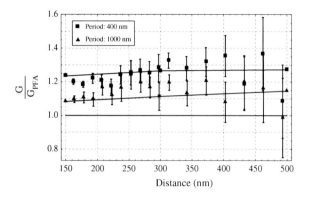

Fig. 4.9 Comparisons between experimental measurements and exact calculations for the Casimir force gradient between a gold sphere and two types of silicon gratings. Squares and circles correspond to data points provided by Ho Bun Chan for a grating period of 400 nm and 1 µm respectively. The solid curves are calculated data obtained using the scattering approach for the corresponding experimental parameters

Fig. 4.10 Same plot as Fig. 4.9, but with corrected filling factors

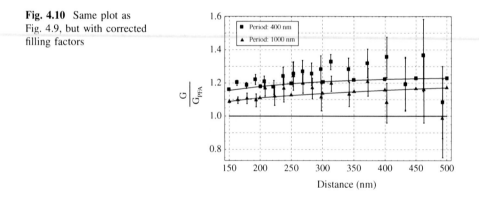

the result for the Casimir energy was found to have converged. The result of our calculation for the filling factors such as originally given in [80] is shown in Fig. 4.9 as the solid green and red curves for the 400 nm ($p = 0.51$) and 1 µm ($p = 0.48$) samples respectively. The theoretical predictions and the experimental data points are in excellent agreement. In particular, due to a new improved numerical code the agreement is better than the one presented in [82].

After submitting this paper we have been informed by Chan et al. that the filling factors in [80] were erroneously interchanged. The correct filling factors are: $p = 0.48$ for the 400 nm grating and $p = 0.51$ for the 1 µm grating. In Fig. 4.10 we show the same calculations as before but with the new filling factors. The agreement between experimental and theoretical data points is slightly degraded for the 400 nm grating, but the overall agreement remains very good.

4.4.3 Exact Calculations in the Plane-sphere Geometry

The plane-sphere geometry is the configuration in which the most precise Casimir force measurements are currently performed [73]. The Casimir interaction in this geometry can also be calculated in a formally exact manner using the general scattering formula (4.19). Such calculations have first been performed for perfectly reflecting mirrors [68, 69] where it was found that the Casimir energy was smaller than expected from the PFA and, furthermore, that the result for electromagnetic fields was departing from PFA more rapidly than was expected from previously existing scalar calculations [65, 66]. It is only very recently that the same calculations have been performed for the more realistic case of metallic mirrors at zero temperature [70] and at arbitrary temperature [71, 72] where both the lossless plasma model dielectric function and the lossy Drude dielectric function have been studied. We will sketch the method in the following.

The set-up of a sphere of radius R above a flat plate is schematically presented in Fig. 4.11. We denote respectively L and $\mathcal{L} \equiv L + R$ the closest approach distance and the center-to-plate distance. In this configuration, the general expression of the Casimir free energy at temperature T may be written as

$$\mathcal{F} = k_{\mathrm{B}} T \sum_m {}' \operatorname{Tr} \ln D(i\xi_m), \quad D \equiv 1 - R_{\mathrm{S}} e^{-K\mathcal{L}} R_{\mathrm{P}} e^{-K\mathcal{L}}. \tag{4.31}$$

The expression contains the reflection operators of the sphere R_{S} and the plate R_{P} which are evaluated with reference points placed at the sphere center and at its projection on the plane of the plate. They are sandwiched in between operators $e^{-K\mathcal{L}}$ describing the propagation between the two reference points.

The upper expression is conveniently written through a decomposition on suitable plane-wave and multipole basis [70]; R_{P} is thus expressed in terms of the Fresnel reflection coefficients r_p with $p = \mathrm{TE}$ and TM for the two electromagnetic polarizations, while R_{S} contains the Mie coefficients a_ℓ, b_ℓ for respectively electric and magnetic multipoles at order $\ell = 1, 2, \dots$. Due to rotational symmetry around the z-axis, each eigenvalue of the angular momentum m gives a separate contribution to the Casimir free energy $\mathcal{F}^{(m)}$, obtained through the same formula as (4.31). The scattering formula is obtained by writing also transformation formulas from the plane waves basis to the spherical waves basis and conversely.

The result takes the form of a multipolar expansion with spherical waves labeled by ℓ and m ($|m| \leq \ell$). It can be considered as an "exact" multipolar series of the Casimir free energy. Of course, the numerical computations of this series can only be done after truncating the vector space at some maximum value $\ell_{\max}$ of the orbital index ℓ.

The effect of this truncation is represented on Fig. 4.12 where the Casimir energy in the plane-sphere geometry divided by its PFA estimation

$$\rho_E = \frac{E}{E^{\mathrm{PFA}}} \tag{4.32}$$

Fig. 4.11 The geometry of a sphere of radius R and a plate at distance L; the center-to-plate distance is $\mathcal{L} \equiv L + R$

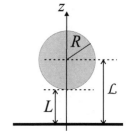

is plotted for various values of ℓ_{max}, in the special case of perfect mirrors in vacuum ($T = 0$). The figure shows that as expected the numerical results are more and more accurate when ℓ_{max} is increased. More precisely the accuracy is significantly degraded when the ratio L/R goes below a minimal value inversely proportional to ℓ_{max}

$$x \equiv \frac{L}{R} > x_{min}, \quad x_{min} \propto \frac{1}{\ell_{max}}. \tag{4.33}$$

To illustrate the effect of the truncation, one may say that the accuracy is degraded by typically more than 0.1% when $x < 0.05$ for a value of $\ell_{max} = 85$. For small values of x, which corresponds to the most precise current experiments, it may be possible to obtain information through an extrapolation of the numerical results. As an example, the dashed line on Fig. 4.12 shows the result of a third degree polynomial fit using accurate numerical evaluations.

As a further step, we show now on Fig. 4.13 the results corresponding to perfect and plasma mirrors, still at zero temperature [70]. We have derived the Casimir energy (4.31) to obtain expressions for the force F and force gradient G, and then formed the ratios of the plane-sphere exact results to the PFA expectations F^{PFA} and G^{PFA} respectively

$$\rho_F = \frac{F}{F^{PFA}}, \quad \rho_G = \frac{G}{G^{PFA}}. \tag{4.34}$$

Using these theoretical evaluations, it is now possible to extract information of interest for a comparison with the experimental study of the PFA in the plane-sphere geometry [73]. In this experiment, the force gradient has been measured for various radii of the sphere and no deviation of the PFA was observed. The authors expressed their result as a constraint on the slope at origin β_G of the function $\rho_G(x)$

$$\rho_G(x) = 1 + \beta_G x + O(x^2), \quad |\beta_G| < 0.4. \tag{4.35}$$

Reasoning along the same lines, we have interpolated our theoretical evaluation of ρ_G at low values of $x = L/R$ [70]. Surprisingly the slope obtained for perfect reflectors was found to lie outside the experimental bound of [73]

$$\beta_G^{perf} \sim -0.48. \tag{4.36}$$

Fig. 4.12 *Upper graph*: the ratio $\rho_E = E/E^{\mathrm{PFA}}$ of the plane-sphere Casimir energy to its PFA estimation is plotted as a function of $\ell_{\max}$ for different values of $L/R = 0.05, 0.1, 0.2$. *Lower graph*: same ratio ρ_E plotted as a function of L/R for different values of $\ell_{\max} = 20, 40, 80$

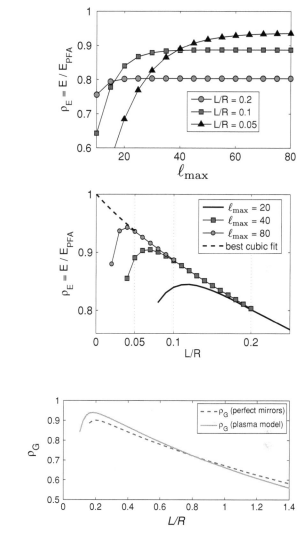

Fig. 4.13 Variation of ρ_G as a function of L/R, for a nanosphere of radius $R = 100$ nm; the solid line corresponds to gold-covered plates ($\lambda_P = 136$ nm) and the dashed line to perfect reflectors. The decrease at low values of L/R represents a numerical inaccuracy due to the limited value of $\ell_{\max}$ (4.24 in this calculation [69])

The consistency with this bound is however recovered for the calculations done for plasma mirrors

$$\beta_G^{\mathrm{plas}} \sim -0.21. \tag{4.37}$$

As a last example of application, we now discuss the effect of a non-zero temperature. To this end we evaluate (4.31) at ambient temperature ($T = 300$ K).

Fig. 4.14 Thermal Casimir
force at $T = 300$ K divided
by the zero temperature force,
computed between perfectly
reflecting sphere and plane
(*upper graph*), and between
Drude metals (*lower graph*),
plotted for $\lambda_P = 136$ nm,
$\lambda_\gamma/\lambda_P = 250$. The solid lines
from bottom to top
correspond to increasing
values of sphere radii. The
dotted curve in the upper
graph is the analytical
asymptotic expression in the
$L \gg R$ limit. The PFA
expressions are given by the
dashed curves

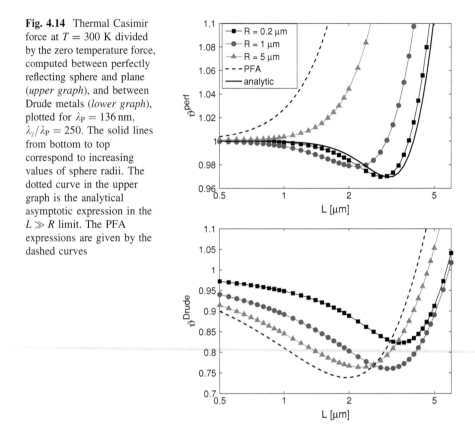

The results of the numerical computations are shown on Fig. 4.14, for the limiting case of perfect reflection (left) and for Drude metals (right) evaluated for $\lambda_P = 136$ nm,$\lambda_\gamma/\lambda_P = 250$ (values corresponding to gold). We have calculated the Casimir force F^{perf} and F^{Drud} between the plane and the sphere at ambient temperature and then plotted the corresponding ratios ϑ^{perf} and ϑ^{Drud} of this force to a reference force corresponding to zero temperature

$$F^{\text{perf}}(L, T) \equiv -\frac{\partial \mathcal{F}^{\text{perf}}}{\partial L}, \quad \vartheta^{\text{perf}} \equiv \frac{F^{\text{perf}}(L, T)}{F^{\text{perf}}(L, 0)},$$

$$F^{\text{Drud}}(L, T) \equiv -\frac{\partial \mathcal{F}^{\text{Drud}}}{\partial L}, \quad \vartheta^{\text{Drud}} \equiv \frac{F^{\text{Drud}}(L, T)}{F^{\text{Drud}}(L, 0)}. \tag{4.38}$$

The various solid curves are drawn for different sphere radii R as a function of the separation L; the dashed curves on Fig. 4.14 represent the quantities $\vartheta_{\text{PFA}}^{\text{perf}}$ and $\vartheta_{\text{PFA}}^{\text{Drud}}$ obtained from (4.38) by using the PFA; the dotted curve in the upper graph is an analytical asymptotic expression discussed below. We do not show the corresponding plots for plasma mirrors as they are very similar to the perfect mirror case.

The comparison of ϑ^{perf} and ϑ^{Drud} reveals surprising features, which could not be expected from an analysis in the parallel-plate geometry. First both ratios ϑ start from unity at small distances $L/R \to 0$. For R small enough, the ratios then decrease below unity with increasing distance, reach a radius-dependent minimum and then increase again at very large distances. This behavior entails that the Casimir force is smaller at $T=300$ K than at $T = 0$, implying a repulsive contribution from thermal fluctuations. The dotted-dashed PFA curve in the upper graph of Fig. 4.14 representing $\vartheta^{\text{perf}}_{\text{PFA}}$ is always larger than unity, excluding such a repulsive contribution from thermal fluctuations in the plane-plane geometry.

A second important feature showing up in Fig. 4.14 is that the PFA expression always overestimates the effect of temperature on the force between perfect (and plasma) mirrors. However between Drude metals, the PFA underestimates this effect at small distances and overestimates it at large distances, the overestimation being however smaller than for perfect mirrors. These results clearly indicate that there is a strong correlation between the effects of plane-sphere geometry, temperature and dissipation.

The calculation of the Casimir free energy may be done analytically for small frequencies corresponding to large plane-sphere separations

$$\mathcal{F}^{\text{perf}}_{\ell=1} = -\frac{3\hbar c R^3}{4\lambda_T L^3}\phi(v), \quad v \equiv \frac{2\pi L}{\lambda_T},$$

$$\phi(v) \equiv \frac{v^2 \cosh v + v \sinh v + \cosh v \sinh^2 v}{2 \sinh^3 v}. \tag{4.39}$$

This simple expression is a good approximation, as proven by the fact that the full expression of ϑ^{perf} tends indeed asymptotically to this simple form for small radii $R \ll L$ (dotted line on upper graph of Fig. 4.14). One can also derive from this expression interesting information about the behavior of the Casimir entropy

$$S^{\text{perf}}_{\ell=1} = -\frac{\partial \mathcal{F}^{\text{perf}}_{\ell=1}}{\partial T} = \frac{3k_B R^3}{4L^4}(\phi(v) + v\phi'(v)). \tag{4.40}$$

This expression takes on negative values for $v \lesssim 1.5$, that is $L \lesssim 1.8\,\mu\text{m}$ at $T = 300$ K, which is in agreement with the behavior observed in the upper graph of Fig. 4.14: in most cases ϑ^{perf} decreases below unity as the distance increases, reaches a minimum and then increases again at long distances. As long as R is not too large, the thermal photons provide a repulsive contribution over a distance range that gets wider as R decreases, to become $L \lesssim \lambda_T/2$ for very small spheres.

We finally will compare the predictions of the dissipationless plasma model and the dissipative Drude model for the thermal Casimir interaction in the plane-sphere geometry. The difference will become particularly clear in the high temperature limit $\mathcal{L} \gg \lambda_T$ where one only needs to take the first Matsubara frequency $\xi_0 = 0$ when computing the Casimir free energy. In the low frequency limit, the Fresnel coefficients (4.16) for the plates are given by $r_{\text{TE}} \approx -r_{\text{TM}} \approx -1$ for the

Fig. 4.15 Ratio of thermal
Casimir force at T=300 K
calculated with the plasma
model and the Drude model,
as a function of surface
separation L for different radii
of the sphere. The solid
curves from bottom to top
correspond to increasing
values of sphere radii. The
dashed curve is the PFA
prediction

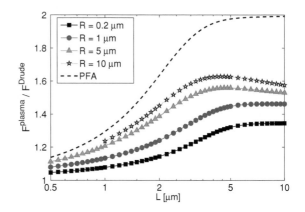

plasma model. The Mie coefficients are easily evaluated [71, 72] and give the
following approximation for the Casimir force within the plasma model

$$\mathcal{F}^{\text{plas}} \approx -\frac{3\hbar c R^3}{8\lambda_T \mathcal{L}^3}\left(1 + \frac{1}{\alpha^2} - \frac{\coth \alpha}{\alpha}\right) \quad \mathcal{L} \gg \lambda_T, R, \lambda_P, \alpha \equiv \frac{2\pi R}{\lambda_P}.$$

This result reproduces, as a particular case, the perfectly-reflecting limit when
$\lambda_P \ll R$.

For the Drude model, the TE Fresnel reflection coefficient has the well-known
low-frequency limit $r_{\text{TE}} \to 0$, whereas the TM coefficient behaves as in the plasma
model: $r_{\text{TM}} \approx 1$. The low-frequency expansion of the Mie coefficients are also
quite different from the plasma case and can be found in [71, 72]. The resulting
high-temperature large-distance limit for the free energy is

$$\mathcal{F}^{\text{Drud}} \approx -\frac{\hbar c R^3}{4\lambda_T \mathcal{L}^3} \quad , \quad \mathcal{L} \gg \lambda_T, R. \tag{4.41}$$

This remarkable result does not depend on the length scales λ_P and λ_γ characterizing the material response, whereas the corresponding plasma result (4.41)
clearly depends on λ_P. One can show that this is always the case in the high-
temperature limit $\lambda_T \ll \mathcal{L}$.

In the case of the Drude model with a non-vanishing relaxation frequency the
free energy for the Drude model turns out to be 2/3 of the expression for perfect
mirrors whereas this ratio is 1/2 in the plane-plane geometry. The latter result is
explained by the fact that the TE reflection coefficient vanishes at zero frequency
so that only the TM modes contribute [42, 45]. The change of the ratio 1/2–2/3 in
the plane-sphere geometry has to be attributed to the redistribution of the TE and
TM contributions into electric and magnetic spherical eigenmodes. The change is
illustrated in Fig. 4.15, where we have plotted the ratio of the thermal Casimir
force F^{plas} calculated with the plasma model to the one F^{Drud} obtained with the
Drude model. Again, the plots correspond to $\lambda_P = 136$ nm and $\lambda_\gamma/\lambda_P = 250$.

The results of our calculations are shown by the solid curves with the sphere radius increasing from bottom to top. The ratio $F^{\text{plas}}/F^{\text{Drud}}$ varies in the plane-sphere geometry as a function of the sphere radius, clearly demonstrating the strong interplay between the effects of temperature, dissipation and geometry. For large spheres ($R \gg \lambda_P$), the ratio converges to the value 3/2, whereas it remains smaller for small spheres (down to 1.2 for $R \sim 100$ nm). The dashed curve gives the variation of the same ratio as calculated within the PFA which leads to a factor of 2 in the limits of large distances or high temperatures, corresponding to the prediction in the parallel-plates geometry. This factor of 2 deduced within the PFA is never approached within the calculations performed in the plane-sphere geometry.

4.5 Conclusion

In this paper we have reviewed the quantum and thermal Casimir interaction between parallel plates, corrugated surfaces and plane and spherical mirrors. To perform our calculations we have extensively used the scattering approach where the objects are characterized by scattering matrices. We have compared our results with predictions obtained within the PFA. When taking the diffraction of the electromagnetic field correctly into account, surprising features appear especially for the thermal Casimir force in the plane-sphere geometry, where the exact results differ substantially from predictions within the PFA. While open problems are still waiting to be tackled, the whole set of presented results clearly illustrates the usefulness and practicality of the scattering approach in Casimir physics.

Acknowledgments The authors thank I. Cavero-Pelaez, D. Dalvit, G.L. Ingold, M.-T. Jaekel, J. Lussange, P.A. Maia Neto, R. Messina, P. Monteiro, I. Pirozenkho and V. Marachevsky for contributions and/or fruitful discussions, H. B. Chan for kindly providing the data of his experiment and the ESF Research Networking Programme CASIMIR (http://www.casimir-network.com) for providing excellent possibilities for discussions and exchange. Financial support from the French Contract ANR-06-Nano-062 and from Capes-Cofecub are gratefully acknowledged.

References

1. Casimir, H.B.G.: Proc. K. Ned. Akad. Wet. **51**, 793 (1948)
2. Milonni, P.W.: The Quantum Vacuum. Academic Press, London (1994)
3. Lamoreaux, S.K.: Resource letter CF-1: Casimir force. Am. J. Phys. **67**, 850 (1999)
4. Reynaud, S., Lambrecht, A., Genet, C., Jaekel, M.T.: Quantum vacuum fluctuations. C. R. Acad. Sci. Paris **IV-2**, 1287 (2001)
5. Bordag, M., Mohideen, U., Mostepanenko, V.M.: New developments in the Casimir effect. Phys. Rep. **353**, 1 (2001)
6. Milton, K.A.: The Casimir effect: recent controversies and progress. J. Phys. A **37**, R209 (2004)
7. Decca, R.S., López, D., Fischbach, E., Klimchitskaya, G.L., Krause, D.E., Mostepanenko, V.M.: Precise comparison of theory and new experiment for the Casimir force leads to

stronger constraints on thermal quantum effects and long-range interactions. Annals Phys. **318**, 37 (2005)

8. Lambrecht, A., Maia Neto, P.A., Reynaud, S.: The Casimir effect within scattering theory. New J. Phys. **8**, 243 (2006)

9. Jaekel, M.T., Reynaud, S.: Movement and fluctuations of the vacuum. Rep. prog. phys. **60**, 863 (1997)

10. Elizalde, E.: Quantum vacuum fluctuations and the cosmological constant. J Phys A **40**, 6647 (2007)

11. Jaekel M.T., and Reynaud S.: In Proceeding of the Orleans School on Mass (2009)

12. Fischbach, E., Talmadge, C.: The Search for Non Newtonian Gravity. AIP Press/Springer, Berlin (1998)

13. Adelberger, E.G., Heckel, B.R., Nelson, A.E.: Tests of the gravitational inverse-square law. Ann. Rev. Nucl. Part. Sci. **53**, 77 (2009)

14. Kapner, D.J., Cook, T.S., Adelberger, E.G., Gundlach, J.H., Heckel, B.R., Hoyle, C.D., Swanson, H.E.: Tests of the gravitational inverse-square law below the dark-energy length scale. Phys. Rev. Lett. **98**, 021101 (2007)

15. Lambrecht A., and Reynaud S.: Poincaré Seminar on Vacuum Energy and Renormalization **1**, 107 (2002) [arXiv quant-ph/0302073]

16. Onofrio, R.: Casimir forces and non-Newtonian gravitation. New J. Phys. **8**, 237 (2006)

17. Decca, R.S., López, D., Fischbach, E., Klimchitskaya, G.L., Krause, D.E., Mostepanenko, V.M.: Novel constraints on light elementary particles and extra-dimensional physics from the Casimir effect. Eur. Phys. J. C **51**, 963 (2007)

18. Lepoutre, S., Jelassi, H., Lonij, V.P.A., Trenec, G., Buchner, M., Cronin, A.D., Vigue, J.: Dispersive atom interferometry phase shifts due to atom-surface interactions. Europhys. Lett. **88**, 20002 (2009)

19. Nesvizhevsky, V.V., Pignol, G., Protasov, K.V.: Neutron scattering and extra-short-range interactions. Phys. Rev. D **77**, 034020 (2008)

20. Parsegian, V.A.: Van der Waals Forces: a Handbook for Biologists, Chemists, Engineers, and Physicists. Cambridge University Press, Cambridge (2006)

21. Sparnaay M.J.: In Physics in the Making. Eds Sarlemijn A., and Sparnaay M.J., North-Holland (1989)

22. Lamoreaux, S.K.L.: Demonstration of the casimir force in the 0.6 to 6 mu m range. Phys. Rev. Lett. **78**, 5 (1997)

23. Mohideen, U., Roy, A.: Precision measurement of the Casimir force from 0.1 to 0.9 mu m. Phys. Rev. Lett. **81**, 4549 (1998)

24. Harris, B.W., Chen, F., Mohideen, U.: Precision measurement of the Casimir force using gold surfaces. Phys. Rev. A **62**, 052109 (2000)

25. Ederth, T.: Template-stripped gold surfaces with 0.4 nm rms roughness suitable for force measurements: Application to the Casimir force in the 20–100 nm range. Phys. Rev. A **62**, 062104 (2000)

26. Bressi, G., Carugno, G., Onofrio, R., Ruoso, G.: Measurement of the Casimir force between parallel metallic surfaces. Phys. Rev. Lett **88**, 041804 (2002)

27. Decca, R.S., López, D., Fischbach, E., Krause, D.E.: Measurement of the Casimir force between dissimilar metals. Phys. Rev. Lett. **91**, 050402 (2003)

28. Chen, F., Klimchitskaya, G.L., Mohideen, U., Mostepananko, V.M.: Theory confronts experiment in the Casimir force measurements: Quantification of errors and precision. Phys. Rev. A **69**, 022117 (2004)

29. Decca, R.S., López, D., Fischbach, E., Klimchitskaya, G.L., Krause, D.E., Mostepanenko, V.M.: Tests of new physics from precise measurements of the Casimir pressure between two gold-coated plates. Phys. Rev. D **75**, 077101 (2007)

30. Munday, J.N., Capasso, F.: Precision measurement of the Casimir-Lifshitz force in a fluid. Phys. Rev. A **75**, 060102(R) (2007)

31. van Zwol, P.J., Palasantzas, G., De Hosson, J.T.M.: Influence of random roughness on the Casimir force at small separations. Phys. Rev. B **77**, 075412 (2008)

32. Munday, J.N., Capasso, F., Parsegian, V.A.: Measured long-range repulsive Casimir-Lifshitz forces. Nature **457**, 170 (2009)
33. Jourdan, G., Lambrecht, A., Comin, F., Chevrier, J.: Quantitative non-contact dynamic Casimir force measurements. Europhys. Lett. **85**, 31001 (2009)
34. Masuda, M., Sasaki, M.: Limits on nonstandard forces in the submicrometer range. Phys. Rev. Lett. **102**, 171101 (2009)
35. de Man, S., Heeck, K., Wijngaarden, R.J., Iannuzzi, D.: Halving the casimir force with conductive oxides. Phys. Rev. Lett. **103**, 040402 (2009)
36. Lambrecht, A., Reynaud, S.: Casimir force between metallic mirrors. Euro. Phys. J. D **8**, 309 (2000)
37. Svetovoy, V.B., van Zwol, P.J., Palasantzas, G., De Hosson, J.T.M.: Optical properties of gold films and the Casimir force. Phys. Rev. B **77**, 035439 (2008)
38. Mehra, J.: Temperature correction to the casimir effect. Physica **37**, 145 (1967)
39. Brown, L.S., Maclay, G.J.: Vacuum stress between conducting plates: an image solution. Phys. Rev. **184**, 1272 (1969)
40. Schwinger, J., de Raad, L.L., Milton, K.A.: Casimir effect in dielectrics. Ann. Phys. **115**, 1 (1978)
41. Genet, C., Lambrecht, A., Reynaud, S.: Temperature dependence of the Casimir effect between metallic mirrors. Phys. Rev. A **62**, 012110 (2000)
42. Boström, M., Sernelius, B.E.: Thermal effects on the Casimir force in the 0.1-5 mu m range. Phys. Rev. Lett. **84**, 4757 (2000)
43. Reynaud S., Lambrecht A., and Genet C.: In Quantum Field Theory Under the Influence of External Conditions. Ed. Milton K.A., Rinton Press (2004) [arXiv quant-ph/0312224]
44. Klimchitskaya, G.L., Mostepanenko, V.M.: Experiment and theory in the Casimir effect. Contemp. Phys. **47**, 131 (2006)
45. Brevik, I., Ellingsen, S.E., Milton, K.A.: Thermal corrections to the Casimir effect. New J. Phys. **8**, 236 (2006)
46. Brevik, I., Ellingsen, S.E., Høye, J.S., Milton, K.A.: Analytical and numerical demonstration of how the Drude dispersive model satisfies Nernst's theorem for the Casimir entropy. J. Phys. A **41**, 164017 (2008)
47. Milton, K.A.: Recent developments in the Casimir effect. J. Phys. Conf. Ser. **161**, 012001 (2009)
48. Jancovici, B., Šamaj, L.: Casimir force between two ideal-conductor walls revisited. Europhys. Lett. **72**, 35 (2005)
49. Buenzli, P.R., Martin, P.A.: The Casimir force at high temperature. Europhys. Lett. **72**, 42 (2005)
50. Bimonte, G.: Bohr-van Leeuwen theorem and the thermal Casimir effect for conductors. Phys. Rev. A **79**, 042107 (2009)
51. Bezerra, V.B., Klimchitskaya, G.L., Mostepanenko, V.M.: Correlation of energy and free energy for the thermal casimir force between real metals. Phys. Rev. A **66**, 062112 (2002)
52. Ingold, G.L., Lambrecht, A., Reynaud, S.: Quantum dissipative Brownian motion and the Casimir effect. Phys. Rev. E **80**, 041113 (2009)
53. Deriagin, B.V., Abrikosova, I.I., Lifshitz, E.M.: Quart. Rev. **10**, 295 (1968)
54. Lifshitz, E.M.: The theory of molecular attractive forces between solids. Sov. Phys. JETP **2**, 73 (1956)
55. Dzyaloshinskii, I.E., Lifshitz, E.M., Pitaevskii, L.P.: The general theory of van der Waals forces. Sov. Phys. Uspekhi **4**, 153 (1961)
56. Balian, R., and Duplantier, B.: Electromagnetic waves near perfect conductors. I. Multiple scattering expansions. Distribution of modes. Ann. Phys. NY **104**,300 (1977)
57. Balian, R., and Duplantier, B.: Electromagnetic waves near perfect conductors. II. Casimir effect. Ann. Phys. NY **112**,165 (1978)
58. Balian R.: In Poincaré Seminar 2002 on Vacuum Energy. Eds Duplantier B., and Rivasseau V., Birkhäuser (2003)
59. Balian R., and Duplantier B.: In 15th SIGRAV Conference on General Relativity and Gravitation.[arXiv quant-ph/0408124] (2004)

60. Schaden, M., Spruch, L.: Focusing virtual photons: Casimir energies for some pairs of conductors. Phys. Rev. Lett. **84**, 459 (2000)
61. Jaffe, R.L., Scardicchio, A.: Casimir effect and geometric optics. Phys. Rev. Lett. **92**, 070402 (2004)
62. Schröder, O., Sardicchio, A., Jaffe, R.L.: Casimir energy for a hyperboloid facing a plate in the optical approximation. Phys. Rev. A **72**, 012105 (2005)
63. Reynaud, S., Maia Neto, P.A., Lambrecht, A.: Casimir energy and geometry: beyond the proximity force approximation. J. Phys. A **41**, 164004 (2008)
64. Emig, T., Graham, N., Jaffe, R.L., Kardar, M.: Casimir forces between arbitrary compact objects. Phys. Rev. Lett **99**, 170403 (2007)
65. Bordag, M., Nikolaev, V.: Casimir force for a sphere in front of a plane beyond proximity force approximation. J. Phys. A **41**, 164002 (2008)
66. Wirzba, A.: The Casimir effect as a scattering problem. J. Phys. A **41**, 164003 (2008)
67. Klingmüller, K., Gies, H.: Geothermal Casimir phenomena. J. Phys. A **41**, 164042 (2008)
68. Emig T.: Fluctuation-induced quantum interactions between compact objects and a plane mirror. J. Stat. Mech. Theory Exp. P04007 (2008)
69. Maia Neto, P.A., Lambrecht, A., Reynaud, S.: Casimir energy between a plane and a sphere in electromagnetic vacuum. Phys. Rev. A **78**, 012115 (2008)
70. Canaguier-Durand, A., Maia Neto, P.A., Cavero-Pelaez, I., Lambrecht, A., Reynaud, S.: Casimir interaction between plane and spherical metallic surfaces. Phys. Rev. Lett. **102**, 230404 (2009)
71. Canaguier-Durand, A., MaiaNeto P., A., Lambrecht, A., Reynaud, S.: Thermal Casimir effect in the plane-sphere geometry. Phys. Rev. Lett. **104**, 040403 (2010)
72. Canaguier-Durand A., Maia Neto P.A., Lambrecht A., and Reynaud S. Thermal Casimir effect for Drude metals in the plane-sphere geometry. submitted [arXiv:1005.4294] (2010)
73. Krause, D.E., Decca, R.S., López, D., Fischbacoh, E.: Experimental investigation of the Casimir force beyond the proximity-force approximation. Phys. Rev. Lett. **98**, 050403 (2007)
74. Chen, F., Mohideen, U., Klimchitskaya, G.L., and Mostepanenko, V.M.: Demonstration of the lateral Casimir force. Phys. Rev. Lett. **88**,101801 (2002)
75. Chen, F., Mohideen, U., Klimchitskaya, G.L., and Mostepanenko, V.M.: Demonstration of the lateral Casimir force. Phys. Rev. Lett. **66**, 032113 (2002)
76. Büscher, R., Emig, T.: Geometry and spectrum of Casimir forces. Phys. Rev. Lett. **94**, 133901 (2005)
77. Rodrigues, R.B., Maia Neto, P.A., Lambrecht, A., and Reynaud, S.:Lateral casimir force beyond the proximity-force approximation. Phys. Rev. Lett. **96**,100402 (2006)
78. Rodrigues, R.B., Maia Neto, P.A., Lambrecht, A., and Reynaud, S.:Lateral casimir force beyond the proximity-force approximation. Reply. Phys. Rev. Lett. **98**,068902 (2007)
79. Rodrigues, R.B., Maia Neto, P.A., Lambrecht, A., Reynaud, S.: Lateral Casimir force beyond the proximity force approximation: A nontrivial interplay between geometry and quantum vacuum. Phys. Rev. A **75**, 062108 (2007)
80. Chan, H.B., Bao, Y., Zou, J., Cirelli, R.A., Klemens, F., Mansfield, W.M., Pai, C.S.: Measurement of the Casimir force between a gold sphere and a silicon surface with nanoscale trench arrays. Phys. Rev. Lett. **101**, 030401 (2008)
81. Chiu, H.C., Klimchitskaya, G.L., Marachevsky, V.N., Mostepanenko, V.M., Mohideen, U.: Demonstration of the asymmetric lateral Casimir force between corrugated surfaces in the nonadditive regime. Phys. Rev. B **80**, 121402 (2009)
82. Lambrecht, A., Marachevsky, V.N.: Casimir interaction of dielectric gratings. Phys. Rev. Lett. **101**, 160403 (2008)
83. Lambrecht, A.: Nanotechnology - Shaping the void. Nature **454**, 836 (2008)
84. Jaekel M.T., and Reynaud S.: Casimir force between partially transmitting mirrors. J. Physique I-1 1395 (1991) [arXiv quant-ph/0101067]
85. Genet, C., Lambrecht, A., Reynaud, S.: Casimir force and the quantum theory of lossy optical cavities. Phys. Rev. A **67**, 043811 (2003)

86. Milton, K.A., Wagner, J.: Multiple scattering methods in Casimir calculations. J. Phys. A **41**, 155402 (2008)
87. Bulgac, A., Magierski, P., Wirzba, A.: Scalar Casimir effect between Dirichlet spheres or a plate and a sphere. Phys. Rev. D **73**, 025007 (2006)
88. Bordag, M.: Casimir effect for a sphere and a cylinder in front of a plane and corrections to the proximity force theorem. Phys. Rev. D **73**, 125018 (2006)
89. Kenneth, O., and Klich, I.: Opposites attract: A theorem about the Casimir force. Phys. Rev. Lett. **97**,160401 (2006)
90. Kenneth, O., and Klich, I.: Opposites attract: A theorem about the Casimir force. Phys. Rev. B **78**,014103 (2008)
91. Emig, T., Jaffe R., L.: Casimir forces between arbitrary compact objects. J. Phys. A **41**, 164001 (2008)
92. Rahi, S.J., Emig, T., Graham, N., Jaffe, R.L., Kardar, M.: Scattering theory approach to electrodynamic Casimir forces. Phys. Rev. D **80**, 085021 (2009)
93. Langfeld, K., Moyaerts, L., Gies, H.: Casimir effect on the worldline. J. High En. Phys. **0306**, 018 (2003)
94. Gies, H., Klingmüller, K.: Casimir effect for curved geometries: Proximity-force-approximation validity limits. Phys. Rev. Lett. **96**, 220401 (2006)
95. Emig, T., Jaffe, R.L., Kardar, M., Scardicchio, A.: Casimir interaction between a plate and a cylinder. Phys. Rev. Lett. **96**, 080403 (2006)
96. Dalvit, D.A.R., Lombardo, F.C., Mazzitelli, F.D., Onofrio, R.: Exact Casimir interaction between eccentric cylinders. Phys. Rev. A **74**, 020101 (2006)
97. Mazzitelli, F.C., Dalvit, D.A.R., Lombardo, F.C.: Exact zero-point interaction energy between cylinders. New J. Phys. **8**, 240 (2006)
98. Rodriguez, A., Ibanescu, M., Iannuzzi, D., Capasso, F., Joannopoulos, J.D., Johnson, S.G.: Computation and visualization of Casimir forces in arbitrary geometries: Nonmonotonic lateral-wall forces and the failure of proximity-force approximations. Phys. Rev. Lett. **99**, 080401 (2007)
99. Döbrich B., DeKieviet M., and Gies H.: Nonpertubative access to Casimir-Polder forces. arXiv:0910.5889 (2009)
100. Bordag M., and Nikolaev V.: First analytic correction beyond PFA for the electromagnetic field in sphere-plane geometry. arXiv:0911.0146 (2009)
101. Gies H., and Weber A.: Geometry-temperature interplay in the Casimir effect. arXiv: 0912.0125 (2009)
102. Bordag M., and Pirozhenko I.: Vacuum energy between a sphere and a plane at finite temperature. arXiv:0912.4047 (2010)
103. Emig T.: Casimir physics: geometry, shape and material. arXiv:1003.0192 (2010)
104. Zandi R., Emig T., and Mohideen U.: Quantum and thermal Casimir interaction between a sphere and a plate: Comparison of Drude and plasma models. arXiv:1003.0068 (2010)
105. Weber A., and Gies H.: Non-monotonic thermal Casimir force from geometry-temperature interplay. arXiv:1003.0430 (2010)
106. Weber A., and Gies H.: Geothermal Casimir phenomena for the sphere-plane and cylinder-plane configurations. arXiv:1003.3420 (2010)
107. Maia Neto, P.A., Lambrecht, A., Reynaud, S.: Roughness correction to the Casimir force: Beyond the proximity force approximation. Europhys. Lett. **69**, 924 (2005)
108. Maia Neto, P.A., Lambrecht, A., Reynaud, S.: Casimir effect with rough metallic mirrors. Phys. Rev. A **72**, 012115 (2005)
109. Rodrigues, R.B., Maia Neto, P.A., Lambrecht, A., Reynaud, S.: Vacuum-induced torque between corrugated metallic plates. Europhys. Lett. **76**, 822 (2006)
110. Rodrigues, R.B., Maia Neto, P.A., Lambrecht, A., Reynaud, S.: Phys. Rev. Lett. **100**, 040405 (2008)
111. Dalvit, D.A.R., Maia Neto, P.A., Lambrecht, A., Reynaud, S.: Probing quantum-vacuum geometrical effects with cold atoms. Phys. Rev. Lett. **100**, 040405 (2008)

112. Messina, R., Dalvit, D.A.R., Maia Neto, P.A., Lambrecht, A., Reynaud, S.: Scattering approach to dispersive atom-surface interactions. Phys. Rev. A **80**, 022119 (2009)
113. Johannson, P., Apell, P.: Geometry effects on the van der Waals force in atomic force microscopy. Phys Rev. B **56**, 4159 (1997)
114. Noguez, C., Roman-Velazquez, C.E., Esquivel-Sirvent, R., Villarreal, C.: High-multipolar effects on the Casimir force: The non-retarded limit. Europhys. Lett. **67**, 191 (2004)
115. Feinberg, G., Sucher, J.: General theory of the van der Waals interactions: A model-independent approach. Phys Rev. A **2**, 2395 (1970)
116. Power, E.A., Thirunamachandran, T.: Zero-point energy differences and many-body dispersion forces. Phys Rev. A **50**, 3929 (1994)
117. Jaekel, M.T., Reynaud, S.: Fluctuations and dissipation for a mirror in vacuum. Quantum Opt. **4**, 39 (1992)
118. Jaekel, M.T., Reynaud, S.: Motional Casimir force. Journal de Physique I- **2**, 149 (1992)
119. Jaekel, M.T., Reynaud, S.: Causality, stability and passivity for a mirror in vacuum. Phys. Lett. A **167**, 227 (1992)
120. Jaekel, M.T., Reynaud, S.: Friction and inertia for a mirror in a thermal field. Phy. Lett. A **172**, 319 (1993)
121. Jaekel, M.T., Reynaud, S.: Quantum fluctuations of position of a mirror in vacuum. J. Phys. I-**3**, 1 (1993)
122. Jaekel, M.T., and Reynaud. S.: Inertia of Casimir energy. J. Phys I-3,1093 (1993)
123. Jaekel, M.T., Reynaud, S.: Quantum fluctuations of mass for a mirror in vacuum. Phys. Lett. A **180**, 9 (1993)
124. Lambrecht, A., Jaekel, M.T., Reynaud, S.: Motion induced radiation from a vibrating cavity. Phys. Rev. Lett. **77**, 615 (1996)
125. Lambrecht, A., Jaekel M., T., Reynaud, S.: The Casimir force for passive mirrors. Phys. Lett. A **225**, 188 (1997)
126. Pitaevskii, L.P.: Thermal lifshitz force between an atom and a conductor with a small density of carriers. Phys. Rev. Lett. **101**, 163202 (2008)
127. Geyer, B., Klimchitskaya, G.L., Mohideen, U., Mostepanenko, V.M.: Comment on "Contribution of Drifting Carriers to the Casimir-Lifshitz and Casimir-Polder Interactions with Semiconductor Materials". Phys. Rev. Lett. **102**, 189301 (2009)
128. Pitaevskii, L.P.: Phys. Rev. Lett. **102**, 189302 (2009)
129. Dalvit, D.A.R., Lamoreaux, S.K.: Contribution of drifting carriers to the Casimir-Lifshitz and Casimir-Polder interactions with semiconductor materials. Phys. Rev. Lett. **101**, 163203 (2008)
130. Decca, R.S., Fischbach, E., Geyer, B., Klimchitskaya, G.L., Krause, D.E., Lopez, D., Mohideen, U., Mostepanenko, V.M.: Comment on "Contribution of Drifting Carriers to the Casimir-Lifshitz and Casimir-Polder Interactions with Semiconductor Materials". Phys. Rev. Lett. **102**, 189303 (2009)
131. Dalvit, D.A.R., Lamoreaux, S.K.: Phys. Rev. Lett. **102**, 189304 (2009)
132. Svetovoy, V.B.: Application of the Lifshitz theory to poor conductors. Phys. Rev. Lett. **101**, 163603 (2008)
133. Svetovoy, V.B.: Application of the Lifshitz theory to poor conductors. Phys. Rev. Lett. **102**, 219903(E) (2009)
134. Kats, E.I.: JETP **46**, 109 (1977)
135. Barnett, S.M., Jeffers, J., Gatti, A., Loudon, R.: Quantum optics of lossy beam splitters. Phys. Rev. A **57**, 2134 (1998)
136. Dalvit, D.A.R., Lombardo, F.C., Mazzitelli, F.D., Onofrio, R.: Casimir force between eccentric cylinders. EuroPhys. Lett. **67**, 517 (2004)
137. Mazzitelli, F.D., Dalvit, D.A.R., Lombardo, F.C.: Exact zero-point interaction energy between cylinders. New Journal of Physics **8**, 240 (2006)
138. Bordag, M., Nikolaev, V.: The vacuum energy for two cylinders with one increasing in size. J. Phys. A **42**, 415203 (2009)

139. Zaheer S., Rahi J., Emig T., and Jaffe R. L.: Casimir interactions of an object inside a spherical metal shell. arXiv 0908.3270 (2009)
140. Chiu, H.C., Klimchitskaya, G.L., Marachevsky, V.N., Mostepanenko, V.M., Mohideen, U.: Lateral Casimir force between sinusoidally corrugated surfaces: Asymmetric profiles, deviations from the proximity force approximation, and comparison with exact theory. Phys. Rev. B **81**, 115417 (2010)
141. Munday, J., Ianuzzi, D., Barash, Y., Capasso, F.: Torque on birefringent plates induced by quantum fluctuations. Phys. Rev. A **71**, 042102 (2005)
142. Bergström, L.: Hamaker constants of inorganic materials. Adv. in Colloid. and Interf. Sci. **70**, 125 (1997)

Chapter 5
Geometry and Material Effects in Casimir Physics-Scattering Theory

Sahand Jamal Rahi, Thorsten Emig and Robert L. Jaffe

Abstract We give a comprehensive presentation of methods for calculating the Casimir force to arbitrary accuracy, for any number of objects, arbitrary shapes, susceptibility functions, and separations. The technique is applicable to objects immersed in media other than vacuum, to nonzero temperatures, and to spatial arrangements in which one object is enclosed in another. Our method combines each object's classical electromagnetic scattering amplitude with universal translation matrices, which convert between the bases used to calculate scattering for each object, but are otherwise independent of the details of the individual objects. This approach, which combines methods of statistical physics and scattering theory, is well suited to analyze many diverse phenomena. We illustrate its power and versatility by a number of examples, which show how the interplay of geometry and material properties helps to understand and control Casimir forces. We also examine whether electrodynamic Casimir forces can lead to stable levitation. Neglecting permeabilities, we prove that any equilibrium position of objects subject to such forces is unstable if the permittivities of all objects are higher or lower than that of the enveloping medium; the former being the generic case for ordinary materials in vacuum.

S. J. Rahi (✉)
Department of Physics, MIT, 02139, Cambridge, MA, USA
e-mail: sjrahi@mit.edu; sjrahi@rockefeller.edu

T. Emig
Laboratoire de Physique Théorique et Modèles Statistiques, Université Paris-Sud, CNRS UMR 8626, 91405, Orsay, France
e-mail: emig@lptms.u-psud.fr

R. L. Jaffe
Center for Theoretical Physics, MIT, Cambridge, MA 02139, USA
e-mail: jaffe@mit.edu

S. J. Rahi
Center for Studies in Physics and Biology, The Rockefeller University, 1230 York Ave, 10065 New York, USA

D. Dalvit et al. (eds.), *Casimir Physics*, Lecture Notes in Physics 834,
DOI: 10.1007/978-3-642-20288-9_5, © Springer-Verlag Berlin Heidelberg 2011

5.1 Introduction

Neutral objects exert a force on one another through electromagnetic fields even if they do not possess permanent multipole moments. Materials that couple to the electromagnetic field alter the spectrum of the field's quantum and thermal fluctuations. The resulting change in energy depends on the relative positions of the objects, leading to a fluctuation-induced force, usually called the Casimir force. Alternatively, one can regard the cause of these forces to be spontaneous charges and currents, which fluctuate in and out of existence in the objects due to quantum mechanics. The name 'Van der Waals force' is sometimes used interchangeably but it usually refers to the Casimir force in the regime where objects are close enough to one another that the speed of light is effectively infinite. The Casimir force has been the subject of precision experimental measurements [16–20, 23, 24, 27, 46, 53, 55, 59, 69, 71–73, 88] and can influence the operation of nanoscale devices [13, 16], see Ref. [54] for a review of the experiments.

Casimir and Polder calculated the fluctuation-induced force on a polarizable atom in front of a perfectly conducting plate and between two polarizable atoms, both to leading order at large separation, and obtained a simple result depending only on the atoms' static polarizabilities [15]. Casimir then extended this result to his famous calculation of the pressure on two perfectly conducting parallel plates [14]. Feinberg and Sucher [34, 35] generalized the result of Casimir and Polder to include both electric and magnetic polarizabilities. Lifshitz, Dzyaloshinskii, and Pitaevskii extended Casimir's result for parallel plates by incorporating nonzero temperature, permittivity, and permeability into a general formula for the pressure on two infinite half-spaces separated by a gap [25, 64, 65]. See also the Chap. 2 by Pitaevskii in this volume.

While these early theoretical predictions of the Casimir force applied only to infinite planar geometries (or atoms), the first precision experiments measured the force between a plate and a sphere. This geometry was preferred because keeping two plane surfaces parallel introduces additional challenges for the experimentalist. To compare the measurements with theory, however, a makeshift solution had to be used: known as the Proximity Force Approximation (PFA), it estimates the Casimir force by integrating the Casimir pressure between opposing infinitesmal surface area elements, as if they were parallel plates, over the area that the sphere and the plate expose to one another [74]. In general, this simple approximation does not capture curvature corrections but in many experimental situations, it performs surprisingly well, as can be seen in Fig. 5.1, for example; at the small separations at which the force is typically probed in precision measurements the sphere and the plate surfaces are well approximated by a collection of infinitesimal parallel plates.

Clearly, for larger separations and for surfaces that are not smooth, the PFA must fail. For example, in measurements of the Casimir force between a sphere and a trench array significant discrepancies were found [17]. And even for the regimes in which the PFA yields good estimates it would be desirable to know what the corrections are.

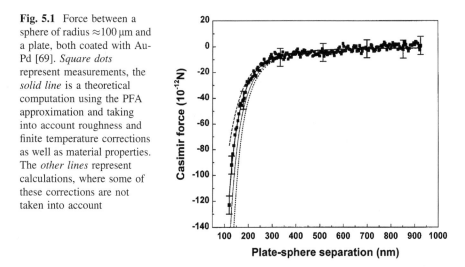

Fig. 5.1 Force between a sphere of radius ≈100 μm and a plate, both coated with Au-Pd [69]. *Square dots* represent measurements, the *solid line* is a theoretical computation using the PFA approximation and taking into account roughness and finite temperature corrections as well as material properties. The *other lines* represent calculations, where some of these corrections are not taken into account

In order to study Casimir forces in more general geometries, it turns out to be advantageous to describe the influence of an arrangement of objects on the electromagnetic field by the way they scatter electromagnetic waves. Here, we derive and apply a representation of the Casimir energy, first developed with various limitations in Refs. [28, 29] and then fully generalized in Ref. [75], that characterizes each object by its on-shell electromagnetic scattering amplitude. The separations and orientations of the objects are encoded in universal translation matrices, which describe how a solution to the source-free Maxwell's equations in the basis appropriate to one object looks when expanded in the basis appropriate to another. The translation matrices depend on the displacement and orientation of coordinate systems, but not on the nature of the objects themselves. The scattering amplitudes and translation matrices are then combined in a simple algorithm that allows efficient numerical and, in some cases, analytical calculations of Casimir forces and torques for a wide variety of geometries, materials, and external conditions. The formalism applies to a wide variety of circumstances, including:

- *n* arbitrarily shaped objects, whose surfaces may be smooth or rough or may include edges and cusps;
- objects with arbitrary linear electromagnetic response, including frequency-dependent, lossy electric permittivity and magnetic permeability tensors;
- objects separated by vacuum or by a medium with uniform, frequency-dependent isotropic permittivity and permeability;
- zero or nonzero temperature;
- and objects outside of one another or enclosed in each other.

These ideas build on a range of previous related work, an inevitably incomplete subset of which is briefly reviewed here: Scattering theory methods were first applied to the parallel plate geometry, when Kats reformulated Lifshitz theory in terms of reflection coefficients [50]. Jaekel and Reynaud derived the Lifshitz formula using reflection coefficients for lossless infinite plates [48] and Genet, Lambrecht, and Reynaud extended this analysis to the lossy case [38]. Lambrecht, Maia Neto, and Reynaud generalized these results to include non-specular reflection [58]. See also the Chap. 4 by Lambrecht et al. in this volume for additional discussions of the scattering approach.

Around the same time as Kats's work, Balian and Duplantier developed a multiple scattering approach to the Casimir energy for perfect metal objects and used it to compute the Casimir energy at asymptotically large separations [4, 5] at both zero and nonzero temperature. In their approach, information about the conductors is encoded in a local surface scattering kernel, whose relation to more conventional scattering formalisms is not transparent, and their approach was not pursued further at the time. One can find multiple scattering formulas in an even earlier article by Renne [81], but scattering is not explicitly mentioned, and the technique is only used to rederive older results.

Another scattering-based approach has been to express the Casimir energy as an integral over the density of states of the fluctuating field, using the Krein formula [6, 56, 57] to relate the density of states to the S-matrix for scattering from the ensemble of objects. This S-matrix is difficult to compute in general. In studying many-body scattering, Henseler and Wirzba connected the S-matrix of a collection of spheres [47] or disks [93] to the objects' individual S-matrices, which are easy to find. Bulgac, Magierski, and Wirzba combined this result with the Krein formula to investigate the scalar and fermionic Casimir effect for disks and spheres [11, 10, 94]. Casimir energies of solitons in renormalizable quantum field theories have been computed using scattering theory techniques that combine analytic and numerical methods [44].

Bordag, Robaschik, Scharnhorst, and Wieczorek [7, 82] introduced path integral methods to the study of Casimir effects and used them to investigate the electromagnetic Casimir effect for two parallel perfect metal plates. Li and Kardar used similar methods to study the scalar thermal Casimir effect for Dirichlet, Neumann, and mixed boundary conditions [62, 63]. The quantum extension was developed further by Golestanian and Kardar [41, 42] and was subsequently applied to the quantum electromagnetic Casimir effect by Emig, Hanke, Golestanian, and Kardar, who studied the Casimir interaction between plates with roughness [31] and between deformed plates [32]. (Techniques developed to study the scalar Casimir effect can be applied to the electromagnetic case for perfect metals with translation symmetry in one spatial direction, since then the electromagnetic problem decomposes into two scalar ones.) Finally, the path integral approach was connected to scattering theory by Emig and Buescher [12].

Closely related to the work we present here is that of Kenneth and Klich, who expressed the data required to characterize Casimir fluctuations in terms of the transition $\mathbb{T}$-operator for scattering of the fluctuating field from the objects [51]. Their abstract representation made it possible to prove general properties of the sign of the Casimir force. In Refs. [28, 29], we developed a framework in which this abstract result can be applied to concrete calculations. In this approach, the $\mathbb{T}$-operator is related to the scattering amplitude for each object individually, which in turn is expressed in an appropriate basis of multipoles. While the $\mathbb{T}$-operator is in general "off-shell," meaning it has matrix elements between states with different spatial frequencies, the scattering amplitudes are the "on-shell" matrix elements of this operator between states of equal spatial frequency.[1] So, it is not the $\mathbb{T}$-operator itself that connects, say, outgoing and standing waves in the case of outside scattering but its on-shell matrix elements, the scattering amplitudes. In this approach, the objects can have any shape or material properties, as long as the scattering amplitude can be computed in a multipole expansion (or measured). The approach can be regarded as a concrete implementation of the proposal emphasized by Schwinger [90] that the fluctuations of the electromagnetic field can be traced back to charge and current fluctuations on the objects. This formalism has been applied and extended in a number of Casimir calculations [40, 52, 67, 68, 80, 91].

The basis in which the scattering amplitude for each object is supplied is typically associated with a coordinate system appropriate to the object. Of course a plane, a cylinder, or a sphere would be described in Cartesian, cylindrical, or spherical coordinates, respectively. However, any compact object can be described, for example, in spherical coordinates, provided that the matrix of scattering amplitudes can be either calculated or measured in that coordinate system. There are a limited number of coordinate systems in which such a partial wave expansion is possible, namely those for which the vector Helmholtz equation is separable. The translation matrices for common separable coordinate systems, obtained from the free Green's function, are supplied in Appendix C of Ref. [75]. For typical cases, the final computation of the Casimir energy can be performed on a desktop computer for a wide range of separations. Asymptotic results at large separation can be obtained analytically.

The primary limitation of the method is on the distance between objects, since the basis appropriate to a given object may become impractical as two objects approach. For small separations, sufficient accuracy can only be obtained if the calculation is taken to very high partial wave order. (Vastly different scales are problematic for numerical evaluations in general.) In the case of two spheres, the scattering amplitude is available in a spherical basis, but as the two spheres approach, the Casimir energy is dominated by waves near the point of closest approach [89]. As the spheres come into contact an infinite number of spherical waves are needed to capture the dominant contribution. A particular basis may also be fundamentally inappropriate at small separations. For instance, if the interaction of two elliptic cylinders is expressed in an ordinary cylindrical basis, when the elliptic cylinders are close enough, the smallest circular cylinder enclosing one may not lie outside the smallest circular cylinder enclosing the other. In that case

---

[1] Because of this relationship, these scattering amplitudes are also referred to as elements of the *T-matrix*. In standard conventions, however, the *T*-matrix differs from the matrix elements of the $\mathbb{T}$-operator by a basis-dependent constant, so we will use the term "scattering amplitude" to avoid confusion.

the cylindrical basis would not "resolve" the two objects (although an elliptic cylindrical basis would). Finally, for a variety of conceptual and computational reasons, we are limited to linear electromagnetic response.

In spirit and in mathematical form our final result resembles similar expressions obtained in surface integral equation methods used in computational electrodynamics [21]. Using such a formulation, in which the unknowns are currents and fields on the objects, one can compute the Casimir energy using more general basis functions, e.g., localized basis functions associated with a grid or mesh, giving rise to finite element and boundary elements methods [80]. See also the Chap. 6 by Johnson in this volume for additional discussions of numerical methods in Casimir physics.

In addition to an efficient computational approach, the scattering formalism has provided the basis for proving general theorems regarding Casimir forces. The seemingly natural question whether the force is attractive or repulsive turns out to be an ill-defined or, at least, a tricky one on closer inspection. When, for example, many bodies are considered, the direction of the force on any one object depends, of course, on which other object's perspective one takes. Even for two objects, "attractive" forces can be arranged to appear as a "repulsive" force, as in the case of two interlocking combs [84]. To avoid such ambiguous situations one can restrict oneself to analyzing two objects that are separable by a plane. Even here, it has turned out that a simple criterion for the direction of the force could not be found. Based on various calculations for simple geometries it was thought that the direction of the force can be predicted based on the relative permittivities and permeabilities of the objects and the medium. Separating materials into two groups, with (i) permittivity higher than the medium or permeability lower than the medium ($\varepsilon > \varepsilon_M$ and $\mu \leq \mu_M$), or (ii) the other way around ($\varepsilon < \varepsilon_M$ and $\mu \geq \mu_M$), Casimir forces had been found to be attractive between members of the same group and repulsive for different types in the geometries considered. However, a recent counterexample [61] shows that this is not always true. A rigorous theorem, which states that Casimir forces are always attractive, exists only for the special case of mirror symmetric arrangements of objects. It was proven first with a $\mathbb{T}$-operator formalism [51], similar to our approach used here, and later using reflection positivity [3]. We have taken an alternative characterization of the force to be fundamental, namely, whether it can produce a stable equilibrium [77]. Here, the categorization of materials into the two groups is meaningful since objects made of materials of the same type cannot produce stable levitation. One practical consequence of this theorem is that it reveals that many current proposals for producing levitation using metamaterials cannot succeed.

To illustrate the general formulation, we provide some sample applications. We include an analysis of the forces between two cylinders or wires [76] and a cylinder and a plate [33, 76, 75]. The Casimir interaction of three bodies is presented subsequently; it reveals interesting multibody effects [78, 76, 85]. The Casimir torque of two spheroids is discussed as well [30]. Furthermore, we analyze the Casimir effect for a parabolic cylinder opposite a plate when both represent perfect metal material boundary conditions [45]. We find that the Casimir force does not vanish in the limit of an infinitesimally thin parabola, where a half plate is arranged above an infinite plate, and we compute the edge effect. Another type of geometries that is treated here consists of a finite sphere or a small spheroid inside a spherical metallic cavity [95].

This chapter is organized as follows: First, we sketch the derivation of the Casimir interaction energy formula (5.39) in Sect. 5.2. Next, the theorem regarding stability is derived in Sect. 5.3. Finally, in Sect. 5.4 sample applications are presented.

5.2 General Theory for Casimir Interactions

This section has been adapted from a longer article, Ref. [75]. Many technical details and extensive appendices have been omitted to fit the format of this book.

5.2.1 Path Integral Quantization

5.2.1.1 Electromagnetic Lagrangian

We consider the Casimir effect for objects without free charges and currents but with nonzero electric and magnetic susceptibilities. The macroscopic electromagnetic Lagrangian density is

$$\mathscr{L} = \frac{1}{2}(\mathbf{E} \cdot \mathbf{D} - \mathbf{B} \cdot \mathbf{H}). \tag{5.1}$$

The electric field $\mathbf{E}(t, \mathbf{x})$ and the magnetic field $\mathbf{B}(t, \mathbf{x})$ are related to the fundamental four-vector potential A^μ by $\mathbf{E} = -c^{-1}\partial_t\mathbf{A} - \nabla A^0$ and $\mathbf{B} = \nabla \times \mathbf{A}$. We treat stationary objects whose responses to the electric and magnetic fields are linear. For such materials, the $\mathbf{D}$ and $\mathbf{B}$ fields are related to the $\mathbf{E}$ and $\mathbf{H}$ fields by the convolutions $\mathbf{D}(t, \mathbf{x}) = \int_{-\infty}^{\infty} dt'\varepsilon(t', \mathbf{x})\mathbf{E}(t - t', \mathbf{x})$ and $\mathbf{B}(t, \mathbf{x}) = \int_{-\infty}^{\infty} dt'\mu(t', \mathbf{x})$ $\mathbf{H}(t - t', \mathbf{x})$ in time, where $\varepsilon(t', \mathbf{x})$ and $\mu(t', \mathbf{x})$ vanish for $t' < 0$. We consider local, isotropic permittivity and permeability, although our derivation can be adapted to apply to non-local and non-isotropic media simply by substituting the appropriate non-local and tensor permittivity and permeability functions. A more formal derivation of our starting point (5.1), which elucidates the causality properties of the permeability and permittivity response functions, is given in Appendix A of Ref. [75].

We define the quantum-mechanical energy through the path integral, which sums all configurations of the electromagnetic fields constrained by periodic boundary conditions in time between 0 and T. Outside of this time interval the fields are periodically continued. Substituting the Fourier expansions of the form $\mathbf{E}(t, \mathbf{x}) = \sum_{n=-\infty}^{\infty} \mathbf{E}(\omega_n, \mathbf{x})e^{-i\omega_n t}$ with $\omega_n = 2\pi n/T$, we obtain the action

$$S(T) = \frac{1}{2} \int_0^T dt \int d\mathbf{x} (\mathbf{E} \cdot \mathbf{D} - \mathbf{B} \cdot \mathbf{H}) = \frac{1}{2} T \sum_{n=-\infty}^{\infty} \int d\mathbf{x} (\mathbf{E}^* \cdot \varepsilon \mathbf{E} - \mathbf{B}^* \cdot \mu^{-1} \mathbf{B}),$$

(5.2)

where ε, $\mathbf{E}$, μ, and $\mathbf{B}$ on the right-hand side are functions of position $\mathbf{x}$ and frequency ω_n, and we have used $\mathbf{D}(\omega, \mathbf{x}) = \varepsilon(\omega, \mathbf{x})\mathbf{E}(\omega, \mathbf{x})$ and $\mathbf{H}(\omega, \mathbf{x}) = \frac{1}{\mu(\omega, \mathbf{x})}\mathbf{B}(\omega, \mathbf{x})$.

From the definition of the fields $\mathbf{E}$ and $\mathbf{B}$ in terms of the vector potential A^μ, we have $\nabla \times \mathbf{E} = i\frac{\omega}{c}\mathbf{B}$, which enables us to eliminate $\mathbf{B}$ in the action,

$$S(T) = \frac{1}{2} T \sum_{n=-\infty}^{\infty} \int d\mathbf{x} \left[\mathbf{E}^* \cdot \left(\mathbb{I} - \frac{c^2}{\omega_n^2} \nabla \times \nabla \times \right) \mathbf{E} - \frac{c^2}{\omega_n^2} \mathbf{E}^* \cdot \mathbb{V}\mathbf{E} \right],$$ (5.3)

where

$$\mathbb{V} = \mathbb{I}\frac{\omega_n^2}{c^2}(1 - \varepsilon(\omega_n, \mathbf{x})) + \nabla \times \left(\frac{1}{\mu(\omega_n, \mathbf{x})} - 1 \right) \nabla \times$$ (5.4)

is the potential operator and we have restored the explicit frequency dependence of ε and μ. The potential operator is nonzero only at those points in space where the objects are located ($\varepsilon \neq 1$ or $\mu \neq 1$).

In the functional integral we will sum over configurations of the field A^μ. This sum must be restricted by a choice of gauge, so that it does not include the infinitely redundant gauge orbits. We choose to work in the gauge $A^0 = 0$, although of course no physical results depend on this choice.

5.2.1.2 Casimir Energy from Euclidean Action

We use standard tools to obtain a functional integral expression for the ground state energy of a quantum field in a fixed background described by $\mathbb{V}(\omega, \mathbf{x})$. The overlap between the initial state $|\mathbf{E}_a\rangle$ of a system with the state $|\mathbf{E}_b\rangle$ after time T can be expressed as a functional integral with the fields fixed at the temporal boundaries [36],

$$\langle \mathbf{E}_b | e^{-iHT\hbar} | \mathbf{E}_a \rangle = \int \mathscr{D}\mathbf{A} \Big|_{\substack{\mathbf{E}(t=0)=\mathbf{E}_a \\ \mathbf{E}(t=T)=\mathbf{E}_b}} e^{\frac{i}{\hbar}S(T)},$$ (5.5)

where $S(T)$ is the action of (5.2) with the time integrals taken between zero and T, and H is the corresponding Hamiltonian.

If the initial and final states are set equal and summed over, the resulting functional integration defines the Minkowski space functional integral

$$\mathscr{Z}(T) \equiv \sum_a \langle \mathbf{E}_a | e^{-iHT/\hbar} | \mathbf{E}_a \rangle = \mathrm{tr}\ e^{-iHT/\hbar} = \int \mathscr{D}\mathbf{A} e^{\frac{i}{\hbar}S(T)},$$ (5.6)

which depends on the time T and the background potential $\mathbb{V}(\omega, \mathbf{x})$. The partition function that describes this system at temperature $1/\beta$ is defined by

$$Z(\beta) = \mathscr{Z}(-i\hbar\beta) = \text{tr } e^{-\beta H}, \tag{5.7}$$

and the free energy F of the field is

$$F(\beta) = -\frac{1}{\beta}\log Z(\beta). \tag{5.8}$$

The limit $\beta \to \infty$ projects the ground state energy out of the trace,

$$\mathscr{E}_0 = F(\beta = \infty) = -\lim_{\beta \to \infty}\frac{1}{\beta}\log Z(\beta). \tag{5.9}$$

The unrenormalized energy $\mathscr{E}_0$ generally depends on an ultraviolet cutoff, but cutoff-dependent contributions arise from the objects individually [43, 44] and do not depend on their separations or orientations. Such terms can remain after ordinary QED renormalization if objects are assumed to constrain electromagnetic waves with arbitrarily high frequencies (for example, if the fields are forced to vanish on a surface). Such boundary conditions should be regarded as artificial idealizations; in reality, when the wavelengths of the electromagnetic waves become shorter than the length scales that characterize the interactions of the material, the influence of the material on the waves vanishes [43]. Accordingly, the potential $\mathbb{V}$ should vanish for real materials in the high-frequency limit. In any event these cutoff dependences are independent of the separation and orientation of the objects, and since we are only interested in energy *differences*, we can remove them by subtracting the ground state energy of the system when the objects are in some reference configuration. In most cases we take this configuration to be when the objects are infinitely far apart, but when calculating Casimir energies for one object inside another, some other configuration must be used. We denote the partition function for this reference configuration by $\overline{Z}$. In this way we obtain the Casimir energy,

$$\mathscr{E} = -\lim_{\beta \to \infty}\frac{1}{\beta}\log Z(\beta)/\overline{Z}(\beta). \tag{5.10}$$

Throughout our calculation of $\mathscr{E}$, we will thus be able to neglect any overall factors that are independent of the relative positions and orientations of the objects.

By replacing the time T by $-i\hbar\beta$, we transform the Minkowski space functional integral $\mathscr{Z}(T)$ into the partition function $Z(\beta)$. In $A^0 = 0$ gauge, the result is simply to replace the frequencies $\omega_n = \frac{2\pi n}{T}$ in (5.4) by $i\frac{2\pi n}{\hbar\beta} = ic\kappa_n$, where κ_n is the nth Matsubara frequency divided by c. (In other gauges the temporal component A^0 of the vector field must be rotated too.)

The Lagrangian is quadratic, so the modes with different κ_n decouple and the partition function decomposes into a product of partition functions for each mode. Since the electromagnetic field is real, we have $\mathbf{E}^*(\omega) = \mathbf{E}(-\omega)$ on the real axis.

We can thus further simplify this decomposition on the imaginary axis by considering $\kappa \geq 0$ only, but allowing $\mathbf{E}$ and $\mathbf{E}^*$ to vary independently in the path integral. Restricting to positive κ is possible because the response functions $\varepsilon(ic\kappa, \mathbf{x})$ and $\mu(ic\kappa, \mathbf{x})$ are invariant under a change of sign in $ic\kappa$, as shown in Appendix A of Ref. [75]. In the limit $\beta \to \infty$, the sum $\sum_{n \geq 0}$ turns into an integral $\frac{\hbar c \beta}{2\pi} \int_0^\infty d\kappa$, and we have

$$\mathcal{E}_0 = -\frac{\hbar c}{2\pi} \int_0^\infty d\kappa \log Z(\kappa), \tag{5.11}$$

where

$$Z(\kappa) = \int \mathcal{D}\mathbf{A} \mathcal{D}\mathbf{A}^* \exp\left[-\beta \int d\mathbf{x}\mathbf{E}^* \cdot \left(\mathbb{I} + \frac{1}{\kappa^2}\nabla \times \nabla\times\right)\mathbf{E} + \frac{1}{\kappa^2}\mathbf{E}^* \cdot \mathbb{V}(ic\kappa, \mathbf{x})\mathbf{E}\right], \tag{5.12}$$

$$\mathbb{V}(ic\kappa, \mathbf{x}) = \mathbb{I}\kappa^2(\varepsilon(ic\kappa, \mathbf{x}) - 1) + \nabla \times \left(\frac{1}{\mu(ic\kappa, \mathbf{x})} - 1\right)\nabla\times. \tag{5.13}$$

The potential $\mathbb{V}(ic\kappa, \mathbf{x})$ is real for real κ, even though ε and μ can have imaginary parts for real frequencies ω. Our goal is now to manipulate $Z(\kappa)$ in (5.12) so that it is computable from the scattering properties of the objects.

5.2.2 Green's Function Expansions and Translation Formulas

The free Green's function and its representations in various coordinate systems are crucial to our formalism. The free electromagnetic field ($\mathbb{V} = 0$) obeys equations of motion obtained by extremizing the corresponding action, (5.2),

$$\left(-\mathbb{I}\frac{\omega^2}{c^2} + \nabla \times \nabla\times\right)\mathbf{E}(\omega, \mathbf{x}) = 0. \tag{5.14}$$

We will employ the electromagnetic dyadic Green's function $\mathbb{G}_0$, defined by

$$\left(-\mathbb{I}\frac{\omega^2}{c^2} + \nabla \times \nabla\times\right)\mathbb{G}_0(\omega, \mathbf{x}, \mathbf{x}') = \mathbb{I}\delta^{(3)}(\mathbf{x} - \mathbf{x}'), \tag{5.15}$$

written here in the position space representation. The Green's function has to be the retarded one, not only on physical grounds, but also as a consequence of the imaginary-frequency formalism, just as is the case for the response functions ε and μ. It is the *retarded* response functions that are analytically continued in the frequency domain to positive imaginary frequency, as shown in Appendix A of Ref. [75].

The representation of the free Green's function, which we need, employs the "regular" and "outgoing" solutions to the differential equation, (5.14),

$$\mathbf{E}_\alpha^{\text{reg}}(\omega, \mathbf{x}) = \langle \mathbf{x} | \mathbf{E}_\alpha^{\text{reg}}(\omega) \rangle, \quad \mathbf{E}_\alpha^{\text{out}}(\omega, \mathbf{x}) = \langle \mathbf{x} | \mathbf{E}_\alpha^{\text{out}}(\omega) \rangle, \qquad (5.16)$$

represented formally by the eigenstate kets $|\mathbf{E}_\alpha^{\text{reg}}(\omega)\rangle$ and $|\mathbf{E}_\alpha^{\text{reg}}(\omega)\rangle$, where the generalized index α labels the scattering channel, including the polarization. For example, for spherical wave functions it represents the angular momentum quantum numbers (l, m) and the polarization E or M. There are six coordinate systems in which the vector wave equation (5.14) can be solved by separation of variables and vector wave functions appropriate to that coordinate system can be constructed [70]. The labels "regular" and "outgoing" denote, respectively, the wave functions' non-singular behavior at the origin or 'outward' direction of energy transport along one of the coordinate system's axes. Let us call the coordinate, along which the latter wave functions are outgoing ξ_1 and the other coordinates ξ_2 and ξ_3. We will usually work on the imaginary ω-axis, in which case we will encounter the corresponding modified special functions.

The free Green's function can be expanded in tensor products of these wave functions,

$$\mathbb{G}_0(\omega, \mathbf{x}, \mathbf{x}') = \sum_\alpha C_\alpha(\omega) \begin{cases} \mathbf{E}_\alpha^{\text{out}}(\omega, \xi_1, \xi_2, \xi_3) \otimes \mathbf{E}_\alpha^{\text{reg}*}(\omega, \xi_1', \xi_2', \xi_3') & \text{if } \xi_1(\mathbf{x}) > \xi_1'(\mathbf{x}') \\ \mathbf{E}_\alpha^{\text{reg}}(\omega, \xi_1, \xi_2, \xi_3) \otimes \mathbf{E}_\alpha^{\text{in}*}(\omega, \xi_1', \xi_2', \xi_3') & \text{if } \xi_1(\mathbf{x}) < \xi_1'(\mathbf{x}') \end{cases},$$

$$(5.17)$$

$\mathbf{E}_\alpha^{\text{in}}$ is the same as $\mathbf{E}_\alpha^{\text{out}}$ except the functional dependence on ξ_1 is complex conjugated, making the wave function 'incoming'. A list of Green's function expansions in various common bases, including the normalization constant, $C_\alpha(\omega)$, is given in Appendix B of Ref. [75]. The wave functions that appear in the series expansion of the free Green's functions in (5.17) satisfy wave equations with frequency ω. As we will see in Sect. 5.2.3, the ability to express the Casimir energy entirely in terms of an "on-shell" partial wave expansion with fixed ω will greatly simplify our calculations.

We will also use the free Green's function in another representation to combine the scattering amplitudes for two different objects. In this calculation the one argument of the Green's function will be located on each object. As long as the pair of objects can be separated in one of the separable coordinate systems by the surface $\xi_1 = \Xi = \text{const.}$, we can distinguish an inside object which lies entirely inside the surface ($\xi_1 < \Xi$) and an outside object ($\xi_1 > \Xi$), see Fig. 5.2. Then, we can expand the free Green's function, when one argument, say $\mathbf{x}$ lies on object i and the other argument, say $\mathbf{x}'$, lies on object j, we expand $\mathbb{G}_0(ic\kappa, \mathbf{x}, \mathbf{x}')$ in terms of coordinates $\mathbf{x}_i$ and $\mathbf{x}_j'$ that describe each point relative to the origin of the body on which it lies. Which of the following expansions is appropriate for a particular

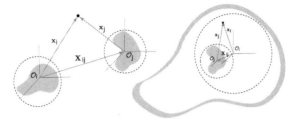

Fig. 5.2 Geometry of the outside (*left*) and inside (*right*) configurations. The dotted lines show surfaces separating the objects on which the radial variable is constant. The translation vector $\mathbf{X}_{ij} = \mathbf{x}_i - \mathbf{x}_j = -\mathbf{X}_{ji}$ describes the relative positions of the two origins

pair of objects depends whether objects i and j are outside of one another, or one object is inside the other,

$$
\mathbb{G}_0(ic\kappa,\mathbf{x},\mathbf{x}') = \sum_{\alpha,\beta} C_\beta(\kappa)
\begin{cases}
\mathbf{E}_\alpha^{\mathrm{reg}}(\kappa,\mathbf{x}_i) \otimes \mathscr{U}_{\alpha\beta}^{ji}(\kappa) \mathbf{E}_\beta^{\mathrm{reg}*}(\kappa,\mathbf{x}'_j) & \text{if } i \text{ and } j \text{ are outside each other}\\[2mm]
\mathbf{E}_\alpha^{\mathrm{reg}}(\kappa,\mathbf{x}_i) \otimes \mathscr{V}_{\alpha\beta}^{ij}(\kappa) \mathbf{E}_\beta^{\mathrm{in}*}(\kappa,\mathbf{x}'_j) &
\begin{cases}
\text{if } i \text{ is inside } j, \text{ or}\\
\text{if } i \text{ is below } j \text{ (plane wave basis)}\}
\end{cases}\\[2mm]
\mathbf{E}_\alpha^{\mathrm{out}}(\kappa,\mathbf{x}_i) \otimes \mathscr{W}_{\alpha}\beta^{ji}(\kappa) \mathbf{E}_\beta^{\mathrm{reg}*}(\kappa,\mathbf{x}'_j) &
\begin{cases}
\text{if } j \text{ is inside } i, \text{ or}\\
\text{if } j \text{ is below } i \text{ (plane wave basis)}\}
\end{cases}
\end{cases}
$$

$$(5.18)$$

where $\mathscr{W}_{\alpha\beta}^{ji}(\kappa) = \mathscr{V}_{\alpha\beta}^{ji,\dagger}(\kappa)\frac{C_\alpha(\kappa)}{C_\beta(\kappa)}$ and C_α is the normalization constant defined in (5.17). The expansion can be written more compactly as

$$
\mathbb{G}_0(ic\kappa) = \sum_{\alpha,\beta} (-C_\beta(\kappa)) \left(|\mathbf{E}_\alpha^{\mathrm{reg}}(\kappa)\rangle \, |\mathbf{E}_\alpha^{\mathrm{out}}(\kappa)\rangle \right) \mathbb{X}_{\alpha\beta}^{ij}(\kappa) \begin{pmatrix} \langle \mathbf{E}_\beta^{\mathrm{reg}}(\kappa)| \\ \langle \mathbf{E}_\beta^{\mathrm{in}}(\kappa)| \end{pmatrix}, \qquad (5.19)
$$

where the $\mathbb{X}$ matrix is defined, for convenience, as the negative of the matrix containing the translation matrices,

$$
\mathbb{X}^{ij}(\kappa) = \begin{pmatrix} -\mathscr{U}^{ji}(\kappa) & -\mathscr{V}^{ij}(\kappa) \\ -\mathscr{W}^{ji}(\kappa) & 0 \end{pmatrix}. \qquad (5.20)
$$

In (5.19) the bras and kets are to be evaluated in position space in the appropriately restricted domains and only one of the three submatrices is nonzero for any pair of objects i and j as given in (5.18). The translation matrices for various geometries are provided in Appendix C of Ref. [75].

5.2.3 Classical Scattering of Electromagnetic Fields

In this section, we summarize the key results from scattering theory needed to compute the scattering amplitude of each body individually. In the subsequent section we will then combine these results with the translation matrices of the previous section to compute $Z(\kappa)$.

By combining the frequency-dependent Maxwell equations, one obtains the vector wave equation

$$(\mathbb{H}_0 + \mathbb{V}(\omega, \mathbf{x}))\mathbf{E}(\omega, \mathbf{x}) = \frac{\omega^2}{c^2}\mathbf{E}(\omega, \mathbf{x}), \qquad (5.21)$$

where $\mathbb{H}_0 = \nabla \times \nabla \times$,

$$\mathbb{V}(\omega, \mathbf{x}) = \mathbb{I}\frac{\omega^2}{c^2}(1 - \varepsilon(\omega, \mathbf{x})) + \nabla \times \left(\frac{1}{\mu(\omega, \mathbf{x})} - 1\right)\nabla \times, \qquad (5.22)$$

which is the same potential operator as the one obtained by rearranging the Lagrangian (see (5.4)).

The Lippmann-Schwinger equation [66]

$$|\mathbf{E}\rangle = |\mathbf{E}_0\rangle - \mathbb{G}_0\mathbb{V}|\mathbf{E}\rangle \qquad (5.23)$$

expresses the general solution to (5.21). Here, $\mathbb{G}_0$ is the free electromagnetic tensor Green's function discussed in Sect. 5.2 and the homogeneous solution $|\mathbf{E}_0\rangle$ obeys $\left(-\frac{\omega^2}{c^2}\mathbb{I} + \mathbb{H}_0\right)|\mathbf{E}_0\rangle = 0$. We can iteratively substitute for $|\mathbf{E}\rangle$ in (5.23) to obtain the formal expansion

$$\begin{aligned}|\mathbf{E}\rangle &= |\mathbf{E}_0\rangle - \mathbb{G}_0\mathbb{V}|\mathbf{E}_0\rangle + \mathbb{G}_0\mathbb{V}\mathbb{G}_0\mathbb{V}|\mathbf{E}\rangle - \cdots \\ &= |\mathbf{E}_0\rangle - \mathbb{G}_0\mathbb{T}|\mathbf{E}_0\rangle,\end{aligned} \qquad (5.24)$$

where the electromagnetic $\mathbb{T}$-operator is defined as

$$\mathbb{T} = \mathbb{V}\frac{\mathbb{I}}{\mathbb{I} + \mathbb{G}_0\mathbb{V}} = \mathbb{V}\mathbb{G}\mathbb{G}_0^{-1}, \qquad (5.25)$$

and $\mathbb{G}$ is the Green's function of the full Hamiltonian, $\left(-\frac{\omega^2}{c^2}\mathbb{I} + \mathbb{H}_0 + \mathbb{V}\right)\mathbb{G} = \mathbb{I}$. We note that $\mathbb{T}$, $\mathbb{G}_0$, and $\mathbb{G}$ are all functions of frequency ω and non-local in space. As can be seen from expanding $\mathbb{T}$ in (5.25) in a power series, $\mathbb{T}(\omega, \mathbf{x}, \mathbf{x}') = \langle\mathbf{x}|\mathbb{T}(\omega)|\mathbf{x}'\rangle$ is zero whenever $\mathbf{x}$ or $\mathbf{x}'$ are not located on an object, *i.e.*, where $\mathbb{V}(\omega, \mathbf{x})$ is zero. This result does not, however, apply to

$$\mathbb{T}^{-1} = \mathbb{G}_0 + \mathbb{V}^{-1}, \qquad (5.26)$$

because the free Green's function is nonlocal. The potential $\mathbb{V}(\omega, \mathbf{x})$ which appears in (5.22) is the coordinate space matrix element of $\mathbb{V}$, $\langle\mathbf{x}|\mathbb{V}|\mathbf{x}'\rangle = \mathbb{V}(\omega, \mathbf{x})\delta(\mathbf{x} - \mathbf{x}')$, which can be generalized to the case where $\mathbb{V}$ is non-local, $\langle\mathbf{x}|\mathbb{V}|\mathbf{x}'\rangle = \mathbb{V}(\omega, \mathbf{x}, \mathbf{x}')$. Note that whether $\mathbb{V}$ is local or non-local, its matrix elements vanish if $\mathbf{x}$ and $\mathbf{x}'$ are on different objects or if either $\mathbf{x}$ or $\mathbf{x}'$ is outside of the objects. The definition of $\mathbb{V}^{-1}$ is natural, $\langle\mathbf{x}|\mathbb{V}^{-1}|\mathbf{x}'\rangle = \mathbb{V}^{-1}(\omega, \mathbf{x})\delta(\mathbf{x} - \mathbf{x}')$ (and

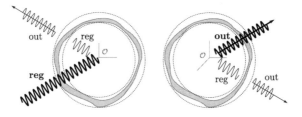

Fig. 5.3 The scattering waves for outside scattering (*left panel*) and inside scattering (*right panel*). In both cases the homogeneous solution $\mathbf{E}_0(\omega)$ is shown in bold. For outside scattering, the homogeneous solution is a regular wave, which produces a regular wave inside the object and an outgoing wave outside the object. For inside scattering, the homogeneous solution is an outgoing wave, which produces a regular wave inside the object and an outgoing wave outside the object

similarly for the non-local case) when $\mathbf{x}$ and $\mathbf{x}'$ are on a single object, which is the only case that enters our analysis.

Next we connect the matrix elements of the $\mathbb{T}$-operator between states with equal ω to the scattering amplitude $\mathscr{F}$. In our formalism, only this restricted subset of $\mathbb{T}$-operator matrix elements is needed in the computation of the Casimir energy.

By the choice of the homogeneous solution, $|\mathbf{E}_0\rangle$, is regular or outgoing, we can distinguish two physically different processes. In the former case, the object scatters the regular wave outward and modifies the amplitude of the imposed regular wave functions inside, a situation we refer to as outside scattering (left panel of Fig. 5.3).[2] In the latter case, the object modifies the amplitude of the transmitted wave and partly reflects it as a regular wave inside (inside scattering, right panel of Fig. 5.3). . 'Outside' and 'inside' are distinguished by surfaces $\xi_1 = \text{constant}$, as before. Here we treat the *outside scattering case*, and refer the reader to Ref. [75] the *inside case* and further details. The expansion in (5.17) allows us to express (5.24) as

$$\mathbf{E}(\omega, \mathbf{x}) = \mathbf{E}_\alpha^{\text{reg}}(\omega, \mathbf{x}) - \sum_\beta \mathbf{E}_\beta^{\text{out}}(\omega, \mathbf{x}) \tag{5.27}$$

at points $\mathbf{x}$ outside a surface $\xi_1 = \text{constant}$ enclosing the object. The equation can be written in Dirac notation, again with the condition that the domain of the functional Hilbert space is chosen appropriately to the type of solution,

$$|\mathbf{E}(\omega)\rangle = |\mathbf{E}_\alpha^{\text{reg}}(\omega)\rangle + \sum_\beta |\mathbf{E}_\beta^{\text{out}}(\omega)\rangle \times \underbrace{(-1) C_\beta(\omega) \langle \mathbf{E}_\beta^{\text{reg}}(\omega) | \mathbb{T}(\omega) | \mathbf{E}_\alpha^{\text{reg}}(\omega) \rangle}_{\mathscr{F}_{\beta,\alpha}^{ee}(\omega)},$$

$$\tag{5.28}$$

---

[2] Alternatively, we can set up asymptotically incoming and outgoing waves on the outside and regular waves inside. The amplitudes of the outgoing waves are then given by the S-matrix, which is related to the scattering amplitude $\mathscr{F}$ by $\mathscr{F} = (S - I)/2$. Although these two matrices carry equivalent information, the scattering amplitude will be more convenient for our calculation.

which defines $\mathscr{F}^{ee}_{\beta,\alpha}$ as the exterior/exterior scattering amplitude (the one evaluated between two regular solutions). We will use analogous notation in the other cases below.

At coordinates $\mathbf{x}$ "far enough inside" a cavity of the object, meaning that $\mathbf{x}$ has smaller ξ_1 than any point on the object, the field $\mathbf{E}$ is given by

$$|\mathbf{E}(\omega)\rangle = |\mathbf{E}^{\mathrm{reg}}_\alpha(\omega)\rangle + \sum_\beta |\mathbf{E}^{\mathrm{reg}}_\beta(\omega)\rangle \times \underbrace{(-1)C_\beta(\omega)\langle \mathbf{E}^{\mathrm{in}}_\beta(\omega)|\mathbb{T}(\omega)|\mathbf{E}^{\mathrm{reg}}_\alpha(\omega)\rangle}_{\mathscr{F}^{ie}_{\beta,\alpha}(\omega)},$$

$$(5.29)$$

where again the free states are only defined over the appropriate domain in position space, and $\mathscr{F}^{ie}$ indicates the interior/exterior scattering amplitude.

We have obtained the scattering amplitude in the basis of free solutions with fixed ω. Since one is normally interested in the scattering of waves outside the object, the scattering amplitude usually refers to $\mathscr{F}^{ee}$. We will use a more general definition, which encompasses all possible combinations of inside and outside. The scattering amplitude is always "on-shell," because the frequencies of both the operator and the states is ω. As a result, it is a special case of the $\mathbb{T}$-operator, which can connect wave functions with different ω.

We find it convenient to assemble the scattering amplitudes for inside and outside into a single matrix,

$$\mathbb{F}(\kappa) = \begin{pmatrix} \mathscr{F}^{ee}(\kappa) & \mathscr{F}^{ei}(\kappa) \\ \mathscr{F}^{ie}(\kappa) & \mathscr{F}^{ii}(\kappa) \end{pmatrix}$$

$$= (-1)C_\alpha(\kappa) \begin{pmatrix} \langle \mathbf{E}^{\mathrm{reg}}_\alpha(\kappa)|\mathbb{T}(ic\kappa)|\mathbf{E}^{\mathrm{reg}}_\beta(\kappa)\rangle & \langle \mathbf{E}^{\mathrm{reg}}_\alpha(\kappa)|\mathbb{T}(ic\kappa)|\mathbf{E}^{\mathrm{out}}_\beta(\kappa)\rangle \\ \langle \mathbf{E}^{\mathrm{in}}_\alpha(\kappa)|\mathbb{T}(ic\kappa)|\mathbf{E}^{\mathrm{reg}}_\beta(\kappa)\rangle & \langle \mathbf{E}^{\mathrm{in}}_\alpha(\kappa)|\mathbb{T}(ic\kappa)|\mathbf{E}^{\mathrm{out}}_\beta(\kappa)\rangle \end{pmatrix}.$$

$$(5.30)$$

where we have set $\omega = ic\kappa$, since this is the case we use. For simplicity we define

$$\mathscr{F}^{ie}_{\beta,\alpha}(\omega)\Big|_{\omega=ic\kappa} \equiv \mathscr{F}^{ee}_{\beta,\alpha}(\kappa).$$

5.2.4 Casimir Free Energy in Terms of the Scattering Amplitudes

With the tools of the previous two sections, we are now able to re-express the Euclidean electromagnetic partition function of (5.12) in terms of the scattering theory results derived in Sect. 5.2.3 for imaginary frequency. We exchange the fluctuating field $\mathbf{A}$, which is subject to the potential $\mathbb{V}(ic\kappa, \mathbf{x})$, for a free field $\mathbf{A}'$, together with fluctuating currents $\mathbf{J}$ and charges $-\frac{i}{\omega} \nabla \cdot \mathbf{J}$ that are confined to the objects.[3]

We multiply and divide the partition function (5.12) by

$$W = \int \mathscr{D}\mathbf{J}\mathscr{D}\mathbf{J}^*\big|_{\text{obj}} \exp\left[-\beta \int d\mathbf{x}\mathbf{J}^*(\mathbf{x}) \cdot \mathbb{V}^{-1}(ic\kappa, \mathbf{x})\mathbf{J}(\mathbf{x})\right] = \det\mathbb{V}(ic\kappa, \mathbf{x}, \mathbf{x}'),$$

(5.31)

where $\big|_{\text{obj}}$ indicates that the currents are defined only over the objects, i.e. the domain where $\mathbb{V}$ is nonzero and therefore $\mathbb{V}^{-1}$ exists.

We then change variables in the integration, $\mathbf{J}(\mathbf{x}) = \mathbf{J}'(\mathbf{x}) + \frac{i}{\kappa}\mathbb{V}(ic\kappa, \mathbf{x})\mathbf{E}(\mathbf{x})$, and a second time, $\mathbf{E}(ic\kappa, \mathbf{x}) = \mathbf{E}'(ic\kappa, \mathbf{x}) - i\kappa \int d\mathbf{x}'\mathbb{G}_0(ic\kappa, \mathbf{x}, \mathbf{x}')\mathbf{J}'(\mathbf{x}')$ and analogously for $\mathbf{J}^*$ and $\mathbf{E}^*$, to obtain

$$Z(\kappa) = \frac{Z_0}{W} \int \mathscr{D}\mathbf{J}'\mathscr{D}\mathbf{J}'^*\big|_{\text{obj}}$$

$$\exp\left[-\beta \int d\mathbf{x}d\mathbf{x}'\mathbf{J}'^*(\mathbf{x}) \cdot \left(\mathbb{G}_0(ic\kappa, \mathbf{x}, \mathbf{x}') + \mathbb{V}^{-1}(ic\kappa, \mathbf{x}, \mathbf{x}')\right)\mathbf{J}'(\mathbf{x}')\right] \quad (5.32)]$$

where

$$Z_0 = \int \mathscr{D}\mathbf{A}'\mathscr{D}\mathbf{A}'^* \exp\left[-\beta \int d\mathbf{x}\mathbf{E}'*(\mathbf{x}) \cdot \left(\mathbb{I} + \frac{1}{\kappa^2}\nabla \times \nabla\times\right)\mathbf{E}'(\mathbf{x})\right], \quad (5.33)$$

is the partition function of the free field, which is independent of the objects. In $Z(\kappa)$, current fluctuations replace the field fluctuations of (5.12). The interaction of current fluctuations on different objects is described by the free Green's function $\mathbb{G}_0(ic\kappa, \mathbf{x}, \mathbf{x}')$ alone. The inverse potential penalizes current fluctuations if the potential is small.

To put the partition function into a suitable form for practical computations, we use the results of the previous sections to re-express the microscopic current fluctuations as macroscopic multipole fluctuations, which then can be connected to the individual objects' scattering amplitudes. This transformation comes about naturally once the current fluctuations are decomposed according to the objects on which they occur and the appropriate expansions of the Green's function are introduced. We begin this process by noticing that the operator in the exponent of the integrand in (5.32) is the negative of the inverse of the $\mathbb{T}$-operator (see (5.26)), and hence

$$Z(\kappa) = Z_0 \det\mathbb{V}^{-1}(ic\kappa, \mathbf{x}, \mathbf{x}')\det\mathbb{T}(ic\kappa, \mathbf{x}, \mathbf{x}') \quad (5.34)$$

---

[3] The sequence of two changes of variables is known as Hubbard-Stratonovich transformation in condensed matter physics.

which is in agreement with a more formal calculation: Since $Z_0 = \det \mathbb{G}_0$ $(ic\kappa, \mathbf{x}, \mathbf{x}')$ and $Z(\kappa) = \det \mathbb{G}(ic\kappa, \mathbf{x}, \mathbf{x}')$, we only need to take the determinant of (5.25) to arrive at the result of (5.34).

Both Z_0 and $\det \mathbb{V}^{-1}(ic\kappa, \mathbf{x})$ are independent of the separation of the objects, since the former is simply the free Green's function, while the latter is diagonal in $\mathbf{x}$. Even a nonlocal potential $\mathbb{V}(ic\kappa, \mathbf{x}, \mathbf{x}')$ only connects points within the same object, so its determinant is also independent of the objects' separation. Because these determinants do not depend on separation, they are canceled by a reference partition function in the final result. We are thus left with the task of computing the determinant of the $\mathbb{T}$-operator.

As has been discussed in Sect. 5.2.3, the $\mathbb{T}$-operator $\mathbb{T}(ic\kappa, \mathbf{x}, \mathbf{x}')$ is not diagonal in the spatial coordinates. Its determinant needs to be taken over the spatial indices $\mathbf{x}$ and $\mathbf{x}'$, which are restricted to the objects because the fluctuating currents $\mathbf{J}(\mathbf{x})$ in the functional integrals are zero away from the objects. This determinant also runs over the ordinary vector components of the electromagnetic $\mathbb{T}$ operator.

A change of basis to momentum space does not help in computing the determinant of the $\mathbb{T}$-operator, even though it does help in finding the determinant of the free Green's function, for example. One reason is that the momentum basis is not orthogonal over the domain of the indices $\mathbf{x}$ and $\mathbf{x}'$, which is restricted to the objects. In addition, a complete momentum basis includes not only all directions of the momentum vector, but also all magnitudes of the momenta. So, in the matrix element $\langle \mathbf{E_k} | \mathbb{T}(\omega) | \mathbf{E_{k'}} \rangle$ the wave numbers k and k' would not have to match, and could also differ from ω/c. That is, the matrix elements could be "off-shell." Therefore, the $\mathbb{T}$-operator could not simply be treated as if it was the scattering amplitude, which is the on-shell representation of the operator in the subbasis of frequency ω (see Sect. 5.2.3), and is significantly easier to calculate. Nonetheless, we will see that it is possible to express the Casimir energy in terms of the on-shell operator only, by remaining in the position basis.

From (5.25), we know that the inverse of the $\mathbb{T}$-operator equals the sum of the free Green's function and the inverse of the potential. Since the determinant of the inverse operator is the reciprocal of the determinant, it is expedient to start with the inverse $\mathbb{T}$-operator. We then separate the basis involving all the objects into blocks for the n objects. In a schematic notation, we have

$$[\langle \mathbf{x} | \mathbb{T}^{-1} | \mathbf{x}' \rangle] = \begin{pmatrix} [\langle \mathbf{x}_1 | \mathbb{T}_1^{-1} | \mathbf{x}_1' \rangle] & [\langle \mathbf{x}_1 | \mathbb{G}_0 | \mathbf{x}_2' \rangle] & \cdots \\ \hline [\langle \mathbf{x}_2 | \mathbb{G}_0 | \mathbf{x}_1' \rangle] & [\langle \mathbf{x}_2 | \mathbb{T}_2^{-1} | \mathbf{x}_2' \rangle] & \cdots \\ \hline \cdots & \cdots & \cdots \end{pmatrix}, \qquad (5.35)$$

where the ij^{th} submatrix refers to $\mathbf{x} \in$ object i and $\mathbf{x}' \in$ object j and $\mathbf{x}_i$ represents a point in object i measured with respect to some fixed coordinate system. Unlike the position vectors in Sect. 5.2.2, at this point the subscript of $\mathbf{x}_i$ does not indicate the origin with respect to which the vector is measured, but rather the object on which the point lies. Square brackets are used to remind us that we are considering the entire matrix or submatrix and not a single matrix element. We note that the operators $\mathbb{T}$ and $\mathbb{G}_0$ are functions of $ic\kappa$, but for simplicity we suppress this argument throughout this derivation. When the two spatial indices lie on different objects, only the free Green's function remains in the off-diagonal submatrices, because $\langle \mathbf{x}_i | \mathbb{V}^{-1} | \mathbf{x}'_j \rangle = 0$ for $i \neq j$.

Next, we multiply $\mathbb{T}^{-1}$ by a reference $\mathbb{T}$-operator $\mathbb{T}_\infty$ without off-diagonal submatrices, which can be interpreted as the $\mathbb{T}$-operator at infinite separation,

$$[\langle \mathbf{x} | \mathbb{T}_\infty \mathbb{T}^{-1} | \mathbf{x}'' \rangle] = \begin{pmatrix} \dfrac{[\langle \mathbf{x}_1 | \mathbf{x}''_1 \rangle]}{[\int d\mathbf{x}'_2 \, \langle \mathbf{x}_2 | \mathbb{T}_2 | \mathbf{x}'_2 \rangle \langle \mathbf{x}'_2 | \mathbb{G}_0 | \mathbf{x}''_1 \rangle]} & \dfrac{[\int d\mathbf{x}'_1 \, \langle \mathbf{x}_1 | \mathbb{T}_1 | \mathbf{x}'_1 \rangle \langle \mathbf{x}'_1 | \mathbb{G}_0 | \mathbf{x}''_2 \rangle] \cdots}{[\langle \mathbf{x}_2 | \mathbf{x}''_2 \rangle]} \begin{matrix} \cdots \\ \cdots \end{matrix} \\ \cdots & \cdots \end{pmatrix}. \tag{5.36}$$

Each off-diagonal submatrix $[\int d\mathbf{x}'_i \langle \mathbf{x}_i | \mathbb{T}_i | \mathbf{x}'_i \rangle \langle \mathbf{x}'_i | \mathbb{G}_0 | \mathbf{x}''_j \rangle].$ is the product of the $\mathbb{T}$-operator of object i, evaluated at two points $\mathbf{x}_i$ and $\mathbf{x}'_i$ on that object, multiplied by the free Green's function, which connects $\mathbf{x}'_i$ to some point $\mathbf{x}''_j$ on object j.

Now we shift all variables to the coordinate systems of the objects on which they lie. As a result, the index on a position vector $\mathbf{x}_i$ now refers to the object i on which the point lies *and* to the coordinate system with origin $\mathcal{O}_i$ in which the vector is represented, in agreement with the notation of Sect. 5.2.2. The off-diagonal submatrices in (5.36) can then be rewritten using (5.19) as,

$$\sum_{\alpha,\beta} \left[\left(\langle \mathbf{x}_i | \mathbb{T}_i | \mathbf{E}^{\text{reg}}_\alpha(\kappa) \rangle \langle \mathbf{x}_i | \mathbb{T}_i | \mathbf{E}^{\text{out}}_\alpha(\kappa) \rangle \right) \mathbb{X}^{ij}_{\alpha\beta} \begin{pmatrix} \langle \mathbf{E}^{\text{reg}}_\beta(\kappa) | \mathbf{x}''_j \rangle \\ \langle \mathbf{E}^{\text{in}}_\beta(\kappa) | \mathbf{x}''_j \rangle \end{pmatrix} (-C_\beta(\kappa)) \right]. \tag{5.37}$$

The matrix $[\langle \mathbf{x} | \mathbb{T}_\infty \mathbb{T}^{-1} | \mathbf{x}'' \rangle]$ has the structure $\mathbb{I} + \mathbb{A}\mathbb{B}$. Using Sylvester's determinant formula $\det(\mathbb{I} + \mathbb{A}\mathbb{B}) = \det(\mathbb{I} + \mathbb{B}\mathbb{A})$, we see that the determinant is unchanged if we replace the off-diagonal submatrices in (5.36) by

$$\left[\sum_\beta (-1) C_\alpha(\kappa) \begin{pmatrix} \langle \mathbf{E}^{\text{reg}}_\alpha(\kappa) | \mathbb{T}_i | \mathbf{E}^{\text{reg}}_\beta(\kappa) \rangle & \langle \mathbf{E}^{\text{reg}}_\alpha(\kappa) | \mathbb{T}_i | \mathbf{E}^{\text{out}}_\beta(\kappa) \rangle \\ \langle \mathbf{E}^{\text{in}}_\alpha(\kappa) | \mathbb{T}_i | \mathbf{E}^{\text{reg}}_\beta(\kappa) \rangle & \langle \mathbf{E}^{\text{in}}_\alpha(\kappa) | \mathbb{T}_i | \mathbf{E}^{\text{out}}_\beta(\kappa) \rangle \end{pmatrix} \mathbb{X}^{ij}_{\beta,\gamma} \right]. \tag{5.38}$$

With this change, the diagonal submatrices in (5.36) become diagonal in the partial wave indices rather than in position space. The matrix elements of the $\mathbb{T}$-operator are the scattering amplitudes, which can be obtained from ordinary scattering calculations, as demonstrated in Sect. 5.2.3. The first matrix in (5.38), including the prefactor $(-1)C_\alpha(\kappa)$, is $\mathbb{F}_i(\kappa)$, the modified scattering amplitude of object i, defined in (5.30).

Putting together (5.11), (5.12), (5.34), and (5.36), we obtain

$$\mathcal{E} = \frac{\hbar c}{2\pi} \int_0^\infty d\kappa \log \det(\mathbb{M}\mathbb{M}_\infty^{-1}), \tag{5.39}$$

where

$$
\mathbb{M} = \begin{pmatrix} \mathbb{F}_1^{-1} & \mathbb{X}^{12} & \mathbb{X}^{13} & \dots \\ \mathbb{X}^{21} & \mathbb{F}_2^{-1} & \mathbb{X}^{23} & \dots \\ \dots & \dots & \dots & \dots \end{pmatrix} \tag{5.40}
$$

and $\mathbb{M}_\infty^{-1}$ is a block diagonal matrix $\mathrm{diag}(\mathbb{F}_1 \mathbb{F}_2 \dots)$.

Using the block determinant identity

$$
\det \begin{pmatrix} \mathbb{A} & \mathbb{B} \\ \mathbb{C} & \mathbb{D} \end{pmatrix} = \det(\mathbb{A})\det(\mathbb{D} - \mathbb{C}\mathbb{A}^{-1}\mathbb{B}) = \det(\mathbb{D})\det(\mathbb{A} - \mathbb{B}\mathbb{D}^{-1}\mathbb{C}), \tag{5.41}
$$

we can simplify this expression for the case of the interaction between two objects,

$$
\mathscr{E} = \frac{\hbar c}{2\pi} \int\limits_0^\infty d\kappa \, \log \det \left(\mathbb{I} - \mathbb{F}_a \mathbb{X}^{ab} \mathbb{F}_b \mathbb{X}^{ba} \right). \tag{5.42}
$$

Usually, not all of the submatrices of $\mathbb{F}$ and $\mathbb{X}$ are actually needed for a computation. For example, if all objects are outside of one another, only the submatrices $\mathscr{F}^{ee}$ of the scattering amplitude that describe outside reflection are needed. If there are only two objects, one inside another, then only the inside reflection submatrix $\mathscr{F}^{ii}$ of the outside object and the outside reflection submatrix $\mathscr{F}^{ee}$ of the inside object are needed.

In order to obtain the free energy at nonzero temperature instead of the ground state energy, we do not take the limit $\beta \to \infty$ in (5.9). Instead, the integral $\frac{\hbar c}{2\pi} \int_0^\infty d\kappa$ is replaced everywhere by $\frac{1}{\beta} \sum_n'$, where $c\kappa_n = \frac{2\pi n}{\hbar \beta}$ with $n = 0, 1, 2, 3 \dots$ is the nth Matsubara frequency. A careful analysis of the derivation shows that the zero frequency mode is weighted by 1/2 compared to the rest of the terms in the sum; this modification of the sum is denoted by a prime on the summation symbol. The factor of 1/2 comes about because the fluctuating charges or currents have to be real for zero frequency. Thus, for κ_0, the expressions on the right hand side of (5.34) should be placed under a square root. (For a complex field, both signs of the integer n would be included separately, and $n = 0$ would be included once, with the normal weight.)

If the medium between the objects is not vacuum but instead has permittivity $\varepsilon_M(ic\kappa)$ and magnetic permeability $\mu_M(ic\kappa)$ different from unity, then the free Green's function is multiplied by $\mu_M(ic\kappa)$, and its argument κ is replaced by $n_M(ic\kappa)\kappa$, where $n_M(ic\kappa) = \sqrt{\varepsilon_M(ic\kappa)\mu_M(ic\kappa)}$ is the medium's index of refraction. Effectively, this change just scales all frequency dependencies in the translation matrices $\mathbb{X}(\kappa)$, which become $\mathbb{X}(n_M(ic\kappa)\kappa)$. Furthermore, the scattering amplitudes absorb the factor $\mu_M(ic\kappa)$ from the free Green's function and change non-trivially, i.e. not just by some overall factor or a scaling of the frequency. They have to be computed with the nonzero electric and magnetic susceptibilities of the medium.

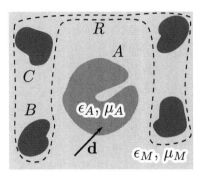

Fig. 5.4 The Casimir energy is considered for objects with electric permittivity $\varepsilon_i(\omega, \mathbf{x})$ and magnetic permeability $\mu_i(\omega, \mathbf{x})$, embedded in a medium with uniform, isotropic, $\varepsilon_M(\omega)$ and $\mu_M(\omega)$. To study the stability of object A, the rest of the objects are grouped in the combined entity R. The stability of the position of object A is probed by displacing it infinitesimally by vector $\mathbf{d}$

5.3 Constraints on Stable Equilibria

Before presenting particular applications of the Casimir energy expression in (5.39), we consider some general properties of electrodynamic Casimir interactions here. This section has been adapted from a Letter co-authored by two of us [77].

As described in the Introduction, some general statements about the attractive or repulsive nature of Casimir forces can be made on the basis of the relative permittivity and permeability of objects and the medium they are immersed in. But the sign of the force is largely a matter of perspective, since attractive forces can be easily arranged to produce repulsion along a specific direction, e.g., as in Ref. [84]. Instead, we focus on the question of stability, see Fig. 5.4, which is more relevant to the design and development of MEMs and levitating devices. We find that interactions between objects within the same class of material (as defined in the Introduction) cannot produce stable configurations.

Let us take a step back and consider the question of stability of mechanical equilibria in the realm of electromagnetism. Earnshaw's theorem [26] states that a collection of charges cannot be held in stable equilibrium solely by electrostatic forces. The charges can attract or repel, but cannot be stably levitated. While the stability of matter (due to quantum phenomena) is a vivid reminder of the caveats to this theorem, it remains a powerful indicator of the constraints to stability in electrostatics. An extension of Earnshaw's theorem to polarizable objects by Braunbek [9, 8] establishes that dielectric and paramagnetic ($\varepsilon > 1$ and $\mu > 1$) matter cannot be stably levitated by electrostatic forces, while diamagnetic ($\mu < 1$) matter can. This is impressively demonstrated by superconductors and frogs that fly freely above magnets [37]. If the enveloping medium is not vacuum, the criteria for stability are modified by substituting the static electric permittivity ε_M and magnetic permeability μ_M of the medium in place of the vacuum value of 1 in the respective inequalities. In fact, if the medium itself has a dielectric constant higher than the objects ($\varepsilon < \varepsilon_M$), stable levitation is possible, as demonstrated for bubbles in liquids (see Ref. [49], and references therein). For dynamic fields the restrictions of electrostatics do not apply; for example, lasers can lift and hold dielectric beads with index of refraction $n = \sqrt{\varepsilon\mu} > 1$ [1]. In addition to the force which keeps the bead in the center of the laser beam there is radiation pressure which pushes the bead along the direction of the Poynting vector. Ashkin and Gordon have proved that no arrangement of lasers can stably levitate an object just based on radiation pressure [2].

We begin our analysis of equilibria of the electrodynamic Casimir force with the precursor of (5.39), which contains the abstract $\mathbb{T}$ and $\mathbb{G}_M$-operators, where $\mathbb{G}_M$ is the electromagnetic Green's function operator for an isotropic, homogeneous medium[4],

$$\mathscr{E} = \frac{\hbar c}{2\pi} \int_0^\infty d\kappa \, \mathrm{tr} \, \ln \mathbb{T}^{-1} \mathbb{T}_\infty, \tag{5.43}$$

where the operator $\left[\mathbb{T}^{-1}(ic\kappa, \mathbf{x}, \mathbf{x}')\right]$ equals

$$\begin{pmatrix} \left[\mathbb{T}_A^{-1}(ic\kappa, \mathbf{x}_1, \mathbf{x}_1')\right] & \left[\mathbb{G}_M(ic\kappa, \mathbf{x}_1, \mathbf{x}_2')\right] & \cdots \\ \left[\mathbb{G}_M(ic\kappa, \mathbf{x}_2, \mathbf{x}_1')\right] & \left[\mathbb{T}_B^{-1}(ic\kappa, \mathbf{x}_2, \mathbf{x}_2')\right] & \\ \cdots & & \cdots \end{pmatrix}, \tag{5.44}$$

and $\mathbb{T}_\infty$ is the inverse of $\mathbb{T}^{-1}$ with $\mathbb{G}_M$ set to zero. The square brackets "[ ]" denote the entire matrix or submatrix with rows indicated by $\mathbf{x}$ and columns by $\mathbf{x}'$.[5] The operator $\left[\mathbb{T}^{-1}(ic\kappa, \mathbf{x}, \mathbf{x}')\right]$ has indices in position space. Each spatial index is limited to lie inside the objects $A, B, \ldots$. For both indices $\mathbf{x}$ and $\mathbf{x}'$ in the same object A the operator is just the inverse $\mathbb{T}$ operator of that object, $\left[\mathbb{T}_A^{-1}(ic\kappa, \mathbf{x}, \mathbf{x}')\right]$. For indices on different objects, $\mathbf{x}$ in A and $\mathbf{x}'$ in B, it equals the electromagnetic Green's function operator $\left[\mathbb{G}_M(ic\kappa, \mathbf{x}, \mathbf{x}')\right]$ for an isotropic, homogeneous medium.

As shown in Sect. 5.2.4, after a few manipulations, the operators $\mathbb{T}_J$ and $\mathbb{G}_M$ turn into the on-shell scattering amplitude matrix, $\mathbb{F}_J$, of object J and the translation matrix $\mathbb{X}$, which converts wave functions between the origins of different objects. While practical computations require evaluation of the matrices in a

---

[4] $\mathbb{G}_M$ satisfies $\left(\nabla \times \mu_M^{-1}(ic\kappa)\nabla \times + \varepsilon_M(ic\kappa)\kappa^2\right)\mathbb{G}_M(ic\kappa, \mathbf{x}, \mathbf{x}') = \delta(\mathbf{x} - \mathbf{x}')\mathbb{I}$, and is related to G_M, the Green's function of the imaginary frequency Helmholtz equation, by $\mathbb{G}_M(ic\kappa, \mathbf{x}, \mathbf{x}') = \mu_M(ic\kappa)\left(\mathbb{I} + (n_M\kappa)^{-2}\nabla \otimes \nabla'\right)G_M(icn_M\kappa, \mathbf{x}, \mathbf{x}')$. Here, $n_M(ic\kappa) = \sqrt{\varepsilon_M(ic\kappa)\mu_M(ic\kappa)}$ is the index of refraction of the medium, whose argument is suppressed to simplify the presentation. Thus $\mathbb{G}_M$, in contrast to $\mathbb{G}_0$, takes into account the permittivity and permeability of the medium when they are different from one.

[5] To obtain the free energy at finite temperature, in place of the ground state energy $\mathscr{E}$, $\int \frac{d\kappa}{2\pi}$ is replaced by the sum $\frac{kT}{\hbar c}\sum'_{\kappa_n \geq 0}$ over Matsubara 'wavenumbers' $\kappa_n = 2\pi nkT/\hbar c$ with the $\kappa_0 = 0$ mode weighted by $1/2$.

particular wave function basis, the position space operators $\mathbb{T}_J$ and $\mathbb{G}_M$ are better suited to our general discussion here.

To investigate the stability of object A, we group the 'rest' of the objects into a single entity R. So, $\mathbb{T}$ consists of 2×2 blocks, and the integrand in (5.43) reduces to $\mathrm{tr}\ln(\mathbb{I} - \mathbb{T}_A\mathbb{G}_M\mathbb{T}_R\mathbb{G}_M)$. Merging the components of R poses no conceptual difficulty given that the operators are expressed in a position basis, while an actual computation of the force between A and R would remain a daunting task. If object A is moved infinitesimally by vector $\mathbf{d}$, the Laplacian of the energy is given by

$$\nabla_{\mathbf{d}}^2\mathscr{E}\big|_{\mathbf{d}=0} = -\frac{\hbar c}{2\pi}\int_0^\infty d\kappa\,\mathrm{tr}\left[2n_M^2(ic\kappa)\kappa^2\frac{\mathbb{T}_A\mathbb{G}_M\mathbb{T}_R\mathbb{G}_M}{\mathbb{I} - \mathbb{T}_A\mathbb{G}_M\mathbb{T}_R\mathbb{G}_M}\right. \tag{5.45}$$

$$+2\mathbb{T}_A\nabla\mathbb{G}_M\mathbb{T}_R(\nabla\mathbb{G}_M)^T\frac{\mathbb{I}}{\mathbb{I} - \mathbb{T}_A\mathbb{G}_M\mathbb{T}_R\mathbb{G}_M} \tag{5.46}$$

$$+2\mathbb{T}_A\nabla\mathbb{G}_M\mathbb{T}_R\mathbb{G}_M\frac{\mathbb{I}}{\mathbb{I} - \mathbb{T}_A\mathbb{G}_M\mathbb{T}_R\mathbb{G}_M} \tag{5.47}$$

$$\left.\cdot\left(\mathbb{T}_A\nabla\mathbb{G}_M\mathbb{T}_R\mathbb{G}_M + \mathbb{T}_A\mathbb{G}_M\mathbb{T}_R(\nabla\mathbb{G}_M)^T\right)\frac{\mathbb{I}}{\mathbb{I} - \mathbb{T}_A\mathbb{G}_M\mathbb{T}_R\mathbb{G}_M}\right].$$

After displacement of object A, the Green's function multiplied by $\mathbb{T}_A$ on the left and $\mathbb{T}_R$ on the right ($\mathbb{T}_A\mathbb{G}_M\mathbb{T}_R$) becomes $\mathbb{G}_M(ic\kappa, \mathbf{x} + \mathbf{d}, \mathbf{x}')$, while that multiplied by $\mathbb{T}_R$ on the left and $\mathbb{T}_A$ on the right ($\mathbb{T}_R\mathbb{G}_M\mathbb{T}_A$) becomes $\mathbb{G}_M(ic\kappa, \mathbf{x}, \mathbf{x}' + \mathbf{d})$. The two are related by transposition, and indicated by $\nabla\mathbb{G}_M(ic\kappa, \mathbf{x}, \mathbf{x}') = \nabla_{\mathbf{d}}\mathbb{G}_M(ic\kappa, \mathbf{x} + \mathbf{d}, \mathbf{x}')\big|_{\mathbf{d}=0}$and-
in the above equation.

In the first line we have substituted $n_M^2(ic\kappa)\kappa^2\mathbb{G}_M$ for $\nabla^2\mathbb{G}_M$; the two differ only by derivatives of δ–functions which vanish since $\mathbb{G}_M(ic\kappa, \mathbf{x}, \mathbf{x}')$ is evaluated with $\mathbf{x}$ in one object and $\mathbf{x}'$ in another. In expressions not containing inverses of $\mathbb{T}$-operators, we can extend the domain of all operators to the entire space: $\mathbb{T}_J(ic\kappa, \mathbf{x}, \mathbf{x}') = 0$ if $\mathbf{x}$ or $\mathbf{x}'$ are not on object $J$and thus operator multiplication is unchanged.

To determine the signs of the various terms in $\nabla_{\mathbf{d}}^2\mathscr{E}\big|_{\mathbf{d}=0}$, an analysis similar to Ref. [51] can be performed. Consequently, the Laplacian of the energy is found to be smaller than or equal to zero as long as both $\mathbb{T}_A$ and $\mathbb{T}_R$ are either positive or negative semidefinite for all imaginary frequencies.[6] The eigenvalues of $\mathbb{T}_J$, defined in (5.25), on the other hand, are greater or smaller than zero depending on the sign s^J of $\mathbb{V}_J$, since

---

[6] In practice, $\mathbb{T}_A$ and $\mathbb{T}_R$ suffice to have the same sign over the frequencies, which contribute most to the integral (or the sum) in (5.43)

$$\mathbb{T}_J = s^J \sqrt{s^J \mathbb{V}_J} \frac{\mathbb{I}}{\mathbb{I} + s^J \sqrt{s^J \mathbb{V}_J} \mathbb{G}_M \sqrt{s^J \mathbb{V}_J}} \sqrt{s^J \mathbb{V}_J}. \qquad (5.48)$$

We are left to find the sign of the potential,

$$\begin{aligned}
\mathbb{V}_J(ic\kappa, \mathbf{x}) &= \mathbb{I}\kappa^2(\varepsilon_J(ic\kappa, \mathbf{x}) - \varepsilon_M(ic\kappa)) \\
&\quad + \nabla \times \left(\mu_J^{-1}(ic\kappa, \mathbf{x}) - \mu_M^{-1}(ic\kappa)\right) \nabla \times,
\end{aligned} \qquad (5.49)$$

of the object A, and the compound object R.[7] The sign is determined by the relative permittivities and permeabilities of the objects and the medium: If $\varepsilon_J(ic\kappa, \mathbf{x}) > \varepsilon_M(ic\kappa)$ and $\mu_J(ic\kappa, \mathbf{x}) \leq \mu_M(ic\kappa)$ hold for all $\mathbf{x}$ in object J, the potential $\mathbb{V}_J$ is positive. If the opposite inequalities are true, $\mathbb{V}_J$ is negative. The curl operators surrounding the magnetic permeability do not influence the sign, as in computing an inner product with $\mathbb{V}_J$ they act symmetrically on both sides. For vacuum $\varepsilon_M = \mu_M = 1$, and material response functions $\varepsilon(ic\kappa, \mathbf{x})$ and $\mu(ic\kappa, \mathbf{x})$ are analytical continuations of the permittivity and permeability for real frequencies [60]. While $\varepsilon(ic\kappa, \mathbf{x}) > 1$ for positive κ, there are no restrictions other than positivity on $\mu(ic\kappa, \mathbf{x})$. (For non-local and non-isotropic response, various inequalities must be generalized to the tensorial operators $\overleftrightarrow{\varepsilon}(ic\kappa, \mathbf{x}, \mathbf{x}')$ and $\overleftrightarrow{\mu}(ic\kappa, \mathbf{x}, \mathbf{x}')$.)

Thus, levitation is not possible for collections of objects characterized by $\varepsilon_J(ic\kappa, \mathbf{x})$ and $\mu_J(ic\kappa, \mathbf{x})$ falling into one of the two classes described earlier, (i) $\varepsilon_J/\varepsilon_M > 1$ and $\mu_J/\mu_M \leq 1$ (positive $\mathbb{V}_J$ and $\mathbb{T}_J$), or (ii) $\varepsilon_J/\varepsilon_M < 1$ and $\mu_J/\mu_M \geq 1$ with (negative $\mathbb{V}_J$ and $\mathbb{T}_J$). (Under these conditions parallel slabs attract.) The frequency and space dependence of the functions has been suppressed in these inequalities. In vacuum, $\varepsilon_M(ic\kappa) = \mu_M(ic\kappa) = 1$; since $\varepsilon(ic\kappa, \mathbf{x}) > 1$ and the magnetic response of ordinary materials is typically negligible [60], one concludes that stable equilibria of the Casimir force do not exist. If objects A and R, however, belong to different categories—under which conditions the parallel plate force is repulsive—then the terms under the trace in lines (45) and (46) are negative. The positive term in line (47) is typically smaller than the first two, as it involves higher powers of $\mathbb{T}$ and $\mathbb{G}_M$. In this case stable equilibrium is possible, as demonstrated recently for a small inclusion within a dielectric filled cavity [79]. For the remaining two combinations of inequalities involving $\varepsilon_J/\varepsilon_M$ and μ_J/μ_M the sign of $\mathbb{V}_J$ cannot be determined a priori. But for realistic distances between objects and the corresponding frequency ranges, the magnetic susceptibility is negligible for ordinary materials, and the inequalities involving μ can be ignored.

In summary, the instability theorem applies to all cases where the coupling of the EM field to matter can be described by response functions ε and μ, which may vary continuously with position and frequency. Obviously, for materials which at a

---

[7] The first curl in the operator $\mathbb{V}_J$ results from an integration by parts. It is understood that it acts on the wave function multiplying $\mathbb{V}_J$ from the left.

microscopic level cannot be described by such response functions, e.g., because of magneto-electric coupling, our theorem is not applicable.

Even complicated arrangements of materials obeying the above conditions are subject to the instability constraint. For example, metamaterials, incorporating arrays of micro-engineered circuitry mimic, at certain frequencies, a strong magnetic response, and have been discussed as candidates for Casimir repulsion across vacuum. (References [87, 86] critique repulsion from dielectric/metallic based metamaterials, in line with our following arguments.) In our treatment, in accord with the usual electrodynamics of macroscopic media, the materials are characterized by $\varepsilon(ic\kappa, \mathbf{x})$ and $\mu(ic\kappa, \mathbf{x})$ at mesoscopic scales. In particular, chirality and large magnetic response in metamaterials are achieved by patterns made from ordinary metals and dielectrics with well-behaved $\varepsilon(ic\kappa, \mathbf{x})$ and $\mu(ic\kappa, \mathbf{x}) \approx 1$ at *short* scales. The interesting EM responses merely appear when viewed as 'effective' or 'coarse grained'.

Clearly, the coarse-grained response functions, which are conventionally employed to describe metamaterials, should produce, in their region of validity, the same scattering amplitudes as the detailed mesoscopic description. Consequently, as long as the metamaterial can be described by $\varepsilon(ic\kappa, \mathbf{x})$ and $\mu(ic\kappa, \mathbf{x}) \approx 1$, the eigenvalues of the $\mathbb{T}$ operators are constrained as described above, and hence subject to the instability theorem. Thus, the proposed use of chiral metamaterials in Ref. [96] cannot lead to stable equilibrium since the structures are composites of metals and dielectrics. Finally, we note that instability also excludes repulsion between two objects that obey the above conditions, if one of them is an infinite flat plate with continuous translational symmetry: Repulsion would require that the energy as a function of separation from the slab should have $\partial_d^2 \mathscr{E} > 0$ at some point since the force has to vanish at infinite separation. A metamaterial does not have continuous translational symmetry at short length scales but this symmetry is approximately valid in the limit of large separations (long wavelengths), where the material can be effectively described as a homogeneous medium. At short separations lateral displacements might lead to repulsion that, however, must be compatible with the absence of stable equilibrium.

5.4 Applications

This section gives an overview on different geometries and shapes that have been studied by the approach that we introduced in Sect. 5.2. A selection of applications has been made to showcase generic situations and important effects that had not been studied in detail before the development of the methods described here. We shall mainly summarize analytical and numerical results for the Casimir interaction in the various systems. For details on their derivation and additional implementations of the scattering approach we refer to the literature.

Fig. 5.5 a Casimir energy for two cylinders of equal radius R as a function of surface-to-surface distance $d - 2R$ (normalized by the radius). The energy is divided by the PFA estimate $E_{PFA}^{cyl-cyl}$ for the energy. The *solid curves* show our numerical results; the *dashed lines* represent the asymptotic results of (5.54). **b** Casimir energy for a cylinder of radius R parallel to a plate as a function of the surface-to-surface distance $H - R$ (normalized by the radius). The energy is divided by the PFA estimate $E_{PFA}^{cyl-plate}$. The *solid curves* reflect our numerical results; the *dashed lines* represent the asymptotic results of (5.58)

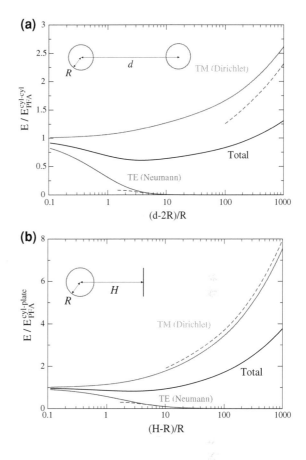

5.4.1 Cylinders, Wires, and Plate

The extent to which EM field fluctuations are correlated depends on the effective dimensionality of the space that can be explored by the fluctuations. Therefore, Casimir interactions are expected to depend strongly on the codimension of the interacting objects. The focus of this subsection is on the particular properties of systems with a codimension of the critical value two. We consider these problems in the context of interactions between cylinders and a cylinder and a plate, both perfect reflectors and dielectric materials. Cylindrical geometries are of recent experimental interest since they are easier to hold parallel than plates and still generate a force that is extensive in one direction.

We consider two cylinders of equal radii R and length $L \to \infty$ with center-to-center separation d, see Fig. 5.5a [76]. (The related configuration where one cylinder is inside another cylinder is treated in Ref. [22].) For this geometry the interaction energy is obtained from the expression

$$\mathcal{E} = \frac{\hbar c}{2\pi} \int_0^\infty d\kappa \log \det\left(\mathcal{I} - \mathcal{F}^{ee}_{cyl}\mathcal{U}^{ba}\mathcal{F}^{ee}_{cyl}\mathcal{U}^{ab}\right). \tag{5.50}$$

with the exterior scattering amplitudes of a cylinder,

$$\mathcal{F}^{ee}_{cyl,k'_z n'E,k_z nM} = \mathcal{F}^{ee}_{cyl,k'_z n'M,k_z nE} = 0,$$

$$\mathcal{F}^{ee}_{cyl,k'_z n'M,k_z nM} = -\frac{2\pi}{L}\delta(k_z - k'_z)\delta_{n,n'}\frac{I'_n(Rp)}{K'_n(Rp)},$$

$$\mathcal{F}^{ee}_{cyl,k'_z n'E,k_z nE} = -\frac{2\pi}{L}\delta(k_z - k'_z)\delta_{n,n'}\frac{I_n(Rp)}{K_n(Rp)}, \tag{5.51}$$

and the matrices $\mathcal{U}^{ab}$, $\mathcal{U}^{ba}$ that translate from cylinder a to b and vice versa. Their elements are summarized in Ref. [75]. The matrix inside the determinant is diagonal in k_z, so the log-determinant over this index turns into an overall integral. A change of variable to polar coordinates converts the integrals over κ and k_z to a single integral over $p = \sqrt{k_z^2 + \kappa^2}$, yielding

$$\mathcal{E} = \frac{\hbar c L}{4\pi} \int_0^\infty p\,dp\left(\log \det \mathcal{N}^M + \log \det \mathcal{N}^E\right), \tag{5.52}$$

where

$$\mathcal{N}^M_{n,n''} = \delta_{n,n''} - \sum_{n'}\frac{I'_n(pR)}{K'_n(pR)}K_{n+n'}(pd)\frac{I'_{n'}(pR)}{K'_{n'}(pR)}K_{n'+n''}(pd)$$

$$\mathcal{N}^E_{n,n''} = \delta_{n,n''} - \sum_{n'}\frac{I_n(pR)}{K_n(pR)}K_{n+n'}(pd)\frac{I_{n'}(pR)}{K_{n'}(pR)}K_{n'+n''}(pd) \tag{5.53}$$

describe magnetic (TE) or Neumann modes and electric (TM) or Dirichlet modes, respectively

For large separations $d \gg R$, the asymptotic behavior of the energy is determined by the matrix elements for $n = n' = 0$ for Dirichlet modes and $n = n' = 0, \pm 1$ for Neumann modes. Taking the determinant of the matrix that consists only of these matrix elements and integrating over p yields straightforwardly the attractive interaction energies

$$\mathcal{E}^E = -\frac{\hbar c L}{d^2}\frac{1}{8\pi \log^2(d/R)}\left(1 - \frac{2}{\log(d/R)} + \cdots\right),$$

$$\mathcal{E}^M = -\hbar c L\frac{7}{5\pi}\frac{R^4}{d^6} \tag{5.54}$$

for electric (Dirichlet) and magnetic (Neumann) modes. The asymptotic interaction is dominated by the contribution from electric (Dirichlet) modes that vanishes for $R \to 0$ only logarithmically.

For arbitrary separations higher order partial waves have to be considered. The number of partial waves has to be increased with decreasing separation. A numerical evaluation of the determinant and the p-integration can be performed easily and reveals an exponentially fast convergence of the energy in the truncation order for the partial waves. Down to small surface-to-surface separations of $(d - 2R)/R = 0.1$ we find that $n = 40$ partial waves are sufficient to obtain precise results for the energy. The corresponding result for the energies of two cylinders of equal radius is shown in Fig. 5.5a. Notice that the minimum in the curve for the total electromagnetic energy results from the scaling by the proximity force approximation (PFA) estimate of the energy. The total energy is monotonic and the force attractive at all separations.

Next we consider a cylinder and an infinite plate, both perfectly reflecting, see Fig. 5.5b. The Casimir energy for this geometry has been computed originally in Ref. [33]. In the limit of perfectly reflecting surfaces, the method of images can be employed to compute the Casimir interaction for this geometry [76]. Here we use a different method that can be also applied to real metals or general dielectrics [75]. We express the scattering amplitude of the cylinder now in a plane wave basis, using

$$\mathscr{F}^{ee}_{\text{cyl},\mathbf{k}_\perp P,\mathbf{k}'_\perp P'} = \sum_{nQ,n'Q'} \frac{C_{\mathbf{k}_\perp P}(\kappa)}{C_Q} D^\dagger_{\mathbf{k}_\perp P,k_z nQ} \mathscr{F}^{ee}_{\text{cyl},k_z nQ,k_z n'Q'} D_{k_z n'Q',\mathbf{k}'_\perp P'}, \quad (5.55)$$

where $\mathbf{k}_\perp$ denotes the vector (k_y, k_z), $C_{\mathbf{k}_\perp P}(\kappa)$ and C_Q are normalization coefficients that can be found together with the matrix elements of the conversion matrix D in Ref. [75]. The elements of the scattering amplitude in the cylindrical basis are given by (5.51). The scattering amplitude of the plate is easily expressed in the plane wave basis as

$$\mathscr{F}^{ee}_{\text{plate},\mathbf{k}'_\perp E,\mathbf{k}_\perp M} = \mathscr{F}^{ee}_{\text{plate},\mathbf{k}'_\perp M,\mathbf{k}_\perp E} = 0,$$

$$\mathscr{F}^{ee}_{\text{plate},\mathbf{k}'_\perp M,\mathbf{k}_\perp M} = \frac{(2\pi)^2}{L^2} \delta^{(2)}(\mathbf{k}_\perp - \mathbf{k}'_\perp) r^M \left(ic\kappa, \sqrt{1 + \mathbf{k}'^2_\perp/\kappa^2} \right)^{-1}, \quad (5.56)$$

$$\mathscr{F}^{ee}_{\text{plate},\mathbf{k}'_\perp E,\mathbf{k}_\perp E} = \frac{(2\pi)^2}{L^2} \delta^{(2)}(\mathbf{k}_\perp - \mathbf{k}'_\perp) r^E \left(ic\kappa, \sqrt{1 + \mathbf{k}'^2_\perp/\kappa^2} \right)^{-1},$$

in terms of the Fresnel coefficients that read for a general dielectric surface

$$r^M(ic\kappa, x) = \frac{\mu(ic\kappa) - \sqrt{1 + (n^2(ic\kappa) - 1)x^2}}{\mu(ic\kappa) + \sqrt{1 + (n^2(ic\kappa) - 1)x^2}},$$

$$r^E(ic\kappa, x) = \frac{\varepsilon(ic\kappa) - \sqrt{1 + (n^2(ic\kappa) - 1)x^2}}{\varepsilon(ic\kappa) + \sqrt{1 + (n^2(ic\kappa) - 1)x^2}}. \quad (5.57)$$

Here, n is the index of refraction, $n(ic\kappa) = \sqrt{\varepsilon(ic\kappa)\mu(ic\kappa)}$. In the limit of a perfectly reflecting plate one has $r^M \to -1$, $r^E \to 1$. The energy given by (5.50) can now be evaluated in the plane wave basis with the translation matrices given by the simple expression $\mathscr{U}^{ab}_{\mathbf{k}_\perp P, \mathbf{k}'_\perp P'} = e^{-\sqrt{\mathbf{k}^2_\perp + \kappa^2} H} \frac{(2\pi)^2}{L^2} \delta^{(2)} (\mathbf{k}_\perp - \mathbf{k}'_\perp) \delta_{P,P'}$.

The asymptotic expression for the attractive interaction energy at large distance $H \gg R$ reads

$$\mathscr{E}^E = -\frac{\hbar c L}{H^2} \frac{1}{16\pi \log(H/R)},$$
$$\mathscr{E}^M = -\hbar c L \frac{5}{32\pi} \frac{R^2}{H^4}.$$

(5.58)

The total electromagnetic Casimir interaction is again dominated by the contribution from the electric (Dirichlet) mode with $n = 0$ which depends only logarithmically on the cylinder radius. The interaction at all separations follows, as in the case of two cylinders, from a numerical computation of the determinant of (5.50) and integration over p. The result is shown in Fig. 5.5b.

The above approach has the advantage that it can be also applied to dielectric objects. The scattering amplitude of a dielectric cylinder can be obtained by solving the wave equation in a cylindrical basis with appropriate continuity conditions [75]. The scattering amplitude is diagonal in k_z and the cylindrical wave index n, but not in the polarization. Here we focus on large distances $H \gg R$. Expanding the log det in (5.50), we obtain for the interaction energy

$$\mathscr{E} = -\frac{3\hbar c L R^2}{128\pi H^4} \int\limits_0^1 dx \frac{\varepsilon_{cyl,0} - 1}{\varepsilon_{cyl,0} + 1} \left[(7 + \varepsilon_{cyl,0} - 4x^2) r^E(0, x) - (3 + \varepsilon_{cyl,0}) x^2 r^M(0, x) \right],$$

(5.59)

if the zero-frequency magnetic permeability $\mu_{cyl,0}$ of the cylinder is set to one. If we do not set $\mu_{cyl,0}$ equal to one, but instead take the perfect reflectivity limit for the plate, we obtain

$$\mathscr{E} = -\frac{\hbar c L R^2}{32\pi H^4} \frac{(\varepsilon_{cyl,0} - \mu_{cyl,0})(9 + \varepsilon_{cyl,0} + \mu_{cyl,0} + \varepsilon_{cyl,0}\mu_{cyl,0})}{(1 + \varepsilon_{cyl,0})(1 + \mu_{cyl,0})}.$$

(5.60)

Finally, if we let ε_{cyl} be infinite from the beginning (the perfect metal limit for the cylinder), only the $n = 0$ TM mode of the scattering amplitude contributes at lowest order. For a plate with zero-frequency permittivity $\varepsilon_{plate,0}$ and permeability $\mu_{plate,0}$, we obtain for the Casimir energy

$$\mathscr{E} = \frac{\hbar c L}{16\pi H^2 \log(R/H)} \phi^E,$$

(5.61)

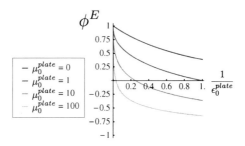

Fig. 5.6 Plots of ϕ^E versus $1/\epsilon_{\text{plate},0}$ for fixed values of $\mu_{\text{plate},0}$: ϕ^E decreases both with increasing $1/\epsilon_{\text{plate},0}$ and increasing $\mu_{\text{plate},0}$. The perfect metal limit ($\phi^E = 1$) is approached slowly for large $\mu_{\text{plate},0}$, as in the case of a sphere opposite a plate. For large $\mu_{\text{plate},0}$ the interaction becomes repulsive, which is expected given similar results for two infinite plates

where

$$\phi^E = \int\limits_0^1 \frac{dx}{1+x}\left[r^E(0,x) - x r^M(0,x)\right]. \tag{5.62}$$

In Fig. 5.6, ϕ^E is plotted as a function of the zero-frequency permittivity of the plate, $\epsilon_{\text{plate},0}$, for various zero-frequency permeability values, $\mu_{\text{plate},0}$.

5.4.2 Three-body Effects

Casimir interactions are not pair-wise additive. To study the consequences of this property, we consider the case of two identical objects near perfectly reflecting walls [78, 76]. Multibody effects were first observed for such a configuration with two rectangular cylinders sandwiched between two infinite plates by Rodriguez et al. [83]. The role of dimension on this effect is studied by considering either cylinders, see Fig. 5.7, or spheres, see Fig. 5.8. While we have given a more detailed description of how the interaction energies follow from the scattering approach in the previous subsection, we mainly provide the final results in this and in the following subsections.

First, we consider the geometry shown in Fig. 5.7 with two cylinders that are placed parallel to one or in-between two parallel plates, where all objects are assumed to be perfectly reflecting. Using the general expression for the Casimir energy of multiple objects, (5.39), the energy can be straightforwardly computed by truncating the matrix $\mathbb{M}$ at a finite partial wave order n. Including up to $n = 35$ partial waves, we obtain for the Casimir force between two cylinders of equal radii in the presence of one or two sidewalls the results shown in Fig. 5.7. In this figure the force at a fixed surface-to-surface distance $d - 2R = 2R$ between the cylinders is plotted as a function of the relative separation $(H - R)/R$ between the plate and cylinder surfaces. Two interesting

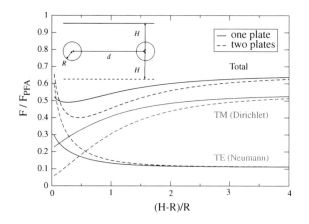

Fig. 5.7 Electromagnetic Casimir force per unit length between two cylinders of radius R and sidewall separation H vs. the ratio of sidewall separation to cylinder radius $(H - R)/R$, at fixed distance $(d - 2R)/R = 2$ between the cylinders, normalized by the total PFA force per unit length between two isolated cylinders, $F_{PFA} = \frac{5}{2}(\hbar c \pi^3/1920)\sqrt{R/(d - 2R)^7}$. The force is attractive. The *solid lines* refer to the case with one sidewall, while *dashed lines* depict the results for two sidewalls. Also shown are the individual TE and TM forces

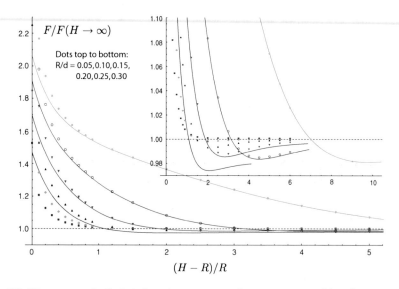

Fig. 5.8 Electromagnetic Casimir force between two spheres next to one sidewall at separation H versus the ratio H/R for different sphere separations d. *Dotted curves* represent numerical results. Shown are also the analytical results of (5.68), including terms up to $j = 10$ for $R/d \leq 0.2$ (*solid curves*) [85]. Inset: Magnification of the nonmonotonicity

features can be observed. First, the attractive total force varies non-monotonically with H: Decreasing for small H and then increasing towards the asymptotic limit between two isolated cylinders for large H, cf. (5.54). The extremum for the one-sidewall case occurs at $H - R \approx 0.27R$, and for the two-sidewall case is at $H - R \approx 0.46R$. Second, the total force for the two-sidewall case in the proximity limit $H = R$ is larger than for $H/R \to \infty$. As might be expected, the H-dependence for one sidewall is weaker than for two sidewalls, and the effects of the two sidewalls are not additive: not only is the difference from the $H \to \infty$ force not doubled for two sidewalls compared to one, but the two curves actually intersect at a separation of $H/R = 1.13$. The non-monotonic sidewall effect arises from a competition between the force from TE and TM modes as demonstrated by the results in Fig. 5.7. The qualitatively different behavior of TE and TM modes can be understood intuitively on the basis of the method of images [76]. The non-monotonicity in H also implies that the force between the cylinders and the sidewalls is not monotonic in d [76].

Second, we replace the two cylinders by two identical, general polarizable compact objects that we specialize later on to spheres [85]. The meaning of the lengths d and H remains unchanged. In the dipole approximation, the retarded limit of the interaction is described by the static electric (α_z, $\alpha_\parallel$) and magnetic (β_z, $\beta_\parallel$) dipole polarizabilities of the objects which can be different in the directions perpendicular (z) and parallel ($\parallel$) to the wall. The well-known Casimir-Polder (CP) potential between two compact objects at large distance is

$$\mathscr{E}_{2,\parallel}(d) = -\frac{\hbar c}{8\pi d^7}\Big[33\alpha_\parallel^2 + 13\alpha_z^2 - 14\alpha_\parallel\beta_z + (\alpha \leftrightarrow \beta)\Big]. \tag{5.63}$$

When a sidewall is added, the energy changes. Its d-dependent part is then

$$\mathscr{E}_{\infty}(d, H) = \mathscr{E}_{2,\parallel}(d) + \mathscr{E}_{2,\backslash}(D, d) + \mathscr{E}_3(D, d) \tag{5.64}$$

with $D = \sqrt{d^2 + 4H^2}$. The change in the relative orientation of the objects with $\ell = d/D$ leads to a modification of the 2-body CP potential

$$\begin{aligned} \mathscr{E}_{2,\backslash}(D, d) = -\frac{\hbar c}{8\pi D^7}\Big[&26\alpha_\parallel^2 + 20\alpha_z^2 - 14\ell^2(4\alpha_\parallel^2 - 9\alpha_\parallel\alpha_z + 5\alpha_z^2) \\ &+ 63\ell^4(\alpha_\parallel - \alpha_z)^2 - 14\big(\alpha_\parallel\beta_\parallel(1-\ell^2) + \ell^2\alpha_\parallel\beta_z\big) + (\alpha \leftrightarrow \beta)\Big]. \end{aligned} \tag{5.65}$$

The three-body energy $\mathscr{E}_3(D, d)$ describes the collective interaction between the two objects and one image object. It is given by

$$\begin{aligned} \mathscr{E}_3(D, d) = \frac{4\hbar c}{\pi}\frac{1}{d^3 D^4(\ell+1)^5}\Big[&\big(3\ell^6 + 15\ell^5 + 28\ell^4 + 20\ell^3 + 6\ell^2 - 5\ell - 1\big) \\ \times &\big(\alpha_\parallel^2 - \beta_\parallel^2\big) - \big(3\ell^6 + 15\ell^5 + 24\ell^4 - 10\ell^2 - 5\ell - 1\big)\big(\alpha_z^2 - \beta_z^2\big) \\ &+ 4\big(\ell^4 + 5\ell^3 + \ell^2\big)\big(\alpha_z\beta_\parallel - \alpha_\parallel\beta_z\big)\Big]. \end{aligned} \tag{5.66}$$

It is instructive to consider the two limits $H \ll d$ and $H \gg d$. For $H \ll d$, $\mathscr{E}_{\underline{\infty}}$ turns out to be the CP potential of (5.63) with the replacements $\alpha_z \to 2\alpha_z$, $\alpha_\parallel \to 0$, $\beta_z \to 0$, $\beta_\parallel \to 2\beta_\parallel$. The two-body and three-body contributions add constructively or destructively, depending on the relative orientation of a dipole and its image which together form a dipole of zero or twice the original strength [85].

For $H \gg d$ the leading correction to the CP potential of (5.63) comes from the three-body energy. The energy then becomes (up to order H^{-6})

$$\mathscr{E}_{\underline{\infty}}(d, H) = \mathscr{E}_{2,\parallel}(d) + \frac{\hbar c}{\pi}\left[\frac{\alpha_z^2 - \alpha_\parallel^2}{4d^3 H^4} + \frac{9\alpha_\parallel^2 - \alpha_z^2 - 2\alpha_\parallel \beta_z}{8dH^6} - (\alpha \leftrightarrow \beta)\right]. \qquad (5.67)$$

The signs of the polarizabilities in the leading term $\sim H^{-4}$ can be understood from the relative orientation of the dipole of one object and the image dipole of the other object [85].

Next, we study the case where the two objects are perfectly reflecting spheres of radius R. Now we consider arbitrary distances and include higher order multipole contributions. For $R \ll d, H$ and arbitrary H/d the result for the force can be written as

$$F = \frac{\hbar c}{\pi R^2} \sum_{j=6}^{\infty} f_j(H/d)\left(\frac{R}{d}\right)^{j+2}. \qquad (5.68)$$

The functions f_j can be computed exactly and their full form is given for $j = 6, 7, 8$ in Ref. [85]. For $H \gg d$ one has $f_6(h) = -1001/16 + 3/(4h^6) + \mathcal{O}(h^{-8})$, $f_8(h) = -71523/160 + 39/(80h^6) + \mathcal{O}(h^{-8})$ so that the wall induces weak repulsive corrections. For $H \ll d$, $f_6(h) = -791/8 + 6741h^2/8 + \mathcal{O}(h^4)$, $f_8(h) = -60939/80 + 582879h^2/80 + \mathcal{O}(h^4)$ so that the force amplitude decreases when the spheres are moved a small distance away from the wall. This proves the existence of a minimum in the force amplitude as a function of H for fixed, sufficiently small R/d.

To obtain the interaction at smaller separations or larger radius, the energy $\mathscr{E}_{\underline{\infty}}$ and force $F = -\partial\mathscr{E}_{\underline{\infty}}/\partial d$ between the spheres has been computed numerically [85]. In order to show the effect of the sidewall, the energy and force between the spheres, normalized to the results for two spheres without a wall, is shown in Fig. 5.8 for fixed d. When the spheres approach the wall, the force first decreases slightly if $R/d \lesssim 0.3$ and then increases strongly under a further reduction of H. For $R/d \gtrsim 0.3$ the force increases monotonically as the spheres approach the wall. This agrees with the prediction of the large distance expansion. The expansion of (5.68) with $j = 10$ terms is also shown in Fig. 5.8 for $R/d \leq 0.2$. Its validity is limited to large d/R and not too small H/R; it fails completely for $R/d > 0.2$ and hence is not shown in this range.

5.4.3 Orientation Dependence

In this subsection we describe the shape and orientation dependence of the Casimir force using (5.39), first reported in Ref. [30]. We consider the orientation dependent force between two spheroids, and between a spheroid and a plane. For two anisotropic objects, the CP potential of (5.63) must be generalized. In terms of the Cartesian components of the standard electric (magnetic) polarizability matrix α (β), the asymptotic large distance potential of two objects (with the $\hat{z}$ axis pointing from one object to the other), can be written as

$$
\mathcal{E} = -\frac{\hbar c}{d^7} \frac{1}{8\pi} \left\{ 13 \left(\alpha_{xx}^1 \alpha_{xx}^2 + \alpha_{yy}^1 \alpha_{yy}^2 + 2\alpha_{xy}^1 \alpha_{xy}^2 \right) \right.
$$
$$
+ 20\alpha_{zz}^1 \alpha_{zz}^2 - 30 \left(\alpha_{xz}^1 \alpha_{xz}^2 + \alpha_{yz}^1 \alpha_{yz}^2 \right) + (\alpha \to \beta)
$$
$$
\left. - 7 \left(\alpha_{xx}^1 \beta_{yy}^2 + \alpha_{yy}^1 \beta_{xx}^2 - 2\alpha_{xy}^1 \beta_{xy}^2 \right) + (1 \leftrightarrow 2) \right\}. \tag{5.69}
$$

For the case of an ellipsoidal object with static electric permittivity ε and magnetic permeability μ, the polarizability tensors are diagonal in a basis oriented to its principal axes, with elements (for $i \in \{1,2,3\}$)

$$
\alpha_{ii}^0 = \frac{V}{4\pi} \frac{\varepsilon - 1}{1 + (\varepsilon - 1)n_i}, \quad \beta_{ii}^0 = \frac{V}{4\pi} \frac{\mu - 1}{1 + (\mu - 1)n_i}, \tag{5.70}
$$

where $V = 4\pi r_1 r_2 r_3 / 3$ is the ellipsoid's volume. In the case of spheroids, for which $r_1 = r_2 = R$ and $r_3 = L/2$, the so-called depolarizing factors, n_j, can be expressed in terms of elementary functions, $n_1 = n_2 = \frac{1-n_3}{2}, n_3 = \frac{1-e^2}{2e^3}$ ($\log \frac{1+e}{1-e} - 2e$), where the eccentricity $e = (1 - \frac{4R^2}{L^2})^{1/2}$ is real for a prolate spheroid ($L > 2R$) and imaginary for an oblate spheroid ($L < 2R$). The polarizability tensors for an arbitrary orientation are then obtained as $\alpha = \mathcal{R}^{-1} \alpha^0 \mathcal{R}$, where $\mathcal{R}$ is the matrix that orients the principal axis of the spheroid relative to a fixed Cartesian basis. Note that for rarefied media with $\varepsilon \simeq 1$, $\mu \simeq 1$ the polarizabilities are isotropic and proportional to the volume. Hence, to leading order in $\varepsilon - 1$ the interaction is orientation independent at asymptotically large separations, as we would expect, since pairwise summation is valid for $\varepsilon - 1 \ll 1$. In the following we focus on the interesting opposite limit of two identical perfectly reflecting spheroids. We first consider prolate spheroids with $L \gg R$. The orientation of each "needle" relative to the line joining them (the initial z-axis) is parameterized by the two angles (θ, ψ), as depicted in Fig. 5.9a. Then the energy is

Fig. 5.9 a Orientation of a prolate (*cigar-shaped*) spheroid: The symmetry axis (initially the z-axis) is rotated by θ about the x-axis and then by ψ about the z-axis. For two such spheroids, the energy at large distances is give by (5.71). The latter is depicted at fixed distance d, and for $\psi_1 = \psi_2$, by a contour plot as function of the angles θ_1, θ_2 for the x-axis rotations . Minima (maxima) are marked by filled (*open*) dots. **b** As in **a** for oblate (*pancake-shaped*) spheroids, with a contour plot of energy at large separations

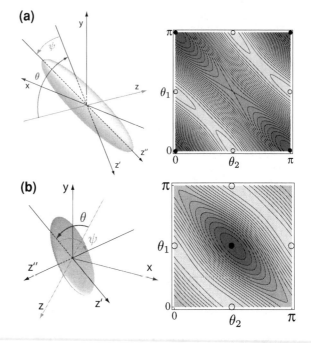

$$\mathcal{E}(\theta_1, \theta_2, \psi) = -\frac{\hbar c}{d^7} \left\{ \frac{5L^6}{1152\pi \left(\ln \frac{L}{R} - 1 \right)^2} \left[\cos^2 \theta_1 \cos^2 \theta_2 \right.\right.$$

$$\left.\left. + \frac{13}{20} \cos^2 \psi \sin^2 \theta_1 \sin^2 \theta_2 - \frac{3}{8} \cos \psi \sin 2\theta_1 \sin 2\theta_2 \right] + \mathcal{O}\left(\frac{L^4 R^2}{\ln \frac{L}{R}} \right) \right\},$$

(5.71)

where $\psi \equiv \psi_1 - \psi_2$. It is minimized for two needles aligned parallel to their separation vector. At almost all orientations the energy scales as L^6, and vanishes logarithmically slowly as $R \to 0$. The latter scaling changes when one needle is orthogonal to $\hat{z}$ (i.e. $\theta_1 = \pi/2$), while the other is either parallel to $\hat{z}$ ($\theta_2 = 0$) or has an arbitrary θ_2 but differs by an angle $\pi/2$ in its rotation about the z-axis (i.e. $\psi_1 - \psi_2 = \pi/2$). In these cases the energy comes from the next order term in (5.71), and takes the form

$$\mathcal{E}\left(\frac{\pi}{2}, \theta_2, \frac{\pi}{2} \right) = -\frac{\hbar c}{1152\pi d^7} \frac{L^4 R^2}{\ln \frac{L}{R} - 1} (73 + 7 \cos 2\theta_2),$$

(5.72)

which shows that the least favorable configuration corresponds to two needles orthogonal to each other and to the line joining them.

For perfectly reflecting oblate spheroids with $R \gg L/2$, the orientation of each "pancake" is again described by a pair of angles (θ, ψ), as depicted in Fig. 5.9b. To leading order at large separations, the energy is given by

$$\mathscr{E} = -\frac{\hbar c}{d^7} \left\{ \frac{R^6}{144\pi^3} \left[765 - 5(\cos 2\theta_1 + \cos 2\theta_2) + 237 \cos 2\theta_1 \cos 2\theta_2 \right.\right.$$
$$\left.\left. + 372 \cos 2\psi \sin^2 \theta_1 \sin^2 \theta_2 - 300 \cos \psi \sin 2\theta_1 \sin 2\theta_2 \right] + \mathcal{O}\left(R^5 L\right) \right\}. \quad (5.73)$$

The leading dependence is proportional to R^6, and does not disappear for any choice of orientations. Furthermore, this dependence remains even as the thickness of the pancake is taken to zero ($L \to 0$). This is very different from the case of the needles, where the interaction energy vanishes with thickness as $\ln^{-1}(L/R)$. The lack of L dependence is due to the assumed perfectly reflectivity. The energy is minimal for two pancakes lying on the same plane ($\theta_1 = \theta_2 = \pi/2$, $\psi = 0$) and has energy $-\hbar c(173/18\pi^3)R^6/d^7$. When the two pancakes are stacked on top of each other, the energy is increased to $-\hbar c(62/9\pi^3)R^6/d^7$. The least favorable configuration is when the pancakes lie in perpendicular planes, i.e., $\theta_1 = \pi/2$, $\theta_2 = 0$, with an energy $-\hbar c(11/3\pi^3)R^6/d^7$.

For an anisotropic object interacting with a perfectly reflecting mirror, at leading order the CP potential is given by

$$\mathscr{E} = -\frac{\hbar c}{d^4} \frac{1}{8\pi} \text{tr}(\alpha - \beta) + \mathcal{O}(d^{-5}), \quad (5.74)$$

which is clearly independent of orientation. Orientation dependence in this system thus comes from higher multipoles. The next order also vanishes, so the leading term is the contribution from the partial waves with $l = 3$ for which the scattering matrix is not known analytically. However, we can obtain the preferred orientation by considering a distorted sphere in which the radius R is deformed to $R + \delta f(\vartheta, \varphi)$. The function f can be expanded into spherical harmonics $Y_{lm}(\vartheta, \varphi)$, and spheroidal symmetry can be mimicked by choosing $f = Y_{20}(\vartheta, \varphi)$. The leading orientation dependent part of the energy is then obtained as

$$\mathscr{E}_f = -\hbar c \frac{1607}{640\sqrt{5}\pi^{3/2}} \frac{\delta R^4}{d^6} \cos(2\theta). \quad (5.75)$$

A prolate spheroid ($\delta > 0$) thus minimizes its energy by pointing towards the mirror, while an oblate spheroid ($\delta < 0$) prefers to lie in a plane perpendicular to the mirror. (It is assumed that the perturbative results are not changed for large distortions.) These configurations are also preferred at small distances d, since (at fixed distance to the center) the object reorients to minimize the closest separation. Interestingly, the latter conclusion is not generally true. In Ref. [30] it has been shown that there can be a transition in preferred orientation as a function of d in the simpler case of a scalar field with Neumann boundary conditions. The separation at which this transition occurs varies with the spheroid's eccentricity.

Fig. 5.10 **a** Energy $EH^2/(\hbar cL)$ versus H/R for $\theta = 0$ and $R = 1$ on a log-linear scale for the parabolic cylinder-plane geometry. The *dashed line* gives the $R = 0$ limit and the *solid curve* gives the PFA result. **b** The coefficient $c(\theta)$ as a function of angle for $R = 0$. The exact result at $\theta = \pi/2$ is marked with a *cross*. Inset: Dirichlet (*circles*) and Neumann (*squares*) contributions to the full electromagnetic result

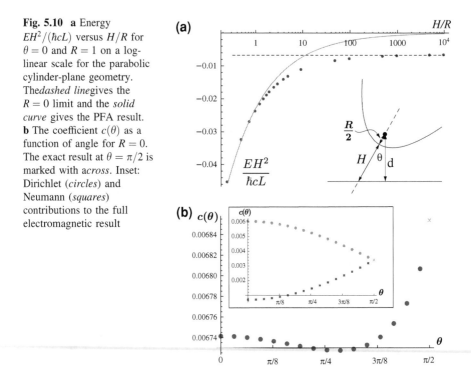

5.4.4 Edge and Finite Size Effects

In this subsection, based on work reported in Ref. [45], it is demonstrated that *parabolic* cylinders provide another example were the scattering amplitudes can be computed exactly. We use the exact results for scattering from perfect mirrors to compute the Casimir force between a parabolic cylinder and a plate. In the limiting case when the curvature at its tip vanishes, the parabolic cylinder becomes a semi-infinite plate (a knife's edge), and we can consider how edges and finite size effects influence the Casimir energy.

The surface of a parabolic cylinder in Cartesian coordinates is described by $y = (x^2 - R^2)/2R$ for all z, as shown in Fig. 5.10a, where R is the curvature at the tip. In parabolic cylinder coordinates, defined through $x = \mu\lambda$, $y = (\lambda^2 - \mu^2)/2$, $z = z$, the surface is simply $\mu = \mu_0 = \sqrt{R}$ for $-\infty < \lambda, z < \infty$. Since sending $\lambda \to -\lambda$ and $\mu \to -\mu$ returns us to the same point, we restrict our attention to $\mu \geq 0$ while considering all values of λ. Then μ plays the role of the "radial" coordinate in scattering theory and one can again define regular and outgoing waves [45]. Since both objects are perfect mirrors, translational symmetry along the z-axis enables us to decompose the electromagnetic field into two scalar fields, as in the case of circular cylinders in Sect. 5.4.1. Each scalar field, describing E (Dirichlet) or M (Neumann) modes, can then be treated independently, with the sum of their contributions giving the full electromagnetic result.

The scattering amplitude of the plate is expressed in a plane wave basis and is given by (5.56) with $r^M = -1$ and $r^E = 1$. The scattering amplitude of the parabolic cylinder for E and M polarization is obtained in a parabolic cylinder wave basis as [45]

$$
\mathscr{F}^{ee}_{\text{para},k_z \nu E, k'_z \nu' E} = -\frac{2\pi}{L}\delta(k_z - k'_z)\delta_{\nu,\nu'}f_{k_z \nu E}, \quad f_{k_z \nu E} = i^\nu \frac{D_\nu(i\tilde{\mu}_0)}{D_{-\nu-1}(\tilde{\mu}_0)}
$$

$$
\mathscr{F}^{ee}_{\text{para},k_z \nu M, k'_z \nu' M} = -\frac{2\pi}{L}\delta(k_z - k'_z)\delta_{\nu,\nu'}f_{k_z \nu M}, \quad f_{k_z \nu M} = i^{\nu+1} \frac{D'_\nu(i\tilde{\mu}_0)}{D'_{-\nu-1}(\tilde{\mu}_0)},
$$

(5.76)

with $\tilde{\mu}_0 = \sqrt{2R\sqrt{\kappa^2 + k_z^2}}$ and the parabolic cylinder function $D_\nu(u)$ for integer ν.

For the present geometry, the general formula for the Casimir energy per unit length can be expressed explicitly as

$$
\frac{\mathscr{E}}{\hbar c L} = \int_0^\infty \frac{d\kappa}{2\pi} \int_{-\infty}^\infty \frac{dk_z}{2\pi} \log\det\left(\delta_{\nu,\nu'} - f_{k_z \nu P}\int_{-\infty}^\infty dk_x \mathscr{U}_{\nu k_x k_z}(d,\theta)r^P \mathscr{U}_{\nu' k_x k_z}(d,-\theta)\right)
$$

(5.77)

for polarization $P=E$ or M. Here the matrix $\mathscr{U}$ with elements

$$
\mathscr{U}_{\nu k_x k_z}(d,\theta) = \sqrt{\frac{i}{2k_y \nu!\sqrt{2\pi}}}\frac{\left(\tan\frac{\phi+\theta}{2}\right)^\nu}{\cos\frac{\phi+\theta}{2}}e^{ik_y d}
$$

(5.78)

with $k_y = i\sqrt{\kappa^2 + k_x^2 + k_z^2}$ and $\tan\phi = k_x/k_y$ describes the translation from parabolic cylinder to plane waves over the distance d from the focus of the parabola to the plane where θ is the angle of inclination of the parabolic cylinder.

Numerical computations of the energy are performed by truncating the determinant at index $\nu_{\max}$. For the numbers quoted below, we have computed for $\nu_{\max}$ up to 200 and then extrapolated the result for $\nu_{\max} \to \infty$, and in Fig. 5.10 we have generally used $\nu_{\max} = 100$. The dependence of the energy on the separation $H = d - R/2$ for $\theta = 0$ is shown in Fig. 5.10a. At small separations ($H/R \ll 1$) the proximity force approximation, given by

$$
\frac{\mathscr{E}_{\text{pfa}}}{\hbar c L} = -\frac{\pi^2}{720}\int_{-\infty}^\infty \frac{dx}{[H + x^2/(2R)]^3} = -\frac{\pi^3}{960\sqrt{2}}\sqrt{\frac{R}{H^5}},
$$

(5.79)

should be valid. The numerical results in Fig. 5.10a indeed confirm this expectation.

A more interesting limit is obtained when $R/H \to 0$, corresponding to a semi-infinite plate for which the PFA energy vanishes. The exact result for the energy for $R = 0$ and $\theta = 0$ is

$$\frac{\mathscr{E}}{\hbar c L} = -\frac{C_\perp}{H^2}, \qquad (5.80)$$

where $C_\perp = 0.0067415$ is obtained by numerical integration. When the semi-infinite plate is tilted by an angle θ, dimensional analysis suggest for the Casimir energy [39, 92]

$$\frac{\mathscr{E}}{\hbar c L} = -\frac{C(\theta)}{H^2}. \qquad (5.81)$$

The function $c(\theta) = \cos(\theta)C(\theta)$ is shown in Fig. 5.10b. A particularly interesting limit is $\theta \to \pi/2$, when the two plates are parallel. In this case, the leading contribution to the Casimir energy should be proportional to the area of the half-plane according to the parallel plate formula, $E_\parallel/(\hbar c A) = -c_\parallel/H^3$ with $c_\parallel = \pi^2/720$, plus a subleading correction due to the edge. Multiplying by $\cos \theta$ removes the divergence in the amplitude $C(\theta)$ as $\theta \to \pi/2$. As in [39], we assume $c(\theta \to \pi/2) = c_\parallel/2 + (\theta - \pi/2)c_{\text{edge}}$, although we cannot rule out the possibility of additional non-analytic forms, such as logarithmic or other singularities. With this assumption, we can estimate the edge correction $c_{\text{edge}} = 0.0009$ from the data in Fig. 5.10b. From the inset in Fig. 5.10b, we estimate the Dirichlet and Neumann contributions to this result to be $c_{\text{edge}}^D = -0.0025$ and $c_{\text{edge}}^N = 0.0034$, respectively. For extensions to other geometries with edges, inclusion of thermal fluctuations and experimental implications, see Ref. [45].

5.4.5 Interior Configurations

In this last subsection we consider so-called interior configurations where one object is contained within another that can be also studied with the methods introduced in Sect. 5.2. Specifically, we obtain the electrodynamic Casimir interaction of a conducting or dielectric object inside a perfectly conducting spherical cavity [95]. In the case where an object, i, lies inside a perfectly conducting cavity, the outer object o, the Casimir energy of (5.42) becomes

$$\mathscr{E} = \frac{\hbar c}{2\pi} \int\limits_0^\infty d\kappa \log \frac{\det(\mathscr{I} - \mathscr{F}_o^{ii} \mathscr{W}^{io} \mathscr{F}_i^{ee} \mathscr{V}^{io})}{\det(\mathscr{I} - \mathscr{F}_o^{ii} \mathscr{F}_i^{ee})}, \qquad (5.82)$$

where $\mathscr{F}_o^{ii}$ is the scattering amplitude for interior scattering of the conducting cavity, a sphere in our case, and $\mathscr{F}_i^{ee}$ the scattering amplitude of the interior object. The amplitude matrix for interior scattering is the inverse of the corresponding exterior matrix. These scattering amplitudes are evaluated in a spherical vector wave basis with respect to appropriately chosen origins within each object. The translation matrices, $\mathscr{W}^{io}$ and $\mathscr{V}^{io}$, relate regular wave functions between the coordinate systems of the interior object and the spherical cavity, see Ref. [75] for details. The determinant in the denominator of (5.82) subtracts the Casimir energy when the origins of the two objects coincide. This way of normalizing the Casimir energy differs from the exterior cases considered before, where the objects are removed to infinite separation; a choice that would be unnatural in the interior case.

First, we determine the forces and torques on a small object, dielectric or conducting, well separated from the cavity walls. This is the interior analogue of the famous Casimir-Polder force on a polarizable molecule near a perfectly conducting plate [15]. In this case the first term in a multiple scattering expansion, where the integrand of (5.82) is replaced by $-\mathrm{Tr}(\mathscr{F}_o^{ii}\mathscr{W}^{io}\mathscr{F}_i^{ee}\mathscr{V}^{io})$, already gives an excellent approximation to the energy. Since the object is small, the scattering amplitude $\mathscr{F}_{i,lmP,l'm'P'}^{ee}$, (where l and m are angular momentum indices and P labels M or E polarization) can be expanded in powers of κ. Only the following terms contribute to lowest order: $\mathscr{F}_{i,1mP,1m'P}^{ee}(\kappa) = 2\kappa^3\alpha_{mm'}^P/3 + O(\kappa^4)$, where $\alpha_{mm'}^P$ is the static electric ($P = E$) or magnetic ($P = M$) polarizability tensor of the inner object. We consider an exterior spherical shell of radius R and define a to be the displacement of the center of the interior object from the center of the shell. Using the dipole approximation for the inner object but including all multipoles of the exterior shell, we find for the Casimir energy to leading order in r/R (where r is the typical length scale of the interior object), the energy

$$\frac{3\pi R^4}{\hbar c}\mathscr{E}(a/R) = \left[f^E(a/R) - f^E(0)\right]\mathrm{Tr}\alpha^E$$
$$+ g^E(a/R)(2\alpha_{zz}^E - \alpha_{xx}^E - \alpha_{yy}^E) + (E \leftrightarrow M). \quad (5.83)$$

The z-axis is oriented from the center of the shell to the innterior object, and α_{ij}^P represent the interior object's static polarizability tensors in a Cartesian basis. The coefficient functions f^P and g^P can be obtained in terms of an integral over modified Bessel functions, see Ref. [95]. f^E is negative and decreasing with a/R, while f^M is positive and increasing. There are important differences between (5.83) and the classic Casimir-Polder result: first, the energy depends in a non-trivial way on a/R; second, at any non-zero distance from the center, the interior object experiences a torque; and third, the force between the two bodies depends on the interior object's orientation.

To explore the orientation dependence of (5.83) assume, for simplicity, there is a single frame in which both α^E and α^M are diagonal. In this body-fixed frame, write $\alpha_{xx}^0 - \alpha_{yy}^0 = \beta$ and $\alpha_{zz}^0 - \frac{1}{2}(\alpha_{xx}^0 + \alpha_{yy}^0) = \gamma$ (where we have suppressed the M/E label). The polarizability in the "lab frame" is obtained by $\alpha = \mathscr{R}\alpha^0\mathscr{R}^{-1}$, where $\mathscr{R}$ is a rotation matrix that orients the principal axes of the inner object with

Fig. 5.11 a The ratio g^P/f^P, which determines the preferred orientation of the interior object, plotted versus $x = a/R$ showing the change in preferred orientation from interior $(a/R < 1)$ to exterior $(a/R > 1)$ (displayed by two small ellipses as described in the text). The solid curves are fits of the form $c_1(1 - x) + c_2(1 - x)^2$ to these data points. **b** PFA correction coefficients for spheres. r/R ranges from -1 (interior concentric), to zero (sphere-plane), to $+1$ (exterior, equal radii). The data points correspond to the exact values of θ_1 calculated numerically, while the solid curve is a fit (see text). Inset: "interior" and "exterior" geometrical configurations

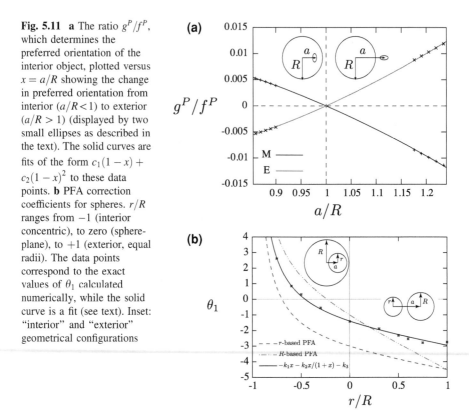

respect to the lab frame. This procedure leaves $\mathrm{Tr}\alpha^0$ invariant, and gives for the second line in (5.83),

$$\sum_{P=M,E} g^P(a/R)\left(\frac{3\beta^P}{2}\sin^2\theta\cos 2\phi + \gamma^P(3\cos^2\theta - 1)\right),$$

where ϕ corresponds to the azimuthal rotation of the object about its principal z-axis, and θ is the angle between the object's principal z-axis and the "laboratory" z-axis connecting the center of the sphere to the origin of the interior object.

If $\beta \neq 0$ then the object held at fixed inclination, θ, experiences a torque that causes it to rotate about the body-fixed z-axis. If, however, the object has axial symmetry $(\beta = 0)$, then the only torque on the object tries to align it either parallel or perpendicular to the displacement axis.

A "cigar shaped" object $(\gamma > 0)$ prefers to orient so as to point perpendicular to the z axis, and a "pancake" $(\gamma < 0)$ tries to align its two large axes perpendicular to the z axis. The small ellipse inside the sphere in Fig. 5.11a illustrates a side view of both the cigar and the pancake in their preferred orientation. It is interesting to note that g^E and g^M are both positive. So, in contrast to the force, the contributions to the torque from magnetic and electric polarizabilities are in the same direction, if they have the same sign. More complicated behavior is possible if, for example, the electric and magnetic polarizabilities are not diagonal in the same body-fixed coordinate system. Note that our results cannot be compared to the PFA approximation since the the size of the inner object, not the separation of surfaces, d, has been assumed to be the smallest scale in the analysis.

An identical analysis can be performed for a polarizable object outside a metallic sphere where $a/R > 1$. It turns out that the analogous exterior function $g(a/R) < 0$ for both polarizations. Therefore, the preferred orientation of a polarizable object outside a metallic sphere is opposite of that in the interior case (see the small ellipse outside the large sphere in Fig. 5.11a). The continuation of the functions f and g from "interior" to "exterior" is displayed in Fig. 5.11a, where the transition from one orientation to the other is clear.

Second, we compute numerically from (5.82) the interaction energy of a finite-size metal sphere with the cavity walls when the separation, d, between their surfaces tends to zero. In this limit the Casimir force F between two conducting spheres, which is attractive, is proportional in magnitude to d^{-3}, where $d = R - r - a$ is the separation of surfaces. The coefficient of d^{-3} is given by the PFA,

$$\lim_{d \to 0} d^3 F = -\frac{\pi^3 \hbar c}{360} \frac{rR}{r+R}. \tag{5.84}$$

This result holds for both the interior and the exterior configuration of two spheres. For fixed r we formally distinguish the cases: $R > 0$ for the exterior, $R \to \infty$ for the plate-sphere, and $R < 0$ for the interior configuration, see Fig. 5.11b. All possible configurations are taken into account by considering $-1 \leq r/R \leq 1$. Although we know of no derivation of the functional form of the Casimir force beyond the leading term in the PFA, our numerical results are well fit by a power series in d/r,

$$F = -\frac{\pi^3 \hbar c}{360 d^3} \frac{rR}{r+R} \left(1 + \theta_1(r/R) \frac{d}{2r} - \theta_2(r/R) \frac{d^2}{2r^2} + \cdots \right) \tag{5.85}$$

We have used this functional form to extract the coefficient $\theta_1(r/R)$.

Although the PFA is accurate only in the limit $d/r \to 0$, it can be extended in various ways to the whole range of d, r, and R. Depending on the surface O from which the normal distance to the other surface is measured, one obtains the "O-based" PFA energy. Clearly, the result depends on which object one chooses as O, but the various results do agree to leading order in d/r. We can choose either of the two spheres to arrive at the "r-based PFA" or the "R-based PFA", see Fig. 5.11b. Either one yields a 'correction' to the leading order PFA,

$$\theta_{1,r}^{\mathrm{PFA}}(x) = -\left(x + \frac{x}{1+x} + 3 \right), \quad \theta_{1,R}^{\mathrm{PFA}} = -\left(3x + \frac{x}{1+x} + 1 \right),$$

where $x = r/R$. In Fig. 5.11b we plot the values of θ_1 extracted from a numerical evaluation of the force from (5.82) for various values of $r/R < 0$. For reference, the two PFA estimates are also shown.

The numerical data in Fig. 5.11b show a smooth transition from the interior to the exterior configuration. Although the PFA estimates do not describe the data, the r-based PFA has a similar functional form and divergence as $x \rightarrow -1$. Therefore, we fit the data in Fig. 5.11b to a function, $\theta_1(x) = -(k_1 x + k_2 x/(1 + x) + k_3)$ and find, $k_1 = 1.05 \pm 0.14, k_2 = 1.08 \pm 0.08, k_3 = 1.38 \pm 0.06$. Notice, however, that the actual function $\theta_1(x)$ is not known analytically and that the fit represents a reasonable choice which may not be unique. Our results show that the correction to the PFA has a significant dependence on ratio of curvatures of the two surfaces.

Acknowledgements The research presented here was conducted together with Noah Graham, Steven G. Johnson, Mehran Kardar, Alejandro W. Rodriguez, Pablo Rodriguez-Lopez, Alexander Shpunt, and Saad Zaheer, whom we thank for their collaboration. This work was supported by the National Science Foundation (NSF) through grant DMR-08-03315 (SJR), by the DFG through grant EM70/3 (TE) and by the U. S. Department of Energy (DOE) under cooperative research agreement #DF-FC02-94ER40818 (RLJ).

References

1. Ashkin, A.: Acceleration and trapping of particles by radiation pressure. Phys. Rev. Lett. **24**, 156–159 (1970)
2. Ashkin, A., Gordon, J.P.: Stability of radiation-pressure particle traps: an optical Earnshaw theorem. Opt. Lett. **8**, 511–513 (1983)
3. Bachas, C.P.: Comment on the sign of the Casimir force. J. Phys. A: Math. Theor. **40**, 9089–9096 (2007)
4. Balian, R., Duplantier, B.: Electromagnetic waves near perfect conductors. II. Casimir effect. Ann. Phys., NY **104**, 300–335 (1977)
5. Balian, R., Duplantier, B.: Electromagnetic waves near perfect conductors. I. Multiple scattering expansions. Distribution of modes. Ann. Phys., NY **112**, 165–208 (1978)
6. Birman, M.S., Krein, M.G.: On the theory of wave operators and scattering operators. Sov. Math.-Dokl. **3**, 740–744 (1962)
7. Bordag, M., Robaschik, D., Wieczorek, E.: Quantum field theoretic treatment of the Casimir effect. Ann. Phys., NY **165**, 192–213 (1985)
8. Braunbek, W.: Freies Schweben diamagnetischer Körper im magnetfeld. Z. Phys. **112**, 764–769 (1939)
9. Braunbek, W.: Freischwebende Körper im elektrischen und magnetischen Feld. Z. Phys. **112**, 753–763 (1939)
10. Bulgac, A., Magierski, P., Wirzba, A.: Scalar Casimir effect between Dirichlet spheres or a plate and a sphere. Phys. Rev. D **73**, 025007 (2006)
11. Bulgac, A., Wirzba, A.: Casimir Interaction among objects immersed in a fermionic environment. Phys. Rev. Lett. **87**, 120404 (2001)
12. Büscher, R., Emig, T.: Geometry and spectrum of Casimir forces. Phys. Rev. Lett. **94**, 133901 (2005)
13. Capasso, F., Munday, J.N., Iannuzzi, D., Chan, H.B.: Casimir forces and quantum electrodynamical torques: Physics and Nanomechanics. IEEE J. Sel. Top. Quant. **13**, 400–414 (2007)

14. Casimir, H.B.G.: On the attraction between two perfectly conducting plates. Proc. K. Ned. Akad. Wet. **51**, 793–795 (1948)
15. Casimir, H.B.G., Polder, D.: The Influence of retardation on the London-van der Waals forces. Phys. Rev. **73**, 360–372 (1948)
16. Chan, H.B., Aksyuk, V.A., Kleiman, R.N., Bishop, D.J., Capasso, F.: Quantum mechanical actuation of microelectromechanical systems by the Casimir force. Science **291**, 1941–1944 (2001)
17. Chan, H.B., Bao, Y., Zou, J., Cirelli, R.A., Klemens, F., Mansfield, W.M., Pai, C.S.: Measurement of the Casimir force between a gold sphere and a silicon surface with nanoscale trench arrays. Phys. Rev. Lett. **101**, 030401 (2008)
18. Chen, F., Klimchitskaya, G.L., Mostepanenko, V.M., Mohideen, U.: Demonstration of the difference in the Casimir force for samples with different charge-carrier densities. Phys. Rev. Lett. **97**, 170402 (2006)
19. Chen, F., Klimchitskaya, G.L., Mostepanenko, V.M., Mohideen, U.: Control of the Casimir force by the modification of dielectric properties with light. Phys. Rev. B **76**, 035338 (2007)
20. Chen, F., Mohideen, U., Klimchitskaya, G.L., Mostepanenko, V.M.: Demonstration of the lateral Casimir force. Phys. Rev. Lett. **88**, 101801 (2002)
21. Chew, W.C., Jin, J.M., Michielssen, E., Song, J.M. (eds.): Fast and Efficient Algorithms in Computational Electrodynamics. Artech House, Norwood, MA (2001)
22. Dalvit, D.A.R., Lombardo, F.C., Mazzitelli, F.D., Onofrio, R.: Exact Casimir interaction between eccentric cylinders. Phys. Rev. A **74**, 020101(R) (2006)
23. Decca, R.S., López, D., Fischbach, E., Klimchitskaya, G.L., Krause, D.E., Mostepanenko, V.M.: Tests of new physics from precise measurements of the Casimir pressure between two gold-coated plates. Phys. Rev. D **75**, 077101 (2007)
24. Druzhinina, V., DeKieviet, M.: Experimental observation of quantum reflection far from threshold. Phys. Rev. Lett. **91**, 193202 (2003)
25. Dzyaloshinskii, I.E., Lifshitz, E.M., Pitaevskii, L.P.: The general theory of van der Waals forces. Adv. Phys. **10**, 165–209 (1961)
26. Earnshaw, S.: On the nature of the molecular forces which regulate the constitution of the luminiferous ether. Trans. Camb. Phil. Soc. **7**, 97–112 (1842)
27. Ederth, T.: Template-stripped gold surfaces with 0.4-nm rms roughness suitable for force measurements: Application to the Casimir force in the 20–100 nm range. Phys. Rev. A **62**, 062104 (2000)
28. Emig, T., Graham, N., Jaffe, R.L., Kardar, M.: Casimir forces between arbitrary compact objects. Phys. Rev. Lett. **99**, 170403 (2007)
29. Emig, T., Graham, N., Jaffe, R.L., Kardar,M.: Casimir forces between compact objects: The scalar objects. Phys. Rev. D **77**, 025005 (2008)
30. Emig, T., Graham, N., Jaffe, R.L., Kardar, M.: Orientation dependence of Casimir forces. Phys. Rev. A **79**, 054901 (2009)
31. Emig, T., Hanke, A., Golestanian, R., Kardar, M.: Probing the strong boundary shape dependence of the Casimir force. Phys. Rev. Lett. **87**, 260402 (2001)
32. Emig, T., Hanke, A., Golestanian, R., Kardar, M.: Normal and lateral Casimir forces between deformed plates. Phys. Rev. A **67**, 022114 (2003)
33. Emig, T., Jaffe, R.L., Kardar, M., Scardicchio, A.: Casimir interaction between a plate and a cylinder. Phys. Rev. Lett. **96**, 080403 (2006)
34. Feinberg, G., Sucher, J.: General form of the retarded van der Waals potential. J. Chem. Phys. **48**, 3333–3334 (1698)
35. Feinberg, G., Sucher, J.: General theory of the van der Waals interaction: A model-independent approach. Phys. Rev. A **2**, 2395–2415 (1970)
36. Feynman, R.P., Hibbs, A.R.: Quantum mechanics and path integrals. McGraw-Hill, New York (1965)
37. Geim, A.: Everyone's magnetism. Phys. Today **51**(9), 36–39 (1998)
38. Genet, C., Lambrecht, A., Reynaud, S.: Casimir force and the quantum theory of lossy optical cavities. Phys. Rev. A **67**, 043811 (2003)

39. Gies, H., Klingmüller, K.: Casimir edge effects. Phys. Rev. Lett. **97**, 220405 (2006)
40. Golestanian, R.: Casimir-Lifshitz interaction between dielectrics of arbitrary geometry: A dielectric contrast perturbation theory. Phys. Rev. A **80**, 012519 (2009)
41. Golestanian, R., Kardar, M.: Mechanical response of vacuum. Phys. Rev. Lett. **78**, 3421–3425 (1997)
42. Golestanian, R., Kardar, M.: Path-integral approach to the dynamic Casimir effect with fluctuating boundaries. Phys. Rev. A **58**, 1713–1722 (1998)
43. Graham, N., Jaffe, R.L., Khemani, V., Quandt, M., Scandurra, M., Weigel, H.: Casimir energies in light of quantum field theory. Phys. Lett. B **572**, 196–201 (2003)
44. Graham, N., Quandt, M., Weigel, H.: Spectral methods in quantum field theory. Springer, Berlin (2009)
45. Graham, N., Shpunt, A., Emig, T., Rahi, S.J., Jaffe, R.L., Kardar, M.: Casimir force at a knife's edge. Phys. Rev. D **81**, 061701(R) (2010)
46. Harber, D.M., Obrecht, J.M., McGuirk, J.M., Cornell, E.A.: Measurement of the Casimir-Polder force through center-of-mass oscillations of a Bose-Einstein condensate. Phys. Rev. A **72**, 033610 (2005)
47. Henseler, M., Wirzba, A., Guhr, T.: Quantization of HyperbolicN-Sphere scattering systems in three dimensions. Ann. Phys., NY **258**, 286–319 (1997)
48. Jaekel, M.T., Reynaud, S.: Casimir force between partially transmitting mirrors. J. Physique I **1**, 1395–1409 (1991)
49. Jones, T.B.: Electromechanics of Particles. Cambridge University Press, Cambridge (1995)
50. Kats, E.I.: Influence of nonlocality effects on van der Waals interaction. Sov. Phys. JETP **46**, 109 (1997)
51. Kenneth, O., Klich, I.: Opposites Attract: A theorem about the Casimir force. Phys. Rev. Lett. **97**, 160401 (2006)
52. Kenneth, O., Klich, I.: Casimir forces in a T-operator approach. Phys. Rev. B **78**, 014103 (2008)
53. Kim, W.J., Brown-Hayes, M., Dalvit, D.A.R., Brownell, J.H., Onofrio, R.: Anomalies in electrostatic calibrations for the measurement of the Casimir force in a sphere-plane geometry.Phys. Rev. A **78**,020101(2008)
54. Klimchitskaya, G.L., Mohideen, U., Mostepanenko, V.M.: The Casimir force between real materials: Experiment and theory. Rev. Mod. Phys. **81**, 1827–1885 (2009)
55. Krause, D.E., Decca, R.S., López, D., Fischbach, E.: Experimental investigation of the Casimir force beyond the proximity-force approximation. Phys. Rev. Lett. **98**, 050403 (2007)
56. Krein, M.G.: On the trace formula in perturbation theory. Mat. Sborn. (NS) **33**, 597–626 (1953)
57. Krein, M.G.: Perturbation determinants and a formula for the trace of unitary and selfadjoint operators. Sov. Math.-Dokl. **3**, 707–710 (1962)
58. Lambrecht, A., Neto, P.A.M., Reynaud, S.: The Casimir effect within scattering theory. New J. Phys. **8**, 243 (2006)
59. Lamoreaux, S.K.: Demonstration of the Casimir force in the 0.6 to 6 μm range. Phys. Rev. Lett. **78**, 5–8 (1997)
60. Landau, L.D., Lifshitz, E.M.: Electrodynamics of continuous media. Pergamon Press, Oxford (1984)
61. Levin M., McCauley A.P., Rodriguez A.W., Reid M.T.H., Johnson S.G. (2010) Casimir repulsion between metallic objects in vacuum. arXiv:1003.3487
62. Li, H., Kardar, M.: Fluctuation-induced forces between rough surfaces. Phys. Rev. Lett. **67**, 3275–3278 (1991)
63. Li, H., Kardar, M.: Fluctuation-induced forces between manifolds immersed in correlated fluids. Phys. Rev. A **46**, 6490–6500 (1992)
64. Lifshitz, E.M.: The theory of molecular attractive forces between solids. Sov. Phys. JETP **2**, 73–83 (1956)
65. Lifshitz, E.M., Pitaevskii, L.P.: Statistical physics Part 2. Pergamon Press, New York (1980)

66. Lippmann, B.A., Schwinger, J.: Variational principles for scattering processes. i. Phys. Rev. **79**, 469–480 (1950)
67. Milton, K.A., Parashar, P., Wagner, J.: Exact results for Casimir interactions between dielectric bodies: The weak-coupling or van der waals limit. Phys. Rev. Lett. **101**, 160402 (2008)
68. Milton K.A., Parashar P., Wagner J. (2008) From multiple scattering to van der waals interactions: exact results for eccentric cylinders. arXiv:0811.0128
69. Mohideen, U., Roy, A.: Precision measurement of the Casimir force from 0.1–0.9 μm. Phys. Rev. Lett. **81**, 4549–4552 (1998)
70. Morse, P.M., Feshbach, H.: Methods of theoretical physics. McGraw-Hill, New York (1953)
71. Munday, J.N., Capasso, F.: Precision measurement of the Casimir-Lifshitz force in a fluid. Phys. Rev. A **75**, 060102(R) (2007)
72. Munday, J.N., Capasso, F., Parsegian, V.A.: Measured long-range repulsive Casimir-Lifshitz forces. Nature **457**, 170–173 (2009)
73. Palasantzas, G., van Zwol, P.J., De Hosson, J.T.M.: Transition from Casimir to van der Waals force between macroscopic bodies. Appl. Phys. Lett. **93**, 121912 (2008)
74. Parsegian, V.A.: van der Waals Forces. Cambridge University Press, Cambridge (2005)
75. Rahi, S.J., Emig, T., Graham, N., Jaffe, R.L., Kardar, M.: Scattering theory approach to electrodynamic Casimir forces. Phys. Rev. D **80**, 085021 (2009)
76. Rahi, S.J., Emig, T., Jaffe, R.L., Kardar, M.: Casimir forces between cylinders and plates. Phys. Rev. A **78**, 012104 (2008)
77. Rahi S.J., Kardar M., Emig T. Constraints on stable equilibria with fluctuation-induced forces. Phys. Rev. Lett. **105**, 070404 (2010)
78. Rahi, S.J., Rodriguez, A.W., Emig, T., Jaffe, R.L., Johnson, S.G., Kardar, M.: Nonmonotonic effects of parallel sidewalls on Casimir forces between cylinders. Phys. Rev. A **77**, 030101 (2008)
79. Rahi, S.J., Zaheer, S.: Stable levitation and alignment of compact objects by Casimir spring forces. Phys. Rev. Lett. **104**, 070405 (2010)
80. Reid, M.T.H., Rodriguez, A.W., White, J., Johnson, S.G.: Efficient computation of Casimir interactions between arbitrary 3D objects. Phys. Rev. Lett. **103**, 040401 (2009)
81. Renne, M.J.: Microscopic theory of retarded Van der Waals forces between macroscopic dielectric bodies. Physica **56**, 125–137 (1971)
82. Robaschik, D., Scharnhorst, K., Wieczorek, E.: Radiative corrections to the Casimir pressure under the influence of temperature and external fields. Ann. Phys., NY **174**, 401–429 (1987)
83. Rodriguez, A., Ibanescu, M., Iannuzzi, D., Capasso, F., Joannopoulos, J.D., Johnson, S.G.: Computation and visualization of Casimir forces in arbitrary geometries: Non-monotonic lateral-wall forces and failure of proximity force approximations. Phys. Rev. Lett. **99**, 080401 (2007)
84. Rodriguez, A.W., Joannopoulos, J.D., Johnson, S.G.: Repulsive, nonmonotonic Casimir forces in a glide-symmetric geometry. Phys. Rev. A **77**, 062107 (2008)
85. Rodriguez-Lopez, P., Rahi, S.J., Emig, T.: Three-body Casimir effects and nonmonotonic forces. Phys. Rev. A **80**, 022519 (2009)
86. Rosa, F.S.S.: On the possibility of Casimir repulsion using metamaterials. J. Phys.: Conf. Ser. **161**, 012039 (2009)
87. Rosa, F.S.S., Dalvit, D.A.R., Milonni, P.W.: Casimir-Lifshitz theory and metamaterials. Phys. Rev. Lett. **100**, 183602 (2008)
88. Roy, A., Lin, C.Y., Mohideen, U.: Improved precision measurement of the Casimir force. Phys. Rev. D **60**, 111101(R) (1999)
89. Schaden, M., Spruch, L.: Infinity-free semiclassical evaluation of Casimir effects. Phys. Rev. A **58**, 935–953 (1998)
90. Schwinger, J.: Casimir effect in source theory. Lett. Math. Phys. **1**, 43–47 (1975)
91. Ttira, C.C., Fosco, C.D., Losada, E.L.: Non-superposition effects in the Dirichlet–Casimir effect. J. Phys. A: Math. Theor. **43**, 235402 (2010)

92. Weber, A., Gies, H.: Interplay between geometry and temperature for inclined Casimir plates. Phys. Rev. D **80**, 065033 (2009)
93. Wirzba, A.: Quantum mechanics and semiclassics of hyperbolic n-disk scattering systems. Phys. Rep. **309**, 1–116 (1999)
94. Wirzba, A.: The Casimir effect as a scattering problem. J. Phys. A: Math. Theor. **41**, 164003 (2008)
95. Zaheer, S., Rahi, S.J., Emig, T., Jaffe, R.L.: Casimir interactions of an object inside a spherical metal shell. Phys. Rev. A **81**, 030502 (2010)
96. Zhao, R., Zhou, J., Koschny, T., Economou, E.N., Soukoulis, C.M.: Repulsive Casimir force in chiral metamaterials. Phys. Rev. Lett. **103**, 103602 (2009)

Chapter 6
Numerical Methods for Computing Casimir Interactions

Steven G. Johnson

Abstract We review several different approaches for computing Casimir forces and related fluctuation-induced interactions between bodies of arbitrary shapes and materials. The relationships between this problem and well known computational techniques from classical electromagnetism are emphasized. We also review the basic principles of standard computational methods, categorizing them according to three criteria—choice of problem, basis, and solution technique—that can be used to classify proposals for the Casimir problem as well. In this way, mature classical methods can be exploited to model Casimir physics, with a few important modifications.

6.1 Introduction

Thanks to the ubiquity of powerful, general-purpose computers, large-scale numerical calculations have become an important part of every field of science and engineering, enabling quantitative predictions, analysis, and design of ever more complex systems. There are a wide variety of different approaches to such calculations, and there is no single "best" method for all circumstances—not only are some methods better suited to particular situations than to others, but there are also often severe trade-offs between generality/simplicity and theoretical efficiency. Even in relatively mature areas like computational classical electromagnetism (EM), a variety of techniques spanning a broad range of sophistication and

S. G. Johnson (✉)
Department of Mathematics, Massachusetts Institute of Technology, Cambridge, MA 02139, USA
e-mail: stevenj@math.mit.edu

D. Dalvit et al. (eds.), *Casimir Physics*, Lecture Notes in Physics 834,
DOI: 10.1007/978-3-642-20288-9_6, © Springer-Verlag Berlin Heidelberg 2011

generality remain in widespread use (and new variations are continually developed) [1–8]. Semi-analytical approaches also remain important, especially perturbative techniques to decompose problems containing widely differing length scales (the most challenging situation for brute-force numerics). Nevertheless, many commonalities and guiding principles can be identified that seem to apply to a range of numerical techniques.

Until a few years ago, Casimir forces and other EM fluctuation-induced interactions occupied an unusual position in this tableau. Realistic, general numerical methods to solve for Casimir forces were simply unavailable; solutions were limited to special high-symmetry geometries (and often to special materials like perfect metals) that are amenable to analytical and semi-analytical approaches. This is not to say that there were not, *in principle*, decades-old theoretical frameworks capable of describing fluctuations for arbitrary geometries and materials, but practical techniques for *evaluating* these theoretical descriptions on a computer have only been demonstrated in the last few years [9–27]. In almost all cases, these approaches turn out to be closely related to computational methods from *classical* EM, which is fortunate because it means that Casimir computations can exploit decades of progress in computational classical EM once the relationship between the problems becomes clear. The long delay in developing numerical methods for Casimir interactions, from the time the phenomenon was first proposed in 1948 [28], can be explained by three factors. First, accurate measurements of Casimir forces were first reported only in 1997 [29] and experimental interest in complex Casimir geometries and materials has only recently experienced dramatic growth due to the progress in fabricating nanoscale mechanical devices. Second, even the simplest numerical prediction of a single force requires the equivalent of a large number of classical EM simulations, a barrier to casual numerical experimentation. Third, there have historically been many equivalent theoretical formulations of Casimir forces, but some formulations are much more amenable to computational solution than others, and these formulations are often couched in a language that is opaque to researchers from classical computational EM.

The purpose of this review is to survey the available and proposed numerical techniques for evaluating Casimir forces, energies, torques, and related interactions, emphasizing their relationships to standard classical-EM methods. Our goal is not to identify a "best" method, but rather to illuminate the strengths and weaknesses of each approach, highlighting the conclusions that can be gleaned from the classical experience. We will review an intellectual framework in which to evaluate different numerical techniques, comparing them along several axes for which quasi-independent choices of approach can be made. We will also emphasize a few key departures of Casimir problems from ordinary classical EM, such as the necessity of imaginary- or complex-frequency solutions of Maxwell's equations and the need for wide-bandwidth analyses, that impact the adaptation of off-the-shelf computational methods.

6.2 Characterization of Numerical Methods: Three Axes

Numerical methods from distinct groups or research papers often differ in several ways simultaneously, complicating the task of directly comparing or even describing them. In order to organize one's understanding of numerical approaches, it is useful to break them down along three axes of comparison, representing (roughly) independent choices in the design of a method:

- What **problem** does the method solve—even within a single area such as classical EM, there are several conceptually different *questions* that one can ask and several *ways of asking* them that lead to different categories of methods.
- What **basis** is used to express the unknowns—how the *infinite* number of unknowns in the exact partial differential equation (PDE) or integral equation are reduced to a *finite* number of unknowns for solution on a computer.
- What **solution technique** is used to determine these unknowns—even with the same equations and the same unknowns, there are vast differences among the types of direct, sparse, and iterative methods that can be used to attack the problem, and the efficient application of a particular solution technique to a particular problem is sometimes a research task unto itself.

In this section, we briefly summarize the available problems, basis choices, and solution techniques for Casimir problems. In subsequent sections, we then discuss in more detail the specific approaches that have currently been demonstrated or proposed.

6.2.1 Posing Casimir Problems

In classical EM, there are several types of problems that are typically posed [6, Appendix D], such as computing source-free time-harmonic eigensolutions $\mathbf{E}, \mathbf{H} \sim e^{-i\omega t}$ and eigenfrequencies ω, computing time-harmonic fields resulting from a time-harmonic current source $\mathbf{J} \sim e^{-i\omega t}$, or computing the time-dependent fields created by an arbitrary time-dependent source $\mathbf{J}(t)$ starting at $t = 0$. Although these are all closely mathematically related, and in some sense the solution of one problem can give solutions to the other problems, they lead to very different types of numerical simulations.

In a similar way, despite the fact that different formulations of Casimir-interaction problems are ultimately mathematically equivalent (although the equivalencies are often far from obvious)—and are usually answering the same conceptual question, such as what is the force or interaction energy for some geometry—each one leads most naturally to distinct classes of computational methods. Here, we exclude formulations such as proximity-force ("parallel-plate") approximations [30–32], pairwise summation of Casimir–Polder forces (valid in the dilute-gas limit) [33–35], and ray optics [36–39], that are useful in special cases but represent uncontrolled approximations if they are applied to arbitrary geometries. Although at some point the distinctions are blurred by the mathematical equivalencies, we can crudely categorize the approaches as:

- Computing the eigenfrequencies ω_n and summing the zero-point energy $\sum_n \frac{\hbar\omega_n}{2}$ [28, 40]. See Sect. 6.3.
- Integrating the mean energy density or force density (stress tensor), by evaluating field correlation functions $\langle E_i E_j \rangle_\omega$ and $\langle H_i H_j \rangle_\omega$ in terms of the classical EM Green's functions at ω via the fluctuation–dissipation theorem [11–13, 19, 20, 23, 27]. See Sect. 6.5.
- Evaluating a path-integral expression for the interaction energy (or its derivative), constrained by the boundary conditions—usually, portions of the path integrals are performed analytically to express the problem in terms of classical scattering matrices or Green's functions at each ω [9, 10, 14–18, 21, 22, 24–26]. See Sect. 6.6.

In each case, the result must be summed/integrated over all frequencies ω to obtain the physical result (corresponding to thermodynamic/quantum fluctuations at all frequencies). The relationship of the problem to *causal* Green's functions (fields appear *after* currents) means that the integrand is analytic for $\mathrm{Im}\,\omega \geq 0$ [41]. As a consequence, there is a choice of *contours* of ω integration in the upper-half complex plane, which is surprisingly important—it turns out that the integrands are wildly oscillatory on the real-ω axis and require accurate integration over a huge bandwidth, whereas the integrands are much better-behaved along the imaginary-ω axis ("Wick-rotated" or "Matsubara" frequencies). This means that Casimir calculations almost always involve classical EM problems evaluated at *complex or imaginary frequencies*, as is discussed further in Sect. 6.4. The nonzero-temperature case, where the integral over imaginary frequencies becomes a sum (numerically equivalent to a trapezoidal-rule approximation), is discussed in Sect. 6.8.

There is also another way to categorize the problem to be solved: whether one is solving a partial differential equation (**PDE**) or an **integral equation**. In a PDE, one has *volumetric unknowns*: fields or other functions at every point in space, related to one another *locally* by derivatives and so on. In an integral equation, one typically has *surface unknowns*: the fields or currents on the *boundaries* between piecewise-homogeneous regions, related to one another *non-locally* by the Green's functions of the homogeneous regions (typically known analytically) [1, 3] (described further in Sect. 6.5.3). The key point is to take advantage of the common situation in which one has piecewise-constant materials, yielding a *surface integral equation*. (There are also *volume integral equations* for inhomogeneous media [42], as well as hybrid integral/PDE approaches [1], but these are less common.) There are other hybrid approaches such as *eigenmode expansion* [43–45], also called *rigorous coupled-wave analysis* (RCWA) [46, 47] or a *cross-section* method [48]: a structure is broken up along one direction into piecewise-constant cross-sections, and the unknown fields at the *interfaces* between cross-sections are propagated in the uniform sections via the *eigenmodes* of those cross-sections (computed analytically or numerically by solving the PDE in the cross-section). Eigenmode expansion is most advantageous for geometries in which the cross-section is constant over substantial regions, just as integral-equation methods are most advantageous to exploit large homogeneous regions.

6.2.2 Choices of Basis

Casimir problems, for the most part, reduce to solving classical EM linear PDEs or integral equations where the unknowns reside in an infinite-dimensional vector space of functions. To *discretize* the problem approximately into a finite number N of unknowns, these unknown functions must be expanded in some finite *basis* (that converges to the exact solution as $N \to \infty$). There are three typical types of basis:

- **Finite differences** [2, 49, 50] (**FD**): approximate a function $f(x)$ by its values on some uniform grid with spacing Δx, approximate derivatives by some difference expression [e.g. second-order center differences $f'(x) \approx \frac{f(x+\Delta x)-f(x-\Delta x)}{2\Delta x} + O(\Delta x^2)$] and integrals by summations (e.g. a trapezoidal rule).
- **Finite-element methods** [1, 3, 4, 7] (**FEM**): divide space into geometric *elements* (e.g. triangles/tetrahedra), and expand an unknown $f(x)$ in a simple *localized* basis expansion for each element (typically, low-degree polynomials) with some continuity constraints. (FD methods are viewable as special cases of FEMs for uniform grids.) For an integral-equation approach, where the unknowns are functions on surfaces, the same idea is typically called a **boundary-element method** (**BEM**) [1, 3, 7, 51, 52].[1]
- **Spectral** methods [53]: expand functions in a non-localized complete basis, truncated to a finite number of terms. Most commonly, Fourier series or related expansions are used (cosine series, Fourier–Bessel series, spherical or spheroidal harmonics, Chebyshev polynomials, etc.).

Finite differences have the advantage of simplicity of implementation and analysis, and the disadvantages of uniform spatial resolution and relatively low-order convergence (errors typically $\sim \Delta x^2$ [2] or even $\sim \Delta x$ in the presence of discontinuous materials unless special techniques are used [54, 55]). FEMs can have nonuniform spatial resolution to resolve disparate feature sizes in the same problem, at a price of much greater complexity of implementation and solution techniques, and can have high-order convergence at the price of using complicated curved elements and high-order basis functions. Spectral methods can have very high-order or possibly exponential ("spectral") convergence rates [53] that can

---

[1] The name *method of moments* is also commonly applied to BEM techniques for EM. However, this terminology is somewhat ambiguous, and can refer more generally to Galerkin or other weighted-residual methods (and historically referred to monomial test functions, yielding statistical "moments") [53].

even suit them to analytical solution—hence, spectral methods were the dominant technique before the computer era and are typically the first class of methods that appear in any field, such as in Mie's classic solution of wave scattering from a sphere [56]. However, exponential convergence is usually obtained only if all discontinuities and singularities are taken explicitly into account in the basis [53]. With discontinuous materials, this is typically only practical for very smooth, high-symmetry geometries like spheres, cylinders, and so on; the use of a generic Fourier/spectral basis for arbitrary geometries reduces to a brute-force method that is sometimes very convenient [57], but may have unremarkable convergence rates [53, 57, 58]. BEMs require the most complicated implementation techniques, because any nontrivial change to the Green's functions of the homogeneous regions (e.g. a change in dimensionality, boundary conditions, or material types) involves tricky changes to the singular-integration methods required to assemble the matrix [59–61] and to the fast-solver methods mentioned in Sect. 6.2.3.

Given FEM/BEM or spectral basis functions $b_n(x)$ and a linear equation $\hat{A}u(x) = v(x)$ for an unknown function u in terms of a linear differential/integral operator $\hat{A}$, there are two common ways [53] to obtain a finite set of N equations to determine the N unknown coefficients c_n in $u(x) \approx \sum_n c_n b_n(x)$. One is a **collocation** method: require that $(\hat{A}u - v)|_{x_n} = 0$ be satisfied at N collocation points x_n. The other is a **Galerkin** method: require that $\langle b_k, \hat{A}u - v \rangle = 0$ be satisfied for $k = 1, \ldots, N$, where $\langle \cdot, \cdot \rangle$ is some inner product on the function space. Both approaches result in an $N \times N$ matrix equation of the form $A\mathbf{u} = \mathbf{v}$. A Galerkin method has the useful property that if $\hat{A}$ is Hermitian and/or definite then the matrix $A_{kn} = \langle b_k, \hat{A}b_n \rangle$ has the same properties.

The specific situation of vector-valued unknowns in EM creates additional considerations for the basis functions. In order to obtain center-difference approximations for all the field components, FD methods for EM typically use a staggered **Yee grid** [2, 49], in which each component of the EM fields is offset onto its own $\frac{\Delta x}{2}$-shifted grid. In FEMs for EM, in order to maintain the appropriate continuity conditions for curl or divergence operators, one uses special classes of *vector-valued* basis functions such as Nédélec elements [7, 62]. In BEMs for EM, vector-valued **RWG** (Rao, Wilton, and Glisson) basis functions [63] (or generalizations thereof [64]) are used in order to enforce a physical continuity condition on surface currents (to preclude accumulation of charge at element edges); see also Fig. 6.3 in Sect. 6.5.3. A spectral integral-equation method for EM with cylindrical or spherical scatterers is sometimes called a **multipole-expansion** method [5], since the obvious spectral basis is equivalent to expanding the scattered fields in terms of multipole moments.

6.2.3 Solution Techniques for Linear Equations

Given a particular problem and basis choice, one at the end obtains some $N \times N$ set of linear equations $A\mathbf{x} = \mathbf{b}$ to solve (or possibly eigenequations

$A\mathbf{x} = \lambda B\mathbf{x}$).[2] Note also that a *single* Casimir-force calculation requires the solution of *many* such equations, at the very least for an integral over frequencies (see Sect. 6.4). There are essentially three ways to solve such a set of equations:

- **Dense-direct** solvers: solve $A\mathbf{x} = \mathbf{b}$ using direct matrix-factorization methods (e.g. Gaussian elimination),[3] requiring $O(N^2)$ storage and $O(N^3)$ time [65].
- **Sparse-direct** solvers [66]: if A is *sparse* (mostly zero entries), use similar direct matrix-factorization methods, but cleverly re-arranged in an attempt to preserve the sparsity. Time and storage depend strongly on the *sparsity pattern* of A (the pattern of nonzero entries).
- **Iterative methods** [65, 67, 68]: repeatedly improve a guess for the solution $\mathbf{x}$ (usually starting with a random or zero guess), only referencing A via repeated matrix–vector multiplies. Time depends strongly on the properties of A and the iterative technique, but typically requires only $O(N)$ storage. Exploits any fast way [ideally $O(N)$ or $O(N \log N)$] to multiply A by any arbitrary vector.

If the number N of degrees of freedom is small, i.e. if the basis converges rapidly for a given geometry, dense-direct methods are simple, quick, and headache-free (and have a standard state-of-the-art implementation in the free LAPACK library [69]). For example, $N = 1000$ problems can be solved in under a second on any modern computer with a few megabytes of memory. Up to $N \sim 10^4$ is reasonably feasible, but $N = 10^5$ requires almost 100 GB of memory and days of computation time without a large parallel computer. This makes dense-direct solvers the method of choice in simple geometries with a rapidly converging spectral basis, or with BEM integral-equation methods for basic shapes that can be accurately described by a few thousand triangular panels, but they rapidly become impractical for larger problems involving many and/or complex objects (or for moderate-size PDE problems even in two dimensions).

In PDE methods with a localized (FD or FEM) basis, the matrices A have a special property: they are *sparse* (mostly zero). The locality of the operators in a typical PDE means that each grid point or element directly interacts only with a bounded number of neighbors, in which case A has only $O(N)$ nonzero entries and can be stored with $O(N)$ memory. The process of solving $A\mathbf{x} = \mathbf{b}$, e.g. computing the LU factorization $A = LU$ by Gaussian elimination [65], unfortunately, ordinarily destroys this sparsity: the resulting L and U triangular matrices are generally not sparse. However, the pattern of nonzero entries that arises from a PDE is not random, and it turns out that clever re-orderings of the rows and columns during factorization can partially preserve sparsity for typical patterns; this insight leads

---

[2] This applies equally well, if somewhat indirectly, to the path-integral expressions of Sect. 6.6 where one evaluates a log determinant or a trace of an inverse, since this is done using either eigenvalues or the same matrix factorizations that are used to solve $A\mathbf{x} = \mathbf{b}$.

[3] Technically, all eigensolvers for $N > 4$ are necessarily iterative, but modern dense-eigensolver techniques employ direct factorizations as steps of the process [65].

to *sparse-direct* solvers [66], available via many free-software packages implementing different sparsity-preserving heuristics and other variations [68]. The sparsity pattern of A depends on the dimensionality of the problem, which determines the number of neighbors a given element interacts with. For meshes/grids having nearest-neighbor interactions, a sparse-direct solver typically requires $O(N)$ time and storage in 1d (where the matrices are band-diagonal), $O(N^{3/2})$ time with $O(N \log N)$ storage in 2d, and $O(N^2)$ time with $O(N^{4/3})$ storage in 3d [66, 70]. The practical upshot is that sparse-direct methods work well for 1d and 2d PDEs, but can grow to be impractical in 3d. For BEM and spectral methods, the interactions are not localized and the matrices are not sparse, so sparse-direct methods are not directly applicable (but see below for an indirect technique).

For the largest-scale problems, or for problems lacking a sparse A, the remaining possibility is an *iterative* method. In these methods, one need only supply a fast way to multiply A by an arbitrary vector y, and the trick is to use this Ay operation on a clever sequence of vectors in such a way as to make an arbitrary initial guess x_0 converge as rapidly as possible to the solution x, ideally using only $O(N)$ storage. Many such techniques have been developed [65, 67, 68]. The most favorable situation for $Ax = b$ occurs when A is Hermitian positive-definite, in which case an ideal Krylov method called the *conjugate-gradient* method can be applied, with excellent guaranteed convergence properties [65, 67], and fortunately this is precisely the case that usually arises for the imaginary-frequency Casimir methods below. There are two wrinkles that require special attention, however. First, one must have a fast way to compute Ay. If A is sparse (as for PDE and FD methods), then only $O(N)$ nonzero entries of A need be stored (as above) and Ay can be computed in $O(N)$ operations. In a spectral method, A is generally dense, but for spectral PDE methods there are often fast $O(N \log N)$ techniques to compute Ay using only $O(N)$ storage (A is stored implicitly), based on fast Fourier transform (FFT) algorithms [53, 57]. In a BEM, where A is again dense, a variety of sophisticated methods that require only $O(N \log N)$ computation time and $O(N)$ storage to compute Ay (again storing A implicitly) have been developed [1, 3, 7, 71], beginning with the pioneering fast-multipole method (FMM) [72]. These fast BEMs exploit the localized basis and the decaying, convolutional nature of the Green's function to approximate long-range interactions (to any desired accuracy). FMMs can be viewed as an approximate factorizations into sparse matrices, at which point sparse-direct methods are also applicable [73]. A second wrinkle is that the convergence rates of iterative methods depend on the condition number of A (the ratio of largest and smallest singular values) [65, 67], and condition numbers generally worsen as the ratio of the largest and smallest lengthscales in the problem increases. To combat this, users of iterative methods employ **preconditioning** techniques: instead of solving $Ax = b$, one solves $KAx = Kb$ or similar, where the preconditioning matrix K is some crude *approximate inverse* for A (but much simpler to compute than A^{-1}!) such that the condition number of KA is reduced [67]. The difficulty with this approach is that good preconditioners tend to be highly problem-dependent, although a variety of useful approaches such as incomplete factorization and coarse-grid/multigrid approximations have been identified [65, 67]. The upshot is that, while the largest-scale solvers almost invariably use iterative techniques, for any given class of physical problems it sometimes takes significant research before the iterative approach becomes well-optimized.

6.3 The Impracticality of Eigenmode Summations

Perhaps the simplest way to express the Casimir energy, at zero temperature, is as a sum of zero-point energies of all oscillating EM modes in the system:

$$U = \sum_{\omega} \frac{\hbar \omega}{2}, \tag{6.1}$$

where ω is the frequency of the mode ($\sim e^{-i\omega t}$) [28, 74]. That is, when the electromagnetic field is quantized into photons with energy $\hbar \omega$, it turns out that the vacuum state in the absence of photons is not empty, but rather has the energy equivalent of "half a photon" in each mode. The computational strategy is then straightforward, in principle: compute the EM eigenfrequencies ω in the problem by some numerical method (many techniques are available for computing eigenfrequencies [1, 57]) and sum them to obtain U. Forces are then given by the derivative of U with respect to changes in the geometry, which could be approximated by finite differences or differentiated analytically with a Hellman–Feynman technique [75] (more generally, derivatives of any computed quantity can be computed efficiently by an adjoint method [76]).

Of course, U in (6.1) has the disadvantage of being formally infinite, but this is actually a minor problem in practice: as soon as one discretizes the problem into a finite number of degrees of freedom (e.g., a finite number of grid points), the number of eigenfrequencies becomes finite (with the upper bound representing a Nyquist-like frequency of the grid). This is the numerical analogue [12] of analytical regularization techniques that are applied to truncate the same sum in analytical computations [28]. (These regularizations do not affect energy *differences* or forces for rigid-body motions.) Matters are also somewhat subtle for dissipative or open systems [77]. But the most serious problem is that, even in the lossless case, this sum is badly behaved: even when one differentiates with separation a to obtain a finite force $F = -\frac{\hbar}{2} \sum \frac{d\omega}{da}$, the summand is wildly oscillatory and includes substantial contributions from essentially *every* frequency, which mostly cancel to leave a tiny result [12, 78]. Numerically, therefore, one must ostensibly compute *all* of the modes, to high precision, which requires $O(N^3)$ time and $O(N^2)$ storage (for a dense-direct eigensolver [65]) given N degrees of freedom. This is possible in simple 1d problems [12, 40], but is impractical as a general approach.

Because of the mathematical equivalence of the different approaches to the Casimir problem, the mode-summation method is sometimes useful as a starting point to derive alternative formulations, but the end result is invariably quite different in spirit from computing the eigenfrequencies one by one and summing them. For example, if one has a function $z(\omega)$ whose roots are the eigenfrequencies, then one can equivalently write U, via the residue theorem of complex analysis, as $U = \frac{1}{2\pi i} \oint_C \frac{\hbar \omega}{2} \frac{d[\ln z(\omega)]}{d\omega} d\omega$, where C is any closed contour in the complex-ω plane that encloses the roots [79]. However, finding functions whose roots are the eigenfrequencies naturally points towards Green's functions (to relate different boundary conditions), and the contour choices typically involve Wick rotation as in Sect. 6.4, so this approach leads directly to imaginary-frequency scattering-matrix techniques as in Sect. 6.6 [18]. A similar contour integral arises from a zeta-function regularization of (6.1) [80].

6.4 The Complex-Frequency Plane and Contour Choices

In order to better understand the frequency integration/summation in Casimir problems, it is illustrative to examine the analytical formula for the simple case of two perfect-metal plates in vacuum separated by a distance a, in which case it can be derived in a variety of ways that the attractive force F is given by [81]:

$$
F = \frac{\hbar}{\pi^2 c^3} \mathrm{Re} \left[\int_0^\infty d\omega \int_1^\infty dp \frac{p^2 \omega^3}{e^{2ip(\omega + i0^+)a/c} - 1} \right]
$$

$$
= \mathrm{Re} \left[\int_0^\infty f(\omega) d\omega \right] = \mathrm{Im} \left[\int_0^\infty f(i\xi) d\xi \right] = \frac{\hbar c}{240 a^4}, \tag{6.2}
$$

where $f(\omega)$ is the contribution of each frequency ω to the force and p is related to the plate-parallel momentum of the contributing modes/fluctuations. In this special case, the entire integral can be performed analytically, but for parallel plates of some finite permittivity ε the generalization (the *Lifshitz formula* [81]) must be integrated numerically. In practice, however, the formula and its generalizations are never integrated in the form at left—instead, one uses the technique of contour integration from complex analysis to *Wick rotate* the integral to imaginary frequencies $\omega = i\xi$, integrating over ξ. (In fact, the formula is typically derived *starting* in imaginary frequencies, via a Matsubara approach [81].) In this section, we review why a trick of this sort is both *possible* and *essential* in numerical computations for all of the methods described below.

Wick rotation is always *possible* as a consequence of causality. It turns out that the frequency contributions $f(\omega)$ for arbitrary materials and geometries, for all of the different formulations of the Casimir force below, are ultimately expressed in terms of classical EM *Green's functions* at ω: the EM fields in response to time-harmonic currents $\mathbf{J} \sim e^{-i\omega t}$. As a consequence of the causality of Maxwell's equations and physical materials—EM fields always arise *after* the source currents, not before—it mathematically follows that the Green's functions must be analytic

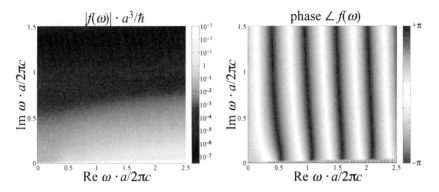

Fig. 6.1 Contributions $f(\omega)$ to the Casimir force, from each fluctuation/mode frequency ω, for two perfect-metal plates with separation a, in the complex-ω plane. *Left*: magnitude $|f(\omega)|$. *Right*: phase $\angle f(\omega)$. [The magnitude is truncated at $10^3 \hbar/a^3$, as it diverges towards the real-ω axis, and some numerical artifacts (rapid oscillations) are visible near the real-ω axis in the phase due to difficulty in evaluating $f(\omega)$.] The key point is that $f(\omega)$ is badly behaved (oscillatory and non-decaying) along contours parallel to the real-ω axis, whereas $f(\omega)$ is nicely behaved (non-oscillatory and exponentially decaying) along contours parallel to the imaginary-ω axis

functions (no poles or other singularities) when $\text{Im } \omega > 0$ (the upper-half complex plane) [41]. Poles in the Green's function correspond to eigenfrequencies or resonances of the source-free Maxwell's equations, and must lie at $\text{Im } \omega \leq 0$ for any physical system with dissipative materials (with the poles approaching $\text{Im } \omega = 0^-$ in the idealized lossless limit). [One can easily see explicitly that this is true for the $f(\omega)$ above: the poles result from a vanishing denominator in the p integrand, which only occurs for purely real ω corresponding to the real-frequency modes trapped between two perfect-metal plates.] As an elementary consequence of complex analysis, this analyticity means that the $\int d\omega$ can be arbitrarily deformed to any contour in the upper-half complex-ω plane without changing the integration result.

Wick rotation is *essential* for computation because the frequency contributions $f(\omega)$ to the force (or interaction energy or other related quantities) are extremely ill-behaved near to the real-ω axis: they are wildly oscillatory and slowly decaying. For example, the magnitude and phase of the function $f(\omega)$ are plotted in the complex ω plane in Fig. 6.1, where the p integral was evaluated numerically with a high-order Clenshaw–Curtis quadrature scheme [82]. Merely evaluating $f(\omega)$ along the real-ω axis is difficult because of singularities (which ultimately reduce the integral to a summation over eigenfrequency-contributions as in Sect. 6.3); in physical materials with dissipation, the real-ω axis is non-singular but is still badly behaved because of poles (lossy modes) located just below the axis. Along any contour parallel to the real-ω axis, the integrand is oscillatory (as can be seen from the phase plot) and non-decaying (as can be seen from the magnitude plot): formally, just as with the infinite summation over eigenmodes in Sect. 6.3, one must integrate over an infinite bandwidth, regularized in some way (e.g. by the Nyquist frequency placing an upper bound on ω for a finite grid), where the oscillations almost entirely cancel to leave a tiny remainder (the force).

(Any physical materials must cease to polarize as $\omega \to \infty$ where the susceptibility vanishes [41], which will make the force contributions eventually vanish as $\omega \to \infty$ even in 1d, but a very wide-bandwidth oscillatory integral is still required.) This is a disaster for any numerical method—even when one is only integrating an analytical expression such as the Lifshitz formula, mere roundoff errors are a severe difficulty for real ω. Along the imaginary-ω axis, on the other hand (or any sufficiently vertical contour), $f(\omega)$ is exponentially decaying and mostly non-oscillatory—an ideal situation for numerical integration.

Therefore, in order for classical EM solvers to be used for Casimir problems, they need to be adapted to solve Maxwell's equations at complex or imaginary ω. Although this sounds strange at first, the frequency-domain problem actually becomes numerically *easier* in every way at imaginary ω; this is discussed in more detail in Sect. 6.5.1.2. In fact, one can even identify an exact mathematical equivalence between a particular complex-ω contour and a *real*-frequency system where an *artificial dissipation* has been introduced, as discussed in Sect. 6.5.5 below—using this trick, one can actually use classical EM solvers with no modification at all, as long as they handle dissipative media. In any case, one needs an integral over frequencies to compute a physically meaningful quantity, which means that solvers and material models, not to mention any physical intuition used for guidance, must be valid for more than just a narrow real-ω bandwidth (unlike most problems in classical EM).

Numerically, it should be pointed out that the $f(i\xi)$ integrand is smooth and exponentially decaying, and so the ξ integral can be approximated to high accuracy by an exponentially convergent quadrature (numerical integration) scheme using evaluations at relatively few points ξ. For example, one can use Gauss–Laguerre quadrature [83], Gaussian quadrature with an appropriate change of variables [84], or Clenshaw–Curtis quadrature with an appropriate change of variables [82].

6.5 Mean Energy/Force Densities and the Fluctuation–Dissipation Theorem

Another, equivalent, viewpoint on Casimir interactions is that they arise from geometry-dependent fluctuations of the electromagnetic fields $\mathbf{E}$ and $\mathbf{H}$, which on average have some nonzero energy density and exert a force. If we can compute these average fields, we can integrate the resulting energy density, stress tensors, and so on, to obtain the energy, force, or other quantities of interest. The good news is that there is a simple expression for those fluctuations in terms of the **fluctuation–dissipation** theorem of statistical physics: the correlation function of the fields is related to the corresponding *classical* Green's function [81]. Ultimately, this means that any standard classical EM technique to compute Green's functions (fields from currents) can be applied to compute Casimir forces, with the caveat that the techniques must be slightly modified to work at imaginary or complex frequencies as described below.

6.5.1 Background

The temperature-T correlation function for the fluctuating electric field at a given frequency ω is given by [81]:

$$\langle E_j(\mathbf{x})E_k(\mathbf{x}')\rangle_\omega = -\frac{\hbar}{\pi}\mathrm{Im}\left[\omega^2 G_{jk}^E(\omega;\mathbf{x},\mathbf{x}')\right]\coth(\hbar\omega/2k_BT), \qquad (6.3)$$

where $G_{jk}^E = (\mathbf{G}_k^E)_j$ is the classical dyadic "photon" Green's function, proportional[4] to the relationship between an electric-dipole current in the k direction at $\mathbf{x}'$ to the electric field at $\mathbf{x}$, and solves

$$\left[\nabla\times\mu(\omega,\mathbf{x})^{-1}\nabla\times-\omega^2\varepsilon(\omega,\mathbf{x})\right]\mathbf{G}_k^E(\omega,\mathbf{x},\mathbf{x}') = \delta^3(\mathbf{x}-\mathbf{x}')\hat{e}_k, \qquad (6.4)$$

where ε is the electric permittivity tensor, μ is the magnetic permeability tensor, and $\hat{e}_k$ is a unit vector in direction k. Similarly, the magnetic-field correlation function is

$$\langle H_j(\mathbf{x})H_k(\mathbf{x}')\rangle_\omega = -\frac{\hbar}{\pi}\mathrm{Im}\left[\omega^2 G_{jk}^H(\omega;\mathbf{x},\mathbf{x}')\right]\coth(\hbar\omega/2k_BT). \qquad (6.5)$$

The magnetic Green's function $\mathbf{G}^H$ can be defined in two essentially equivalent ways. The first is as derivatives $\frac{1}{\omega^2\mu(\mathbf{x})}\nabla\times\mathbf{G}^E\times\nabla'\frac{1}{\mu(\mathbf{x}')}$ of the electric Green's function $G_{jk}^E(\mathbf{x},\mathbf{x}')$, where ∇ and ∇' denote derivatives with respect to $\mathbf{x}$ and $\mathbf{x}'(\nabla'$ acting to the left), respectively [81]. The second way to define $\mathbf{G}^H$ is proportional to the magnetic field in response to a magnetic-dipole current, analogous to (6.4):

$$\left[\nabla\times\varepsilon(\omega,\mathbf{x})^{-1}\nabla\times-\omega^2\mu(\omega,\mathbf{x})\right]\mathbf{G}_k^H(\omega,\mathbf{x},\mathbf{x}') = \delta^3(\mathbf{x}-\mathbf{x}')\hat{e}_k, \qquad (6.6)$$

which can be more convenient for numerical calculation [13]. These two definitions are related [85] by $\mathbf{G}^H = \frac{1}{\omega^2\mu(\mathbf{x})}\nabla\times\mathbf{G}^E\times\nabla'\frac{1}{\mu(\mathbf{x}')} - \frac{1}{\omega^2\mu(\mathbf{x}')}\delta(\mathbf{x}-\mathbf{x}')I$ (with I being the 3×3 identity matrix),[5] where the second (diagonal) term has no effect on energy differences or forces below and is therefore irrelevant. Now, these equations are rather nasty along the real-ω axis: not only will there be poles in $\mathbf{G}$ just below the axis corresponding to lossy modes, but in the limit where the dissipative losses vanish (ε and μ become real), the combination of the poles

---

[4] The electric field $\mathbf{E}(\mathbf{x})$ from a dipole current $\mathbf{J} = \delta^3(\mathbf{x}-\mathbf{x}')\hat{e}_k e^{-i\omega t}$ is $\mathbf{E}(\mathbf{x}) = i\omega G_k^E$ $(\omega,\mathbf{x},\mathbf{x}')e^{-i\omega t}$.

[5] This can be seen more explicity by substituting $\mathbf{G}^H = \frac{1}{\omega^2}\frac{1}{\mu}\nabla\times\mathbf{G}^E\times\nabla'\frac{1}{\mu'} - \frac{1}{\omega^2\mu}\delta$ into (6.6), with δ denoting $\delta(\mathbf{x}-\mathbf{x}')I$ and μ or μ' denoting $\mu(\mathbf{x})$ or $\mu(\mathbf{x}')$, respectively. In particular, $[\nabla\times\frac{1}{\varepsilon}\nabla\times-\omega^2\mu](\frac{1}{\omega^2}\frac{1}{\mu}\nabla\times\mathbf{G}^E\times\nabla'\frac{1}{\mu'} - \frac{1}{\omega^2\mu}\delta)$ yields $\nabla\times[\frac{1}{\omega^2\varepsilon}\nabla\times\frac{1}{\mu}\nabla\times\mathbf{G}^E - \mathbf{G}^E]\times$ $\nabla'\frac{1}{\mu'} - \nabla\times\frac{1}{\omega^2\mu'\varepsilon}\nabla\times\delta + \delta$, which via (6.4) gives $+\nabla\times\frac{1}{\omega^2\varepsilon}\delta\times\nabla'\frac{1}{\mu'} - \nabla\times\frac{1}{\omega^2\mu'\varepsilon}\nabla\times\delta + \delta = \delta$ as desired, where in the last step we have used the fact that $\delta\times\nabla' = \nabla\times\delta$ [since $\nabla\times$ is antisymmetric under transposition and $\nabla'\delta(\mathbf{x}-\mathbf{x}') = -\nabla\delta(\mathbf{x}-\mathbf{x}')$].

approaching the real axis with the Im in the correlation function results in a delta function at each pole[6] and integrals of the correlation functions turn into sums over modes as in Sect. 6.3 However, the saving grace, as pointed out in Sect. 6.4, is that Green's functions are *causal*, allowing us to transform any integral over all real fluctuation frequencies into an integral over *imaginary* fluctuation frequencies $\omega = i\xi$. The coth factor has poles that alter this picture, but we will eliminate those for now by considering only the $T = 0^+$ case where $\coth(+\infty) = 1$, returning to nonzero temperatures in Sect. 6.8.

6.5.1.1 Energy Density

In particular, to compute the Casimir energy U, we merely integrate the classical energy density in the EM field [41] over all positions and all fluctuation frequencies, Wick-rotated to an integral over imaginary frequencies, resulting in the expression:

$$U = \int\limits_{0}^{\infty} d\xi \int \frac{1}{2}\left[\frac{d(\xi\varepsilon)}{d\xi}\langle|\mathbf{E}|^2\rangle_{i\xi} + \frac{d(\xi\mu)}{d\xi}\langle|\mathbf{H}|^2\rangle_{i\xi}\right]d^3\mathbf{x}, \tag{6.7}$$

where we have simplified to the case of isotropic materials (scalar ε and μ). At thermodynamic equilibrium, this expression remains valid even for arbitrary dissipative/dispersive media thanks to a direct equivalence with a path-integral expression [87], which is not obvious from the classical viewpoint in which the energy density is usually only derived in the approximation of negligible absorption [41]. (Thanks to the relationship between the Green's function and the local density of states [88], there is also a direct equivalence between this energy integral and eigenmode summation [12].) In the common case where μ has negligible frequency dependence (magnetic responses are usually negligible at the short wavelengths where Casimir forces are important, so that $\mu \approx \mu_0$), we can use the identity[7] that $\int \varepsilon|\mathbf{E}|^2 = \int \mu|\mathbf{H}|^2$ for fields at any given frequency [6] to simplify this expression to [12]:

---

[6] This follows from the standard identity that the limit $\mathrm{Im}[(x + i0^+)^{-1}]$, viewed as a distribution, yields $-\pi\delta(x)$ [86].

[7] Lest the application of this field identity appear too glib, we can also obtain the same equality directly from the Green's functions in the correlation functions. We have $\int \mu\langle|\mathbf{H}|^2\rangle = \frac{\hbar}{\pi}\mathrm{tr}\int \xi^2\mu\mathbf{G}^H(\mathbf{x},\mathbf{x})$, and from the identity after (6.6) we know that $\xi^2\mu\mathbf{G}^H = -\nabla \times \mathbf{G} \times \nabla' \frac{1}{\mu'} + \delta$. However, because $\nabla\times$ is self-adjoint [6], we can integrate by parts to move $\nabla\times$ from the first argument/index of $\mathbf{G}^E$ to the second, obtaining $-\mathbf{G}^E \times \nabla' \frac{1}{\mu'} \times \nabla' = \xi^2\varepsilon'\mathbf{G}^E - \delta$ from the first term under the integral. (Here, we employ the fact that $\mathbf{G}^E$ is real-symmetric at imaginary $\omega = i\xi$, from Sect. 6.5.1.2, to apply (6.10) to the second index/argument instead of the first.) This cancels the other delta from $\xi^2\mu\mathbf{G}^H$ and leaves $\xi^2\varepsilon\mathbf{G}^E$, giving $\varepsilon\langle|\mathbf{E}|^2\rangle$ as desired.

$$U = \int_0^\infty d\xi \int \frac{1}{2\xi} \frac{d(\xi^2 \varepsilon)}{d\xi} \langle |\mathbf{E}|^2 \rangle_{i\xi} d^3\mathbf{x}. \tag{6.8}$$

Here, the zero-temperature imaginary-frequency mean-square electric field is given by:

$$\langle |\mathbf{E}(\mathbf{x})|^2 \rangle_{i\xi} = \frac{\hbar}{\pi} \xi^2 \text{tr} \mathbf{G}^E(i\xi; \mathbf{x}, \mathbf{x}), \tag{6.9}$$

where tr denotes the trace $\sum_j G^E_{jj}$ and the Im has disappeared compared to (6.3) because $\mathbf{G}^E(i\xi)$ is real and the Im cancels the i in $d\omega \rightarrow id\xi$.

Equation (6.13) may at first strike one as odd, because one is evaluating the Green's function (field) at $\mathbf{x}$ from a source at $\mathbf{x}$, which is formally infinite. This is yet another instance of the formal infinities that appear in Casimir problems, similar to the infinite sum over modes in Sect. 6.3. In practice, this is not a problem either analytically or numerically. Analytically, one typically regularizes the problem by subtracting off the vacuum Green's function (equivalent to only looking at the *portion* of the fields at $\mathbf{x}$ which are reflected off of inhomogeneities in ε or μ) [81]. Numerically, in an FD or FEM method with a finite grid, the Green's function is everywhere finite (the grid is its own regularization) [12]. In a BEM, the Green's function is explicitly written as a sum of the vacuum field and scattered fields, so the former can again be subtracted analytically [12]. As in Sect. 6.3, these regularizations do not affect physically observable quantities such as forces or energy differences, assuming rigid-body motion.

6.5.1.2 The Remarkable Imaginary-Frequency Green's Function

This imaginary-frequency Green's function is actually a remarkably nice object. Wick-rotating (6.4), it satisfies:

$$\left[\nabla \times \mu(i\xi, \mathbf{x})^{-1} \nabla \times + \xi^2 \varepsilon(i\xi, \mathbf{x}) \right] \mathbf{G}^E_k(i\xi, \mathbf{x}, \mathbf{x}') = \delta^3(\mathbf{x} - \mathbf{x}') \hat{e}_k. \tag{6.10}$$

Because of causality, it turns out that ε and μ are strictly real-symmetric and positive-definite (in the absence of gain) along the imaginary-frequency axis, even for dissipative/dispersive materials [41]. Furthermore, the operator $\nabla \times \mu^{-1}\nabla\times$ is real-symmetric positive-semidefinite for a positive-definite real-symmetric μ [6]. Thus, the entire bracketed operator $[\cdots]$ in (6.10) is *real-symmetric positive-definite* for $\xi > 0$, which lends itself to some of the best numerical solution techniques (Cholesky decomposition [65], tridiagonal QR [65], conjugate gradients [65, 67], and Rayleigh-quotient methods [68]). (This definiteness is also another way of seeing the lack of poles or oscillations for $\omega = i\xi$.) It follows that the integral operator whose kernel is $\mathbf{G}^E$, i.e. the inverse of the $[\cdots]$ operator in (6.10), is also real-symmetric positive-definite, which is equally useful for integral-equation methods.

In vacuum, the 3d real-ω Green's function $\sim e^{i\omega|x-x'|/c}/|\mathbf{x} - \mathbf{x}'|$ [41] is Wick-rotated to $\sim e^{-\xi|x-x'|/c}/|\mathbf{x} - \mathbf{x}'|$, an *exponentially decaying, non-oscillatory* function. This is yet another way of understanding why, for $\omega = i\xi$, there are no interference effects and hence no "modes" (poles in $\mathbf{G}$), and integrands tend to be non-oscillatory and exponentially decaying (as $\xi \to \infty$, $\mathbf{G}$ becomes exponentially short-ranged and does not "see" the interacting objects, cutting off the force contributions). (It also means, unfortunately, that a lot of the most interesting phenomena in classical EM, which stem from interference effects and resonances, may have very limited consequences for Casimir interactions.)

One other property we should mention is that the operator becomes semidefinite for $\xi = 0$, with a nullspace encompassing any static field distribution ($\nabla\phi$ for any scalar ϕ). This corresponds to the well-known singularity of Maxwell's equations at zero frequency [89, 90], where the electric and magnetic fields decouple [41]. Since we eventually integrate over ξ, the measure-zero contribution from $\xi = 0$ does not actually matter, and one can use a quadrature scheme that avoids evaluating $\xi = 0$. However, in the nonzero-temperature case of Sect. 6.8 one obtains a sum over discrete-ξ contributions, in which case the zero-frequency term is explicitly present. In this case, $\xi = 0$ can be interpreted if necessary as the limit $\xi \to 0^+$ (which can be obtained accurately in several ways, e.g. by Richardson extrapolation [91], although some solvers need special care to be accurate at low frequency [89, 90]); note, however, that there has been some controversy about the zero-frequency contribution in the unphysical limit of perfect/dissipationless metals [92].

6.5.1.3 Stress Tensor

In practice, one often wants to know the Casimir force (or torque) on an object rather than the energy density. In this case, instead of integrating an electromagnetic energy density over the volume, one can integrate an electromagnetic *stress tensor* over a surface enclosing the object in question, schematically depicted in Fig. 6.2 [81]. The mean stress tensor for the Casimir force is [81]:

$$\langle T_{jk}(\mathbf{x})\rangle_\omega = \varepsilon(\mathbf{x}, \omega)\left[\langle E_j(\mathbf{x})E_k(\mathbf{x})\rangle_\omega - \frac{\delta_{jk}}{2}\sum_\ell\langle E_\ell(\mathbf{x})^2\rangle_\omega\right]$$
$$+ \mu(\mathbf{x}, \omega)\left[\langle H_j(\mathbf{x})H_k(\mathbf{x})\rangle - \frac{\delta_{jk}}{2}\sum_\ell\langle H_\ell(\mathbf{x})^2\rangle_\omega\right]. \tag{6.11}$$

As above, the field correlation functions are expressed in terms of the classical Green's function, and the integral of the contributions over all ω is Wick-rotated to imaginary frequencies $\omega = i\xi$:

$$\mathbf{F} = \int_0^\infty d\xi \oiint_{\text{surface}} \langle \mathbf{T}(\mathbf{x})\rangle_{i\xi} \cdot d\mathbf{S}, \tag{6.12}$$

Fig. 6.2 Schematic
depiction of two objects
whose Casimir interaction is
desired. One computational
method involves integrating a
mean stress tensor around
some closed surface (*dashed
line*) surrounding an object,
yielding the force on that
object

with (zero-temperature) correlation functions

$$\langle E_j(\mathbf{x})E_k(\mathbf{x})\rangle_{i\xi} = \frac{\hbar}{\pi}\xi^2 G^E_{jk}(i\xi; \mathbf{x}, \mathbf{x}), \tag{6.13}$$

$$\langle H_j(\mathbf{x})H_k(\mathbf{x})\rangle_{i\xi} = \frac{\hbar}{\pi}\xi^2 G^H_{jk}(i\xi; \mathbf{x}, \mathbf{x}), \tag{6.14}$$

corresponding to the fields on the stress-integration surface in response to currents placed on that surface. To compute a Casimir torque around an origin $\mathbf{r}$, one instead uses $(\mathbf{x} - \mathbf{r}) \times \langle \mathbf{T}(\mathbf{x})\rangle_{i\xi} \cdot d\mathbf{S}$ [93].

The derivation of this stress tensor (6.11) is not as straightforward as it might at first appear. If the stress-integration surface lies entirely in vacuum $\varepsilon \approx \varepsilon_0$ and $\mu \approx \mu_0$, then one can interpret (6.11) as merely the ordinary EM stress tensor from the microscopic Maxwell equations [41], albeit integrated over fluctuations. If the stress-integration surface lies in a dispersive/dissipative medium such as a fluid, however, then the classical EM stress tensor is well known to be problematic [41] and (6.11) may seem superficially incorrect. However, it turns out that these problems disappear in the context of thermodynamic equilibrium, where a more careful free-energy derivation of the Casimir force from fluctuations indeed results in (6.11) [81, 94],[8] which has also proved consistent with experiments [95, 96]. Note also that, while (6.11) assumes the special case of isotropic media at $\mathbf{x}$, it can still be used to evaluate the force on objects made of anisotropic materials, as long as the stress-integration surface lies in an isotropic medium (e.g. vacuum or most fluids).

This formulation is especially important for methods that use an iterative solver for the Green's functions as discussed below, because it only requires solving for the response to currents on the stress-integration surface, rather than currents at every point in space to integrate the energy density, greatly reducing the number of right-hand sides to be solved for the linear (6.10) [12]; additional reductions in the number of right-hand sides are described in Sect. 6.5.6.

---

[8] If compressibility of the fluid and the density-dependence of ε is not neglected, then there is an additional $\partial\varepsilon/\partial\rho$ term in (6.11) resulting from fluctuations in the density ρ [81].

6.5.2 Finite-Difference Frequency-Domain (FDFD)

In order to determine the Casimir energy or force, one evaluates the Green's function by solving (6.10) and then integrates the appropriate energy/force density over volume/surface and over the imaginary frequency ξ. The central numerical problem is then the determination of the Green's function by solving a set of linear equations corresponding to (6.10), and probably the simplest approach is based on a finite-difference (FD) basis: space is divided into a uniform grid with some resolution Δx, derivatives are turned into differences, and one solves (6.10) by some method for every desired right-hand side. In classical EM (typically finding the field from a given current at real ω), this is known as a finite-difference frequency-domain (FDFD) method, and has been widely used for many years [5, 49].

For example, in one dimension for z-directed currents/fields with $\mu = 1$, (6.10) becomes $\left(-\frac{d^2}{dx^2} + \xi^2\right)\varepsilon G_{zz}^E = \delta(x - x')\hat{z}$. If we approximate $G_{zz}^E(n\Delta x, x') \approx G_n$, then the corresponding finite-difference equation, with a standard center-difference approximation for d^2/dx^2 [50], is

$$-\frac{G_{n+1} - 2G_n + G_{n-1}}{\Delta x^2} + \xi^2 \varepsilon_n G_n = \frac{\delta_{nn'}}{\Delta x}, \tag{6.15}$$

replacing the $\delta(x - x')$ with a discrete equivalent at n'. Equation (6.15) is a tridiagonal system of equations for the unknowns G_n. More generally, of course, one has derivatives in the y and z directions and three unknown $\mathbf{G}$ (or $\mathbf{E}$) components to determine at each grid point. As mentioned in Sect. 6.2.2, it turns out that accurate center-difference approximations for the $\nabla \times \nabla\times$ operator in three dimensions are better suited to a "staggered" grid, called a *Yee* grid [2, 49], in which different field components are discretized at points slightly offset in space: e.g., $E_x([n_x + \frac{1}{2}]\Delta x, n_y\Delta y, n_z\Delta z)$, $E_y(n_x\Delta x, [n_y + \frac{1}{2}]\Delta y, n_z\Delta z)$, and $E_z(n_x\Delta x, n_y\Delta y, [n_z + \frac{1}{2}]\Delta z)$ for the $\mathbf{E}$ field components. Note that any arbitrary frequency dependence of ε is trivial to include, because in frequency domain one is solving each ξ separately, and a perfect electric conductor is simply the $\varepsilon(i\xi) \to \infty$ limit.

One must, of course, somehow truncate the computational domain to a finite region of space in order to obtain a finite number N of degrees of freedom. There are many reasonable ways to do this because Casimir interactions are rapidly decaying in space (force $\sim 1/a^{d+1}$ or faster with distance a in d dimensions, at least for zero temperature). One could simply terminate the domain with Dirichlet or periodic boundary conditions, for example, and if the cell boundaries are far enough away from the objects of interest then the boundary effects will be negligible (quite different from classical EM problems at real ω!) [12]. In classical EM, one commonly uses the more sophisticated approach of a **perfectly matched layer** (PML), an artificial reflectionless absorbing material placed adjacent to the boundaries of the computational domain to eliminate outgoing waves [2]. Mathematically, a PML in a direction x is equivalent to a complex "coordinate stretching" $\frac{d}{dx} \to (1 + i\sigma/\omega)^{-1}\frac{d}{dx}$ for an artificial PML "conductivity" $\sigma(x) > 0$ [2, 97–99], where the $1/\omega$ factor is introduced to give an equal attenuation rate at all frequencies. However, at imaginary frequencies $\omega = i\xi$, a PML therefore results simply in a *real* coordinate stretching $\frac{d}{dx} \rightarrow (1 + \sigma/\xi)^{-1}\frac{d}{dx}$: in a PDE with decaying, non-oscillatory solutions (such as the imaginary-ω Maxwell equations), it is well known that a reasonable approach to truncating infinite domains is to perform a (real) coordinate transformation that compresses space far away where the solution is small [53]. A convenience of Maxwell's equations is that any coordinate transformation (real or complex) can be converted merely into a change of ε and μ [100], so any PML can be expressed simply as a change of materials while keeping the same PDE and discretization [98, 99].

Such a center-difference scheme is nominally second-order accurate, with discretization errors that vanish as $O(\Delta x^2)$ [2, 50]. One can also construct higher-order difference approximations (based on more grid points per difference). As a practical matter, however, the accuracy is limited instead by the treatment of material interfaces where ε changes discontinuously. If no special allowance is made for these interfaces, the method still converges, but its convergence rate is reduced by the discontinuity to $O(\Delta x)$ [54, 55, 101, 102] (unless one has **E** polarization completely parallel to all interfaces so that there is no field discontinuity). There are various schemes to restore second-order (or higher) accuracy by employing specialized FD equations at the interfaces [54, 103], but an especially simple scheme involves unmodified FD equations with modified materials: it turns out that, if the discontinuous ε is smoothed in a particular way (to avoid introducing first-order errors by the smoothing itself), then second-order accuracy can be restored [55, 101, 102].[9]

Given the FD equations, one must then choose a solution technique to solve the resulting linear equations $A\mathbf{x} = \mathbf{b}$, where $\mathbf{x}$ is the Green's function (or field), $\mathbf{b}$ is the delta-function (or current) right-hand side, and A is the discretized $\nabla \times \mu^{-1}\nabla \times + \xi^2\varepsilon$ operator. Note that A is a real-symmetric positive-definite matrix at imaginary frequencies, as discussed in Sect. 6.5.1.2. Because A is sparse [only $O(N)$ nonzero entries], one can utilize a sparse-direct Cholesky factorization $A = R^T R$ (R is upper-triangular) [66] (for which many software packages are available [68]). Given this factorization, any right-hand side can be solved quickly by backsubstitution, so one can quickly sum the energy density over all grid points (essentially computing the trace of A^{-1}) to find the Casimir energy, or alternatively sum the stress tensor over a stress-integration surface to find the force. Precisely such a sparse-direct FD method for the Casimir energy was suggested by Pasquali and Maggs [23], albeit derived by a path-integral log det expression that is mathematically equivalent to summing the energy density

---

[9] Even if the ε discontinuities are dealt with in this way, however, one may still fail to obtain second-order accuracy if the geometry contains sharp corners, which limit the accuracy to $O(\Delta x^p)$ for some $1 < p < 2$ [101]. This is an instance of *Darboux's principle*: the convergence rate of a numerical method is generally limited by the strongest singularity in the solution that has not been explicitly compensated for [53].

(6.8) [87]. The alternative is an iterative technique, and in this case A's Hermitian definiteness means that an ideal Krylov method, the conjugate-gradient method [65, 67] can be employed [12]. The conjugate-gradient method requires $O(N)$ storage and time per iteration, and in the absence of preconditioning requires a number of iterations in d dimensions proportional to the diameter $O(N^{1/d})$ of the grid for each right-hand side [104]. The stress-tensor approach reduces the number of right-hand sides to be solved compared to energy-density integration: one only needs to evaluate the Green's function for sources on a stress-integration surface, which has $O(N^{\frac{d-1}{d}})$ points in d dimensions. This gives a total time complexity of $O(N) \cdot O(N^{1/d}) \cdot O(N^{\frac{d-1}{d}}) = O(N^2)$ for an unpreconditioned iterative method; an ideal multigrid preconditioner can in principle reduce the number of iterations to $O(1)$ [4, 105] (when N is increased by improving spatial resolution), yielding an $O(N^{2-\frac{1}{d}})$ time complexity. Substantial further improvements are obtained by realizing that one does not, in fact, need to sum over every point on the stress-integration surface, instead switching to a different spatial integration scheme described in Sect. 6.5.6.

6.5.3 Boundary-Element Methods (BEMs)

In some sense, a volume discretization such as an FD method is too general: in most physical situations, the medium is piecewise-constant, and one might want to take advantage of this fact. In particular, for the basic problem of finding the field in response to a current source at a given frequency, one can instead use a surface integral-equation approach: the unknowns are *surface currents* on the interfaces between homogeneous materials, and one solves for the surface currents so that the total field (source + surface currents) satisfies the appropriate boundary conditions at the interfaces [1, 3, 7]. For example, in the case of a perfect electric conductor, the surface-current unknowns can be the physical electric currents $\mathbf{J}$ at the interface, and the boundary condition is that of vanishing tangential $\mathbf{E}$ field.[10] In the case of permeable media ε and μ, the physical (bound) currents are volumetric within the medium [e.g., the electric bound current is $\mathbf{J} = -i\omega(\varepsilon - \varepsilon_0)\mathbf{E}$], not surface currents [41]. However, it turns out that one can introduce *fictitious* surface electric and magnetic currents at all interfaces to provide enough degrees of freedom to satisfy the boundary condition of continuous tangential $\mathbf{E}$ and $\mathbf{H}$, and thus to fully solve Maxwell's equations. The application of this *equivalence*

---

[10] This is known as an *electric-field integral equation* (EFIE); one can also express the equations for perfect conductors in terms of boundary conditions enforced on *magnetic* fields (MFIE) or some linear *combination* of the two (CFIE), and the most effective formulation is still a matter of debate [90].

Fig. 6.3 Example triangular mesh of the surfaces of two objects for a BEM solver [9]. Associated with each edge k is an "RWG" basis function $\mathbf{b}_k$ [63], schematically represented in the inset, which vanishes outside the adjacent two triangles

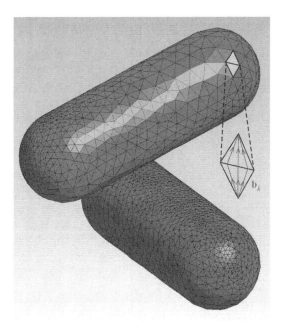

principle[11] to obtain surface integral equations for BEM is known as the PMCHW approach (Poggio, Miller, Chang, Harrington, and Wu) [110–112]. In either case, one has surface (electric and/or magnetic) currents $\mathbf{J}_s$, plus an external current source $\mathbf{J}$ [e.g., the right-hand side of (6.4)], so one can express the $\mathbf{E}$ or $\mathbf{H}$ field at any point $\mathbf{x}$ as a convolution of $\mathbf{J} + \mathbf{J}_s$ with the *analytically known* Green's function $\mathbf{G}_0(\mathbf{x} - \mathbf{x}')$ of the corresponding *homogeneous* medium at $\mathbf{x}$. In BEM, one expresses $\mathbf{J}_s$, in turn, as a sum of *localized* basis functions $\mathbf{b}_k$ associated with some discrete mesh approximation of the surface. For example, Fig. 6.3 depicts a standard triangular-type mesh of two objects, where there is a localized basis function $\mathbf{b}_k$ (inset) associated with each *edge* of this mesh such that $\mathbf{b}_k$ is nonzero only on the adjacent two triangles [63]; this is the RWG basis mentioned in Sect. 6.2.2. Abstractly, the resulting equations for the fields could then be written in the following form:

$$\text{field}(\mathbf{x}) = \mathbf{G}_0 * (\mathbf{J} + \mathbf{J}_s) = \mathbf{G}_0 * \mathbf{J} + \sum_{k=1}^{N} \mathbf{G}_0 * \mathbf{b}_k c_k, \qquad (6.16)$$

where $\mathbf{G}_0*$ denotes convolution with the (dyadic) analytical Green's function of the homogeneous medium at $\mathbf{x}$, and c_k are the unknown coefficients of each basis function.[12] (More generally, $\mathbf{G}_0 * \mathbf{J}$ could be replaced by any arbitrary incident

---

[11] The idea of solving scattering problems by introducing fictitous boundary currents had its origins [106–109] many years before its application to BEM by Harrington [110] and subsequent refinements.

[12] Technically, only currents from surfaces bordering the medium of $\mathbf{x}$ contribute to this sum.

field, regardless of how it is created.) In a Galerkin method (see Sect. 6.2.2), one obtains N equations for the N unknowns c_k by taking the inner product of both sides of this equation (substituted into the appropriate boundary condition) with the same basis functions $\mathbf{b}_j$ (since they work just as well as a basis for the tangential field as for the tangential surface currents). This ultimately results in a set of linear equations $A\mathbf{c} = \mathbf{d}$, where the matrix A multiplying the unknown coefficients c_k is given by

$$A_{jk} = \iint \bar{\mathbf{b}}_j(\mathbf{x}) \cdot \mathbf{G}_0(\mathbf{x} - \mathbf{x}') \cdot \mathbf{b}_k(\mathbf{x}') d^2\mathbf{x} d^2\mathbf{x}'. \qquad (6.17)$$

[For the case of a perfect conductor with a vanishing tangential $\mathbf{E}$, the right-hand-side $\mathbf{d}$ is given by $d_j = -\langle \mathbf{b}_j, \mathbf{G}_0 * \mathbf{J} \rangle = -\iint \bar{\mathbf{b}}_j(\mathbf{x}) \cdot \mathbf{G}_0(\mathbf{x} - \mathbf{x}') \cdot \mathbf{J}(\mathbf{x}') d^2\mathbf{x} d^2\mathbf{x}'.$] One then solves the linear system for the unknown coefficients c_k, and hence for the unknown surface currents $\mathbf{J}_s$. Implementing this technique is nontrivial because the A_{jk} integrands (6.17) are singular for $j = k$ or for j adjacent to k, necessitating specialized quadrature techniques for a given form of $\mathbf{G}_0$ [60, 61], but substantial guidance from the past decades of literature on the subject is available.

Given these currents, one can then evaluate the electric or magnetic field at any point $\mathbf{x}$, not just on the surface, by evaluating (6.16) at that point. In particular, one can evaluate the field correlation functions via the fluctuation-dissipation theorem (6.3): $\langle E_j(\mathbf{x}) E_k(\mathbf{x}) \rangle_\omega$ is given in terms of the electric field in the j direction at $\mathbf{x}$ from a delta-function current $\mathbf{J}$ in the k direction at $\mathbf{x}$. Of course, as noted previously, this is infinite because the $\mathbf{G}_0 * \mathbf{J}$ term (the field from the delta function) blows up at $\mathbf{x}$, but in the Casimir case one is only interested in the *change* of the correlation functions due to the geometry—so, one can use the standard trick [81] of subtracting the vacuum contribution $\mathbf{G}_0 * \mathbf{J}$ and only computing the surface-current contribution $\mathbf{G}_0 * \mathbf{J}_s$ to the field at $\mathbf{x}$. In this way, one can compute the stress tensor, the energy density, and so on, as desired.

As explained in Sect. 6.4, the integral of contributions over all frequencies is best performed at imaginary frequencies, so all of the above must use $\omega = i\xi$. This only has the effect of Wick-rotating the homogeneous-medium dyadic Green's function $\mathbf{G}_0$ to the $\sim e^{-\xi|\mathbf{x}-\mathbf{x}'|}/|\mathbf{x} - \mathbf{x}'|$ imaginary-frequency Green's function. This makes the problem *easier*, in principle. First, the exponential decay cuts off long-range interactions, making fast-solver techniques (see Sect. 6.2.3) potentially even more effective. Second, the matrix A is now real-symmetric and positive-definite, which allows the use of more efficient linear solvers as noted previously. Fortunately, the $1/|\mathbf{x} - \mathbf{x}'|$ singularity of $\mathbf{G}_0$ is the same at real and imaginary frequencies, allowing existing techniques for the integration of (6.17) to be leveraged.

At first glance, this approach seems most straightforwardly applicable to the stress-tensor technique, as suggested in Ref. [12]: one uses the BEM solution to evaluate the mean stress tensor $\langle \mathbf{T} \rangle$ on any integration surface around a body, integrating via some quadrature technique to obtain the force. If one uses a dense-direct solver (when N is not too big), the Cholesky factorization of A can be computed once for a given ξ and then many right-hand sides can be solved quickly via backsubstitution [65] in order to integrate $\langle \mathbf{T} \rangle$ over the stress-integration surface. Precisely such a dense-direct BEM stress-tensor method was recently demonstrated to compute Casimir forces in two and three dimensions [19, 20]. As described in Sect. 6.2.3, fast-solver techniques can be applied to multiply A by a vector in $O(N \log N)$ time with $O(N)$ storage; given a good preconditioner, this implies that an iterative method such as conjugate-gradient (applicable since A is real-symmetric positive-definite) could find $\langle \mathbf{T} \rangle$ at a single $\mathbf{x}$ and ξ in $O(N \log N)$ time. The remaining question is the number of points required for the surface integral of $\langle \mathbf{T} \rangle$, which depends on *why* one is increasing N: either to increase accuracy for a fixed geometry or to increase the complexity of the geometry for a fixed accuracy. In the former case, the smoothness of $\langle \mathbf{T} \rangle$ in $\mathbf{x}$ means that exponentially convergent quadrature techniques are applicable, which converge much faster than the (polynomial) BEM basis for the surface currents, so that ultimately the number of stress-quadrature points[13] should be independent of N and the overall complexity becomes $O(N \log N)$. In the latter case, for a fixed accuracy and increasingly complex geometry (or smaller feature sizes), it appears likely that the number of stress-quadrature points will increase with N, but detailed studies of this scaling are not yet available.

It turns out that this BEM approach is closely related to the BEM path-integral approach described in Sect. 6.6.3. Both approaches end up solving linear equations with exactly the same matrix A of (6.17), with the same degrees of freedom. The path-integral approach shows, however, that this same matrix can be applied to compute the Casimir interaction energy as well as the force, with comparable computational cost for dense solvers. Moreover, as explained below, expressing the force in terms of the derivative of the path-integral energy results in a trace expression that is conceptually equivalent to integrating a stress tensor over the surface of an object, where the number of "quadrature points" is now exactly equal to N. An unanswered question, at this point, is whether a fast solver can be more efficiently (or more easily) exploited in the stress-tensor approach or in the path-integral approach.

6.5.4 Other Possibilities: FEM and Spectral Methods

There are of course, many other frequency-domain techniques from classical EM that could potentially be used to solve for the Green's function and hence the

---

[13] Numeric integration (*quadrature*) approximates an integral $\int f(x) dx$ by a sum $\sum_i f(x_i) w_i$ over quadrature points x_i with weights w_i. There are many techniques for the selection of these points and weights, and in general one can obtain an error that decreases exponentially fast with the number of points for analytic integrands [53, 83, 84, 113]. Multidimensional quadrature, sometimes called *cubature*, should be used to integrate the stress tensor over a 2d surface, and numerous schemes have been developed for low-dimensional cubature [114, 115] (including methods that adaptively place more quadrature points where they are most needed [116]). For spherical integration surfaces (or surfaces that can be smoothly mapped to spheres), specialized methods are available [117, 118].

energy/force density. For example, one could use spectral integral-equation methods, such as multipole expansions for spheres and cylinders [5], to compute responses to currents, although the advantages of this approach compared to the spectral path-integral approach in Sect. 6.6.2 are unclear. One can also solve the PDE formulation of the Green's function (6.10) using a finite-element (FEM) approach with some general mesh; in principle, existing FEM techniques from classical EM [1, 3, 4, 7] are straightforwardly applicable. One subtlety that arises in FEM methods with a nonuniform resolution is the regularization, however [12]. In principle, as mentioned above, one needs to subtract the vacuum Green's function contribution from the field correlation functions in order to get a physical result [since the vacuum Green's function $\mathbf{G}(\mathbf{x}, \mathbf{x}')$ diverges as $\mathbf{x}' \to \mathbf{x}$, although the divergence is cut off by the the finite mesh resolution]. With a uniform mesh, this vacuum contribution is the same everywhere in the mesh and hence automatically integrates to zero in the force (when the stress tensor is integrated over a closed surface or the energy is differentiated). For a nonuniform mesh, however, the vacuum contribution varies at different points in space with different resolution, so some "manual" regularization seems to be required (e.g., subtracting a calculation with the same mesh but removing the objects). These possibilities currently remain to be explored for Casimir physics.

6.5.5 Finite-Difference Time-Domain (FDTD) Methods

Casimir effects are fundamentally broad-bandwidth, integrating contributions of fluctuations at all frequencies (real or imaginary), although the imaginary-frequency response is dominated by a limited range of imaginary frequencies. In classical EM, when a broad-bandwidth response is desired, such as a transmission or reflection spectrum from some structure, there is a well-known alternative to computing the contributions at each frequency separately—instead, one can simulate the same problem in *time*, Fourier-transforming the response to a short pulse excitation in order to obtain the broad-bandwidth response in a single *time-domain* simulation [6, 119]. The same ideas are applicable to the Casimir problem, with a few twists, yielding a practical method [13, 27] that allows Casimir calculations to exploit off-the-shelf time-domain solvers implementing the standard *finite-difference time-domain* (FDTD) method [2]. There are two key components of this approach [13]: first, converting the frequency integral to a time integral and, second, finding a time-domain equivalent of the complex-fequency idea from Sect. 6.4.

As reviewed above, the mean fluctuations in the fields, such as $\langle E^2(\mathbf{x}) \rangle_\omega$, can be expressed in terms of the fields at $\mathbf{x}$ from a frequency-ω current at $\mathbf{x}$. If, instead of a frequency-ω current, one uses a current with $\delta(t)$ time dependence, it follows by linearity of (6.4) that the Fourier transform of the resulting fields must yield exactly the same $\langle E^2(\mathbf{x}) \rangle_\omega$. Roughly, the procedure could be expressed as follows: First, we compute some function $\Gamma(t)$ of the time-domain fields from a sequence of simulations with $\delta(t)$ sources, e.g. where $\Gamma(t)$ is the result of spatially integrating the fields making up the mean stress tensor $\langle \mathbf{T}(\mathbf{x}) \rangle$ [noting that each point $\mathbf{x}$ involves several separate $\delta(t)$-response simulations]. Second, we Fourier transform $\Gamma(t)$ to obtain $\tilde{\Gamma}(\omega)$. Third, we obtain the force (or energy, etcetera) by integrating $\int \tilde{\Gamma}(\omega) \tilde{g}(\omega) d\omega$ with appropriate frequency-weighting factor $\tilde{g}(\omega)$ (which may come from the frequency dependence of ε in $\langle \mathbf{T} \rangle$, a Jacobian factor from below, etcetera). At this point, however, it is clear that the Fourier transform of Γ was entirely unnecessary: because of the unitarity of the Fourier transform (the Plancherel theorem), $\int \tilde{\Gamma}(\omega) \tilde{g}(\omega) d\omega = \int \Gamma(t) g(-t) dt$. That is, we can compute the force (or energy, etcetera) by starting with $\delta(t)$ sources and simply integrating the response $\Gamma(t)$ in time (accumulated as the simulation progresses) multiplied by some (precomputed, geometry-independent) kernel $g(t)$ (which depends on temperature if the coth factor is included for $T > 0$). The details of this process, for the case of the stress tensor, are described in Refs. [13, 27].

Although it turns out to be possible to carry out this time-integration process as-is, we again find that a transformation into the complex-frequency plane is desirable for practical computation (here, to reduce the required simulation time) [13]. Transforming the *frequency* in a *time*-domain method, however, requires an indirect approach. The central observation is that, in (6.4) for the electric-field Green's function $\mathbf{G}^E$, the frequency only appears explicitly in the $\omega^2 \varepsilon$ term, together with ε. So, any transformation of ω can equivalently be viewed as a transformation of ε. In particular suppose that we wish to make some transformation $\omega \rightarrow \omega(\xi)$ to obtain an ω in the upper-half complex plane, where ξ is a real parameter (e.g. $\omega = i\xi$ for a Wick rotation). Equivalently, we can view this as a calculation at a *real* frequency ξ for a transformed *complex material*: $\omega^2 \varepsilon(\omega, \mathbf{x}) \rightarrow \xi^2 \varepsilon_c(\xi, \mathbf{x})$ where the transformed material is [13, 120]

$$\varepsilon_c(\xi, \mathbf{x}) = \frac{\omega^2(\xi)}{\xi^2} \varepsilon(\omega(\xi), \mathbf{x}). \tag{6.18}$$

For example, a Wick rotation $\omega \rightarrow i\xi$ is equivalent to operating at a real frequency ξ with a material $\varepsilon(\omega) \rightarrow -\varepsilon(i\xi)$. However, at this point we run into a problem: multiplying ε by -1 yields exponentially growing solutions at negative frequencies [13, 120], and this will inevitably lead to exponential blowup in a time-domain simulation (which cannot avoid exciting negative frequencies, if only from roundoff noise). In order to obtain a useful time-domain simulation, we must choose a contour $\omega(\xi)$ that yields a *causal, dissipative* material ε_c, and one such choice is $\omega(\xi) = \xi \sqrt{1 + i\sigma/\xi}$ for any constant $\sigma > 0$ [13, 120]. This yields $\varepsilon_c = (1 + i\sigma/\omega)\varepsilon$, where the $i\sigma/\omega$ term behaves exactly like an artificial *conductivity* added everywhere in space. In the frequency-domain picture, we would say from Sect. 6.4 that this $\omega(\xi)$ contour will improve the computation by moving away from the real-ω axis, transforming the frequency integrand into something exponentially decaying and less oscillatory. In the time-domain picture, the σ term adds a *dissipation* everywhere in space that causes $\Gamma(t)$ to *decay exponentially in time*, allowing us to truncate the simulation after a short time. As long as we include the appropriate Jacobian factor $\frac{d\omega}{d\xi}$ in our frequency integral, absorbing it into $g(t)$, we will obtain the *same result* in a much shorter time. The computational details of this transformation are described in Refs. [13, 27]. More generally, this equivalence between the Casimir force and a relatively narrow-bandwidth real-frequency response of a dissipative system potentially opens other avenues for the understanding of Casimir physics [120].

The end result is a computational method for the Casimir force in which one takes an off-the-shelf time-domain solver (real time/frequency), adds an artificial conductivity σ everywhere, and then accumulates the response $\Gamma(t)$ to short pulses multiplied by a precomputed (geometry independent) kernel $g(t)$. The most common time-domain simulation technique in classical EM is the FDTD method [2]. Essentially, FDTD works by taking the same spatial Yee discretization as in the FDFD method above, and then also discretizing time with some time step Δt. The fields are then marched through time in steps of Δt, where each time step requires $O(N)$ work for N spatial grid points. Because the complex-ω contour is implemented entirely as a choice of materials ε_c, existing FDTD software can be used without modification to compute Casimir forces, and one can exploit powerful existing software implementing parallel calculations, various dimensionalities and symmetries, general dispersive and anisotropic materials, PML absorbing boundaries, and techniques for accurate handling of discontinuous materials. One such FDTD package is available as free/open-source software from our group [119], and we have included built-in facilities to compute Casimir forces [121].

6.5.6 Accelerating FD Convergence

Finally, we should mention a few techniques that accelerate the convergence and reduce the computational cost of the finite-difference approaches. These techniques are not *necessary* for convergence, but they are simple to implement and provide significant efficiency benefits.

The simplest technique is extrapolation in Δx: since the convergence rate of the error with the spatial resolution Δx is generally known *a priori*, one can fit the results computed at two or more resolutions in order to extrapolate to $\Delta x \to 0$. The generalization of this approach is known as *Richardson extrapolation* [91], and it can essentially increase the convergence order cheaply, e.g., improving $O(\Delta x)$ to $O(\Delta x^2)$ [122].

Second, suppose one is computing the force between two objects A and B surrounded by a homogeneous medium. If one of the objects, say B, is removed, then (in principle) there should be no net remaining force on A. However, because of discretization asymmetry, a computation with A alone will sometimes still give a small net force, which converges to zero as $\Delta x \to 0$. If this "error" force is subtracted from the A–B force calculation, it turns out that the net error is reduced. More generally, the error is greatly reduced if one computes the A–B force and then subtracts the "error" forces for A alone and for B alone, tripling the number of computations but greatly reducing the resolution that is required for an accurate result [12].

Third, when integrating the stress tensor $\langle \mathbf{T}(\mathbf{x}) \rangle_{i\xi}$ over $\mathbf{x}$ to obtain the net force (6.12), the most straightforward technique in FD is to simply sum over all the grid points on the integration surface—recall that each point $\mathbf{x}$ requires a linear solve (a different right-hand side) in frequency domain, or alternatively a separate time-domain simulation (a separate current pulse). This is wasteful, however, because $\langle \mathbf{T}(\mathbf{x}) \rangle_{i\xi}$ is conceptually smoothly varying in space—if one could evaluate it at arbitrary points $\mathbf{x}$ (as is possible in the BEM approach), an exponentially convergent quadrature scheme could be exploited to obtain an accurate integral with just a few $\mathbf{x}$'s. This is not directly possible in an FD method, but one can employ a related approach. If the integration surface is a box aligned with the grid, one can expand the fields on each side of the box in a cosine series (a discrete cosine transform, or DCT, since space is discrete)—this generally converges rapidly, so only a small number terms from each side are required for an accurate integration. But instead of putting in point sources, obtaining the responses, and expanding the response in a cosine series, it is equivalent (by linearity) to put in cosine sources directly instead of point sources. [Mathematically, we are exploiting the fact that a delta function can be expanded in any orthonormal basis $b_n(\mathbf{x})$ over the surface, such as a cosine series, via: $\delta(\mathbf{x} - \mathbf{x}') = \sum_n \bar{b}_n(\mathbf{x}') b_n(\mathbf{x})$. Substituting this into the right-hand side of (6.10), each $b_n(\mathbf{x})$ acts like a current source and $\bar{b}_n(\mathbf{x}')$ scales the result, which is eventually integrated over $\mathbf{x}'$.] The details of this process and its convergence rate are described in Ref. [27], but the consequence is that many fewer linear systems (fewer right-hand sides) need be solved (either in frequency or time domain) than if one solved for the stress tensor at each point individually.

6.6 Path Integrals and Scattering Matrices

Another formulation of Casimir interactions is to use a derivation based on path integrals. Although the path-integral derivation itself is a bit unusual from the perspective of classical EM, and there are several slightly different variations on this idea in the literature, the end result is straightforward: Casimir energies and forces are expressed in terms of log determinants and traces of classical scattering matrices [10, 14–18, 21, 22, 24–26], or similarly the interaction matrices (6.17) that arise in BEM formulations [9]. Here, we omit the details of the derivations and focus mainly on the common case of piecewise-homogeneous materials, emphasizing the relationship of the resulting method to surface-integral equations from classical EM via the approach in Ref. [9].

Path integrals relate the Casimir interaction energy U of a given configuration to a functional integral over all possible vector-potential fields $\mathbf{A}$. Assuming

piecewise-homogeneous materials, the constraint that the fields in this path integral must satisfy the appropriate boundary conditions can be expressed in terms of
auxiliary fields $\mathbf{J}$ at the interfaces (a sort of Lagrange multiplier) [123].[14] At this
point, the original functional integral over $\mathbf{A}$ can be performed analytically,
resulting in an energy expression involving a functional integral $Z(\xi)$ over only the
auxiliary fields $\mathbf{J}$ at each imaginary frequency ξ, of the form (at zero temperature):

$$U = -\frac{\hbar c}{2\pi} \int_0^\infty \log \det \frac{Z(\xi)}{Z_\infty(\xi)} d\xi, \tag{6.19}$$

$$Z(\xi) = \int \mathcal{D}\mathbf{J} e^{-\frac{1}{2}\iint d^2\mathbf{x} \iint d^2\mathbf{x}' \mathbf{J}(\mathbf{x}) \cdot \mathbf{G}_\xi(\mathbf{x}-\mathbf{x}') \cdot \mathbf{J}(\mathbf{x}')}. \tag{6.20}$$

Here, Z_∞ denotes Z when the objects are at infinite separation (non-interacting),
regularizing U to just the (finite) interaction energy (See also the Chap. 5 by S.J.
Rahi et al. in this volume for additional discussion of path integrals and Casimir
interactions.) In the case of perfect electric conductors in vacuum, $\mathbf{J}$ can be
interpreted as a surface current on each conductor (enforcing the vanishing tangential $\mathbf{E}$ field), and $\mathbf{G}_\xi$ is the vacuum Green's function in the medium outside the
conductors [9]. For permeable media (finite ε and μ), it turns out that a formulation
closely related to the standard PMCHW integral-equation model (see Sect. 6.5.3)
can be obtained: $\mathbf{J}$ represents fictitious surface electric and magnetic currents on
each interface (derived from the continuity of the tangential $\mathbf{E}$ and $\mathbf{H}$ fields), with
$\mathbf{G}_\xi$ again being a homogeneous Green's function (with one Z factor for each
contiguous homogeneous region) [124]. Alternatively, because there is a direct
correspondence between surface currents and the outgoing/scattered fields from a
given interface, "currents" $\mathbf{J}$ can be replaced by scattered fields, again related at
different points $\mathbf{x}$ and $\mathbf{x}'$ by the Green's function of the homogeneous medium; this
is typically derived directly from a T-matrix formalism [14, 22, 25, 26]. Here, we
will focus on the surface-current viewpoint, which is more common in the classical-EM integral-equation community.

The path integral (6.20) is somewhat exotic in classical EM, but it quickly
reduces to a manageable expression once an approximate (finite) basis $\mathbf{b}_k$ is chosen
for the currents $\mathbf{J}$. Expanding in this basis, $\mathbf{J} \approx \sum c_k \mathbf{b}_k(\mathbf{x})$ and the functional
integral $\mathcal{D}\mathbf{J}$ is replaced by an ordinary integral over the basis coefficients
$dc_1 \cdots dc_N$. Equation (6.20) is then a Gaussian integral that can be performed
analytically to obtain $Z(\xi) = \#/\sqrt{\det A(\xi)}$ for a proportionality constant $\#$ [9],
where $A_{jk} = \int \bar{\mathbf{b}}_j \cdot \mathbf{G}_\xi \cdot \mathbf{b}_k$ is essentially the same as the BEM matrix (6.17), albeit

---

[14] Alternatively, the path integral can be performed directly in $\mathbf{A}$, resulting in an expression
equivalent to the sum over energy density in Sect. 6.5 [87] and which in an FD discretization
reduces in the same way to repeated solution of the Green's-function diagonal at every point in
space [23].

here in an arbitrary basis. In the log det of (6.19), proportionality constants and exponents cancel, leaving:

$$U = +\frac{\hbar c}{2\pi} \int\limits_0^\infty \log \det \left[A_\infty(\xi)^{-1} A(\xi) \right] d\xi. \tag{6.21}$$

Just as in Sect. 6.7, the use of a real-symmetric positive-definite homogeneous Green's function $\mathbf{G}_\xi$ at imaginary frequencies means that $A(\xi)$ is also real-symmetric and positive-definite, ensuring positive real eigenvalues and hence a real log det. Several further simplifications are possible, even before choosing a particular basis. For example, let $\mathbf{p}$ be the position of some object for which the force $\mathbf{F}$ is desired. The components F_i of the force (in direction p_i) can then be expressed directly as a trace [9, 125]:

$$F_i = -\frac{dU}{dp_i} = -\frac{\hbar c}{2\pi} \int\limits_0^\infty \mathrm{tr}\left(A^{-1} \frac{\partial A}{\partial p_i} \right) d\xi. \tag{6.22}$$

Equivalently, this trace is the sum of eigenvalues λ of the generalized eigenproblem $\frac{\partial A}{\partial p_i} \mathbf{v} = \lambda A \mathbf{v}$; again, these λ are real because A is real-symmetric positive-definite and $\partial A / \partial p_i$ is real-symmetric. (If dense-direct solvers are used, computing $A^{-1} \frac{\partial A}{\partial p_i}$ via Cholesky factorization is much more efficient than computing eigenvalues, however [65].) The matrix A can be further block-decomposed in the usual case where one is computing the interactions among two or more disjoint objects (with disjoint surface currents $\mathbf{J}$). For example, suppose that one has two objects 1 and 2, in which case one can write

$$A = \begin{pmatrix} A_{11} & A_{12} \\ A_{12}^\mathsf{T} & A_{22}, \end{pmatrix} \tag{6.23}$$

where A_{11} and A_{22} couple currents on each object to other currents on the same object, and A_{12} and $A_{12}^\mathsf{T} = A_{21}$ couple currents on object 1 to object 2 and vice versa. In the limit of infinite separation for A_∞, one obtains $A_{12} \to 0$ while A_{11} and A_{22} are unchanged, and one can simplify the log det in (6.21) to

$$\log \det \left[A_\infty(\xi)^{-1} A(\xi) \right] = \log \det \left[I - A_{22}^{-1} A_{12}^\mathsf{T} A_{11}^{-1} A_{21} \right]. \tag{6.24}$$

Computationally, only A_{12} depends on the relative positions of the objects, and this simplification immediately allows several computations to be re-used if the energy or force is computed for multiple relative positions.

6.6.1 Monte-Carlo Path Integration

Before we continue, it should be noted that there also exists a fundamentally different approach for evaluating a path-integral Casimir formulation. Instead of reducing the problem to surface/scattering unknowns and analytically integrating Z to obtain a matrix log det expression, it is possible to retain the original path-integral expression, in terms of a functional integral over vector potentials $\mathbf{A}$ in the volume, and perform this functional integral numerically via Monte-Carlo methods [126, 127]. This reduces to a Monte-Carlo integration of an action over all possible closed-loop paths ("worldlines"), discretized into some number of points per path. Because this technique is so different from typical classical EM computations, it is difficult to directly compare with the other approaches in this review. Evaluating its computational requirements involves a statistical analysis of the scaling of the necessary number of paths and number of points per path with the desired accuracy and the complexity of the geometry [12], which is not currently available. A difficulty with this technique is that it has currently only been formulated for scalar fields with Dirichlet boundary conditions, not for the true Casimir force of vector electromagnetism.

6.6.2 Spectral Methods

One choice of basis functions $\mathbf{b}_k$ for the path-integral expressions above is a spectral basis, and (mirroring the history of integral equations in classical EM) this was the first approach applied in the Casimir problem. With cylindrical objects, for example, the natural spectral basis is a Fourier-series $e^{im\phi}$ in the angular direction ϕ. For planar surfaces the natural choice is a Fourier transform, for spheres it is spherical harmonics $Y_{\ell m}$ (or their vector equivalents [41]), and for spheroids there are spheroidal harmonics [128]. Equivalently, instead of thinking of surface currents expanded in a Fourier-like basis, one can think of the scattered fields from each object expanded in the corresponding Fourier-like basis (e.g. plane, cylindrical, or spherical waves), in which case A relates the incoming to outgoing/scattered waves for each object; this has been called a "scattering-matrix" or "T-matrix" method and is the source of many pioneering results for Casimir interactions of non-planar geometries [14, 22, 25, 26]. Even for nonspherical/spheroidal objects, one can expand the scattered waves in vector spherical harmonics [22], and a variety of numerical techniques have been developed to relate a spherical-harmonic basis to the boundary conditions on nonspherical surfaces [58]. These spectral scattering methods have their roots in many classical techniques for EM scattering problems [56, 129] (See also the Chap. 5 by S.J. Rahi et al. and Chap. 4 by A. Lambrecht et al. in this volume for additional discussions of scattering techniques and Casimir interactions.) Here, we will use the surface-current viewpoint rather than the equivalent scattered-wave viewpoint.

Many simplifications occur in the interaction matrix A of (6.23) for geometries with highly symmetrical objects and a corresponding spectral basis [22]. Consider, for example, the case of spherical objects, with surface currents expressed in a vector spherical-harmonic basis (spherical harmonics for two polarizations [22]). In the interaction matrix $A_{jk} = \iint \mathbf{b}_j(\mathbf{x}) \cdot \mathbf{G}(\mathbf{x} - \mathbf{x}') \cdot \mathbf{b}_k(\mathbf{x}')$, the convolution $\int G(\mathbf{x} - \mathbf{x}') \cdot \mathbf{b}_k(\mathbf{x}')$ of a Green's function $\mathbf{G}$ with $\mathbf{b}_k$ is known analytically: it is just the outgoing spherical wave produced by a spherical-harmonic current. If $\mathbf{b}_j$ is another spherical-harmonic current on the same sphere, then the orthogonality of the spherical harmonics means that the $\mathbf{x}$ integral of $\mathbf{b}_j(\mathbf{x})$ against the spherical wave is zero unless $j = k$. Thus, the self-interaction blocks A_{11} and A_{22} of (6.23), with an appropriate normalization, are simply identity matrices. The A_{12} entries are given by the coupling of a spherical wave from $\mathbf{b}_k$ on sphere 2 with a spherical-harmonic basis function $\mathbf{b}_j$ on sphere 1, but again this integral can be expressed analytically, albeit as an infinite series: the spherical wave from sphere 2 can be re-expressed in the basis of spherical waves centered on sphere 1 via known translation identities of spherical waves, and as a result A_{12} takes the form of a "translation matrix" [22]. Furthermore, if there are only two spheres in the problem, then their spherical harmonics can be expressed with respect to a common z axis passing through the centers of the spheres, and a $Y_{\ell m}$ on sphere 1 will only couple with a $Y_{\ell' m'}$ on sphere 2 if $m = m'$, greatly reducing the number of nonzero matrix elements. Related identities are available for coupling cylindrical waves around different origins, expanding spherical/cylindrical waves in terms of planewaves for coupling to planar surfaces, and so on [22].

As was noted in Sect. 6.2.2, such a spectral basis can converge exponentially fast if there are no singularities (e.g. corners) that were not accounted for analytically, and the method can even lend itself to analytical study. Especially for cylinders and spheres, the method is simple to implement and allows rapid exploration of many configurations; the corresponding classical "multipole methods" are common in classical EM for cases where such shapes are of particular interest [5]. On the other hand, as the objects become less and less similar to the "natural" shape for a given basis (e.g. less spherical for spherical harmonics), especially objects with corners or cusps, the spectral basis converges more slowly [58]. Even for the interaction between two spheres or a sphere and a plate, as the two surfaces approach one another the multipole expansion will converge more slowly [25, 130, 131]—conceptually, a spherical-harmonic basis has uniform angular resolution all over the sphere, whereas for two near-touching surfaces one would rather have more resolution in the regions where the surfaces are close (e.g. by using a nonuniform BEM mesh). This exponential convergence of a spectral (spherical harmonic [22]) Casimir calculation is depicted in Fig. 6.4 for the case of the Casimir interaction energy U between two gold spheres of radius $R = 1\,\mu\text{m}$, for various surface-to-surface separations a. The error $\Delta U/U$ decreases exponentially with the maximum spherical-harmonic order ℓ [corresponding to $N = 4\ell(\ell + 2)$ degrees of freedom for two spheres], but the exponential rate slows as a/R decreases. (On the other hand, for small a/R a perturbative expansion or extrapolation may become applicable [130].)

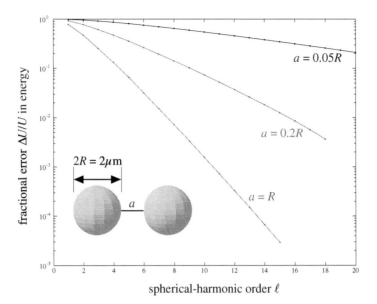

Fig. 6.4 Fractional error $\Delta U/U$ in the Casimir interaction energy U between two gold spheres of radius $R = 1\,\mu m$, for various surface-surface separations a, as a function of the maximum spherical-harmonic order ℓ of the spectral path-integral (scattering-matrix/multipole) method. The error converges exponentially with ℓ, but the exponential rate slows as a/R shrinks. (Calculations thanks to A. Rodriguez.)

6.6.3 Boundary-Element Methods (BEMs)

In a BEM, one meshes the interfaces, say into triangles, and uses a set of localized basis functions $\mathbf{b}_k$ as discussed in Sect. 6.5.3. In this case, the interaction matrix A that arises in the path-integral formulation is exactly the same as the interaction matrix that arises in classical BEM methods (albeit at an imaginary frequency), and is the same as the matrix A that arises in a BEM stress-tensor approach as described in Sect. 6.5.3 . The main difference, compared to the stress-tensor approach, lies in how one *uses* the matrix A: instead of solving a sequence of linear equations to find the mean stress tensor $\langle \mathbf{T} \rangle$ at various points on a surface around an object, one computes $\log \det A$ or $\operatorname{tr}\left[A^{-1} \frac{\partial A}{\partial p_i}\right]$ to obtain the energy (6.21) or force (6.22). We have demonstrated this approach for several three-dimensional geometries, such as the crossed capsules of Fig. 6.3 [9].

If one is using dense-matrix techniques, the advantage of this approach over the stress-tensor technique seems clear [9]: it avoids the complication of picking a stress-integration surface and an appropriate surface-integration technique, and allows the size of the linear system to be easily reduced via blocking as in (6.24). The situation is less clear as one moves to larger and larger problems, in which dense-matrix solvers become impractical and one requires an iterative method.

In that case, computing $\text{tr}\left[A^{-1}\frac{\partial A}{\partial p_i}\right]$ straightforwardly requires N linear systems to be solved; if each linear system can be solved in $O(N\log N)$ time with a fast solver (as discussed in Sects. 6.2.3 and 6.5.3), then the overall complexity is $O(N^2\log N)$ [with $O(N)$ storage], whereas it is possible that the stress-tensor surface integral may require fewer than N solves. On the other hand, there may be more efficient ways to compute the trace (or log det) via low-rank approximations: for example, if the trace (or log det) is dominated by a small number of extremal eigenvalues, then these eigenvalues can be computed by an iterative method [68] with the equivalent of $\ll N$ linear solves. The real-symmetric property of A, as usual, means that the most favorable iterative methods can be employed, such as a Lanczos or Rayleigh-quotient method [68]. Another possibility might be sparse-direct solvers via a fast-multipole decomposition [73]. The most efficient use of a fast $O(N\log N)$ BEM solver in Casimir problems, whether by stress-tensor or path-integral methods, remains an open question (and the answer may well be problem-dependent).

In the BEM approach with localized basis functions, the $\text{tr}\left[A^{-1}\frac{\partial A}{\partial p_i}\right]$ expression for the force corresponds to a sum of a diagonal components for each surface element, and in the exact limit of infinite resolution (infinitesimal elements) this becomes an integral over the object surfaces. Expressing the force as a surface integral of a quantity related to Green's-function diagonals is obviously reminiscent of the stress-tensor integration from Sect. 6.5.1.3, and it turns out that one can prove an exact equivalence using only vector calculus [124]. (At least one previous author has already shown the algebraic equivalence of the stress tensor and the derivative of the path-integral energy for forces between periodic plates [132].)

6.6.4 Hybrid BEM/Spectral Methods

It is possible, and sometimes very useful, to employ a hybrid of the BEM and spectral techniques in the previous two sections. One can discretize a surface using boundary elements, and use this discretization to solve for the scattering matrix A_{kk} of each object in a spectral basis such as spherical waves. That is, for any given incident spherical wave, the outgoing field can be computed with BEM via (6.16) and then decomposed into outgoing spherical waves to obtain one row/column of A_{kk} at a time; alternatively, the multipole decomposition of the outgoing wave can be computed directly from the multipole moments of the excited surface currents $\mathbf{J_s}$ [41]. This approach appears to be especially attractive when one has complicated objects, for which a localized BEM basis works well to express the boundary conditions, but the interactions are only to be computed at relatively large separations where the Casimir interaction is dominated by a few low-order multipole moments. One can perform the BEM computation once per object and re-use the resulting scattering matrix many times via the analytical translation matrices, allowing one to efficiently compute interactions for many rearrangements of the same objects and/or for "dilute" media consisting of many copies of the same objects [133]. (Essentially, this could be viewed as a form of low-rank approximation of the BEM matrix, capturing the essential details relevant to moderate-range Casimir interactions in a much smaller matrix.) Such a hybrid approach is less attractive for closer separations, however, in which the increasing number of relevant multipole moments will eventually lead to an impractically large matrix to be computed.

6.6.5 Eigenmode-Expansion/RCWA Methods

Consider the case of the interaction between two corrugated surfaces depicted in Fig 6.5, separated in the z direction. From the scattering-matrix viewpoint, it is natural to consider scattering off of each object by planewaves. In this case, the self-interaction matrices A_{11}^{-1} and A_{22}^{-1} can be re-expressed in terms of reflection matrices R_1 and R_2 for each surface, relating the amplitudes of incident waves at some plane (dashed line) above each surface to the reflected (specular and non-specular) planewave amplitudes. The matrices A_{12} and A_{21} are replaced by a diagonal matrix $D_{12} = D_{21}^T$ that relates the planewave amplitudes at the planes for objects 1 and 2, separated by a distance a—at real frequencies, this would be a phase factor, but at imaginary frequencies it is an exponential decay as discussed below. This results in the following expression for the Casimir interaction energy:

$$U = \frac{\hbar c}{2\pi} \int_0^\infty \log \det[I - R_2 D_{12} R_1 D_{12}] d\xi. \tag{6.25}$$

Alternatively, instead of viewing it as a special case of the T-matrix/scattering-matrix idea [22], the same expression can be derived starting from an eigenmode-summation approach [18].

The problem then reduces to computing the scattering of an incident planewave off of a corrugated surface, with the scattered field decomposed into outgoing planewaves. For this problem, one could use any of the tools of computational EM (such as BEM, FD, and so on), but there is a notable method that is often well-suited to the case of periodic surfaces, especially periodic surfaces with piecewise-constant cross-sections[15] (as in object 2 of Fig. 6.5). This method is called *eigenmode expansion* [43–45] or *rigorous coupled-wave analysis* (RCWA) [46, 47], or alternatively a *cross-section* method [48]. RCWA has a long history because it is closely tied to semi-analytical methods to study waveguides with slowly/weakly varying cross-sections [48, 134]. An analogous method was recently applied to Casimir problems [18]. In RCWA, one computes reflection and scattering matrices at a given frequency ω along some direction z by expanding the

---

[15] See also the Chap. 4 by A. Lambrecht et al. in this volume for additional discussion of Casimir interactions among periodic structures.

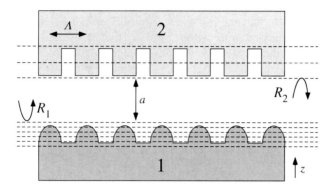

Fig. 6.5 Schematic problem for which eigenmode-expansion is well suited: the interaction between two corrugated surfaces, with period Λ. The Casimir problem reduces to computing the reflection matrices R_1 and R_2 for each individual surface, in a planewave basis. Eigenmode expansion works by expanding the field in each cross-section (*dashed lines*) in the basis of eigenmodes of a z-invariant structure with that cross-section, and then matching boundary conditions whenever the cross-section changes

fields at each z in the basis of the eigenmodes of the cross-section at that z (waves with z dependence $e^{i\beta z}$ at a given ω, where β is called the *propagation constant* of the mode). Along regions of uniform cross-section, the z dependence $e^{i\beta z}$ of each mode is known analytically and no computation is required (the mode amplitudes are multiplied by a diagonal propagation matrix D). Regions of continuously varying cross-section are approximated by breaking them up into a finite number of constant–cross-section layers (as in object 1 of Fig. 6.5). At any z where the cross-section changes, a change of basis is performed by matching boundary conditions (the xy components of the fields must be continuous), yielding a *transfer matrix* at that interface. All these transfer and propagation matrices can then be combined to compute scattering/reflection matrices for an entire structure.

The main difference here from classical RCWA computations is that the modes are computed at imaginary frequencies ξ. As in Sect. 6.5.1.2, this actually simplifies the problem. At an imaginary frequency $\omega = i\xi$, the modes of a given cross-section $\varepsilon(i\xi, x, y)$ and $\mu(i\xi, x, y)$ with z dependence $e^{i\beta z} = e^{-\gamma z} (\gamma = -i\beta)$ satisfy the eigenequation (for isotropic materials) [135, 136]:

$$\begin{pmatrix} \xi\varepsilon + \nabla_{xy} \times \frac{1}{\xi\mu}\nabla_{xy}\times & \\ & \xi\mu + \nabla_{xy} \times \frac{1}{\xi\varepsilon}\nabla_{xy}\times \end{pmatrix} \begin{pmatrix} \mathbf{E}_{xy} \\ \mathbf{H}_{xy} \end{pmatrix} = \gamma \begin{pmatrix} & & 1 \\ & -1 & \\ -1 & & \\ 1 & & \end{pmatrix} \begin{pmatrix} \mathbf{E}_{xy} \\ \mathbf{H}_{xy} \end{pmatrix},$$

(6.26)

where the xy subscript indicates a two-component vector with xy (transverse) components. The operators on both the left- and right-hand sides are real-symmetric, while the operator on the left-hand side is positive-definite, and as a result the eigenvalues γ are purely real. This means that the propagation constants β are purely imaginary (all of the imaginary-frequency modes are evanescent in z), and the analogues of incoming/outgoing waves are those that are exponentially decaying towards/away from the surface. Moreover, the numerical problem of solving for these eigenmodes in a given cross-section reduces to a positive-definite generalized eigenvalue problem (a definite matrix pencil [69]), to which the most desirable numerical solvers apply [68, 69] (unlike the classical real-ω problem in which there are both propagating and evanescent modes because the problem is indefinite [135, 136]). For homogeneous cross-sections (as in the space between the two objects), the solutions are simply planewaves of the form $e^{ik_x x + ik_y y - \gamma z + \xi t}$, where for vacuum $\gamma = \pm\sqrt{|\mathbf{k}_{xy}|^2 + \xi^2/c^2}$.

For sufficiently simple cross-sections, especially in two-dimensional or axisymmetric geometries, it is possible to solve for the modes analytically and hence obtain the scattering matrices, and this is how the technique was first applied to the Casimir problem [18]. For more general geometries, one can solve for the modes numerically by a variety of techniques, such as by a transfer-matrix method in two dimensions [43] or by a planewave expansion (in the xy cross-section) in three dimensions [46]. Of course, one truncates to a finite number of modes via some cutoff $|\gamma|$ (which follows automatically from discretizing the cross-section in a finite grid, for example), and convergence is obtained in the limit as this cutoff increases. Given a basis of eigenmodes with some cutoff, the process of constructing the scattering/reflection matrices is thoroughly discussed elsewhere [43–47], so we do not review it here.

The strength of RCWA is that regions of uniform cross-section are handled with at most a 2d discretization of the cross-section, independent of the thickness of the region, so very thick or very thin layers can be solved efficiently. The main limitation of RCWA methods is that the transfer matrices (and the resulting reflection matrices R_1 and R_2) are dense $N \times N$ matrices, where N is the number of modes required for convergence. If N is large, as in complicated three-dimensional structures, the problem can quickly become impractical because of the $O(N^2)$ storage and $O(N^3)$ computation requirements. The most favorable case is that of periodic structures with relatively simple unit cells, in which case the problem can be reduced to that of computing the modes of each periodic unit cell (with Bloch-periodic boundary conditions) as discussed below, and RCWA can then be quite practical even in three dimensions. Non-periodic structures, such as compact objects, can be handled by perfectly matched layer (PML) absorbing boundaries [44], albeit at greater computational cost because of the increased cross-section size.

6.7 Periodicity and Other Symmetries

In this section, we briefly discuss the issue of periodicity and other symmetries, which can be exploited to greatly reduce the computational effort in Casimir calculations just as for classical EM calculations.

If a structure is periodic in the x direction with period Λ, as in Fig. 6.5, the problem simplifies considerably because one can reduce the computation to a single unit cell of thickness Λ. In particular, one imposes Bloch-periodic boundary conditions—the fields at $x = \Lambda$ equal the fields at $x = 0$ multiplied by a phase factor $e^{ik_x\Lambda}$—and computes the Casimir energy or force for each Bloch wavevector k_x separately, then integrates the result over k_x via $\int_{-\pi/\Lambda}^{\pi/\Lambda}(\cdots)dk_x$. This can be derived in a variety of ways, for example by applying Bloch's theorem [6] to decompose the eigenmodes into Bloch-wave solutions for each k_x, or by expanding the delta functions of the fluctuation–dissipation approach in a Fourier series [12]. More generally, for any periodic unit cell, one can perform the Casimir energy/ force computation for the unit cell with Bloch periodic boundaries and then integrate the Bloch wavevector **k** over the irreducible Brillouin zone (multiplied by the volume ratio of the Brillouin zone and the irreducible Brillouin zone).

The specific case of continuous translational symmetry, say in the x direction, corresponds $\Lambda \to 0$ and one must integrate over all k_x (the Brillouin zone is infinite). Certain additional simplifications apply in the case of a perfect-metal structure with continuous translational symmetry, in which case the fields decompose into two polarizations and the k integration can be performed implicitly [12].

Rotational symmetry can be handled similarly: the fields can be decomposed into fields with $e^{im\phi}$ angular dependence, and the total force or energy is the sum over all integers m of the contributions for each m [27]. More generally, the Casimir contributions can be decomposed into a sum of contributions from irreducible representations of the symmetry group of the structure (e.g. all eigenmodes can be classified into these representations [137, 138]); translational and rotational symmetries are merely special cases. As another example, in a structure with a mirror symmetry one could sum even- and odd-symmetry contributions (in fact, this is the underlying reason for the TE/TM polarization decomposition in two dimensions [6]).

6.8 Nonzero-Temperature Corrections

In the preceding sections, we discussed only the computation of Casimir interactions at zero temperature $T = 0^+$. However, the modification of any imaginary-frequency expression for a Casimir interaction from $T = 0$ to $T > 0$ is almost trivial: one simply performs a sum instead of an integral. If the $T = 0$ interaction (energy, force, etc.) is expressed as an integral $\int_0^\infty C(\xi)d\xi$ of some contributions $C(\xi)$ at each imaginary frequency ξ, then the $T > 0$ interaction is well known to be simply [81]:

$$\int_0^\infty C(\xi)d\xi \to \frac{2\pi k_B T}{\hbar} \sum_{n=0}^\infty {}' C\left(\frac{2\pi k_B T}{\hbar}n\right), \qquad (6.27)$$

where k_B is Boltzmann's constant and $\sum'$ indicates a sum with weight $\frac{1}{2}$ for the $n = 0$ term. The frequencies $\xi_n = 2\pi k_B T n/\hbar$ are known as Matsubara frequencies, and the corresponding (imaginary) Matsubara wavelengths are $\lambda_n = 2\pi/\xi_n = \lambda_T/n$ where $\lambda_T = \hbar/k_B T$. The conversion of the $T = 0$ integral into a summation can be derived in a variety of ways, most directly by considering thermodynamics in the Matsubara formalism [81]. Physically, this arises from the $\coth(\hbar\omega/2kT)$ Bose–Einstein distribution factor that appears in the fluctuation–dissipation expressions (6.3) for nonzero temperatures. When the contour integration is performed over ω, the coth introduces poles at $\hbar\omega/2kT = i\pi n$ that convert the integral into a sum via the residue theorem (with the $n = 0$ residue having half weight because it lies on the real-ω axis) [77]. As explained in Sect. 6.5.1.2, some care must be applied in evaluating the $n = 0$ term because of the well known singularity of Maxwell's equations at $\omega = 0$ (where the **E** and **H** fields decouple), and one may need to take the limit $\xi \to 0^+$ (although there is some controversy in the unphysical case of perfect metals [92]).

Mathematically, the sum of (6.27) is exactly the same as a trapezoidal-rule approximation for the $T = 0$ integral, with equally spaced abscissas $\Delta\xi = 2\pi/\lambda_T$ [53, 91, 139]. Thanks to the $O(\Delta\xi^2)$ convergence of the trapezoidal rule [53], this means that the $T > 0$ result is quite close to the $T = 0$ result unless $C(\xi)$ varies rapidly on the scale of $2\pi/\lambda_T$. In particular, suppose that $C(\xi)$ varies on a scale $2\pi/a$, corresponding to some lengthscale a in the problem (typically from a surface–surface separation). In that case, assuming $C(\xi)$ has nonzero slope[16] at $\xi = 0^+$ (typical for interactions between realistic metal surfaces), then the nonzero-T correction should be of order $O(a^2/\lambda_T^2)$. At room temperature ($T = 300\,\mathrm{K}$), $\lambda_T \approx 7.6\,\mu\mathrm{m}$, and the temperature corrections to Casimir interactions are typically negligible for submicron separations [81, 140]. On the other hand, it is possible that careful material and geometry choices may lead to larger temperature effects [139]. There is also the possibility of interesting effects in nonequilibrium situations (objects at different temperatures) [141, 142], but such situations are beyond the scope of this review.

6.9 Concluding Remarks

The area of numerical Casimir computations remains rich with opportunities. Relatively few geometry and material combinations have as yet been explored, and thus many newly answerable questions remain regarding the ways in which Casimir phenomena can be modified by exploiting the degrees of freedom available in modern nanofabrication. In the regime of computational techniques, while several effective methods have already been proposed and demonstrated, the

---

[16] If $C(\xi)$ has zero slope at $\xi = 0^+$, then the trapezoidal rule differs from the integral by $O(\Delta\xi^4)$ or less, depending upon which derivative is nonzero at $\xi = 0^+$ [53].

parallels with computational electromagnetism lead us to anticipate ongoing improvements and developments for some time to come. The same parallels also caution against any absolute "rankings" of the different approaches, as different numerical techniques have always exhibited unique strengths and weaknesses in both theory and practice. And because computer time is typically much less expensive than programmer time, there is something to be said for methods that may be theoretically suboptimal but are easy to implement (or are available off-the-shelf) for very general geometries and materials. Nor is the value of analytical and semi-analytical techniques diminished, but rather these approaches are freed from the tedium of hand computation to focus on more fundamental questions.

Acknowledgements This work was supported in part by the Army Research Office through the ISN under contract W911NF-07-D-0004, by the MIT Ferry Fund, and by the Defense Advanced Research Projects Agency (DARPA) under contract N66001-09-1-2070-DOD. We are especially grateful to our students, A. W. Rodriguez, A. P. McCauley, and H. Reid for their creativity and energy in pursuing Casimir simulations. We are also grateful to our colleagues F. Capasso, D. Dalvit, T. Emig, R. Jaffe, J. D. Joannopoulos, M. Kardar, M. Levin, M. Lončar, J. Munday, S. J. Rahi, and J. White, for their many suggestions over the years.

References

1. Chew, W.C., Jian-Ming, J., Michielssen, E., Jiming, S.: Fast and Efficient Algorithms in Computational Electromagnetics. Artech, Norwood, MA (2001)
2. Taflove, A., Hagness, S.C.: Computational Electrodynamics: The Finite-Difference Time-Domain Method. Artech, Norwood, MA (2000)
3. Volakis, J.L., Chatterjee, A., Kempel, L.C.: Finite Element Method Electromagnetics: Antennas, Microwave Circuits, and Scattering Applications. IEEE Press, New York (2001)
4. Zhu, Y., Cangellaris, A.C.: Multigrid Finite Element Methods for Electromagnetic Field Modelling. John Wiley and Sons, Hooboke, NJ (2006)
5. Yasumoto, K. (ed.): Electromagnetic Theory and Applications for Photonic Crystals. CRC Press, Boca Raton, FL (2005)
6. Joannopoulos, J.D., Johnson, S.G., Winn, J.N., Meade, R.D.: Photonic Crystals: Molding the Flow of Light, 2nd edn. Princeton University Press, Princeton, NJ (2008)
7. Jin, J.: The Finite Element Method in Electromagnetics, 2nd edn. Wiley, New York (2002)
8. Rao, S.M., Balakrishnan, N.: Computational electromagnetics. Curr. Sci. **77**(10), 1343–1347 (1999)
9. Reid, M.T.H., Rodriguez, A.W., White, J., Johnson, S.G.: Efficient computation of three-dimensional Casimir forces. Phys. Rev. Lett. **103**(4), 040–401 (2009)
10. Reynaud, S., Maia Neto, P.A., Lambrecht, A.: Casimir energy and geometry: Beyond the proximity force approximation. J. Phys. A: Math. Theor. **41**, 164–004 (2008)
11. Rodriguez, A., Ibanescu, M., Iannuzzi, D., Capasso, F., Joannopoulos, J.D., Johnson, S.G.: Computation and visualization of Casimir forces in arbitrary geometries: Non-monotonic lateral-wall forces and failure of proximity force approximations. Phys. Rev. Lett. **99**(8), 080–401 (2007)
12. Rodriguez, A., Ibanescu, M., Iannuzzi, D., Joannopoulos, J.D., Johnson, S.G.: Virtual photons in imaginary time: Computing Casimir forces in arbitrary geometries via standard numerical electromagnetism. Phys. Rev. **76**(3), 032–106 (2007)
13. Rodriguez, A.W., McCauley, A.P., Joannopoulos, J.D., Johnson, S.G.: Casimir forces in the time domain: Theory. Phys. Rev. A **80**(1), 012–115 (2009)

14. Emig, T., Graham, N., Jaffe, R.L., Kardar, M.: Casimir forces between arbitrary compact objects. Phys. Rev. Lett. **99**, 170–403 (2007)
15. Emig, T., Hanke, A., Golestanian, R., Kardar, M.: Probing the strong boundary shape dependence of the Casimir force. Phys. Rev. Lett. **87**, 260–402 (2001)
16. Emig, T., Jaffe, R.L., Kardar, M., Scardicchio, A.: Casimir interaction between a plate and a cylinder. Phys. Rev. Lett. **96**, 080–403 (2006)
17. Lambrecht, A., Maia Neto, P.A., Reynaud, S.: The Casimir effect within scattering theory. New J. Phys. **8**, 243 (2008)
18. Lambrecht, A., Marachevsky, V.I.: Casimir interactions of dielectric gratings. Phys. Rev. Lett. **101**, 160–403 (2008)
19. Xiong, J.L., Chew, W.C.: Efficient evaluation of Casimir force in z-invariant geometries by integral equation methods. Appl. Phys. Lett. **95**, 154–102 (2009)
20. Xiong, J.L., Tong, M.S., Atkins, P., Chew, W.C.: Efficient evaluation of Casimir force in arbitrary three-dimensional geometries by integral equation methods. Phys. Lett. A **374**(25), 2517–2520 (2010)
21. Rahi, S.J., Emig, T., Jaffe, R.L., Kardar, M.: Casimir forces between cylinders and plates. Phys. Rev. A **78**, 012–104 (2008)
22. Rahi, S.J., Emig, T., Jaffe, R.L., Kardar, M.: Scattering theory approach to electrodynamic casimir forces. Phys. Rev. D **80**, 085–021 (2009)
23. Pasquali, S., Maggs, A.C.: Numerical studies of Lifshitz interactions between dielectrics. Phys. Rev. A. **79**, 020–102 (2009)
24. Maia Neto, P.A., Lambrecht, A., Reynaud, S.: Roughness correction to the Casimir force: Beyond the proximity force approximation. Europhys. Lett. **69**, 924–930 (2005)
25. Maia Neto, P.A., Lambrecht, A., Reynaud, S.: Casimir energy between a plane and a sphere in electromagnetic vacuum. Phys. Rev. A **78**, 012–115 (2008)
26. Kenneth, O., Klich, I.: Casimir forces in a T-operator approach. Phys. Rev. B **78**, 014–103 (2008)
27. McCauley, A.P., Rodriguez, A.W., Joannopoulos, J.D., Johnson, S.G.: Casimir forces in the time domain: Applications. Phys. Rev. A **81**, 012–119 (2010)
28. Casimir, H.B.G.: On the attraction between two perfectly conducting plates. Proc. K. Ned. Akad. Wet. **51**, 793–795 (1948)
29. Lamoreaux, S.K.: Demonstration of the Casimir force in the 0.6 to 6μm range. Phys. Rev. Lett. **78**, 5–8 (1997)
30. Derjaguin, B.V., Abrikosova, I.I., Lifshitz, E.M.: Direct measurement of molecular attraction between solids separated by a narrow gap. Q. Rev. Chem. Soc. **10**, 295–329 (1956)
31. Bordag, M.: Casimir effect for a sphere and a cylinder in front of a plane and corrections to the proximity force theorem. Phys. Rev. D **73**, 125–018 (2006)
32. Bordag, M., Mohideen, U., Mostepanenko, V.M.: New developments in the Casimir effect. Phys. Rep. **353**, 1–205 (2001)
33. Casimir, H.B.G., Polder, D.: The influence of retardation on the London-van der Waals forces. Phys. Rev. **13**(4), 360–372 (1948)
34. Sedmik, R., Vasiljevich, I., Tajmar, M.: Detailed parametric study of Casimir forces in the casimir polder approximation for nontrivial 3d geometries. J. Comput. Aided Mater. Des. **14**(1), 119–132 (2007)
35. Tajmar, M.: Finite element simulation of Casimir forces in arbitrary geometries. Intl. J. Mod. Phys. C **15**(10), 1387–1395 (2004)
36. Jaffe, R.L., Scardicchio, A.: Casimir effect and geometric optics. Phys. Rev. Lett. **92**, 070–402 (2004)
37. Hertzberg, M.P., Jaffe, R.L., Kardar, M., Scardicchio, A.: Casimir forces in a piston geometry at zero and finite temperatures. Phys. Rev. D **76**, 045–016 (2007)
38. Zaheer, S., Rodriguez, A.W., Johnson, S.G., Jaffe, R.L.: Optical-approximation analysis of sidewall-spacing effects on the force between two squares with parallel sidewalls. Phys. Rev. A **76**(6), 063–816 (2007)

39. Scardicchio, A., Jaffe, R.L.: Casimir effects: An optical approach I. Foundations and examples. Nucl. Phys. B **704**(3), 552–582 (2005)
40. Van Enk, S.J.: The Casimir effect in dielectrics: A numerical approach. J. Mod. Opt. **42**(2), 321–338 (1995)
41. Jackson, J.D.: Classical Electrodynamics, 3rd edn. Wiley, New York (1998)
42. Schaubert, D.H., Wilton, D.R., Glisson, A.W.: A tetrahedral modeling method for electromagnetic scattering by arbitrarily shaped inhomogeneous dielectric bodies. IEEE Trans. Antennas Propagat. **32**, 77–85 (1984)
43. Bienstman, P., Baets, R.: Optical modelling of photonic crystals and VCSELs using eigenmode expansion and perfectly matched layers. Opt. Quantum Electron. **33**(4–5), 327–341 (2001)
44. Bienstman, P., Baets, R.: Advanced boundary conditions for eigenmode expansion models. Opt. Quantum Electron. **34**, 523–540 (2002)
45. Willems, J., Haes, J., Baets, R.: The bidirectional mode expansion method for two dimensional waveguides. Opt. Quantum Electron. **27**(10), 995–1007 (1995)
46. Moharam, M.G., Grann, E.B., Pommet, D.A., Gaylord, T.K.: Formulation for stable and efficient implementation of the rigorous coupled-wave analysis of binary gratings. J. Opt. Soc. Am. A **12**, 1068–1076 (1995)
47. Moharam, M.G., Pommet, D.A., Grann, E.B., Gaylord, T.K.: Stable implementation of the rigorous coupled-wave analysis for surface-relief gratings: Enhanced transmittance matrix approach. J. Opt. Soc. Am. A **12**, 1077–1086 (1995)
48. Katsenelenbaum, B.Z., Mercaderdel Río, L., Pereyaslavets, M., Sorolla Ayza, M., Thumm, M.: Theory of Nonuniform Waveguides: The Cross-Section Method (1998)
49. Christ, A., Hartnagel, H.L.: Three-dimensional finite-difference method for the analysis of microwave-device embedding. IEEE Trans. Microwave Theory Tech. **35**(8), 688–696 (1987)
50. Strikwerda, J.: Finite Difference Schemes and Partial Differential Equations. Wadsworth and Brooks/Cole, Pacific Grove, CA (1989)
51. Bonnet, M.: Boundary Integral Equation Methods for Solids and Fluids. Wiley, Chichester (1999)
52. Hackbush, W., Verlag, B.: Integral Equations: Theory and Numerical Treatment. Birkhauser Verlag, Basel, Switzerland (1995)
53. Boyd, J.P.: Chebychev and Fourier Spectral Methods, 2nd edn. Dover, New York (2001)
54. Ditkowski, A., Dridi, K., Hesthaven, J.S.: Convergent Cartesian grid methods for Maxwell's equations in complex geometries. J. Comp. Phys. **170**, 39–80 (2001)
55. Oskooi, A.F., Kottke, C., Johnson, S.G.: Accurate finite-difference time-domain simulation of anisotropic media by subpixel smoothing. Opt. Lett. **34**, 2778–2780 (2009)
56. Stratton, J.A.: Electromagnetic Theory. McGraw-Hill, New York (1941)
57. Johnson, S.G., Joannopoulos, J.D.: Block-iterative frequency-domain methods for Maxwell's equations in a planewave basis. Opt. Express **8**(3), 173–190 (2001)
58. Kuo, S.H., Tidor, B., White, J.: A meshless, spectrally accurate, integral equation solver for molecular surface electrostatics. ACM J. Emerg. Technol. Comput. Syst. **4**(2), 1–30 (2008)
59. Sladek, V., Sladek, J.: Singular Integrals in Boundary Element Methods. WIT Press, Southampton, UK (1998)
60. Taylor, D.J.: Accurate and efficient numerical integration of weakly singular integrals in Galerkin EFIE solutions. IEEE Trans. Antennas Propagat. **51**, 1630–1637 (2003)
61. Tong, M.S., Chew, W.C.: Super-hyper singularity treatment for solving 3d electric field integral equations. Microwave Opt. Tech. Lett. **49**, 1383–1388 (2006)
62. Nédélec, J.C.: Mixed finite elements in $\mathbb{R}^3$. Numerische Mathematik. **35**, 315–341 (1980)
63. Rao, S.M., Wilton, D.R., Glisson, A.W.: Electromagnetic scattering by surfaces of arbitrary shape. IEEE Trans. Antennas Propagat. **30**, 409–418 (1982)
64. Cai, W., Yu, Y., Yuan, XC.: Singularity treatment and high-order RWG basis functions for integral equations of electromagnetic scattering. Intl. J. Numer. Meth. Eng. **53**, 31–47 (2002)
65. Trefethen, L.N., Bau, D.: Numerical Linear Algebra. SIAM, Philadelphia (1997)

66. Davis, T.A.: Direct Methods for Sparse Linear Systems. SIAM, Philadelphia (2006)
67. Barrett, R., Berry, M., Chan, T.F., Demmel, J., Donato, J., Dongarra, j., Eijkhout, V., Pozo, R., Romine, C., Vander Vorst, H.: Templates for the Solution of Linear Systems: Building Blocks for Iterative Methods, 2nd edn. SIAM, Philadelphia (1994)
68. Bai, Z., Demmel, J., Dongarra, J., Ruhe, A., VanDer Vorst, H.: Templates for the Solution of Algebraic Eigenvalue Problems: A Practical Guide. SIAM, Philadelphia (2000)
69. Anderson, E., Bai, Z., Bischof, C., Blackford, S., Demmel, J., Dongarra, J., Du Croz, J., Greenbaum, A., Hammarling, S., McKenney, A., McKenney, A.: LAPACK Users' Guide, 3rd edn. SIAM, Philadelphia (1999)
70. Duff, I.S., Erisman, A.M., Reid, J.K.: On George's nested dissection method. SIAM J. Numer. Anal. **13**, 686–695 (1976)
71. Phillips, J.R., White, J.K.: A precorrected-FFT method for electrostatic analysis of complicated 3-D structures. IEEE Trans. Comput. Aided Des. **16**, 1059–1072 (1997)
72. Greengard, L., Rokhlin, V.: A fast algorithm for particle simulations. J. Comp. Phys. **73**, 325–348 (1987)
73. Greengard, L., Gueyffier, D., Martinsson, P.G., Rokhlin, V.: Fast direct solvers for integral equations in complex three-dimensional domains. Acta Numer **18**, 243–275 (2009)
74. Milonni, P.W.: The Quantum Vacuum: An Introduction to Quantum Electrodynamics. Academic Press, San Diego (1993)
75. Cohen-Tannoudji, C., Din, B., Laloë, F.: Quantum Mechanics. Hermann, Paris (1977)
76. Strang, G.: Computational Science and Engineering. Wellesley-Cambridge Press, Wellesley, MA (2007)
77. Lamoreaux, S.K.: The Casimir force: background, experiments, and applications. Rep. Prog. Phys. **68**, 201–236 (2005)
78. Ford, L.H.: Spectrum of the Casimir effect and the Lifshitz theory. Phys. Rev. A **48**, 2962–2967 (1993)
79. Nesterenko, V.V., Pirozhenko, I.G.: Simple method for calculating the Casimir energy for a sphere. Phys. Rev. D **57**, 1284–1290 (1998)
80. Cognola, G., Elizalde, E., Kirsten, K.: Casimir energies for spherically symmetric cavities. J. Phys. A **34**, 7311–7327 (2001)
81. Lifshitz, E.M., Pitaevskii, L.P.: Statistical Physics: Part 2. Pergamon, Oxford (1980)
82. Boyd, J.P.: Exponentially convergent Fourier–Chebyshev quadrature schemes on bounded and infinite intervals. J. Sci. Comput. **2**, 99–109 (1987)
83. Stroud, A.H., Secrest, D.: Gaussian Quadrature Formulas. Prentice-Hall, Englewood Cliffs, NJ (1966)
84. Piessens, R., de Doncker-Kapenga, E., Uberhuber, C., Hahaner, D.: QUADPACK: A Subroutine Package for Automatic Integration. Springer-Verlag, Berlin (1983)
85. Buhmann, S.Y., Scheel, S.: Macroscopic quantum electrodynamics and duality. Phys. Rev. Lett. **102**, 140–404 (2009)
86. Gel'fand, I.M., Shilov, G.E.: Generalized Functions. Academic Press, New York (1964)
87. Milton, K.A., Wagner, J., Parashar, P., Brevek, I.: Casimir energy, dispersion, and the Lifshitz formula. Phys. Rev. D **81**, 065–007 (2010)
88. Economou, E.N.: Green's Functions in Quantum Physics, 3rd edn. Springer, Heidelberg (2006)
89. Zhao, J.S., Chew, W.C.: Integral equation solution of Maxwell's equations from zero frequency to microwave frequencies. IEEE Trans. Antennas Propagat. **48**, 1635–1645 (2000)
90. Epstein, C.L., Greengard, L.: Debye sources and the numerical solution of the time harmonic Maxwell equations. Commun. Pure Appl. Math. **63**, 413–463 (2009)
91. Press, W.H., Teukolsky, S.A., Vetterling, W.T., Flannery, B.P.: Numerical Recipes in C: The Art of Scientific Computing, 2nd edn. Cambridge Univ. Press, (1992)
92. Hoye, J.S., Brevik, I., Aarseth, J.B., Milton, K.A.: What is the temperature dependence of the casimir effect? J. Phys. A: Math. Gen. **39**(20), 6031–6038 (2006)

93. Rodriguez, A.W., Munday, J., Dalvit, D.A.R., Capasso, F., Joannopoulos, J.D., Johnson, S.G.: Stable suspension and dispersion-induced transition from repulsive Casimir forces between fluid-separated eccentric cylinders. Phys. Rev. Lett. **101**(19), 190–404 (2008)
94. Pitaevskii, L.P.: Comment on Casimir force acting on magnetodielectric bodies in embedded in media. Phys. Rev. A **73**, 047–801 (2006)
95. Munday, J., Capasso, F., Parsegian, V.A.: Measured long-range repulsive Casimir-lifshitz forces. Nature **457**, 170–173 (2009)
96. Munday, J.N., Capasso, F.: Precision measurement of the Casimir-Lifshitz force in a fluid. Phys. Rev. A **75**, 060–102 (2007)
97. Chew, W.C., Weedon, W.H.: A 3d perfectly matched medium from modified Maxwell's equations with stretched coordinates. Microwave Opt. Tech. Lett. **7**(13), 599–604 (1994)
98. Zhao, L., Cangellaris, A.C.: A general approach for the development of unsplit-field time-domain implementations of perfectly matched layers for FDTD grid truncation. IEEE Microwave and Guided Wave Lett. **6**(5), 209–211 (1996)
99. Teixeira, F.L., Chew, W.C.: General closed-form PML constitutive tensors to match arbitrary bianisotropic and dispersive linear media. IEEE Microwave and Guided Wave Lett. **8**(6), 223–225 (1998)
100. Ward, A.J., Pendry, J.B.: Refraction and geometry in Maxwell's equations. J. Mod. Opt. **43**(4), 773–793 (1996)
101. Farjadpour, A., Roundy, D., Rodriguez, A., Ibanescu, M., Bermel, P., Joannopoulos, J.D., Johnson, S.G., Burr, G.: Improving accuracy by subpixel smoothing in FDTD. Opt. Letters. **31**, 2972–2974 (2006)
102. Kottke, C., Farjadpour, A., Johnson, S.G.: Perturbation theory for anisotropic dielectric interfaces, and application to sub-pixel smoothing of discretized numerical methods. Phys. Rev. E **77**, 036–611 (2008)
103. Zhao, S.: High order matched interface and boundary methods for the Helmholtz equation in media with arbitrarily curved interfaces. J. Comp. Phys. **229**, 3155–3170 (2010)
104. Golub, G.H., Loan, C.F.V.: Matrix Computations, 3rd edn. Johns Hopkins University Press, Baltimore, MD (1996)
105. Trottenberg, U., Ooseterlee, C., Schüller, A.: Multigrid. Academic Press, London (2001)
106. Rengarajan, S.R., Rahmat-Samii, Y.: The field equivalence principle: Illustration of the establishment of the non-intuitive null fields. IEEE Antennas Propag. Mag. **42**, 122–128 (2000)
107. Schelkunoff, S.A.: Some equivalence theorems of electromagnetic waves. Bell Syst. Tech. J. **15**, 92–112 (1936)
108. Stratton, J.A., Chu, L.J.: Diffraction theory of electromagnetic waves. Phys. Rev. **56**, 99–107 (1939)
109. Love, A.E.H.: The integration of equations of propagation of electric waves. Phil. Trans. Roy. Soc. London A **197**, 1–45 (1901)
110. Harrington, R.F.: Boundary integral formulations for homogeneous material bodies. J. Electromagn. Waves Appl. **3**, 1–15 (1989)
111. Umashankar, K., Taflove, A., Rao, S.: Electromagnetic scattering by arbitrary shaped three-dimensional homogeneous lossy dielectric objects. IEEE Trans. Antennas Propagat. **34**, 758–766 (1986)
112. Medgyesi-Mitschang, L.N., Putnam, J.M., Gedera, M.B.: Generalized method of moments for three-dimensional penetrable scatterers. J. Opt. Soc. Am. A **11**, 1383–1398 (1994)
113. Trefethen, L.N.: Is Gauss quadrature better than clenshaw–curtis. SIAM Review **50**, 67–87 (2008)
114. Cools, R.: Advances in multidimensional integration. J. Comput. Appl. Math **149**, 1–12 (2002)
115. Cools, R.: An encyclopaedia of cubature formulas. J. Complexity **19**, 445–453 (2003)
116. Berntsen, J., Espelid, T.O., Genz, A.: An adaptive algorithm for the approximate calculation of multiple integrals. ACM Trans. Math. Soft. **17**, 437–451 (1991)
117. Atkinson, K., Sommariva, A.: Quadrature over the sphere. Elec. Trans. Num. Anal. **20**, 104–118 (2005)

118. Le Gia, Q.T., Mhaskar, H.N.: Localized linear polynomial operators and quadrature formulas on the sphere. SIAM J. Num. Anal. **47**, 440–466 (2008)
119. Oskooi, A.F., Roundy, D., Ibanescu, M., Bermel, P., Joannopoulos, J.D., Johnson, S.G.: Meep: A flexible free-software package for electromagnetic simulations by the FDTD method. Comp. Phys. Comm. **181**, 687–702 (2010)
120. Rodriguez, A.W., McCauley, A.P., Joannopoulos, J.D., Johnson, S.G.: Theoretical ingredients of a Casimir analog computer. Proc. Nat. Acad. Sci. **107**, 9531–9536 (2010)
121. McCauley, A.P., Rodriguez, A.W., Johnson, S.G.: Casimir Meep wiki. http://ab-initio. mit.edu
122. Werner, G.R., Cary, J.R.: A stable FDTD algorithm for non-diagonal anisotropic dielectrics. J. Comp. Phys. **226**, 1085–1101 (2007)
123. Li, H., Kardar, M.: Fluctuation-induced forces between rough surfaces. Phys. Rev. Lett. **67**, 3275–3278 (1991)
124. Reid, M.T.H.: Fluctuating surface currents: A new algorithm for efficient prediction of Casimir interactions among arbitrary materials in arbitrary geometries. Ph.D. Thesis, Department of Physics, Massachusetts Institute of Technology (2010)
125. Emig, T.: Casimir forces: An exact approach for periodically deformed objects. Europhys. Lett. **62**, 466 (2003)
126. Gies, H., Klingmuller, K.: Worldline algorithms for Casimir configurations. Phys. Rev. D **74**, 045–002 (2006)
127. Gies, H., Langfeld, K., Moyaerts, L.: Casimir effect on the worldline. J. High Energy Phys. **6**, 018 (2003)
128. Emig, T., Graham, N., Jaffe, R.L., Kardar, M.: Orientation dependence of Casimir forces. Phys. Rev. A **79**, 054–901 (2009)
129. Waterman, P.C.: The T-matrix revisited. J. Opt. Soc. Am. A **24**(8), 2257–2267 (2007)
130. Emig, T.: Fluctuation-induced quantum interactions between compact objects and a plane mirror. J. Stat. Mech. p. P04007 (2008)
131. Emig, T., Jaffe, R.L.: Casimir forces between arbitrary compact objects. J. Phys. A **41**, 164–001 (2008)
132. Bimonte, G.: Scattering approach to Casimir forces and radiative heat transfer for nanostructured surfaces out of thermal equilibrium. Phys. Rev. A **80**, 042–102 (2009)
133. McCauley, A.P., Zhao, R., Reid, M.T.H., Rodriguez, A.W., Zhao, J., Rosa, F.S.S., Joannopoulos, J.D., Dalvit, D.A.R., Soukoulis, C.M., Johnson, S.G.: Microstructure effects for Casimir forces in chiral metamaterials. Phys. Rev. B **82**, 165108 (2010)
134. Marcuse, D.: Theory of Dielectric Optical Waveguides, 2nd edn. Academic Press, San Diego (1991)
135. Johnson, S.G., Bienstman, P., Skorobogatiy, M., Ibanescu, M., Lidorikis, E., Joannopoulos, J.D.: Adiabatic theorem and continuous coupled-mode theory for efficient taper transitions in photonic crystals. Phys. Rev. E. **66**, 066–608 (2002)
136. Skorobogatiy, M., Yang, J.: Fundamentals of Photonic Crystal Guiding. Cambridge University Press, Cambridge (2009)
137. Inui, T., Tanabe, Y., Onodera, Y.: Group Theory and Its Applications in Physics. Springer, Heidelberg (1996)
138. Tinkham, M.: Group Theory and Quantum Mechanics. Dover, New York (2003)
139. Rodriguez, A.W., Woolf, D., McCauley, A.P., Capasso, F., Joannopoulos, J.D., Johnson, S.G.: Achieving a strongly temperature-dependent Casimir effect. Phys. Rev. Lett. **105**, 060–401 (2010)
140. Milton, K.A.: The Casimir effect: recent controveries and progress. J. Phys. A: Math. Gen. **37**, R209–R277 (2004)
141. Najafi, A., Golestanian, R.: Forces induced by nonequilibrium fluctuations: The soret–Casimir effect. Europhys. Lett. **68**, 776–782 (2004)
142. Obrecht, J.M., Wild, R.J., Antezza, M., Pitaevskii, L.P., Stringari, S., Cornell, E.A.: Measurement of the temperature dependence of the Casimir-Polder force. Phys. Rev. Lett. **98**(6), 063–201 (2007)

Chapter 7
Progress in Experimental Measurements of the Surface–Surface Casimir Force: Electrostatic Calibrations and Limitations to Accuracy

Steve K. Lamoreaux

Abstract Several new experiments have extended studies of the Casimir force into new and interesting regimes. This recent work will be briefly reviewed. With this recent progress, new issues with background electrostatic effects have been uncovered. The myriad of problems associated with both patch potentials and electrostatic calibrations are discussed and the remaining open questions are brought forward.

7.1 Introduction

Nowadays, it is unclear what it means to write a review article, or a review chapter for a book, on a particular subject. This unclarity results simply from the ease with which modern digital reference and citation resources can be used; with a mere typing of a keyword or two into a computer hooked up to the internet, one has an instant review of any field of interest. As such, at the present time, review articles tend to be op-ed pieces that tend to be less than scientifically enlightening. Rather than continue in the tradition of collecting up a series of electronic database searches, I will give an overview of some recent experiments and also describe how anomalous electrostatic effects might have affected the results of these experiments. This Chapter is not meant to be a review of every paper in the Casimir force experimental measurement field, but a review of what I consider are the credible experiments, that have carried the field forward, that were performed over the last decade or so. As such, there will be little mention of experimental studies that have claimed 1% or better agreement, simply because it

S. K. Lamoreaux (✉)
Physics Department, Yale University, New Haven, CT 06520-8120, USA
e-mail: steve.lamoreaux@yale.edu

D. Dalvit et al. (eds.), *Casimir Physics*, Lecture Notes in Physics 834,
DOI: 10.1007/978-3-642-20288-9_7, © Springer-Verlag Berlin Heidelberg 2011

is unclear to me what these experiments really mean. If the reader is interested, a recent review of this 1% level work is presented in [1]. Of course I admit freely that my review presented here reflects my own opinions, however I hope the reader accepts or rejects my points based on verifiable facts and an independent scientific analysis. It must be remembered that simply because a paper appears in print, in a credible and leading journal, it is not necessarily scientifically correct or accepted by the community at large. Neither does the fact that work is funded by the DOE, NSF, or DARPA (or other funding agencies beyond the realm of the U.S.A.) guarantee its validity or broad acceptance in the scientific community. And perhaps most interestingly as a remark on the general history of science, the "consensus opinion" is not necessarily correct either. In particular, in the surface–surface Casimir force measurement field, there have been more than a few "Comments" on various papers; the interested reader would do well to ignore most, but not all, of these "Comments" as they are confusing, if not bogus, but certainly inflammatory.

Watching the field develop since my 1997 experimental result [2], which served as a watershed for new interest in surface–surface Casimir force measurements, has been fascinating. I had no preconceived notions as to how large or small the effect should be relative to the case of assumed simple perfect conductors (e.g., ignoring effects like surface plasmons), but I had no illusions as to the accuracy of my work, hence the words "Demonstration of the Casimir force" in the title of my paper. I simply did not have the time or resources to perform a study of possible systematic effects that likely limited the accuracy of my result; the precision was at the 5% level, at the point of closest approach. Again the accuracy of my result was, and remains, an open question, as it does for any experiment.

At the time the work reported in [2] was performed, there were no precision calculations of the Casimir force for real materials. Describing the metal plates with the simplest plasma model, for parallel plates, the correction to the force compared to the perfect conducting case is [3, 4]

$$\eta(d) = 1 - \frac{16}{3}\frac{c}{\omega_p d}, \tag{7.1}$$

where $\eta(d)$ is a force correction factor which varies with plate separation d, c is the velocity of light, and ω_p is the plasma frequency, where the form of the permittivity of the metal is

$$\epsilon(\omega) = 1 - \frac{\omega_p^2}{\omega^2}, \tag{7.2}$$

which is valid at high frequency. As ω approaches zero, (7.2) become invalid, and in addition the effect of static conductivity must be included also. Equation (7.1) can be easily modified for a sphere-plane geometry [2]. However, the magnitude of this correction was certainly outside what was reasonable based on the precision of my experiment, which appeared to be best described by plates with perfect conductivity. There was some skepticism regarding the lack of a finite conductivity correction in my result, and although several theorists expressed interest in performing a more accurate calculation, none did. Eventually I attempted the calculations myself, with mixed results. My calculations were based on published optical properties of Au and Cu, with the Cu calculations intended as a test case. These calculations showed roughly 10–15% (for Cu) and 20–30% (for Au) reductions in force, compared to perfect conductors, for distances of order one micron; I eventually found an error in the radius of curvature of the spherical-surface plate used in my experiment [5, 6] that lowered the experimentally measured force by 10%, but did not bring the experimental result into agreement with my Au calculation. Later work showed that Au and Cu are nearly identical, with my Cu result being the more accurate; the discrepancy was due to the way I interpolated between data points in the tabulated optical data [7]. With the refined calculation, my experiment and theory appeared to be in agreement, however by this time I was skeptical of my results, as stated in the ensuing discussion, in [8]. Interestingly enough, I had spent considerable effort trying to find corrections that would bring my experimental result into agreement with my original inaccurate calculation, so I felt that I was prepared to comment against a new theoretical result, obtained by Boström and Sernelius [9, 10], that leads to a major correction to the Casimir force between real, non-superconducting materials. This correction reduces the force by a full factor of two at large separations. More will be said of this correction later in this review; in particular, in light of new electrostatic systematic effects that have recently been discovered, the rhetoric against the result of Boström and Sernelius no longer appears as certain. In addition, all of the 1% work that was reported before [9, 10] does not show the predicted correction, nor does subsequent 1% level work. So we are faced with the possibility that the degree of precision isn't as high as stated in the 1% work, or that the theory is not at all understood. Instead of questioning experimental accuracy, new fantastic theoretical suggestions have been made, regarding the low frequency permittivity of metals, that eliminate the new correction. This remains a major open topic in the field.

There is a tendency among workers in this field to confuse precision with accuracy, of which I am guilty myself. *Precision* relates to the number of significant figures a measurement device or system provides; lots of digits can be useful for detecting small changes in some "large" parameter, assuming that the system is stable. *Accuracy* is the assignment of meaning to precision, it is the connection between accepted definitions of, for example, lengths, voltages, and forces, and the measurements that come out of an experimental apparatus. As an example, for Casimir force measurements using the sphere-plane geometry, an essential parameter is the radius of curvature of the sphere. A radius of curvature accuracy of 0.5% for a sphere of 0.2 mm diameter corresponds to 1000 nm, a bit larger than the wavelength of visible light. Thus optical measurements of adequate accuracy appear as hopeless; can electron microscopy attain this level of precision? The answer is not obvious. Of course, an experiment can be designed that does not require a high accuracy radius of curvature measurement, e.g., when the ratio of Casimir to electrostatic force is measured. Nonetheless, the attention to this problem in those works reporting 1% or better accuracy does not appear as sufficient to warrant such accuracy claims. The precision might be that level, but the cross checks required for accurate work are missing.

In general, to attain a given experimental accuracy, say 1%, requires that the calibrations and force measurements must be done to much better than 1% accuracy, particularly for comparisons between theory and experiment with no adjustable parameters. As there are possibly five or more absolute measurements that must be made to interpret an experiment, a reasonable requirement for the average calibration accuracy is 0.5%, assuming that the uncertainties can be added in quadrature (this point is open to debate; many precision measurement experts insist that the uncertainties be simply added, which bring the required average accuracy to the 0.2% level). Some of the required calibrations are as follows: The optical properties of the surfaces must be adequately characterized to allow calculation of the force to 0.5% accuracy; the radius of curvature of the spherical surface (for a sphere-plane experiment) needs to be measured to 0.5% accuracy; the absolute separation must be determined to high accuracy. This last point is perhaps the most difficult, as

$$\left| \frac{\delta F}{F} \right| = \left| n \frac{\delta d}{d} \right|, \tag{7.3}$$

where n is the exponent in the power law. For a sphere-plane geometry where $n \approx -3$ we see immediately that if we want 0.5% force accuracy as limited by the distance measurement, at the point of closest approach, say 100 nm, then the fractional error must be 0.5%/3 or about 0.17%, and when $d = 100$ nm this corresponds to $\delta d = 0.17$ nm $= 1.7$ Å. This is at the level where, in the atomic force microscopy (AFM) community, the definition of the surface location is agreed as controversial. So we see immediately that it is pointless to include any discussion of experiments that claim 1% accuracy as the radius measurement is not discussed in sufficient detail in any of the papers making such claims. My statements here should be considered as a call for details.

The general experimental techniques used in all Casimir experiments to date are rather straightforward. Many experiments employ AFM or micromechanical techniques drawn from fields that enjoy tremendous engineering support. The trick of Casimir force measurements lies in the attainment of very high force measurement sensitivity subjected to precise and rigorous calibrations, and in the elimination of long-range background electrostatic effects that can mask or distort the now-well-studied AFM signals extrapolated to very large distances. At large distances, the attractive force between two surfaces, "the" Casimir force, becomes a property of the bulk material(s) that the plates comprise, and is viewed as a fundamental physical effect arising from the quantum vacuum, as opposed to AFM signals used to detect surface roughness, for example. Experimental rigor is required to transform precision into accuracy on the fundamental vacuum effect.

Because the measurement techniques are largely borrowed from other fields, I will not give a nuts and bolts discussion of measurements in this review, for the simple reason that I know nothing about AFM techniques. Nowadays one can simply buy an AFM system from Veeco, for example, and adapt it to the samples and longer distance ranges required for Casimir measurement. There are companies that commercially produce bare cantilevers, and most engineering schools have fabrication facilities where NEMs and MEMs systems can be produced with just about any desired properties in configurations limited only by the imagination. Alternatively, my own work employs torsion balances, and the interested reader can refer to Cavendish's experiment for most details of such systems. An analysis of the force sensitivity of a torsion pendulum can be found in [11].

The principle advantage to AFM type or torsion pendulum type measurements (in fact there is no fundamental difference between them, it's a matter of scale) is elimination of stiction associated with the fulcrum type balances used in practically all earlier experiments. The proliferation of high accuracy mechanical and opto-mechanical translation stages, together with high quality digital data acquisition systems has made precision Casimir force measurement possible; the questions of accuracy are now the central theme, not the simple detection of the force.

This is not to say that the experiments are easy or simple; again, the art of the experiments lies in the attainment of high force measurement sensitivity, reliable calibrations, the production of well-characterized optical surfaces, and the elimination of background effects due to, for example, electrostatic effects. The electrostatic effects are common to all experiments, either in regard to system calibrations or systematic background effect, or both. Given the importance of electrostatic effects, I will discuss them at length in this review.

It is often said that the Casimir force is simply the retarded van der Waals potential. This view strikes me as fundamentally flawed, as the Casimir force does not depend on the properties of the individual atoms of the plates, but on their bulk properties. Indeed, the non-additivity of the van der Waals effect has been discussed at length in the literature (see [12] for a discussion and references). It is more profitable to think of the Casimir force as the zero point electromagnetic field stress on a parallel plate waveguide. This force is apparently largest when the waveguide is constructed from perfectly conducting material(s). The effects of imperfect conductivity can be calculated provided the optical constants of the material(s) are known over an adequate wavelength range. Furthermore, most of the surface–surface Casimir effect is due to conduction electrons. It is meaningless to assign a retarded van der Waals force between the individual electrons in a conductor. Likewise, if the Casimir force was simply the retarded van der Waals force, it would make little sense to consider modifying the Casimir force, in a fundamental way, by altering the mode structure imposed by specially tailored boundaries.

7.2 Motivation for the Experimental Study of the Casimir Force: Some Recent Results

The Casimir force is of fundamental interest in that it is taken as evidence for the existence of the fluctuations associated with the quantum vacuum [13]. One can almost as easily derive the Casimir force by treating the electromagnetic field classically, with the field fluctuation due to dissipation in the material bodies; this is the Lifshitz approach [14]. A principal controversy associated with the quantum vacuum interpretation lies in the fact that the zero point electromagnetic field energy, when integrated to the Planck scale (which is the natural cutoff), leads to a cosmological energy density some 130 orders of magnitude larger than observed. This is an open problem in modern physics.

There are three principal motivations for studying the Casimir force. One question is how well do we understand the basic underlying physics? This relates to the second motivation which lies in the testing for the existence of short range corrections to gravity, or a new force associated with axion exchange, for example. For such tests, the Casimir force represents a systematic background effect that must be characterized or physically eliminated by employing a shield. The third motivation comes from interest in modifying the Casimir force to eliminate stiction, for example, or make it useful in nanodevices. These categories are not mutually exclusive, and of course overlap considerably as the questions all have a fundamental element.

7.2.1 Progress in Understanding the Fundamental Casimir Force

In 2000, Boström and Sernelius [9, 10] put forward the first fundamentally new idea relating to the surface–surface Casimir effect in over 40 years, since Lifshitz's paper [14], which lies in the treatment of material permittivities in the zero-frequency limit. The problem of finite conductivity was addressed earlier by Hargreaves and later by Schwinger et al. [3, 4] who proposed a possible means to deal with it, that is, to let the surface material permittivity diverge before setting the frequency to zero. The point is that in calculating the Casimir force at finite temperature, the integral includes a Boltzmann's factor which accounts for the thermal population of the electromagnetic modes,

$$N(\omega) + \frac{1}{2} = \frac{1}{e^{\hbar\omega/k_bT} - 1} + \frac{1}{2} = \frac{1}{2}\coth\frac{\hbar\omega}{2k_bT}, \tag{7.4}$$

where $\hbar\omega$ is the energy of a photon, k_b is Boltzmann's constant, and T is the absolute temperature. Because $\coth x$ has simple poles at $x = \pm in\pi$, the integral over frequency in calculating the Casimir force can be replaced by a sum of the residues at the poles of (7.4), or Matsubara frequencies,

$$\omega_n \equiv \frac{n\pi k_bT}{\hbar}. \tag{7.5}$$

Analytic continuation of the permittivity function allows the transformation of the integral from over real frequencies to a contour integral on the complex frequency plane, and it is valid to replace the integral over frequency with a sum over the poles. The upshot is that the transverse electric (TE) mode with $n=0$ does not contribute to the force at all if the permittivity diverges slower than ω^{-2} in the limit as ω goes to zero. It is generally assumed that for metals with a finite conductivity, at zero frequency the permittivity goes as

$$\epsilon(\omega) = \frac{4\pi i\sigma}{c\omega},\tag{7.6}$$

in which case the TE $n=0$ mode does not contribute at all to the force. This is important because at room temperature, at distances greater than about 10 microns, this mode accounts for roughly half of the force. The implied correction at separations of 1 micron is about 30%. This appears to be at odds with a number of experiments, including my own. In particular, I had spent much effort in finding a correction to my experiment that would bring the results into agreement with my own incorrect calculation for Au. Thus I was well-equipped to reject this result outright, as did a number of others.

One possible solution is that the permittivity diverges as ω^{-2} as the frequency goes to zero. This has led to the proposal of a generalized plasma model [15],

$$\epsilon_{gp}(i\xi) = \epsilon(i\xi) + \frac{\omega_p^2}{\xi^2},\tag{7.7}$$

where $i\xi$ represents the frequency along the imaginary axis, ϵ is the usual Drude model permittivity, for example, and ω_p is the so-called plasma frequency due to free electrons. Normally this expansion is assumed to be valid at very high frequencies, much above the resonances in the system of atoms and charges that comprise the plates. However assuming the permittivity of this form brings back the contribution of the TE $n=0$ mode, and apparently improves the agreement between theory and experiment.

There are consequences in a broader complex of phenomena when this generalized plasma model is introduced. In particular, if we consider the interaction of a low-frequency magnetic field with a material surface, by use of Maxwell's equation, it is straightforward to show that [16]

$$-\nabla^2 H = \frac{\omega^2}{c}\epsilon(\omega)H,\tag{7.8}$$

which represents so-called eddy current effects, and can be easily extended to the complex frequency plane. We see immediately that if ϵ diverges as ω^{-2} that at zero frequency,

$$-\nabla^2 H \propto H,\tag{7.9}$$

which predicts that a static magnetic field will interact with an ordinary conductor in a manner different from universal diamagnetism. Such an extra effect is not experimentally observed, as (7.8) together with (7.6) is known to describe the non-diamagnetic interaction of low frequency fields with conductors. So we are faced with discarding over a century of electrical engineering knowledge in order to explain a few 1% level Casimir force experiments of questionable accuracy, and my own. This is not acceptable.

The crux of the problem lies in the fact that at equilibrium, all electric fields at a surface of a conductor must terminate normal to the surface [17]. An electric field parallel to a surface implies a flowing current; such currents can exist in a transitory fashion as associated with a fluctuation as required for generating the Casimir force, but such fluctuations cannot occur with zero frequency. For the *TE* modes, the electric field is parallel to the surface, so at zero frequency *TE* modes simply cannot be supported, assuming that equilibrium and zero frequency are equivalent. We will return to this problem later in this review in relation to electrostatic calibrations.

This issue is, however, not yet settled as new precise experiments are required. It is interesting that this effect becomes less pronounced at smaller separations, simply because the $n = 0$ modes contribute a relatively smaller fraction to the total force. For my own experiment [2] the possibility of a systematic error is becoming more and more apparent. It should be emphasized, however, that AFM type experiments probe an order of magnitude smaller distance scale than the torsion pendulum experiments, and the relative contributions of various effects are rapidly varying.

Work with AFMs and MEM type systems have demonstrated the difficulty of producing metal and other films, together with their characterization, that allows a comparison between experiment and theory at a level of better than 10%. For example, Svetovoy et al. [18] show that the prediction of the Casimir force between metals with a precision better than 10% must be based on the material optical response measured from visible to mid-infrared range, that the tabulated data is generally not good enough for precision work better than 10% accuracy. The issues of roughness are well-discussed in [1], however, additional new work by van Zwol et al. [19] amplifies the problems of surface roughness particularly in determining the absolute separation. See also the Chap. 10 of van Zwol et al. in this volume for additional discussions of roughness in Casimir physics. It appears that the best prospect for determining the correct form of the permittivity function at zero frequency is to do a measurement at very large separations. Indeed, problems of surface roughness correction virtually disappear for typical optical finishes at distances about 500 nm. Above 2–3 microns, the difference between the force with and without the *TE* $n = 0$ mode approaches a factor of two. Recent experimental work on Au films at Yale show that the Boström–Sernelius analysis is likely correct, but this work is at a very preliminary stage.

7.2.2 The Detection of New Long Range Forces

In the mid-1980s, the question of the possible existence of a new so-called fifth force was suggested based on data from Eötvos-type experiments [20]. Presently, interest in such forces is greater than ever due to possible modification of gravity as allowed by String Theory, and due to the observation of dark energy in the Universe which might be due to particles associated with new long range forces that could manifest themselves on many different length scales [21]. The basic idea is that our four dimensional Universe is embedded in a space of more than 10 dimensions. Leakage of lines of force between the larger space and our four dimensional world could lead to a modification of the inverse square law, for example. Although there is no specific prediction from a String theory, the possibility does exist in its context.

With the publication of my 1997 experimental result, I received many suggestions to analyze my experiment in light of an additional force that would appear along with the Casimir force, however I rejected these suggestions because my experiment was intended as a demonstration and any limit would be at the level of 100% of the Casimir force. Taken as a fraction of the gravitational field, my result was not particularly spectacular. Nonetheless, others analyzed my experiment. Among the first to do so, in the context of a general review of limits on subcentimeter forces, was Long et al. [22] and earlier, with a more detailed analysis, was Klimchitskaya et al. [23].

The most ambitious recent work on this subject is by Decca et al. [24] who achieved an astounding accuracy without observing any anomalous effects. Use of the proximity force theorem, to be discussed later in this review, to calculate the limits on a possible new force has been criticized. The issue is that the proximity force theorem really only applies to a force that depends on the location of the body surfaces; the approximation is not valid for the volume integral required for calculating the anomalous force. The applicability is addressed by Dalvit and Onofrio [25] where corrections to the calculation in [26] are pointed out.

Earlier work by Decca et al. [27] appears as more reliable at constraining new forces. The technique developed here, a so-called isoelectronic method, relied on the properties of an Au film being independent of the substrate. For different materials coated with Au films of identical optical characteristics and of sufficient thickness, the Casimir force should be the same. In this work Au/Au and Au/Ge composites are compared, and the result is "Casimir-less." Techniques such as this appear as the most likely way to achieve the best sensitivity to new forces, however, unfortunately the minimum separation is limited by the Au film thickness, hence the later work [24]. See also Chap. 9 by Decca et al. in this volume. It should be noted that use of a screening film to eliminate electrostatic forces and other background effects have been used in other "fifth force" experiments for separations at the mm scale, but clearly the trick can be scaled down to distances limited only by the skill of the experimenter (Luther, G.: Private Communication (1997)).

7.2.3 Modification of the Casimir Force

The possibility of modification of the Casimir force is a topic of current great interest. With the rising of nanotechnology, the need to control, modify, or make good use of the Casimir force is imperative as it is among the dominant forces affecting MEMs and NEMs. At very short distances, at the atomic scale, the large-scale geometrical aspects of the surfaces become irrelevant, and the force becomes dominated by the van der Waals force between atoms comprising the plates; the atom-atom force along with roughness leads to stiction and friction. At such short distances, the treatment of the plates in a continuum fashion fails. Any possibility to control either the short range or long range force can have enormous techno-logical benefits. These issues have generated renewed interest in measuring the Casimir force with improved precision, in applying it to nano-mechanical devices, and in controlling it. In many instances, the attractive nature of the force leads to more problems than to solutions because, for example, it leads to irreversible sticking of the components in a nano-device. There have been proposals to develop "metamaterials" which provide a boundary condition that makes the force repulsive, but the extremely large frequency range of electromagnetic field modes that contribute to the force suggests that this is not possible [28].

The internal sticking problem of MEMs, however, might be slightly overstated. Recent commentary relating to this possible problem has been based on the work of Buks and Roukes [29] where irreversible stiction was observed in MEMs devices. In this work, the mechanical motion was monitored by use of an electron beam which caused the components of the MEMs to become highly charged. Whether the irreversibility is really due to the Casimir force, or if it is due to charge surface interactions, remains an open question. Nonetheless, it is agreed that a full understanding of the Casimir force, and its possible control, are central to the future of MEMs and NEMs engineering. See also Chap. 8 of Capasso et al. and Chap. 9 of Decca et al. in this volume for discussions of the use of MEMS and NEMS in Casimir force measurements.

The prospects of engineering a coating that can significantly modify the Casimir force appear as dismal. This is because the Casimir force is a "broad-band" phenomenon. Use of magnetic films has been suggested, but unfortunately ferromagnetic response does not extended into the near-infrared and visible spectrum that would be required to modify the Casimir force.

Recently, it has been demonstrated experimentally that a conductive oxide film, Indium–Tin Oxide (ITO) produces a Casimir force about half of that due to metals [30]. ITO has a number of interesting features, including transparency over the optical spectrum and chemical inertness. Thus it appears as an interesting material from a nanoengineering viewpoint.

Casimir himself attempted to apply his namesake force to the electron, spe-cifically to calculate the fine structure constant. Casimir modelled the electron as a conducting ball of uniform charge that would contract due to the zero point energy of the external electromagnetic modes. This force would be balanced by the space charge repulsion of the uniform charge density, when the conducting sphere of constant total charge was just the right diameter. The fine structure constant $\alpha \approx 1/137$, which relates to the electron diameter, could then be determined from fundamental parameters along with a calculation of how the electromagnetic mode zero point energy changes as the sphere contracts [12]. However, Boyer subsequently found that the exterior spherical modes cause the sphere to expand [31]. Boyer's result was interesting enough that it led to the exploration of the effects of geometry on the Casimir force.

The change in boundary conditions that had been considered cannot be realized experimentally; for example, if one cuts a conducting sphere in half and tries to measure the force between the hemispheres, the force is different from the stress outside the continuous conducting sphere—simply because the two halves are now separated by a vacuum gap and there will be an attraction there, and because the structure of the surface modes is altered by the gap. Nonetheless, several experiments aimed at directly modifying the Casimir force have been performed in the last decade or so, and are continuing.

7.2.4 Hydrogen Switchable Mirror

An experiment with a surprising result employed a hydrogen switchable mirror, and a change in the Casimir force was sought when the mirror was switched between its low reflectivity and high reflectivity states [32]. The surprise was that no significant change in the Casimir force was observed with the switching, despite the rather dramatic change in the mirror from nearly transparent to highly reflecting.

The explanation of the null result likely lies in the construction of the mirror which has a very thin (5 nm) palladium layer to protect the underlying sensitive structure. This layer tends to dominate the Casimir effect, even though the layer is about one-half of a skin depth for the frequencies that are affect by the hydrogen switching. Other complications include the narrow spectral width of the mirror state which reduces the effect further, and the layered structure of the mirror—it is possible that the principal activity occurs in the deeper layers. In spite of these problems, hope remains that an effect on the Casimir force will be detectable [33].

7.2.5 Geometrical Boundary Effects

Until now, no significant or non-trivial corrections to the Casimir force due to boundary modifications have been observed experimentally. As mentioned above, for the systems that had previously been considered such as the conducting sphere, it is not clear that an experimental measurement of the external stress is even possible. Cutting a sphere in half clearly changes the boundary value problem; it is unlikely that the two halves of such a sliced sphere will be repelled with a force that is given by the external stress on the sphere.

However, there are other possible ways to generate a geometrical influence on the Casimir force. A conceptually straightforward way is to contour the surfaces of the plates at a length scale comparable to the mode wavelengths that contribute

Fig. 7.1 An approximately
scaled schematic
representation of the
experiment of Chan et al. The
trench arrays, of varying
width and depth, were made
from the same doped p-type
Si substrate. (Public Domain,
by S. K. Lamoreaux)

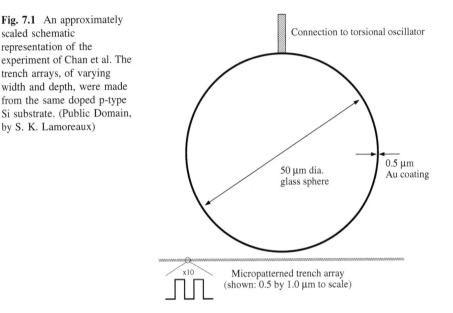

Connection to torsional oscillator

50 μm dia.
glass sphere

0.5 μm
Au coating

x10

Micropatterned trench array
(shown: 0.5 by 1.0 μm to scale)

most to the net Casimir force. For a plate separation d, the wavelengths that
contribute most are $\approx \pi d$. This means that a surface nano-patterned at 400 nm
length scale should show significant geometrical effects for separations below
1 μm. Using such a system, Chan et al. have produced a convincing measurement
of a non-trivial geometrical influence on the Casimir force [34].

These measurements, between a nanostructured silicon surface and a Au coated
sphere, were made using a micromechanical torsional oscillator. The change in
resonant frequency of the oscillator, as a function of separation between the Au
sphere and the surface, provided a measure of the gradient of the Casimir force. The
sphere, of radius 50 μm coated with 400 nm of gold, was attached to one side of the
oscillator that comprised a 3.5 μm thick, 500 μm square silicon plate suspended by
two tiny torsion rods. The sphere and oscillator were moved toward the nanostructured surface by use of a piezoelectric actuator (see Fig. 7.1 of this chapter, and also
Fig. 8.5 of the chapter of Capasso et al. for a sketch scheme of this experiment).

Two different nanostructured plates, compared with a smooth plate, were
measured in this work. The geometry of the nanostructures, rectangular trenches
etched in the surface of highly p-doped silicon, was chosen because the effects are
expected to be large in such a geometry. Emig and Büscher had previously calculated the effective modification of the Casimir force due to such a geometry, but
for the case of perfect conductors [35]. Even though the calculations were not for
real materials, these theoretical results appeared as a reasonable starting point for a
comparison with an experiment.

Although much progress has recently been made toward a realistic and
believable accuracy and precision with which the Casimir force can be calculated
for real materials [18], the problems associated with the well-known experimental

variability of sputtered or evaporated films were avoided in the work of Chan et al. by comparing two different nanostructured plates with a smooth plate, all made from the same silicon substrate, and all using the same Au coated sphere. The trick is comparable to the isoelectronic method described in Sect. 7.2.2. So even though ab initio calculations of the Casimir force for real material using tabulated optical properties cannot be accurate to better than 10%, this problem was simply circumvented by the comparison technique.

The geometric modification of the Casimir force was detected by measuring a deviation from that expected by use of the Proximity Force Approximation (PFA), or the Pairwise Additive Approximation (PAA), both of which will be described later in this review. The success of the PFA is so good that it suggests a means of detecting a geometrical effect. Basically, the surface is divided into infinitesimal units, and it is assumed that the total Casimir force can be determined by adding the Casmir force, appropriately scaled by area, between surface unit pairs in opposite surfaces; this is the PAA. Thus, for the nanostructured surfaces, a 50% reduction in force would be expected by the PAA, because the very deep trenches (depth $t = 2a \approx 1\,\mu m$), etched as a regular array, were designed to remove half of the surface. As mentioned, two different trench spacings λ were fabricated and measured, such that $\lambda/a = 1.87$ (sample A) and 0.82 (sample B), and compared to a smooth surface. The Casimir force between the gold sphere and the smooth plate, as calculated from the tabulated properties of gold and silicon, taking into account the conductivity due to the doping, agree with the experimental results to about 10% accuracy. For sample A, the force is 10% larger than expected by the PAA, using the measured smooth surface force, and for sample B, it is 20% larger, in the range $150 < z < 250$ nm. The deviation increases as λ/a decreases, as expected.

The theory of Emig and Büscher predicts deviations from the PAA twice as large as were observed. Nonetheless, the results of Chan et al. indicate a clear effect of geometry on the Casimir force. However, much theoretical work remains to be done toward gaining a complete understanding of the experimental observations. The already difficult calculations are made more so by the finite conductivity effects of the plates, and the sharp features of the trenches as opposed to the smooth simple sinusoidal corrugations. New calculational techniques have been developed that will allow reasonable accuracy calculations. See Chap. 4 of Lambrecht et al. in this volume for related discussions. Also a number of possible systematics associated with electrostatic effects were not fully investigated.

7.2.6 Repulsive Casimir Effect

The generalized Liftshitz formulation of the Casimir force allows for a material between the plates. The force is thus altered from the case of a vacuum between the plates, and the effect can be calculated. It is easy to envision filling the space between the sphere and plate of a Casimir setup with a liquid and measuring the effects of replacing the vacuum. A first experiment using alcohol between the plates was done by Munday et al. [36] where a substantial reduction in the force was observed compared to what is expected with vacuum between the plates. The effects of Debye screening and other electrostatic effects were also thoroughly studied [37].

Munday et al. extended their studies to a very interesting situation where the Casimir force becomes repulsive, by suitably choosing the permittivities of the plates and liquids. If the plates' material dielectric permittivities are ϵ_1 and ϵ_2, and the liquid between has ϵ_3, the force will be repulsive when $\epsilon_1 > \epsilon_3 > \epsilon_2$. Of course, the permittivities are frequency dependent, so this relationship must hold over a sufficiently broad range of frequencies.

Perhaps a more familiar problem is the wetting of a material surface by a liquid. In this case, one plate is replaced by air or vacuum so $\epsilon_2 = 1$, and if the liquid permittivity is less than that of the remaining plate, the liquid spreads out in a thin film rather than forming droplets. For example, liquid helium, which has a very small permittivity, readily forms a thin film because it is "repelled" by the vacuum ($\epsilon_1 > \epsilon_3 > \epsilon_2 = 1$), and we say that the liquid wets the surface. On the other hand, liquid mercury which has a high effective permittivity does not wet glass ($\epsilon_1 < \epsilon_3 > \epsilon_2 = 1$).

Although there are many liquids that wet glass or fused silica, there are only a few sets of materials that will satisfy the requirement for a repulsive force between material plates. The set employed by Munday et al. was fused silica and gold, with bromobenzene as the liquid. The experimental setup was based on an atomic force microscope (AFM) that was modified slightly for the detection of average surface forces rather than atomic-scale point forces. For measuring the Casimir force, the sharp tip was replaced by a gold coated microsphere (diameter = 39.8 microns) which serves as the gold plate. Using a spherical surface for one plate simplifies the system geometry, which is completely defined by the sphere radius and distance of closest approach from the flat fused silica plate.

A problem that all Casimir force experiments face is the system force calibration. For this work and related work, a most clever calibration technique was devised. Because the fluid produces a hydrodynamic force when the sphere/plate separation is changed, and this force is linear with velocity, subtracting the force when the separation is changed at two different speeds produces the hydrodynamic force without any contribution from the Casimir force. The hydrodynamic force thus measured, which can be calculated to high accuracy, provided the calibration. In addition, this force, scaled to the appropriate velocity, was then subtracted from the force vs. distance measurement, yielding a clean measurement of the Casimir force. The measurements spanned a range of 20 nm to several hundred nm, with the minimum distance limited by surface roughness, and the maximum distance limited by system sensitivity. Various spurious effects were accounted for and shown to have no significant contribution within the statistical accuracy of the measurement.

Showing that it is indeed possible to produce and measure a repulsive Casimir force is important to both fundamental physics and to nanodevice engineering. There has been much discussion of such forces as they will provide a means of quantum levitation of one material above another. Even in a fluid, it will be possible to suppress mechanical stiction and make ultra-low friction sensors and devices. It might be possible to "tune" the liquid (e.g., by use of a mixture) so that at sufficiently large distances, the force becomes attractive, while being repulsive at short distances. This would allow objects to levitate above a liquid covered surface, for example. See Chap. 8 of Capasso et al. in this volume for related discussions.

7.3 Approximations, Electrostatic Calibrations, and Background Effects

Wittingly or unwittingly, many approximations have been included in all Casimir force experiments to date. For example, most experiments employ the use of an electrostatic force from accurately measured applied voltage for calibrations and the detection of spurious contact potentials between the plates. The force is assumed to follow the form

$$F(d) = \frac{1}{2} \frac{\partial C(d)}{\partial d} V^2, \tag{7.10}$$

where $C(d)$ is the capacitance between the Casimir plates, as a function of distance d between them. An exact calculation exists between a sphere and a plane, however, for most situations the so-called Proximity Force Approximation (PFA) can be used. In the case of a plate with spherical surface with curvature R, with a distance d at the point of closest approach to a plane surface, the force between the two plates is

$$F(d) = 2\pi R \mathcal{E}(d), \tag{7.11}$$

where $\mathcal{E}(d)$ is the energy per unit area between plane parallel surfaces that leads to the attractive force.

Briefly, the PFA was introduced by Deryagiun [39] to describe the Casimir force between curved surfaces, and this approximation is known to be extremely accurate when the curvature is much less than the separation between the surfaces. The PFA can be used beyond the Casimir force and has quite general applicability [40,41]. The PFA is a special case of the Pairwise Additive Approximation (PAA) where the plate surfaces are divided into infinitesimal area elements, and the force is determined through a pairwise addition of corresponding elements. The PFA and PAA work very well for electrostatic effects because, for a conductor (even poor) in equilibrium, the electric lines of force must be normal to the surface, otherwise currents would flow in contradiction to the assumption that the system is in equilibrium.

The use of a sphere and a flat plate vastly simplifies an experiment because the system is fully mechanically defined in terms of the point of closest approach and the radius of curvature of the sphere. For two flat plates the system is specified by two tilt angles, the areas, long-scale smoothness, and a separation, which all need to be defined, measured, and controlled. It is interesting to note that if the force is measured as a function of applied voltage in the sphere-plane configuration that the result should be

$$F(d) = \frac{\pi \epsilon_0 R}{d} V^2 = \alpha V^2, \tag{7.12}$$

where ϵ_0 is the permittivity of free space, and R is the radius of curvature of the spherical surface. The absolute distance between the sphere and the plane surface is proportional to α^{-1} and this provides a means of determining the distance.

Even when the full form of the sphere-plane capacitance is used in (7.10), approximations still exist. Specifically, there are additional terms to the force given by (7.10) because the capacitance is in fact a tensor. This can be easily seen, as when a charged sphere is bisected, the two halves repel each other, with a force

$$F = \frac{q^2}{8R},$$

where q is the charge on the sphere [17] (Prob. 2, Sect. 5). Note that this is the force for a fixed charge, which must be modified for a fixed voltage. The point is that the two halves experience a force, even though their potential difference is zero; there are apparently additional terms that need to be added to (7.10). As the geometry is not critical in this argument, we can conclude that if the two plates of a Casimir experiment are at the same non-zero potential, there will be an additional repulsive force between them. This sort of effect has not been considered at all.

The other problem that has received significant attention only recently is the effect of patch potentials on a conducting surface. The effect is well-known, and *is largest with clean samples* because when dirt is present, ions tend to accumulate at the boundaries between the patches, shielding the effect [17] (Sect. 23).

To date, every Casimir experiment that has bothered measuring the contact potential as a function of distance has shown an apparent distance dependence of that potential. Various experiments are nicely reviewed in [42]. The basic essential problem manifests itself in anomalous behavior in the electrostatic calibration of an experiment, for example, as experienced in [43]. It was suggested that the anomalous effects that were observed are due to irregularities of the spherical surface. Roughness effects [44] certainly can cause problems at short distances, but the possibility that the anomalous effects are due to simple geometrical effects is credibly discarded in [42].

The contact potential is simply measured by finding a voltage potential difference V_m between the two plates that minimizes the force given by (7.10). V_m is manifest as an asymmetry in the force between $\pm V$ applied between the plates.

In Figs. 7.2 and 7.3 we show a picture of our experiment [45] with Germanium (Ge) plates and a general scheme of the control system of the torsion pendulum apparatus. We were initially confused because a $1/d^{1.2}$ to $1/d^{1.5}$ force persisted

Fig. 7.2 A photograph of the apparatus, in operation, used to measure the attractive force between Ge plates. The glass bell jar introduces some distortion; visible are the "compensating plates" on the left of the torsion pendulum, and the plates (2.54 cm diameter) between which the Casimir force is measured, on the right. A ThorLab T25XYZ translation stage is used to position the "fixed" plate. The fine tungsten torsion wire is not visible. (Public Domain, by S.K. Lamoreaux)

Fig. 7.3 A schematic drawing of the control system of the torsion pendulum apparatus. (Public Domain, by S.K. Lamoreaux)

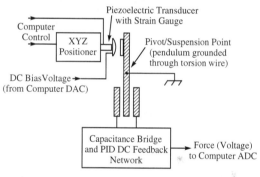

when the electrostatic force was minimized at each distance. Our initial conclusion was that there was a distance offset, as described in the next section, together with an uncompensated voltage offset. de Man et al. [46] have also observed a distance dependence of the contact potential, and concluded that it did not lead to any anomalies in their electrostatic calibrations, however, the measurements are at shorter distances than were used in the Ge experiment. In general, the relative electrostatic effect, compared to the Casimir force, should scale roughly as $(1/d)/(1/d^3) = d^2$. I will now tell the story of how we came to understand the results of our measurements using Ge plates.

7.3.1 Inclusion of the Debye Screening Length?

In the early calibrations of our Ge plate Casimir experiment [45], we had a long-range background force that depended on distance not quite as $1/d$, as described above. Our initial guess was that there was a distance offset in our calibrations due to penetration into the plates of the calibration electric field. The problem is that a quasi-static electric field can propagate a finite distance into a semiconductor (see, e.g., [47]); this distance is determined by the combined consideration of diffusion and field driven electric currents, leading to an effective field penetration length (Debye–Hückel length)

$$\lambda = \sqrt{\frac{\epsilon \epsilon_0 kT}{e^2 c_t}}, \tag{7.13}$$

where $c_t = c_h + c_e$ is the total carrier concentration, which for an intrinsic semiconductor, $c_e = c_h$. For intrinsic Ge $\lambda \approx 0.6$ μm, while for a good conductor, it is less than 1 nm. λ is independent of the applied field so long as the applied field E times λ is less than the thermal energy, $k_b T$ where k_b is Boltzmann's constant. In this limit, and at sufficiently low frequencies and wavenumbers, thermal diffusion dominates the field penetration into the material. A sufficiently low frequency for Ge would be $v_c/\lambda \sim 10$ GHz, where v_c is a typical thermal velocity of a carrier.

The potential in a plane semiconductor, if the potential is defined on a surface $x = 0$ is

$$V(x) = V(0)e^{-|x|/\lambda}, \tag{7.14}$$

where λ is the Debye–Hückel screening length, defined previously.

We are interested in finding the electrostatic energy between two thick Ge plates separated by a distance d, with voltages $+V/2$ and $-V/2$ applied to the back surfaces of the plates. In this case, the field is normal to the surface. After we find the energy per unit area, we can use the proximity force approximation to get the attractive force between a spherical and flat plate.

Let $x = 0$ refer to the surface of the plate 1, and $x = d$ refer to the surface of plate 2. By symmetry, the potential at the center position between the plates is zero. The potential in plate 1 can be written as

$$V_1(x) = V/2 - (V/2 - V_s)e^{-|x|/\lambda}, \tag{7.15}$$

and for the space between the plates

$$V_0(x) = -2V_s x/d + V_s,$$

where we assume the field is uniform. V_s, the surface potential, is to be determined.

We need only consider the boundary conditions in plate 1, which are

$$V_1(-\infty) = V/2,$$

$$V_0(0) = V_1(0),$$

(which has already been used)

$$\epsilon \frac{dV_1(x)}{dx}\Big|_{x=0} = \frac{dV_0(x)}{dx}\Big|_{x=0},$$

where the last two imply that $D = \epsilon E$ is continuous across the boundary.
The solution is

$$V_s = \frac{V}{2}\left(\frac{1}{1 + 2\lambda/\epsilon d}\right). \tag{7.16}$$

With this result, it is straightforward to calculate the total field energy per unit area
in both plates and in the space between the plates. The result is

$$\mathcal{E} = \frac{1}{2}\frac{\epsilon_0 V^2}{d}\left[\frac{y + y^2}{(y + 2)^2}\right], \tag{7.17}$$

where the dimensionless length $y = \epsilon d/\lambda$ has been introduced. By expanding this
result for small y, it can be easily seen that the effect appears as an apparent offset
in the distance that is determined by measuring the capacitance between the plates.
For small voltages, this offset is approximately $\lambda/\epsilon = 0.68/16 \approx 0.05$ μm.

If $V - V_s$ is large compared to $k_b T$, the effective penetration depth increases
because the charge density is modified in the vicinity of the surface. The potential
in the plates is no longer a simple exponential, however one can define an effective
shielding length [47]

$$\frac{\lambda'}{\lambda} = \frac{|\phi|}{\sqrt{e^\phi + e^{-\phi} - 2}}, \tag{7.18}$$

where

$$\phi = \frac{V - V_s}{k_b T}. \tag{7.19}$$

Given that $k_b T = 30$ meV, at plate separations of order 1 μm for Ge this begins to
be a large correction when voltages larger than 60 mV are applied between the
plates, however the potentials used in our experiment were far smaller.

We eventually realized that this effect is not present at very low frequencies; the
lifetime of Ge surface states is on the order of milliseconds. The lack of penetration of quasi-static fields into semiconductors was first observed in the development of the field effect transistor, and explained by Bardeen [48] as shielding
due to surface states. Again, in equilibrium, the electric field must enter normal to
the plate surfaces, otherwise a current would be flowing in contradiction to the
assumption of equilibrium. Therefore, even on very poor conductors, charges
rearrange to force any applied field to be perpendicular to the surface; when this
situation is attained, the electric field terminates at the surface. The boundary
condition is that of a perfect conductor.

The presence of time-dependent surface states might be responsible for some
of the anomalous electrostatic calibration effects observed by Kim et al. [43].
Particulary if there is a slight oxide coating on a metal surface, the surface states
might not have enough time to reach equilibrium in the dynamic measurement
system that was employed. The relaxation times for trapped surface states can be

many milliseconds. However, the possibility that these sorts of states contribute to the anomalous effect is very speculative, and it is difficult to come up with an experiment to check this hypothesis.

As an aside, our consideration of this effect led us to the realization that the usual permittivity treatment of materials with non-degenerate conduction electrons is not correct, but must be solved in a different way than simply assigning a conductivity to the material [49, 50]. The discussion of this theoretical point is beyond the scope of this review.

7.3.2 Variable Contact Potential

It was recognized that a distance dependence of the minimizing potential would lead to extra electrostatic forces that are not necessarily zero at the minimizing potential [51]. The force at the voltage which minimizes the force at each separation was thought to represent the pure "Casimir" force between the plates. However, the applied voltage $V_a(d)$ required to minimize the (electrostatic) force is observed to depend on d, and is of the form (in the 1–50 μm range)

$$V_a(d) = a \log d + b, \tag{7.20}$$

where a and b are constants with magnitude of a few mV. This variation leads to a long-range $1/d$ -like potential for the minimized force. An analysis suggests that this force is better described as $1/d^m$ where $m \approx 1.2-1.4$.

As we show here, the variation in $V_a(d)$ implies an additional force that increases as $1/d^{1.25}$, assuming that the voltage variation is due to the potential of the plates actually changing with distance. Such changes could come about due to external fixed fields or potential variations associated with the plate translation mechanism, and is equivalent to having an adjustable battery in series with the plates. We were unable to come up with a model that can give a sufficiently large effect based on interactions between, for example, the charge carriers in the plates. However, at sufficient sensitivity, it is likely that such effects will be important.

This analysis, while it predicts the correct form of the extra force, predicts that this force is negative or repulsive. However, it is enlightening to go through the analysis, and this work will never be published elsewhere. An understanding of the specific origin of the variation of applied minimizing potential $V_a(d)$ is not necessary to correct for the additional force that it causes, we simply need the experimentally determined $V_a(d)$, and assume it is tied to the plate positions.

We note further that $V_a(d)$ is not a measure of the contact potential, but the voltage which minimizes the force. We call the "true" contact potential $V_c(d)$, which might depend on distance.

In performing our experiment, at each separation d, V_a is varied and its value that minimizes the attractive force is determined. That is, we assume the force is proportional to the derivative of the capacitance between the plate, times the square of the potential difference between them. However, this assumption is not necessarily correct, as we observe that the force minimizing potential varies with distance, with value $V_a(d)$. In order to assess the implied effects of $V_a(d)$, let us first consider the energy as a function of position, assuming that the two plates are equipotential surfaces (not necessarily true), let us determine a relationship between $V_a(d)$ and $V_c(d)$ (which is unknown). In order to do this, we assume that V_a is an independent variable:

$$\mathcal{E}(d) = \frac{1}{2} C(d)(V_a + V_c(d))^2, \tag{7.21}$$

where $C(d)$ is the capacitance between the plates, V_a is the applied potential which can be varied, and $V_c(d)$ is the contact potential between the plates (fixed and unknown), assuming equipotential surfaces.

The force between the plates is given by the derivative of $\mathcal{E}$,

$$F(d) = \frac{\partial \mathcal{E}(d)}{\partial d} = \frac{1}{2} \frac{\partial C(d)}{\partial d}(V_a + V_c(d))^2 + C(d)(V_a + V_c(d))\frac{\partial V_c(d)}{\partial d}. \tag{7.22}$$

Now the minimum in the force is determined by the derivative with V_a:

$$\frac{\partial F(d)}{\partial V_a} = \frac{\partial C(d)}{\partial d}(V_a + V_c(d)) + C(d)\frac{\partial V_c(d)}{\partial d} = 0, \tag{7.23}$$

which determines $V_a(d)$, no longer an independent variable. Thus,

$$\frac{\partial V_c(d)}{\partial d} = -\frac{1}{C(d)}\frac{\partial C(d)}{\partial d}(V_a(d) + V_c(d)), \tag{7.24}$$

which allows the determination of $V_c(d)$ when $V_a(d)$ is known. The differential equation can be solved numerically, noting that at long distances $V_a(d) = -V_c(d)$, and that $V_c(d)$ becomes constant.

The electrostatic force between the plates at the minimized potential is given by

$$F(d) = -\frac{1}{2}\frac{\partial}{\partial d}\left[C(d)(V_a(d) + V_c(d))^2\right]. \tag{7.25}$$

There are some nice features to this result. First, if we apply a constant offset V_0 to $V_c(d)$, this effect is compensated by $V_a(d) - V_0$ which is easily seen as the relationship is linear.

Unfortunately, the sign of the effect indicates that it is repulsive, and thus is not the explanation of the long range attractive force that persists at the minimizing potential, after correcting for $1/d$ force, as observed in our Ge experiment. That is, the variation in contact potential, parameterized as $V_c(d)$ is not due to a long-range effect between the plates, affecting for example the surface charge densities.

Clearly, however, a position variation in V_a will lead to extra terms in the force. It should be emphasized that any precision measurement of the Casimir force requires verification that the contact potential is not changing as a function of distance, and if it is, some correction to the force as described here might very well exist. In the next subsection, we will explore another model that produces a variation in V_a, and describes well the results obtained with our Ge measurements.

7.3.3 Patch Potential Effects

It is often assumed that the surface of a conductor is an equipotential. While this would be true for a perfectly clean surface of a homogeneous conductor cut along one of its crystalline planes, it is not the case for any real surface which can be polycrystalline, stressed, or chemically contaminated. Experiments show that even with precautions for extreme cleanliness, typical surface potential variations are on the order of at least a few millivolts [52]. This is most likely due to local variations in surface crystalline structure, giving rise to varying work functions and hence varying-potential patches. It is well known that the work function of a metal surface depends on the crystallographic plane along which it lies; as an example, for gold the work functions are 5.47, 5.37, and 5.31 eV for surfaces in the $\langle 100 \rangle$, $\langle 110 \rangle$, and $\langle 111 \rangle$ directions respectively. This variation is most likely due to different effective electron masses, hence Fermi energies, for the different axes.

The means by which surface potential patches form is described in [17], Sect. 22. Briefly, when two conductors, A and B, of different work functions are brought into contact, electrons flow until the chemical potential (i.e., the Fermi energy) in both conductors equalizes. If we consider moving an electron in a closed path that moves from inside conductor A, across the boundary to inside conductor B, through the surface of B into the vacuum, back through surface A, and to the starting point, the total work must be zero in equilibrium. If we take the contact potential difference between the conductors as ϕ_{ab}, and the surface work functions as W_a and W_b, for the total work to be zero we must have

$$\phi_{ab} = W_b - W_a,$$

implying that the contact potential is simply the difference in the surface work functions.

It is straightforward to calculate the electric field energy of random patches, as has been done by Speake and Trenkel [53]. Consider two plane and parallel surfaces separated by a distance d. Assume a potential $V = 0$ at $x = 0$, while at $x = d$, $V = V_0 \cos ky$. It is easy to show that, in the region between the plates,

$$V(x, y) = V_0 \cos ky \frac{e^{kx} - e^{-kx}}{e^{kd} - e^{-kd}}.$$

The field energy, per unit area is given by

$$\mathcal{E} = \int_0^d \left[\left(\frac{\partial V}{\partial x} \right)^2 + \left(\frac{\partial V}{\partial y} \right)^2 \right] dx,$$

where we have used the fact that $\langle \cos^2 ky \rangle = \langle \sin^2 ky \rangle = 1/2$ to do the y integral. Letting

$$u = e^{kx} - e^{-kx} \quad dv = e^{kx} - e^{-kx},$$

so

$$du = k[e^{kx} + e^{-kx}] \quad v = \frac{1}{k}[e^{kx} + e^{-kx}],$$

and integrating by parts

$$\int_0^d [e^{kx} - e^{-kx}]^2 dx = \frac{1}{k}[e^{2kx} - e^{-2kx}]\big|_0^d - \int_0^d [e^{kx} + e^{-kx}]^2 dx.$$

The LHS is proportional to the field energy for E_y while the last term on the RHS is proportional to (minus) the field energy for E_y. We thus have

$$\mathcal{E} = k\frac{V_0^2}{2}\frac{e^{2kd} - e^{-2kd}}{[e^{kd} - e^{-kd}]^2}.$$

By use of the proximity force approximation, the (attractive) force between a flat surface and spherical surface is $F(d) = 2\pi R \mathcal{E}(d)$ where R is the radius of curvature, where d is the point of closest approach between the surfaces. In the limit $kd \to 0$,

$$F = 2\pi R \frac{V_0^2}{4d} \propto \frac{1}{d}.$$

This shows that when $kd \ll 1$ or $d \ll \lambda/2\pi$ where λ is a characteristic length of a potential patch, the force goes as $1/d$. This is what we expect from the PAA when the surfaces are very close.

There is an intermediate range where the force transforms from $1/d$ to exponential variation; at further distances, the force becomes a constant, as $\mathcal{E}$ does not vary with d. Between parallel plates, at long distances, the force is zero because the field energy does not change with separation. It is interesting to note this significant difference between the PFA result for a spherical surface and the result for parallel plates. As a constant force is in reality unobservable, this long distance force should be subtracted from the PFA result.

It should be noted that the field equations are linear, so we can add other $\cos(k'y)$, $\cos(k'z)$ fluctuations, and the integral over z, y leads to delta functions of $k - k'$. We can therefore rewrite the attractive force as an integral over k_y, k_z where we have $V_{k_y}(y) + V_{k_z}(z)$ representing the amplitude spectrum in k space of the surface fluctuations. If we take $V_{k_y}(y) \sim V_{k_z}(z)$ and assume they are uncorrelated, the integral over k_y, k_z leads to

$$F = \pi R V_{rms}^2 \int_0^\infty (2\pi k \; dk)(kS(k)) \frac{e^{2kd} - e^{-2kd}}{\left[e^{kd} - e^{-kd}\right]^2},$$

where, by use of the Wiener-Khinchine theorem, $S(k)$ is the normalized cosine Fourier transform (in polar coordinates) of the surface potential spatial correlation function.

In order to compute the patch effect on the force in the sphere-plane configuration we make use of the proximity force approximation. Just as in the case of roughness in Casimir physics [44], one must distinguish between two PFAs: one is for the treatment of the curvature of the sphere (valid when $d \ll R$, where R is the radius of curvature), and the other one is the PFA applied to the surface patch distribution (valid when $kd \ll 1$). We assume that we are in the conditions for PFA for the curvature, but we keep kd arbitrary. Then, the electrostatic force in the sphere-plane case is $F_{sp}(d) = 2\pi R \mathcal{E}(d)$, implying

$$F_{sp} = 2\pi\epsilon_0 R \int_0^\infty dk \frac{k^2 e^{-kd}}{\sinh(kd)} S(k). \tag{7.26}$$

There are a number of models that can be used to describe the surface fluctuations. The simplest is to say that the potential autocorrelation function is, for a distance r along a plate surface,

$$\mathcal{R}(r) = \begin{cases} V_0^2 & \text{for } r \leq \lambda, \\ 0 & \text{for } r > \lambda. \end{cases} \tag{7.27}$$

Then, by the Wiener-Khinchin theorem, the power spectral density $S(k)$ can be evaluated as the cosine two-dimensional Fourier transform of the autocorrelation function, which in our notation is [54]

$$S(k) = V_0^2 \lambda^2 \frac{J_1(\lambda k)}{\lambda k}, \tag{7.28}$$

with J_1 the Bessel function of first kind. The plane-sphere force is then given by, using $k = u/\lambda$,

$$F_{sp} = 2\pi\epsilon_0 R \int_0^\infty du \; u \frac{J_1(u)}{e^{2ud/\lambda} - 1}. \tag{7.29}$$

A numerical calculation shows that, for $d < .01\lambda$,

$$F_{sp} \approx \frac{\pi\epsilon_0 R V_0^2}{d}, \tag{7.30}$$

suggesting that $V_{rms}^2 = V_0^2$, as expected. For $50\lambda > d > \lambda$, the force falls with distance as $1/d^3$.

We see immediately that at short distances, there is a residual force due to patches that varies as $1/d$, and there is no minimizing potential that can compensate this effect. It is, in a restricted sense, equivalent to having an oscillating potential between the plates; there is no way for a static field to compensate the oscillating field energy.

As described in the last section, in our own work [45] and in a number of other experiments [43, 46], a distance-variation in the electrical potential minimizing the force between the plates has been observed. It had been suggested already that this variation in contact potential can cause an additional electrostatic force, and an estimate was made for the possible size of the effect [51]. However, further experimental work shows that the model used in [51], where the varying contact potential is considered to be a varying voltage in series with the plates, does not reproduce our experimental results [55, 56].

A model that produces a residual electrostatic force consistent with our observations [45] is shown in Fig. 7.3. In this figure, the two capacitors (short distance, $C_a(d)$, long distance $C_b(d + \Delta)$) create a net force on the lower continuous plate (setting $V_1 = 0$ initially),

$$F(d, V_0) = -\frac{1}{2}C'_a V_0^2 - \frac{1}{2}C'_b (V_0 + V_c)^2, \tag{7.31}$$

where

$$C'_a = \frac{\partial C_a(d)}{\partial d}; \quad C'_b = \frac{\partial C_b(d + \Delta)}{\partial d}, \tag{7.32}$$

and V_0 can be varied, with V_c a fixed property of the plates. The force is minimized when

$$\left.\frac{\partial F(d, V_0)}{\partial V_0}\right|_{V_0 = V_m} = 0 \Rightarrow V_m(d) = -\frac{C'_b V_c}{C'_a + C'_b}, \tag{7.33}$$

implying a residual electrostatic force

$$F_{res}^{el}(d) = F(d, V_0 = V_m(d))$$
$$= -\left[C'_a + \frac{C'^2_a}{C'_b}\right]\frac{V_m^2(d)}{2} = -\left[\frac{C'_a C'_b}{C'_a + C'_b}\right]\frac{V_c^2}{2}. \tag{7.34}$$

It is easy to take a case of parallel plate capacitors ($C'_a = -\epsilon_0 A/d^2$ and $C'_b = -\epsilon_0 A/(d + \Delta)^2$, where A is the area of each of the upper plates in Fig. 7.3, assumed to be equal; hence, the lower continuous plate has area $2A$) and to show that there is a residual electrostatic force at the minimizing potential. Indeed, in such case,

$$V_m(d) = -V_c \frac{d^2}{d^2 + (d + \Delta)^2}, \tag{7.35}$$

Fig. 7.4 A toy model
illustrating the mechanism for
the generation of a distance-
dependent minimizing
electrostatic potential $V_m(d)$
and electrostatic residual
force $F_{res}^{el}(d)$. (Public
Domain, by S.K. Lamoreaux)

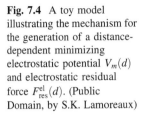
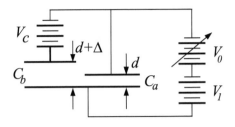

$$F_{res}^{el}(d) = \frac{\epsilon_0 A}{2} \frac{V_c^2}{d^2 + (d+\Delta)^2}. \tag{7.36}$$

Alternatively, in terms of $V_m(d)$ (up to V_1, see below), the force is

$$F_{res}^{el}(d) = \frac{\epsilon_0 A}{2} \frac{V_m^2(d)[d^2 + (d+\Delta)^2]}{d^4}. \tag{7.37}$$

Experimentally, $V_m(d)$ cannot be directly measured; measurements can only determine it up to an overall offset V_1 which arbitrarily depends on the sum of contact potentials in the complete circuit between the plates. Therefore the force should be written as proportional to $(V_m(d) + V_1)^2$ instead of simply $V_m^2(d)$, where V_1 is determined by a fit to experimental data, for example. In the limit $\Delta \gg d$, the residual force is proportional to $1/d^4$ in the plane-offset plane case here considered (see Fig. 7.4).

If we now consider the sphere-plane case, $C_a'(d) = -2\pi\epsilon_0 R/d$, and the denominator of (7.37) becomes d^2. If we further consider the surface divided up into infinitesimal areas, each with a random potential, and integrate over the surface to get the net force, there is a further reduction of the power of d in the denominator (just as in the proximity force approximation), leaving the sphere-plane force proportional to $1/d$. This motivates writing the residual force as

$$F_{res}(d) = \frac{\pi\epsilon_0 R \left[(V_m(d) + V_1)^2 + V_{rms}^2 \right]}{d}, \tag{7.38}$$

where it is understood that $V_m(d)$ is experimentally measured, and V_1 is a fit parameter that represents a sort of surface average potential, plus circuit offsets (this equation is supported both by numerical studies and by our experimental results [45, 55, 56], and is valid when $|V_1| \gg |V_m(d)|$) as observed. The last term in (7.38) is the expected random (i.e., does not contribute to $V_m(d)$) patch potential force, but here should be thought of as a fit parameter that reflects the magnitude of V_{rms}. With this result, the long range force observed in our experiment could be explained, and our work with Ge was completed. The agreement with theory is excellent, however, there is very little difference in theoretical prediction of the force with and without the $TE\ n = 0$ mode, so this work was not able to help with that controversy.

As a final note, the variations in surface potential could be a simple function of position on the conducting surface, for example, due to stresses or impurities within the samples. Alternatively, if there is a slight roughness to the surface, the peaks could have different potentials than the valleys associated with surface irregularities. This latter possibility appears to be a better model as we were unable to detect a variation in V_m when the plates were moved relative to each other, which might be expected for positional surface patches. However, the level of the surface fluctuations is quite small, and for example is beyond the range of state of the art Kelvin probes [57, 58]. These issues need further investigation.

7.4 Conclusions and Outlooks

In many respects, we can consider the measurement of the Casimir force between surfaces as a mature field. However, many open issues remain, particularly in the limits of accuracy that can be expected. In recent years, we have seen a number of experiments claiming 1% precision, but many counter claims that such accuracy is beyond what is possible due to finite knowledge of a plethora of corrections and required absolute calibrations. Some open issues include the effects of finite conductivity on the contribution of the TE $n = 0$ surface mode; the usual Drude model of the permittivity of a metal suggests that this mode does not contribute at all to the force, reducing the force by a factor of two at large separations. It is unclear whether additional short-range AFM type measurements will clear this problem up, as at short distances, the correction is relatively small. Improved measurements at distances above a few microns would appear to offer the best prospects for bringing these issues to closure. Recent work with our torsion pendulum system at Yale seems to be in favor of the no-TE $n = 0$ mode, although the precision is not yet sufficient to make a strong claim. Over the next few months we hope to have new higher accuracy data analyzed.

The effects of patch potentials has not been fully investigated in all experiments to date. For example, in my 1997 experiment [2], an anomalous component to the $1/d$ force would result in an error in the distance determination, which only needed to be 0.1 micron to bring my experiment into agreement with the Boström and Sernelius calculation. Likewise, the boundary modification experiment of Chan et al. [34] did not consider in any obvious way excess forces due to electrostatic patch effects, which might be expected to be substantial due to the sharp features of the etched silicon trenches, and will vary as $1/d^3$ in the limit of the separation much larger than the trench spacing. It is hard to imagine that such an effect is more than 10% of the Casimir force, but some analysis and additional experiments are necessary to eliminate the possibility of such a systematic effect.

In any case, a reasonable ultimate experimental goal is the attainment of 1% agreement between theory and experiment, in terms of true accuracy; this is not a question of simple precision. Hopefully the readers of this review will realize the complexity and difficulty of the challenge presented by this goal.

Acknowledgements I thank my colleagues and collaborators W.-J. Kim, A.O. Sushkov, H.X. Tang, and D.A.R. Dalvit for many fruitful discussions that led to the understanding of our Ge experiment, and to deeper understanding of the Casimir force in general. I also thank R. Onofrio and S. de Man for a number of discussions over the last few years that were helpful in clarifying a number of issues. SKL was supported by the DARPA/MTO Casimir Effect Enhancement project under SPAWAR contract number N66001-09-1-2071.

References

1. Klimchitskaya, G.L., Mohideen, U., Mostepanenko, V.M.: The Casimir force between real materials: Experiment and theory. Rev. Mod. Phys. **81**, 1827 (2009)
2. Lamoreaux, S.K.: Demonstration of the Casimir force in the 0.6 to 6 µm range. Phys. Rev. Lett. **78**, 5 (1997)
3. Hargreaves, C.M.: Corrections to the retarded dispersion force between metal bodies. Proc Kon. Ned. Akad. Wetensh. **68B**, 231 (1965)
4. Schwinger, J., De Raad Jr., L.L., Milton, K.A.: Casimir effect in dielectrics. Ann. Phys. (New York) **115**, 1 (1978)
5. Lamoreaux, S.K.: Calculation of the Casimir force between imperfectly conducting plates. Phys. Rev. A **59**, R3149 (1999)
6. Lamoreaux, S.K.: Erratum: Demonstration of the Casimir force in the 0.6 to 6 µm range. Phys. Rev. Lett. **81**, 5475 (1998)
7. Lambrecht, A., Reynaud, S.: Casimir force between metallic mirrors. Euro. Phys. J. D **8**, 309 (2000)
8. Lambrecht, A., Reynaud, S., Lamoreaux, S.K.: Comment on Demonstration of the Casimir force in the 0.6 to 6 µm range. Phys. Rev. Lett. **84**, 5672 (2000), and references therein
9. Boström, M., Sernelius, B.: Thermal effects on the Casimir force in the 0.1–5 µm range. Phys Rev Lett **84**, 4757 (2000)
10. Boström, M., Sernelius, B.E: Entropy of the Casimir effect between real metal plates. Physica A **339**, 53 (2004)
11. Lamoreaux, S.K., Buttler, W.T.: Thermal noise limitations to force measurements with torsion pendulums: Applications to the measurement of the Casimir force and its thermal correction. Phys. Rev. E **71**, 036109 (2005)
12. Milonni, P.W.: The Quantum Vacuum. Academic Press, San Diego (1994)
13. Casimir, H.B.G.: On the attraction between two perfectly conducting plates. Proc. Kon. Ned. Akad. Wetenschap **51**, 793 (1948)
14. Lifshitz, E.M.: The theory of molecular attractive forces between solids. Sov. Phys. JETP **2**, 73 (1956)
15. See, e.g., Mostepanenko, V.: Theory, and the Casimir effect. J. Phys. Conf. Ser. **161**, 012003 (2009)
16. Jackson, J.D.: Classical Electrodynamics, 2nd ed. Wiley, New York (1975). See Sect. 7.5
17. Landau, L.D., Lifshitz, E.M.: Electrodynamics of Continuous Media. Pergamon, Oxford (1960)
18. Svetovoy, V.B., van Zwol, P.J., Palasantzas, G., Th.M.De Hosson, J.: Optical properties of gold films and the Casimir force. Phys.Rev. B **77**, 035439 (2008)
19. van Zwol, P.J., Svetovoy, V.B., Palasantzas, G.: Distance upon contact: Determination from roughness profile. Phys. Rev. B **80**, 235401 (2009)
20. Fischbach, E., Sudarsky, D., Szafer, A., Talmadge, C., Aronson, S.H.: Reanalysis of the Eotvos experiment. Phys. Rev. Lett. **56**, 3 (1986)
21. Mostepanenko, V.M., Yu. Sokolov, I.: Laboratory tests for the constituents of dark matter. Astronoische Nachrichten **311**, 197 (1990)
22. Long, J.C., Chan, H.W., Price, J.C.: Experimental status of gravitational-strength forces in the sub-centimeter regime. Nucl. Phys. B **539**, 23 (1999)

23. Klimchitskaya, G.L., Bezerrade Mello, E.R., Mostepanenko, V.M.: Constraints on the parameters of degree-type hypothetical forces following from the new Casimir force measurement. Phys. Lett. A **236**, 280 (1997)
24. Decca, R.S., Lopez, D., Fischbach, E., Klimchitskaya, G.L., Kraus, D..E, Mostepanenko, V.M.: Tests of new physics from precise measurements of the Casimir pressure between two gold-coated plates. Phys. Rev. D **75**, 77101 (2007)
25. Dalvit, D.A.R., Onofrio, R.: On the use of the proximity force approximation for deriving limits to short-range gravitational-like interactions from sphere-plane Casimir force experiments. Phys. Rev. D **80**, 064025 (2009)
26. Decca, R.S., Fischbach, E., Klimchitskaya, G.L., Krause, D.E., Lopez, D., Mostepanenko, D.V.M.: Comment on Anomalies in electrostatic calibrations for the measurement of the Casimir force in a sphere-plane geometry. Phys. Rev. D **79**, 124021 (2009)
27. Decca, R.S., Lopez, D., Chan, H.B., Fischbach, E., Krause, D.E., Jamell, C.R.: Constraining new forces in the Casimir regime using the isoelectronic technique. Phys. Rev. Lett. **94**, 240401 (2005)**
28. Rosa, F.S.S., Dalvit, D.A.R., Milonni, P.W.: Casimir interactions for anisotropic magnetodielectric metamaterials. Phys. Rev. A **78**, 032117 (2008)
29. Buks, E., Roukes, M.L.: Metastability and the Casimir effect in micromechanical systems. Europhys. Lett. **54**, 220 (2001)
30. de Man, S., Heeck, K., Wijngaarden, R.J., Iannuzzi, D.: Halving the Casimir force with conductive oxides. Phys. Rev. Lett. **103**, 040402 (2009)
31. Boyer, T.H.: Quantum electromagnetic zero-point energy of a conducting spherical shell and the Casimir model for a charged particle. Phys. Rev. **174**, 1764 (1968)
32. Iannuzzi, D., Lisanti, M., Capasso, F.: Effect of hydrogen-switchable mirrors on the Casimir force. PNAS **101**, 4019 (2004)
33. de Man, S., Iannuzzi, D.: On the use of hydrogen switchable mirrors in Casimir force experiments. New J. Phys. **8**, 235 (2006)
34. Chan, H.B., Bao, Y., Zou, J., Cirelli, R.A., Klemens, F., Mansfield, W.M., Pai, C.S.: Measurement of the Casimir force between a gold sphere and a silicon surface with nanoscale trench arrays. Phys. Rev. Lett. **101**, 03040 (2008)
35. Büscher, R., Emig, T.: Nonperturbative approach to Casimir interactions in periodic geometries. Phys. Rev. A **69**, 062101 (2004)
36. Munday, J.N., Capasso, F.: Precision measurement of the Casimir–Lifshitz force in a fluid. Phys. Rev. A **75**, 060102 (2007)
37. Munday, J.N., Capasso, J.N.F., Parsegian, V.A., Bezrukov, S.M.: Measurements of the Casimir–Lifshitz force in fluids: The effect of electrostatic forces and Debye screening. Phys. Rev. A **78**, 032109 (2008)
38. Munday, J.N., Capasso, F., Parsegian, V.A.: Measured long-range repulsive Casimir–Lifshitz forces. Nature **457**, 170 (2008)
39. Derjaguin, B.V., Abrikosova, I.I.: Direct measurement of the molecular attraction of solid bodies. 1. Statement of the problem and method of measuring forces by using negative feedback. Sov. Phys. JETP **3**, 819 (1957)
40. Blocki, J., Randrup, J., Swiatecki, W.J., Tsang, C.F.: Proximity forces. Annals Phys. **105**, 427 (1977)
41. Blocki, J., Swiatecki, W.J.: Generalization of the proximity force theorem. Annals Phys. **132**, 53 (1981)
42. Kim, W.J., Brown-Hayes, M., Dalvit, D.A.R., Brownell, J.H., Onofrio, R.: Reply to 'Comment on Anomalies in electrostatic calibrations for the measurement of the Casimir force in a sphere-plane geometry'. Phys. Rev. A **79**, 026102 (2009)
43. Kim, W.J., Brown-Hayes, M., Dalvit, D.A.R., Brownell, J.H., Onofrio, R.: Anomalies in electrostatic calibrations for the measurement of the Casimir force in a sphere-plane geometry. Phys. Rev. A **78**, 020101 (2008)
44. Maia Neto, P.A., Lambrecht, A., Reynaud, S.: Casimir effect with rough metallic mirrors. Phys. Rev. A **72**, 012115 (2005)

45. Kim, W.J., Sushkov, A.O., Dalvit, D.A.R., Lamoreaux, S.K.: Measurement of the short-range attractive force between Ge plates using a torsion balance. Phy. Rev. Lett. **103**, 060401 (2009)
46. de Man, S., Heeck, K., Iannuzzi, D.: No anomalous scaling in electrostatic calibrations for Casimir force measurements. Phys. Rev. A **79**, 024102 (2009)
47. Spitsyn, A.I., Vanstan, V.M.: Potential gradient along semiconductor-surface projections in an external electric field. Radiophysics and Quantum Electronics **36**, 752 (1993)
48. Bardeen, J.: Surface states and rectification at metal semi-conductor contact. Phys. Rev. **71**, 717 (1947)
49. Dalvit, D.A.R, Lamoreaux, S.K.: Computation of Casimir forces for dielectrics or intrinsic semiconductors based on the Boltzmann transport equation. J. Phys.: Conference Series **161**, 012009 (2009)
50. Dalvit, D.A.R, Lamoreaux, S.K.: Contribution of drifting carriers to the Casimir–Lifshitz and Casimir–Polder interactions with semiconductor materials. Phys. Rev. Lett. **101**, 163203 (2008)
51. Lamoreaux, S.K.: Electrostatic background forces due to varying contact potentials in Casimir experiments. arXiv:0808.0885
52. Robertson, N.A.: Kelvin Probe Measurements of the Patch Effect Report LIGO-G070481-00-R (available at http://www.ligo.caltech.edu/docs/G/G070481-00.pdf)
53. Speake, C.C., Trenkel, C.: Forces between conducting surfaces due to spatial variations of surface potential. Phys. Rev. Lett. **90**, 160403 (2003)
54. Stein, E., Weiss, G.: Introduction to Fourier Transforms on Euclidean Spaces. Princeton University Press, Princeton (1971)
55. Kim, W.J., Sushkov, A.O., Dalvit,, D.A.R., Lamoreaux,, S.K.: Measurement of the short-range attractive force between Ge plates using a torsion balance. Phy. Rev. Lett. **103**, 060401 (2009)
56. Kim, W.J., Sushkov, A.O, Dalvit, D.A.R, Lamoreaux, S.K.: Surface contact potential patches and Casimir force measurements. Phys. Rev. A **81**, 022505 (2010)
57. Nonnemacher , N., OBoyle , M.P., Wickramasinghe, H.K.: Kelvin Probe Microscopy. Appl. Phys. Lett. **58**, 2921 (1991)
58. Jacobs, H.O., Leuchtmann, P., Homan, O.J., Stemmer, A.: Resolution and contrast in Kelvin probe force microscopy. J. Appl. Phys. **84**, 1168 (1998)

Chapter 8
Attractive and Repulsive Casimir–Lifshitz Forces, QED Torques, and Applications to Nanomachines

Federico Capasso, Jeremy N. Munday and Ho Bun Chan

Abstract This chapter discusses recent developments in quantum electrodynamical (QED) phenomena, such as the Casimir effect, and their use in nanomechanics and nanotechnology in general. Casimir–Lifshitz forces arise from quantum fluctuations of vacuum or more generally from the zero-point energy of materials and their dependence on the boundary conditions of the electromagnetic fields. Because the latter can be tailored, this raises the interesting possibility of designing QED forces for specific applications. After a concise review of the field in the introduction, high precision measurements of the Casimir force using MicroElectroMechanical Systems (MEMS) are discussed. Applications to non-linear oscillators are presented, along with a discussion of their use as nanoscale position sensors. Experiments that have demonstrated the role of the skin-depth effect in reducing the Casimir force are then presented. The dielectric response of materials enters in a non-intuitive way in the modification of the Casimir–Lifshitz force between dielectrics through the dielectric function at imaginary frequencies $\varepsilon(i\xi)$. The latter is illustrated in a dramatic way by experiments on materials that can be switched between a reflective and a transparent state (hydrogen switchable mirrors) and by a large reduction of the Casmir force between a gold sphere and a thick gold film, when the latter is replaced by an indium tin oxide (ITO) thick film. Changing the electromagnetic density of states by altering the shape of the interacting surfaces on a scale comparable to their separation is an effective method to tailor Casimir–Lifshitz forces. Measurements of the latter between a

F. Capasso (✉)
School of Engineering and Applied Sciences, Harvard University, Cambridge, MA 02138, USA

J. N. Munday
Department of Electrical and Computer Engineering, University of Maryland, College Park, Maryland 20742, USA

D. Dalvit et al. (eds.), *Casimir Physics*, Lecture Notes in Physics 834, DOI: 10.1007/978-3-642-20288-9_8, © Springer-Verlag Berlin Heidelberg 2011

silicon surfaces nanostructured with deep trenches and a sphere metalized with thick gold have demonstrated the non-additivity of these forces and the ability to tailor them by suitable surface patterning. Experiments on the Casimir effect in fluids are discussed, including measurements of attractive and repulsive Casimir forces conducted between solids separated by a fluid with $\varepsilon(i\xi)$ intermediate between those of the solids over a large frequency range. Such repulsive forces can be used to achieve quantum levitation in a virtually friction-less environment, a phenomenon that could be exploited in innovative applications to nanomechanics. The last part of the chapter deals with the elusive QED torque between birefringent materials and efforts to observe it. We conclude by highlighting future important directions.

8.1 Introduction

According to QED, quantum fluctuations of the electromagnetic field give rise to a zero-point energy that never vanishes, even in empty space [1]. In 1948, Casimir [2] showed that, as a consequence, two parallel plates, made out of ideal metal (i.e. with unity reflectivity at all wavelengths, or equivalently with infinite plasma frequency), should attract each other in vacuum even if they are electrically neutral, a phenomenon known as the Casimir effect. Because only the electromagnetic modes that have nodes on both walls can exist within the cavity, the zero-point energy depends on the separation between the plates, giving rise to an attractive force. This result in fact can be interpreted as due to the differential radiation pressure associated with zero-point energy (virtual photons) between the "inside" and the "outside" of the plates, which leads to an attraction because the mode density in free space is higher than the density of states between the plates [1]. The interpretation in terms of zero-point energy of the Casimir effect was suggested by Niels Bohr, according to Casimir's autobiography [3]. An equivalent derivation of excellent intuitive value, leading to the Casimir force formula, was recently given by Jaffe and Scardicchio in terms of virtual photons moving along ray optical paths [4, 5]. Between two parallel plates, the Casimir force assumes the form [2]:

$$F_c = -\pi^2 \hbar c A / 240 d^4, \tag{8.1}$$

where c is the speed of light, $\hbar$ is Planck's constant divided by 2π, A is the area of the plates, and d is their separation.

The pioneering experiments of Spaarnay [6] were not able to unambiguously confirm the existence of the Casimir force, due to, among other factors, the large error arising from the difficulty in maintaining a high degree of parallelism between the plates. Clear experimental evidence for the effect was presented by van Blokland and Overbeek in 1978 who performed measurements between a metallic sphere and a metallic plate [7], thus eliminating a major source of uncertainty. Final decisive verification is due to Lamoureaux, who in 1997 reported the first high precision measurements of the Casimir force using a torsional pendulum and sphere-plate configuration [8]. This was followed by several experimental studies, which have produced further convincing confirmation [9–16] for the Casimir effect including the parallel plate geometry [13].

Between a sphere and a plate made of ideal metals the Casimir force reads [17]:

$$F_c = -\pi^3 \hbar c R / 360 d^3, \tag{8.2}$$

where R is the radius of the sphere and d is the minimum distance between the sphere and the plate. In the derivation of (8.2) it was assumed that this distance is much smaller than the sphere diameter (proximity force approximation).

Several reviews on Casimir forces and on the closely related van der Waals forces have recently appeared [18–27]. Both forces are of QED origin. The key physical difference is that in the Casimir case retardation effects due to the finite speed of light cannot be neglected, as in the van der Waals limit, and are actually dominant [1]. This is true for distances such that the propagation time of light between the bodies or two molecules is much greater than the inverse characteristic frequency of the material or of the molecules (for example the inverse plasma frequency in the case of metals and the inverse of the frequency of the dominant transition contributing to the polarizability $\alpha (\omega)$, in the case of molecules) [1]. The complete theory for macroscopic bodies, developed by Lifshitz, Dzyaloshinskii, and Pitaevskii, is valid for any distance between the surfaces and includes in a consistent way both limits [28, 29].

This formulation, a generalization of Casimir's theory to dielectrics, including of course non-ideal metals, is the one which is most often used for comparison with experiments. In this theory, the force between two uncharged surfaces can be derived according to an analytical formula (often called the Lifshitz formula) that relates the zero-point energy to the dielectric functions of the interacting surfaces and of the medium in which they are immersed. This equation for the force between a sphere and plate of the same metal is [28]:

$$F_1(z) = \frac{\hbar}{2\pi c^2} R \int_0^\infty \int_0^\infty p\xi^2 \left\{ \ln \left[1 - \frac{(s-p)^2}{(s+p)^2} e^{-2pz\xi/c} \right] \right.$$

$$\left. + \ln \left[1 - \frac{(s-p\varepsilon)^2}{(s+p\varepsilon)^2} e^{-2pz\xi/c} \right] \right\} dp d\xi, \tag{8.3}$$

where $s = \sqrt{\varepsilon - 1 + p^2}$, $\varepsilon(i\xi)$ is the dielectric function of the dielectric or metal evaluated at imaginary frequency and the integration is over all frequencies and wavevectors of the modes between the plates. The expression for $\varepsilon(i\xi)$ is given by:

$$\varepsilon(i\xi) = 1 + \frac{2}{\pi} \int_0^\infty \frac{\omega \cdot \varepsilon''(\omega)}{\omega^2 + \xi^2} d\omega, \tag{8.4}$$

where $\varepsilon''(\omega)$ is the imaginary part of the dielectric function. The integral in (8.4) runs over all real frequencies, with non-negligible contributions arising from a very wide range of frequencies. (8.3) and (8.4) show that the optical properties of the material influence in a non-intuitive way the Casimir force. The finite conductivity modifications to the Casimir force based on the frequency dependence of the dielectric function can be calculated numerically using the tabulated complex dielectric function of the metal [30–34]. This leads to a reduction in the Casimir force compared to the ideal metal case given by (8.1). Physically this can be understood from the fact that in a real metal the electromagnetic field penetrates by an amount of the order of the skin-depth which leads to an effective increase of the plate separation. See also the Chap. 10 by van Zwol et al. in this volume for a further discussion of the optical properties of materials used in Casimir force measurements.

The second modification, due to the roughness of the metallic surfaces, tends to increase the attraction [35, 36] because the portions of the surfaces that are locally closer contribute much more to the force due its strong nonlinearity with distance.

As previously mentioned, at very short distances, the theory of Lifshitz, Dzyaloshinskii, and Pitaevskii, also provides a complete description of the non-retarded van der Waals force [37, 38]. Recently Henkel et al. [39] and Intravaia et al. [40] have provided a physically intuitive description of the van der Waals limit for real metals with dispersion described by the Drude model. At finite plasma frequency one must include surface plasmons in the counting of electromagnetic modes, i.e. modes associated with surface charge oscillations which exponentially decay away from the surface. At short distances (small compared to the plasma wavelength) the Casimir energy is given by the shift in the zero-point energy of the surface plasmons due to their Coulomb (electrostatic) interaction. The corresponding attractive force between two parallel plates is then given by [41]:

$$F_c = -\frac{\hbar c \pi^2 A}{290 \lambda_p d^3}. \tag{8.5}$$

This formula is an approximation of the short distance limit of the more complete theory [28, 29]. At large separations $(d \gg \lambda_p)$, retardation effects give rise to a long-range interaction that in the case of two ideal metals in vacuum reduces to Casimir's result.

In a number of studies several authors [11, 14, 15] have claimed agreement between Casimir force experiments and theory at the 1% level or better—a claim that has been challenged in some of the literature [12, 42–44]. The authors of [44] have pointed out that the strong non-linear dependence of the force on distance limits the precision in the absolute determination of the force. Uncertainties in the knowledge of the dielectric functions of the thin metallic films used in the experiments and in the models of surface roughness used to correct the Lifshitz theory also typically give rise to errors larger than 1% in the calculation of the expected force [12, 43, 44]. It has also been shown that the calculation of the

Casimir force can vary by as much as 5% depending on which values are chosen for the optical properties of a given material [45]. Another uncertainty is related to the model of surface roughness and in its measurement that translates to an uncertainty in the comparison between theory and experiments. We conclude that claims of agreement between theory and experiment at the 1% level or less are questionable due to experimental errors and uncertainties in the calculations. For a further discussion of modern Casimir force experiments, see the Chap. 7 by Lamoreaux in this volume.

Apart from its intrinsic theoretical interest, the Casimir interaction has recently received considerable attention for its possible technological consequences. The Casimir force, which rapidly increases as the surface separation decreases, is the dominant interaction mechanism between neutral objects at sub-micron distances. In light of the miniaturization process that is moving modern technology towards smaller electromechanical devices, it is reasonable to ask what role the zero-point energy might play in the future development of micro- and nanoelectromechanical systems (MEMS and NEMS) [16, 46, 47].

One of the first experiments was to design a micro-machined torsional device that could be actuated solely by the Casimir force [16]. The results not only demonstrated that this is indeed possible, but also provided one of the most sensitive measurements of the Casimir force between metallized surfaces. In their second experiment [47], the same group showed that the Casimir attraction can also influence the dynamical properties of a micromachined device, changing its resonance frequency, and giving rise to hysteretic behavior and bistability in its frequency response to an ac excitation, as expected for a non-linear oscillator. The authors proposed that this device could serve as a nanometric position sensor. The above developments are covered in Sect. 8.2.

A particularly interesting direction of research on Casimir–Lifshitz forces is the possibility of designing their strength and spatial dependence by suitable control of the boundary conditions of the electromagnetic fields. This can be done by appropriate choice of the materials [48, 49], of the thickness of the metal films [50] and the shape of the interacting surfaces [51–53]. By nanoscale periodic patterning of one of the metallic surfaces and controlling the ratio of the period to the depth of the grooves the Casmir force has been significantly tailored as discussed in Sect. 8.3. This section also discusses experiments aimed at elucidating the role of the skin-depth effect in the Casimir force, by coating one of the surfaces with suitably engineered thin films.

Section 8.3 also covers one of the most interesting features of long-range QED forces: repulsive forces which can arise between suitable surfaces when their dielectric functions and that of the medium separating them satisfy a particular inequality [20, 29, 37, 38]. Measurements of Casimir–Lifshitz forces in fluids are presented, including the measurements of a repulsive force between gold and silicon dioxide separated by bromobenzene. Methods of measuring these forces are discussed in detail and the phenomenon of "quantum levitation" is analyzed along with intriguing applications to nanotechnology such as frictionless bearings and related devices.

QED can give rise to other exotic macroscopic interaction phenomena between materials with anisotropic optical properties such as birefringent crystals. For example a torque due to quantum fluctuations between plates made of uniaxial materials has been predicted but has not yet been observed [54, 55]. Section 8.4 is devoted to discussion of this remarkable effect and related calculations. Specific experiments are proposed along with novel applications.

Section 8.5 provides an outlook on novel directions in this field.

8.2 MEMS Based on the Casimir Force

MEMS are a silicon-based integrated circuit technology with moving mechanical parts that are released by means of etching sacrificial silicon dioxide layers followed by a critical point drying step [56]. They have been finding increasing applications in several areas ranging from actuators and sensors to routers for optical communications. For example the release of the airbag in cars is controlled by a MEMS based accelerometer. In the area of lightwave communications the future will bring about new optical networks with a mesh topology, based on dense wavelength division multiplexing. These intelligent networks will be adaptive and self-healing with capabilities of flexible wavelength provisioning, i.e. the possibility to add and drop wavelengths at specific nodes in response to real time bandwidth demands and rerouting. The *lambda router* [57, 58], a device consisting of an array of thousands of voltage controlled mirrors, which switches an incoming wavelength from one optical fiber to any of many output fibers, is an example of a MEMS technology that might impact future networks.

The development of increasingly complex MEMS will lead to more attention to scaling issues, as this technology evolves towards NanoElecroMechanicalSystems (NEMS). Thus, it is conceivable that a Moore curve for MEMS will develop leading to increasingly complex and compact MEMS having more devices in close proximity [59, 60]. This scenario will inevitably lead to-having to face the issue of Casimir interactions between metallic and dielectric surfaces in close proximity with attention to potentially troublesome phenomena such as stiction, i.e. the irreversible coming into contact of moving parts due to Casimir/van der Waals forces [59]. On the other hand such phenomena might be usable to one's advantage by adding functionality to NEMS based architectures. See also the Chap. 9 by Decca et al. in this volume for additional discussions of MEMS and NEMS based Casimir force experiments.

8.2.1 Actuators

In the first experiment [16], the authors designed and demonstrated a micro-machined torsional device that was actuated by the Casimir force and that

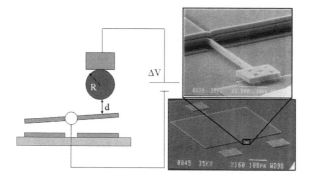

Fig. 8.1 MEMS Casimir force detection setup: schematic of the experiment (not to scale) and scanning electron micrographs of the micromachined torsional device used for the measurement of the Casimir force with a close-up of one of the torsional rods anchored to the substrate. As the metallic sphere approaches the top plate, the Casimir force causes a rotation of the torsional rod

provided a very sensitive measurement of the latter. This device (Fig. 8.1) was subsequently used in a variety of experiments [14, 15, 50, 53]. It consists of a 3.5 μm thick, 500 μm square heavily doped polysilicon plate freely suspended on two of its opposite sides by thin torsional rods. The other ends of the torsional rods are anchored to the substrate via support posts. Two fixed polysilicon electrodes are located symmetrically underneath the plate, one on each side of the torsional rod. Each electrode is half the size of the top plate. There is a 2 μm gap between the top plate and the fixed electrodes created by etching a SiO_2 sacrificial layer. The top plate is thus free to rotate about the torsional rods in response to an external torque.

A schematic of the actuation mechanism based on the Casimir force is shown in Fig. 8.1. A polystyrene sphere with radius $R = 100$ μm is glued on the end of a copper wire using conductive epoxy. A 200 nm thick film of gold with a thin titanium adhesion layer is then evaporated on both the sphere and the top plate of the torsional device. An additional 10 nm of gold is sputtered on the sphere to provide electrical contact to the wire. The micromachined device is placed on a piezoelectric translation stage with the sphere positioned close to one side of the top plate. As the piezo extends, it moves the micromachined device towards the sphere. The rotation of the top plate in response to the attractive Casimir force is detected by measuring the imbalance of the capacitances of the top plate to the two bottom electrodes at different separations between the sphere and the top plate. The measurement is performed at room temperature and at a pressure of less than 1 mTorr. Note that an external bias needs to be applied to the sphere to compensate for the potential V_0 resulting from work function differences between the metallic surfaces and other effects such as contact potentials associated with grounding, patch potential, etc. [7]. The value of V_0 is typically in the 10 to 100 mV range.

Figure 8.2 shows the results of that measurement. One sees that the data points lie above the curve given by (8.2). Two main effects are at work in this

Fig. 8.2 Experimental measurement of the Casimir force from the MEMS torsional apparatus. Angle of rotation of the top plate in response to the Casimir force as a function of distance. The solid line is the predicted Casimir force (8.2) without corrections for surface roughness or finite conductivity. Dots are experimental results

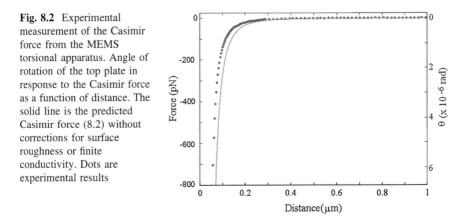

discrepancy. The first one is the finite reflectivity of the metal. This causes virtual photons associated with vacuum fluctuations to penetrate into the metal (skin effect) increasing the effective sphere-plate separation thus decreasing the force. The second effect is the surface roughness, which is estimated from AFM measurements to be a few tens of nanometers depending on the particulars of the experiment. It enhances the Casimir force due to the strong nonlinear dependence with distance. Both effects can be accounted for within the framework of Lifshitz theory, giving a much smaller discrepancy between theory and experiments.

A bridge circuit enables one to measure the change in capacitance to 1 part in 2×10^5, equivalent to a rotation angle of 8×10^{-8} rad, with integration time of 1 s when the device is in vacuum. With a torsional spring constant as small as 1.5×10^{-8} N m rad^{-1}, the device yields a sensitivity of 5 pN Hz$^{-1/2}$ for forces acting at the edge of the plate. Such force sensitivity is comparable to the resolution of conventional atomic force microscopes. The device is insensitive to mechanical noise from the surroundings because the resonant frequency is maintained high enough (~ 2 kHz) due to the small moment of inertia of the plate.

8.2.2 Nonlinear Oscillators

While there is vast experimental literature on the hysteretic response and bistability of nonlinear oscillators in the context of quantum optics, solid-state physics, mechanics, and electronics, the experiment summarized in this section represents to our knowledge, the first observation of bistability and hysteresis caused by a QED effect. A simple model of the Casimir oscillator consists of a movable metallic plate subjected to the restoring force of a spring obeying Hooke's law and the nonlinear Casimir force arising from the interaction with a fixed metallic sphere (Fig. 8.3). For separations d larger than a critical value [61], the system is bistable: the potential energy consists of a local minimum and a global minimum

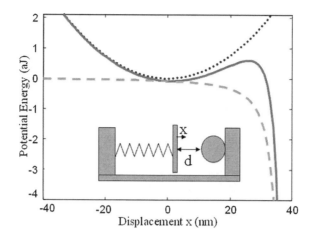

Fig. 8.3 *Inset* a simple model of the nonlinear Casimir oscillator (not to scale). Main figure: elastic potential energy of the spring (*dotted line*, spring constant 0.019 N m^{-1}), energy associated with the Casimir attraction (*dashed line*) and total potential energy (*solid line*) as a function of plate displacement. The distance d between the sphere (100 μm radius) and the equilibrium position of the plate in the absence of the Casimir force, is chosen to be 40 nm

separated by a potential barrier (Fig. 8.3). The local minimum is a stable equilibrium position, about which the plate undergoes small oscillations. The Casimir force modifies the curvature of the confining potential around the minimum, thus changing the natural frequency of oscillation and also introduces higher order terms in the potential, making the oscillations anharmonic.

For this experiment, Fig. 8.1 was used. The torsional mode of oscillation was excited by applying a driving voltage to one of the two electrodes that is fixed in position under the plate. The driving voltage is a small ac excitation V_{ac} with a dc bias V_{dc1} to linearize the voltage dependence of the driving torque. The top plate is grounded while the detection electrode is connected to a dc voltage V_{dc2} through a resistor. Oscillatory motion of the top plate leads to a time varying capacitance between the top plate and the detection electrode. For small oscillations, the change in capacitance is proportional to the rotation of the plate. The detection electrode is connected to an amplifier and a lock-in amplifier measures the output signal at the excitation frequency.

To demonstrate the nonlinear effects introduced by the Casimir force, the piezo was first retracted until the sphere was more than 3.3 μm away from the oscillating plate so that the Casimir force had a negligible effect on the oscillations. The measured frequency response shows a resonance peak that is characteristic of a driven harmonic oscillator (peak I in Fig. 8.4a), regardless of whether the frequency is swept up (hollow squares) or down (solid circles). This ensures that the excitation voltage is small enough so that intrinsic nonlinear effects in the oscillator are negligible in the absence of the Casimir force. The piezo was then extended to bring the sphere close to the top plate while

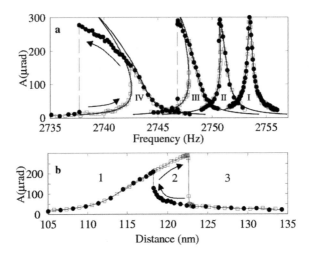

Fig. 8.4 a Hysteresis in the frequency response induced by the Casimir force on an otherwise linear oscillator. *Hollow squares* (*solid circles*) are recorded with increasing (decreasing) frequency. *Solid lines* show the predicted frequency response of the oscillator. The distance z between the oscillator and the sphere is 3.3 μm, 141 nm, 116.5 nm, and 98 nm for peaks I, II, III, and IV, respectively. The excitation amplitude is maintained constant at 55.5 mV for all four separations. The solid lines are the calculated response. The peak oscillation amplitude for the plate is 39 nm at its closest point to the sphere. **b** Oscillation amplitude as a function of distance with excitation frequency fixed at 2748 Hz

maintaining the excitation voltage at fixed amplitude. The resonance peak shifts to lower frequencies (peaks II, III, and IV), by an amount that is consistent with the distance dependence of the force in Fig. 8.2. Moreover, the shape of the resonance peak deviates from that of a driven harmonic oscillator and becomes asymmetric. As the distance decreases, the asymmetry becomes stronger and hysteresis occurs. This reproducible hysteretic behavior is characteristic of strongly nonlinear oscillations [62].

The solid lines in Fig. 8.4a show the predicted frequency response of the oscillator including the first, second, and third spatial derivatives of the Casimir force. Higher orders terms or the full nonlinear potential would need to be included to achieve a better agreement with experiments.

An alternative way to demonstrate the "memory" effect of the oscillator is to maintain the excitation at a fixed frequency and vary the distance between the sphere and the plate (Fig. 8.4b). As the distance changes, the resonance frequency of the oscillator shifts, to first order because of the changing force gradient. In region 1, the fixed excitation frequency is higher than the resonance frequency and vice versa for region 3. In region 2, the amplitude of oscillation depends on the history of the plate position. Depending on whether the plate was in region 1 or region 3 before it enters region 2, the amplitude of oscillation differs by up to a factor of 6. This oscillator therefore acts as a nanometric sensor for the separation between two uncharged metallic surfaces.

8.3 The Design and Control of Casimir Forces

In this section we discuss experiments aimed at tailoring the Casimir–Lifshitz force via control of the boundary conditions of the electromagnetic fields. Several examples will be discussed: (1) control of the geometry of the surfaces by nanostructuring with suitable corrugations; (2) control of the thickness of the metallic layers deposited on the juxtaposed surfaces; (3) choice of materials that can be reversibly switched from metallic to transparent and conductive oxides (4) interleaving fluids between the interacting surfaces; (5) material combinations that give rise to repulsive Casimir–Lifshitz force; (6) devices based on repulsive forces.

8.3.1 Modification of the Casimir Force by Surface Nanostructuring

There exists a close connection between the Casimir force between conductors and the van der Waals (vdW) force between molecules (see the chapter of Henkel et al. Chap. 11 in this volume for a discussion of atom-surface effects). For the former, the quantum fluctuations are often associated with the vacuum electromagnetic field, while the latter commonly refers to the interaction between fluctuating dipoles. In simple geometries such as two parallel planes, the Casimir force can be interpreted as an extension of the vdW force in the retarded limit. The interaction between molecules in the two plates is summed to yield the total force. However, such summation of the vdW force is not always valid for extended bodies because the vdW force is not pairwise additive. The interaction between two molecules is affected by the presence of a third molecule. Recently Chan and coworkers [53] reported measurements of the Casimir force between nanostructured silicon surfaces and a gold sphere (Fig. 8.5). One of the interacting objects consists of a silicon surface with nanoscale, high aspect ratio rectangular corrugations. The other surface is a gold-coated glass sphere attached onto a micromechanical torsional oscillator similar to the one discussed in the previous section. Lateral movements of the surfaces are avoided by positioning the corrugations perpendicular to the torsional axis. The Casimir force gradient is measured from the shifts in the resonant frequency of the oscillator at distances between 150 and 500 nm. Deviations of up to 20% from PAA are observed, demonstrating the strong geometry dependence of the Casimir force.

Figure 8.5a shows a cross section of an array of rectangular corrugations with period of 400 nm (sample B) fabricated on a highly p-doped silicon substrate. Two other samples, one with period 1 µm (sample A) and the other with a flat surface, are also fabricated; the trenches have a depth $t \sim 1$ µm.

The geometry of nanoscale, rectangular trenches was chosen because the Casimir force on such structures is expected to exhibit large deviations from pairwise additive approximation (PAA). Consider the interaction between the

Fig. 8.5 a Cross section of rectangular trenches in silicon, with periodicity of 400 nm and depth of 0.98 μm (sample *B*). **b** *Top view* of the structure. **c** Schematic of the experimental setup (not to scale) including the micromechanical torsional oscillator, gold spheres, and silicon trench array. **d** Measurement scheme with electrical connections. Excitation voltages V_{ac1} and V_{ac2} are applied to the bottom electrodes

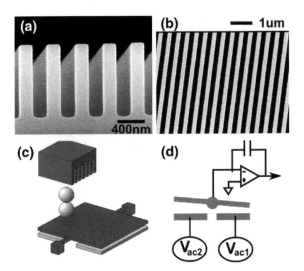

trench array and a parallel flat surface at distance z from the top surface of the trenches. In the pairwise additive picture, this interaction is a sum of two contributions: the volume from the top surface to the bottom of the trench and the volume below the bottom of the trench. The latter component is negligible because the distance to the other surface is more than 1 μm, larger than the distance range at which Casimir forces can be detected in the experiment. For a trench array of 50% duty cycle, the former component yields exactly half of the interaction between two flat surfaces F_{flat} regardless of the periodicity because half of the material is removed [63]. In practice, the trench arrays are created with duty cycle close to but not exactly at 50%. Under PAA, the total force is equal to pF_{flat}, where p is the fraction of solid volume. The calculation of the Casimir force in such corrugated surfaces, in contrast, is highly nontrivial. While perturbative treatments are valid for smooth profiles with small local curvature, they are impractical for the deep, rectangular corrugations.

Using a different approach based on path integrals, Büscher and Emig [63] calculated the Casimir force for the corrugated geometry made of perfect conductors. Strong deviations from PAA were obtained when the ratio z/λ is large, where λ is the pitch. In the limit when λ goes to zero, the force on a trench array approaches the value between flat surfaces, leading to deviations from PAA by a factor of 2. Such large deviations occur because the Casimir force is associated with confined electromagnetic modes with wavelength comparable to the separation between the interacting objects. When $\lambda \ll z$, these modes fail to penetrate into the trenches, rendering the Casimir force on the corrugated surface equal to a flat one.

For these experiments the gradient of the Casimir force on the silicon trench arrays was measured using a gold-coated sphere attached to a micromechanical torsional oscillator similar to the nonlinear Casmir oscillator previously discussed. The oscillator consists of a 3.5 μm thick, 500 μm square silicon plate suspended by two torsional rods. As shown in Fig. 8.5c two glass spheres, each with radius R of 50 μm, are stacked and attached by conductive epoxy onto the oscillator at a distance of $b = 210$ μm from the rotation axis. The large distance (~ 200 μm) between the oscillator plate and the corrugated surface ensures that the attraction between them is negligible and only the interaction between the top sphere and the corrugated surface is measured. Before attachment, a layer of gold with thickness 400 nm is sputtered onto the spheres. Two electrodes are located between the plate and the substrate. Torsional oscillations in the plate are excited when the voltage on one of the electrodes is modulated at the resonant frequency of the oscillator ($f_0 = 1783$ Hz, quality factor $Q = 32,000$). For detecting the oscillations, an additional ac voltage of amplitude 100 mV and frequency of 102 kHz is applied to measure the capacitance change between the top plate and the electrodes. A phase-locked loop is used to track the shifts in the resonance frequency [16, 47] as the sphere approaches the other silicon plate through extension of a closed-loop piezoelectric actuator. As shown in Fig. 8.5c, the movable plate is positioned so that its torsional axis is perpendicular to the trench arrays in the other silicon surface. Such an arrangement eliminates motion of the movable plate in response to possible lateral Casimir forces [64] because the spring constant for translation along the torsional axis is orders of magnitude larger than the orthogonal direction in the plane of the substrate.

The force gradient is measured between the gold sphere and a flat silicon surface [solid circles in Fig. 8.6a] obtained from the same wafer on which the corrugated samples A and B were fabricated. The main source of uncertainty in the measurement (~ 0.64 pN μm^{-1} at $z = 300$ nm) originates from the thermomechanical fluctuations of the micromechanical oscillator. As the distance decreases, the oscillation amplitude is reduced to prevent the oscillator from entering the nonlinear regime. At distances below 150 nm, the oscillation amplitude becomes too small for reliable operation of the phase-locked loop. In Fig. 8.6a, the line represents the theoretical force gradient between the gold sphere and the flat silicon surface, including both the finite conductivity and roughness corrections. Lifshitz's equation is used to take into account the finite conductivity. For the gold surface, tabulated values of the optical properties [32] were used. For the silicon surface, the tabulated values were further modified by the concentration of carriers (2×10^{18} cm^{-3}) determined from the dc conductivity of the wafer (0.028 Ω cm).

Using an atomic force microscope, the main contribution to the roughness was found to originate from the gold surface (~ 4 nm rms) rather than the silicon wafer (~ 0.6 nm rms), which was taken into account using a geometrical averaging method [65]. The Casimir force gradients $F'_{c;A}$ and $F'_{c;B}$ between the same gold sphere and the corrugated samples A and B were then measured and plotted in Figs. 8.6b and c. As described earlier, under PAA, the forces on the trench arrays (where z is measured from the top of the corrugated surface) are equal to the force on a flat surface multiplied by the fractional volumes p_A and p_B. The solid lines in Fig. 8.6b and c represent the corresponding force gradients, $p_A F'_{c,\text{flat}}$ and $p_B' F'_{c,\text{flat}}$, respectively. Measurement of the force gradient was repeated 3 times for each

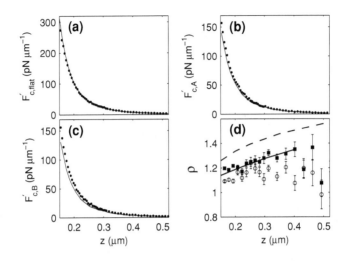

Fig. 8.6 Measured Casimir force gradient between the same gold sphere and **a** a flat silicon surface, $F'_{c,\text{flat}}$, **b** sample A, $F'_{c,A}$ ($\lambda = 1$ μm), and **c** sample B, $F'_{c,B}$ ($\lambda = 400$ nm). In **a**, the *line* represents the theoretical Casimir force gradient including finite conductivity and surface roughness corrections. In **b** and **c**, the *lines* represent the force gradients expected from PAA ($pF'_{c,\text{flat}}$). **d** Ratio ρ of the measured Casimir force gradient to the force gradient expected from PAA, for samples A ($\lambda/a = 1.87$, *hollow circles*) and B ($\lambda/a = 0.82$, *solid squares*), respectively. Theoretical values [63] for perfectly conducting surfaces are plotted as the solid ($\lambda/a = 2$) and dashed lines ($\lambda/a = 1$)

sample, yielding results that are consistent within the measurement uncertainty. To analyze the deviations from PAA, the ratios $\rho_A = F'_{c,A}/p_A F_{c,\text{flat}}$ and $\rho_B = F'_{c,B}/p_B F'_{c,\text{flat}}$ are plotted in Fig. 8.6d. The ratio ρ equals one if PAA is valid. For sample A with $\lambda/a = 1.87$, where a is half the depth of the trenches, the measured force deviates from PAA by $\sim 10\%$. In sample B with $\lambda/a = 0.82$, the deviation increases to $\sim 20\%$. For 150 nm $< z < 250$ nm, the measured Casimir force gradients in both samples show clear deviations from PAA. At larger distances, the uncertainty increases considerably as the force gradient decreases. We compare our experimental results on silicon structures to calculations by Büscher and Emig [63] on perfect conductors [solid and dashed lines in Fig. 8.6d]. In this calculation, the Casimir force between a flat surface and a corrugated structure with $p = 0.5$ was determined for a range of λ/a using a path integral approach. Since $R \gg z$, the proximity force approximation allows a direct comparison of our measured force gradient using a sphere and the predicted force that involved a flat surface. The measured deviation in sample B is larger than sample A, in agreement with the notion that geometry effects become stronger as λ/a decreases. However, the measured deviations from PAA are smaller than the predicted values by about 50%, significantly exceeding the measurement uncertainty for 150 nm $< z < 250$ nm. Such discrepancy is, to a certain extent, expected as a result of the interplay between finite conductivity and geometry effects. The relatively large value of the skin-depth in silicon (~ 11 nm at a wavelength of 300 nm) could reduce the deviations from PAA. See also the Chap. 4 of Lambrecht et al. in this volume for additional information on material and geometry effects.

8.3.2 Modification of the Casimir Force Between Metallic Films Using the Skin-Depth Effect

The use of ultra-thin metallic coatings (i.e. of thickness comparable to the skin-depth at wavelengths comparable to the distance between the surfaces) over transparent dielectrics, as opposed to thick layers, as employed in the experiments of Sect. 8.2, should alter the distance dependence of the force.

At sub-micron distances, the Casimir force critically depends on the reflectivity of the interacting surfaces for wavelengths in the ultraviolet to far infrared [28, 66]. The attraction between transparent materials is expected to be smaller than that between highly reflective mirrors as a result of a less effective confinement of electromagnetic modes inside the optical cavity defined by the surfaces. A thin metallic film can be transparent to electromagnetic waves that would otherwise be reflected by bulk metal. In fact, when their thickness is much less than the skin-depth, most of the light passes through the film. Consequently, the Casimir force between metallic films should be significantly reduced when its thickness is less than the skin-depth at ultraviolet to infrared wavelengths. For most common metals, this condition is reached when the thickness of the layer is ~ 10 nm.

The technique presented in [66] was recently perfected in terms of the calibration method used and allowed the accurate measurement of the Casimir force for different metal film thickness on the sphere [50].

Demonstrating the skin-depth effect requires careful control of the thickness and surface roughness of the films. The sphere was glued to its support and subsequently coated with a 2.92 nm titanium (Ti) adhesion layer and a 9.23 nm film of palladium (Pd). The thickness of the Ti layer and of the Pd film was measured by Rutherford back scattering [67] on a silicon slice that was evaporated in close proximity to the sphere. After evaporation, the sphere was imaged with an optical profiler to determine its roughness and mounted inside the experimental apparatus. After completion of the Casimir force measurements, the sphere was removed from the experimental apparatus, coated with an additional 200 nm of Pd, analyzed with the optical profiler, and mounted back inside the vacuum chamber for another set of measurements. It is important to stress that the surface roughness measured before and after the deposition of the thicker Pd layer was the same within a few percent.

In Fig. 8.7, the results of the thin film measurements are compared with those obtained after the evaporation of the thick layer of Pd. The measurements were repeated 20 times for both the thin and thick films. The results clearly demonstrate the skin-depth effect on the Casimir force. The force measured with the thin film of Pd is in fact smaller than that observed after the evaporation of the thicker film. Measurements were repeated with a similar sphere: the results confirmed the

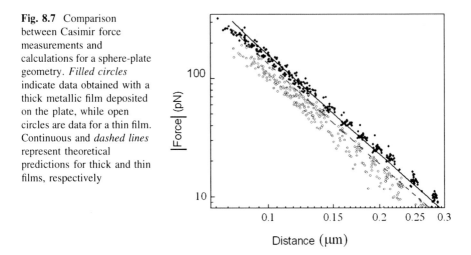

Fig. 8.7 Comparison between Casimir force measurements and calculations for a sphere-plate geometry. *Filled circles* indicate data obtained with a thick metallic film deposited on the plate, while open circles are data for a thin film. Continuous and *dashed lines* represent theoretical predictions for thick and thin films, respectively

skin-depth effect. To rule out possible spurious effects, the data were compared with a theoretical calculation (Fig. 8.7) based on the Lifshitz theory which includes the dielectric function of the metallic coatings and the effects of the surface roughness. The magnitude and spatial distribution of the latter was measured with an optical profilometer and incorporated in the modified Lifshitz equation [36, 50, 68]. The dielectric functions used in the calculation were obtained from Refs. [32–34, 69], and a suitable modification of Lifshitz's theory to account for multiple thin films was used [37].

The discrepancy observed in the case of the thin metallic film is not surprising. The calculation of the force is based on two approximations: (i) the dielectric function for the metallic layers (both titanium and palladium) is assumed to be equal to the one tabulated for bulk-materials, and (ii) the model used to describe the dielectric function of polystyrene is limited to a simplified two-oscillator approximation [69]. These assumptions can lead to significant errors in the estimated force.

8.3.3 Casimir Force Experiments with Transparent Materials

In this section we explore situations in which one of the two noble metal surfaces (typically gold) in Casimir force experiments is replaced by a material that is transparent over a significant range of wavelengths. The expectation would be a large reduction of the Casimir force. Experiments discussed in this section have shown however that to achieve such a reduction the transparency window must be very large. The reason is that, as seen earlier in this chapter, the Casimir–Lifshitz force depends on the dielectric function at imaginary frequencies, which depends on all wavelengths including ones much larger than the plate separation. Here we

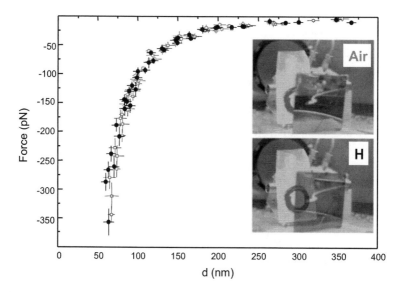

Fig. 8.8 Casimir force between a gold-coated plate and a sphere coated with a Hydrogen Switchable Mirror (HSM) as a function of the distance, in air (*open squares*) and in argon–hydrogen (*filled circles*). *Inset* A HSM in air and in hydrogen. A similar mirror was deposited on the sphere of our experimental apparatus

discuss in detail the case where one of the gold surfaces in the sphere-plate geometry is replaced by a metallic superlattice in which the reflectivity can be tuned by hydrogenation and also include a discussion of a recent experiment in which one of the two surfaces consists of Indium Tin Oxide (ITO), a transparent conductor.

Using the experimental set-up described in Sect. 8.2, the Casimir force between a gold-coated plate and a sphere coated with a Hydrogen Switchable Mirror (HSM) [70] was measured for separations in the 70 to 400 nm range [66]. The HSMs are metallic superlattices obtained by repeating seven consecutive evaporations of alternate layers of magnesium (10 nm) and nickel (2 nm), followed by an evaporation of a thin film of palladium (5 nm). The inset of Fig. 8.8 shows a glass slide coated according to this procedure, both in its as deposited state, and in its hydrogenated state. It is evident that the optical properties of the film are very different in the two situations. The transparency of the film was measured over a wavelength range between 0.5 and 3 μm, and its reflectivity at $\lambda = 660$ nm, keeping the sample in air and in an argon–hydrogen atmosphere (4% hydrogen). The results are in good agreement with the values reported in [71].

The results of Casimir force measurements obtained in air and in a hydrogen-rich atmosphere are shown in Fig. 8.8. It is evident that the force does not change in a discernible way upon hydrogenation of the HSM [66].

In order to explain this apparently surprising result, one should first note that the dielectric properties of the HSMs used in this experiment are known only in a limited range of wavelengths, spanning approximately the range 0.3–2.5 μm [71]. However, because the separation between the sphere and the plate in the experiment is in the 100 nm range, one could expect that it is not necessary to know the dielectric function for wavelengths longer than 2.5 μm, because those modes should not give rise to large contributions to the force. A mathematical analysis carried out using ad hoc models to describe the interacting surfaces has shown that this is not necessarily the case. Because the Casimir force depends on the dielectric function at imaginary frequency (8.4) and the integral in the latter is over all frequencies, long wavelengths compared to the separation between the sphere and the plate can make a significant contribution to the force. Thus, one of the reasons for not having observed a change in the latter upon hydrogenation is likely related to the fact that that the imaginary part of the permittivity might not change significantly at long wavelengths. Recently, a more accurate analysis of the experiment [72] confirmed this result, but also added an important detail: for a correct comparison of data with theory it is necessary take into account also the presence of the 5 nm thick palladium layer that was deposited on top of the HSMs to prevent oxidation and promote hydrogen absorption. Although this layer is fairly transparent to all wavelengths from ultraviolet to infrared, its contribution to the interaction reduces the expected change of the force by nearly a factor of two. It is thus the combination of the effect of the reflectivity at long wavelengths and of the thin palladium film that limits the magnitude of the change of the force following hydrogenation. Still, calculations show that a small change in the Casimir force upon hydrogenation should be observable with improved experimental precision and with the use of HSMs of different composition [72].

Recently the group of Davide Iannuzzi reported a precise measurement in air of the Casimir force between a gold-coated sphere and a glass plate coated with either a thick gold layer or a highly conductive, transparent ITO film [73]. The decrease of the Casimir force due to the different dielectric properties of the reflective gold layer and the transparent oxide film resulted to be as high as 40%–50% at all separations (from 50 to 150 nm). Physically the large reduction of the Casimir force when the Au surface is replaced by ITO is due to the much smaller plasma frequency of ITO (in the near infrared spectrum) compared to that of Au (in the ultraviolet). This experiment shows that, in the presence of a conductive oxide layer, the Casimir force can still be the dominant interaction mechanism even in air, and indicates that, whenever the design might require it, it is possible to tune the Casimir attraction by a factor of 2.

8.3.4 Casimir Forces in a Fluid

To measure the Casimir force in a fluid, a modified atomic force microscopy (AFM) method can be used as shown in Fig. 8.9 [74, 75]. Light from a superluminescent diode is reflected off the back of the cantilever and is detected by a four-quadrant photodetector, which is used to monitor the deflection of the

Fig. 8.9 Experimental setup. A polystyrene sphere is attached to an AFM cantilever and coated with gold. A laser beam is directed through a few millimeter opening in the conductive coating of the cantilever holder and is reflected off the back of the cantilever to monitor its motion

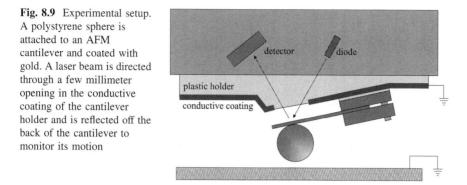

cantilever, as in standard AFM measurements. The difference signal between the top two quadrants and the bottom two quadrants is proportional to the vertical deflection of the cantilever. A piezoelectric column within the AFM is used to advance the cantilever and sphere toward the plate, and the piezoelectric column's advance is detected using a linear variable differential transformer, which minimizes nonlinearities and hysteresis inherent in piezoelectrics. As the sphere approaches the plate, any force between the two will result in a deflection of the cantilever, which will then be detected in the difference signal from the four-quadrant detector. Cleaning and calibration techniques can be used to isolate the Casimir force from other spurious forces (e.g. electrostatic and hydrodynamic) and to convert the deflection signal into a force signal [74–76].

Figure 8.10 shows the Casimir–Lifshitz force in ethanol between the gold-coated sphere and gold-coated plate. The data for 51 runs are shown (dots) along with the average of these data (circles) and Lifshitz theory for ethanol separating the two surfaces and no added salt (solid line). The theory describes the data well, despite the uncertainties in the optical properties. Deviations between the theory and experiment below 30 nm are likely due to the inability of the theory to accurately describe the surface roughness on these scales and the uncertainty in the optical properties.

The Casimir–Lifshitz force for different salt concentrations is shown in Fig. 8.10b along with Lifshitz's theory without corrections due to electrostatics or zero-frequency screening [75]. The data is shown for experiments with no added salt (circles), 0.3 mM NaI (squares), and 30 mM NaI (triangles) and is obtained by averaging 51 data set for each concentration. The inset shows a log–log plot of the data. The difference between the forces due to the modification of the zero-frequency contribution and the Debye screening are greater for smaller separations and both are calculated to be ∼15 pN in the range from 30–40 nm; however, the sensitivity of our apparatus is not adequate to distinguish a significant difference between these curves. Further experimental details can be found in Ref. [75].

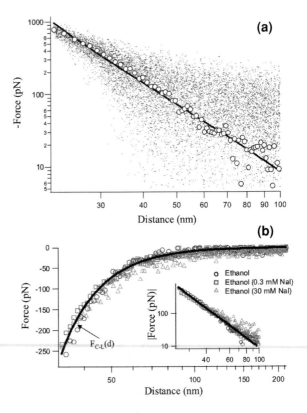

Fig. 8.10 a Measured force between a gold sphere and a gold plate immersed in liquid ethanol is well described by Lifshitz's theory. Dots represent measurements from 51 runs. Circles are average values from the 51 data sets. Solid line is Lifshitz's theory. **b** Comparison of the measured force with different concentrations of salt shows no significant difference. Force data in ethanol with no added salt (*circles*), 0.3 mM NaI (*squares*), and 30 mM NaI (*triangles*). Lifshitz's theory for no added salt is shown as a *solid line*. *Inset* log–log plot of the data

8.3.5 Repulsive Forces and Casimir Levitation

Modification of the Casimir force is of great interest from both a fundamental and an applied point-of-view. It is reasonable to ask whether such modifications can lead to repulsive forces in special cases. In 1968, T.H. Boyer showed that for a perfectly conducting spherical shell the Casimir effect should give rise to an outward pressure [77]. Similar repulsive Casimir forces have also been predicted for cubic and rectangular cavities with specific aspect ratios [78, 79]. However, criticisms concerning these results have been raised [80], and recently the possibility of repulsive forces based on topology for a wide class of systems has been ruled out [81].

The possibility of topological repulsive Casimir forces, i.e. due to the geometrical structure of the interacting metallic bodies in vacuum, is therefore controversial. In this section, we will describe a repulsive force that is due strictly to the optical properties of the materials involved. Such a mechanism is responsible for many phenomena in the non-retarded regime including the surface melting of solids [82] and the vertical ascent of liquid helium within a container (see, for example, the discussion in Refs. [20, 29, 83]).

As was demonstrated by Dzyaloshinskii, Lifshitz, and Pitaevskii in their seminal paper, the sign of the force depends on the dielectric properties of materials involved [29]. See also the Chap. 2 by Pitaevskii in this volume for related discussions. Two plates made out of the same material will always attract, regardless of the choice of the intermediate material (typically a fluid or vacuum); however, between slabs of different materials (here labeled 1 and 2) the force becomes repulsive by suitably choosing the intermediate liquid (labeled 3). Thus, by proper choice of materials, the Casimir–Lifshitz force between slabs 1 and 2 can be either attractive or repulsive. Specifically, the condition for repulsion is:

$$\varepsilon_1 > \varepsilon_3 > \varepsilon_2, \tag{8.6}$$

Here the dielectric functions $\varepsilon_1, \varepsilon_2$ and ε_3, of the materials are evaluated at imaginary frequencies. Because they vary with frequency, it is conceivable that inequality (8.6) may be satisfied for some frequencies and not for others. For various separations between the slabs, different frequencies will contribute with different strengths, which can lead to a change in the sign of the force as a function of separation.

In order to qualitatively understand the origin of these repulsive forces, we consider the following toy model (see Fig. 8.11 and Ref. [84]) for the microscopic interaction of the bodies. To first order, the force between the latter is dominated by the pair-wise summation of the van der Waals forces between all the constituent molecules. This additivity is a good approximation for rarefied media; however, the force between two molecules is affected in general by the presence of a third. Hamaker first used this approach in extending the calculations of London to the short-range interaction (i.e. the non-retarded van der Waals force) between bodies and in particular to those immersed in a fluid. By suitably choosing three materials and their constituent molecules so that their polarizabilities satisfy the inequality $\alpha_1 > \alpha_3 > \alpha_2$, we find the forces between the individual molecules, which are proportional to the product of the polarizabilites integrated over all imaginary frequencies, will obey: $F_{13} > F_{12} > F_{23}$, where the subscript ij represents the interaction between molecules i and j. Thus, it is energetically more favorable for molecule 3 to be near molecule 1 than it is for molecule 2 to be near molecule 1. As more molecules of the same species are added to the system, molecules of type 3 will be strongly attracted to those of type 1, resulting in an increased separation for molecules of type 2 from those of type 1. In this way, Hamaker showed that repulsive forces between two different materials immersed in a liquid are possible by calculating the total interaction energy between the bodies and the fluid as the separation between the bodies is varied. His calculations however were non-rigorous because they neglected non-additivity and retardation effects. When these are included, long-range repulsion between two bodies (materials 1 and 2) separated by a third (material 3) is predicted when their relative dielectric functions obey (8.6). Note that when the fluid has the largest dielectric function, the cohesive van der Waals interaction within the fluid results in an attraction between its molecules that is larger than that between the molecules of the fluid and the plate, which leads to an attractive force between the two plates.

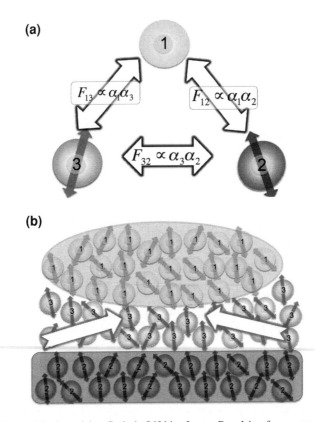

Fig. 8.11 Toy model of repulsive Casimir–Lifshitz forces. Repulsive forces can exist between two materials, schematically represented as an ensemble of molecules separated by a third, typically a liquid, with specific optical properties. **a** Three individual molecules will all experience attractive interactions. **b** For a collection of molecules, with $\alpha_1 > \alpha_3 > \alpha_2$, it is energetically more favorable for the molecules with the largest polarizabilities (α_1 and α_3 for this example) to be close, resulting in an increased separation between molecules of type 1 and type 2. For a condensed system, the net interaction between material 1 and material 2 is repulsive if the corresponding dielectric functions satisfy $\varepsilon_1 > \varepsilon_3 > \varepsilon_2$, as consequence of the similar inequality between polarizabilities. Note that all the α's and the ε's need to be evaluated at imaginary frequencies (see text)

Examples of material systems that obey (8.6) are rare but do exist. One of the earliest triumphs of Lifshitz's equation was the quantitative explanation of the thickening of a superfluid helium film on the walls of a container [29, 83]. In that system, it is energetically more favourable for the liquid to be between the vapour and the container, and the liquid climbs the wall. One set of materials (solid–liquid–solid) that obeys inequality (8.6) over a large frequency range is gold, bromobenzene, and silica (Fig. 8.12).

Using the above-mentioned material combination, we have shown that repulsive Casimir–Lifshitz forces are measureable [76]. Raw deflection versus piezo

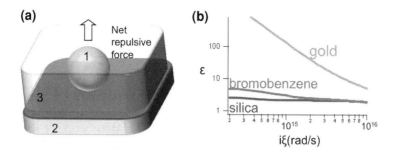

Fig. 8.12 Repulsive quantum electrodynamical forces can exist for two materials separated by a fluid. **a** The interaction between material *1* and material *2* immersed in a fluid (material *3*) is repulsive when $\varepsilon_1(i\xi) > \varepsilon_3(i\xi) > \varepsilon_2(i\xi)$, where the $\varepsilon(i\xi)$'s are the dielectric functions at imaginary frequency. **b** The optical properties of gold, bromobenzene, and silica are such that $\varepsilon_{\text{gold}}(i\xi) > \varepsilon_{\text{bromobenzene}}(i\xi) > \varepsilon_{\text{silica}}(i\xi)$ and lead to a repulsive force between the gold and silica surfaces

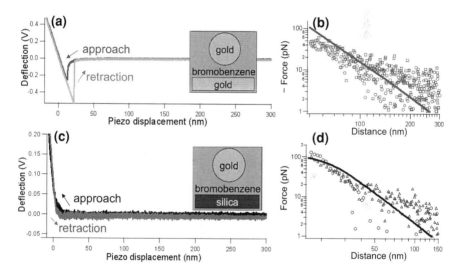

Fig. 8.13 Attractive and repulsive Casimir–Lifshitz force measurements. **a** Deflection data showing attractive interactions between a gold sphere and a gold plate. **c** For the case of the same gold sphere and a silica plate, deflection data show a repulsive interaction evident during both approach and retraction. **d** Measured repulsive force between a gold sphere and a silica plate on a log–log scale (*circles*) and calculated force using Lifshitz's theory (*solid line*) including corrections for the measured surface roughness of the sphere and the plate. Triangles are force data for another gold sphere (nominally of the same diameter)/silica plate pair. **b** Measured attractive force on a log–log scale for two gold sphere/plate pairs (*circles* and *squares*) and calculated force using Lifshitz's theory (*solid line*) including surface roughness corrections corresponding to the data represented by the circles

displacement data show that the force is changed from attractive to repulsive by replacing the gold plate with the silica plate [Fig. 8.13a,c]. The data in Fig. 8.13a,c were acquired with a piezo speed of 45 nm/s. With the gold plate, the cantilever is bent toward the surface during the approach, which corresponds to an attractive force between the sphere and plate until contact [Fig. 8.13a]. Once contact is made, the normal force of the plate pushes against the sphere. Upon retraction, the sphere sticks to the plate for an additional 10 nm, due to stiction between the two gold surfaces, before losing contact with the surface. When the silica plate is used, the cantilever is bent away from the surface during the approach, corresponding to a repulsive interaction [Fig. 8.13c]. During retraction, the sphere continues to show repulsion. This cannot be a result of the hydrodynamic force, because the hydrodynamic force is in a direction that opposes the motion of the sphere and will change sign as the direction is changed. Similarly, the repulsion observed in Fig. 8.13c cannot be due to charge trapped on silica; any charge that does exist on the surface will induce an image charge of opposite sign on the metal sphere and lead to an attractive interaction. Further experimental details can be found in Ref. [76], and a critical analysis of previous experiments in the van der Waals regime are discussed in Ref. [84] and briefly at the end of this section.

The measured forces after calibration show a clear distinction between the attractive and repulsive regimes when the plate is changed from gold to silica (Fig. 8.13b,d). The circles correspond to the average force from 50 runs between the gold sphere and the plate. Histograms of the force data at different distances show a Gaussian distribution and no evidence of systematic errors.

The experiment is repeated with an additional sphere and plate for both configurations. Figure 8.13b shows the measured force for two different spheres of nominally the same diameter and two different gold plates. Similar measurements for two spheres and silica plates are shown in Fig. 8.13d. The solid lines are the temperature dependent Lifshitz's theory including surface roughness corrections for the first sphere/plate pair (circles). Because the second set of measurements are made with spheres and plates of similar surface roughness and size, the corrections are of similar magnitude.

Prior to our work, previous experiments have shown evidence for short-range repulsive forces in the van der Waals regime [85–90]; however, there are many experimental issues that must be considered that, as our analysis below shows, were not adequately addressed in many of these experiments. For separations of a few nm or less, liquid orientation, solvation, and hydration forces become important and should be considered, which are not an issue at larger separations. Surface charging effects are important for all distance ranges. In order to satisfy (8.6), one of the solid materials must have a dielectric function that is lower than the dielectric function of the intermediate fluid. One common choice for this solid material is PTFE (polytetrafluoroethylene), which was used in most experiments [86, 88–90]; however, as was pointed out in Ref. [86], residual carboxyl groups and other impurities can easily be transferred from the PTFE to the other surface, which complicates the detection and isolation of the van der Waals force. In a few experiments, the sign of the force did not agree with the theoretical calculation, which may be attributed to additional electrostatic force contributions [85, 86]. To avoid this problem, Meurk et al.

Fig. 8.14 QED levitation device. A repulsive force develops between the disk immersed in a fluid and the plate, which is balanced by gravity. We show a nano-compass that could be developed to mechanically sense small magnetic fields

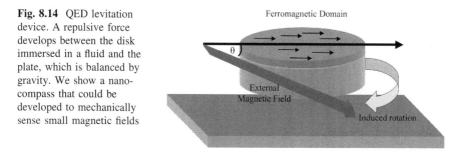

performed experiments with inorganic samples [87]; however, the experimental configuration consisted of a sharp tip and a plate, which limited the surface separations to below 2 nm. For the determination of the cantilever force constant, either the Sader method [91] or the Cleveland method [92] was used in these experiments. The Sader method gives the spring constant of a cantilever based on the geometry of the cantilever and its resonance frequency, and the Cleveland method uses the resonance frequency shift of a cantilever upon the addition of masses to determine the spring constant. These methods lead to an additional 10–20% error in the determination of the force [93], which could be greatly reduced if a calibration method is performed that uses a known force for the calibration [74, 75, 94]. Finally, the determination of the absolute distance was often found by performing a fit of the experimental data to the presumed power law of the van der Waals force [86, 88–90]. Thus, the absolute surface separation could only be determined if one assumed that the measured force was only the van der Waals force and that it was described precisely by a $1/d^2$ force law.

8.3.6 Devices Based on Repulsive Casimir Forces

Repulsive Casimir–Lifshitz forces could be of significant interest technologically as this technique might be used to develop ultra-sensitive force and torque sensors by counterbalancing gravity to levitate an object immersed in fluid above a surface without disturbing electric or magnetic interactions. Because the surfaces never come into direct contact as a result of their mutual repulsion, these objects are free to rotate or translate relative to each other with virtually no static friction. Dynamical damping due to viscosity will put limits on how quickly such a device can respond to changes in its surroundings; however, in principle even the smallest translations or rotations can be detected on longer time scales. Thus, force and torque sensors could be developed that surpass those currently used. Figure 8.14 shows an example of a QED levitation device: a nano-compass sensitive to very small static magnetic fields [95].

Fig. 8.15 A QED torque
develops between two
birefringent parallel plates
with in-plane optical axis
when they are placed in close
proximity

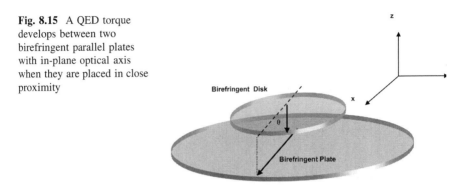

Fig. 8.15 A QED torque develops between two birefringent parallel plates with in-plane optical axis when they are placed in close proximity

8.4 QED Torque

The effect of the zero-point energy between two optically anisotropic materials, as shown in Fig. 8.15, has also been considered [54, 55, 95, 97–103]. In this case, the fluctuating electromagnetic fields have boundary conditions that depend on the relative orientation of the optical axes of the materials; hence, the zero-point energy arising from these fields also has an angular dependence. This leads to a torque that tends to align two of the principal axes of the materials in order to minimize the system's energy. We have recently shown that such torques should indeed be measurable and have suggested experimental configurations to perform these measurements [102, 103].

In 1972 Parsegian and Weiss derived an expression for the short-range, non-retarded van der Waals interaction energy between two semi-infinite dielectrically anisotropic materials immersed within a third material [54]. This result, obtained by the summation of the electromagnetic surface mode fluctuations, showed that the interaction energy was inversely proportional to the separation squared and depended on the angle between the optical axes of the two anisotropic materials. In 1978, Barash independently derived an expression for the interaction energy between two anisotropic materials using the Helmholtz free energy of the electromagnetic modes, which included retardation effects [55]. In the non-retarded limit, Barash's expression confirmed the inverse square distance dependence of Parsegian and Weiss and that the torque, in this limit, varies as $\sin(2\theta)$, where θ is the angle between the optical axes of the materials.

The equations that govern the torque in the general case of arbitrary distances are quite cumbersome and are treated in detail elsewhere [55, 102]. For brevity, we refer the reader to those papers for a more in-depth analysis and simply state a few of the relevant results. First, the torque is proportional to the surface area of the interacting materials and decreases with increasing surface separation. Second it is found that the QED torque at a given distance varies as:

$$M = A \sin(2\theta), \tag{8.7}$$

even in the retarded limit, where A is the value of the torque at $\theta = \pi/4$. Figure 8.16a shows the torque as a function of angle for a 40 µm diameter calcite disk in vacuum above a barium titanate plate at a distance of $d = 100$ nm [102]. The circles correspond to the calculated values of the torque, while the solid line corresponds to a best fit with (8.7).

Experimentally it is difficult to use large disks in close proximity, because at such small separations tolerances in the parallelism of two large surfaces (tens of microns in diameter) are extremely tight; in addition it is difficult to keep them free of dust and contaminants. If the vacuum is replaced by liquid ethanol, the torque remains of the same order of magnitude; however, the three materials (calcite, ethanol, and barium titanate) have dielectric functions that obey (8.6). This will result in a repulsive Casimir–Lifshitz force which will counterbalance the weight of the disk and allow it to float at a predetermined distance above the plate. For a 20 µm thick calcite disk with a diameter of 40 µm above a barium titanate plate in ethanol, the equilibrium separation was calculated to be approximately 100 nm with a maximum torque of $\sim 4 \times 10^{-19}$ N m [102], as shown in Fig. 8.16b.

For the observation of the QED torque, it was suggested in [102] that the disk be rotated by $\theta = \pi/4$ by means of the transfer of angular momentum of light from a polarized beam. The laser could then be shuttered, and one would visually observe the rotation of the disk back to its minimum energy orientation. The amount of angular momentum transfer determines the initial value of the angle of rotation. After the laser beam is blocked the disk can rotate either clockwise or counterclockwise back to the equilibrium position depending on the value of the initial angle, making it possible to verify the $\sin(2\theta)$ dependence of the torque. Procedures to minimize the effect of charges on the plates and other artifacts were also discussed [102].

An alternative scheme involving the statistical analysis of Brownian motion was recently described in [103]. For this situation, the disk size is reduced to the point that Brownian motion causes translation and rotation. When these rotations become comparable to the QED rotation, the disk will no longer rotate smoothly to its minimum energy configuration. Instead the angle between the optical axes will fluctuate to sample all angles. The probability distribution for the observation of the angle θ between the two optical axes is:

$$p(\theta) = p_o \exp[-U(\theta)/k_B T], \tag{8.8}$$

where $U(\theta)$ is the potential energy of the QED orientation interaction, i.e. the energy associated with the torque, $k_B T$ is the thermal energy, p_o is a normalization constant such that $\int p(\theta)\, d\theta = 1$. By observing the angle between the axes as a function of time, one can deduce the probability distribution via a histogram of the angular orientations as shown in Fig. 8.17. This is similar to the determination of the potential energy as a function of distance in Total Internal Reflection Microscopy (TIRM) experiments for optically trapped spherical particles [104, 105].

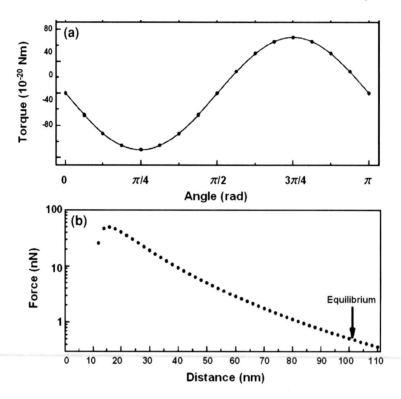

Fig. 8.16 **a** Calculated torque as a function of angle between a 40 μm diameter disk made of calcite and a barium titanate plate separated in vacuum by a distance $d = 100$ nm. The *lines* represent a fit (8.7). **b** Calculated retarded van der Waals force as a function of plate separation for this system at a rotation angle of $\pi/4$. The *arrow* represents the distance at which the retarded van der Waals repulsion is in equilibrium with the weight of the disk

To observe this effect, one needs to levitate a birefringent disk above a birefringent plate at short-range and be able to detect the orientation of the axes. This can be done either by using a repulsive Casimir–Lifshitz force or a double layer repulsion force [38] and video microscopy techniques [106] as described below.

The equilibrium separation occurs when the sum of the forces (Casimir–Lifshitz, double layer, and weight) acting on the particle is zero:

$$\sum F = F_{CL}(d) + D \times \exp[-d/l] - \pi R^2 h \Delta \rho g = 0, \qquad (8.9)$$

where $F_{CL}(d)$ is the Casimir–Lifshitz force at distance d, D is a constant related to the Poisson-Boltzmann potential evaluated at the surface due to charging, l is the Debye length, R is the radius of the disk, h is the thickness of the disk, $\Delta \rho$ is the density difference between the disk and the solution, and g is the acceleration due to gravity. Both the Casimir–Lifshitz force and the weight of the disks are set by the geometry of the system and the materials chosen; however, the double layer force

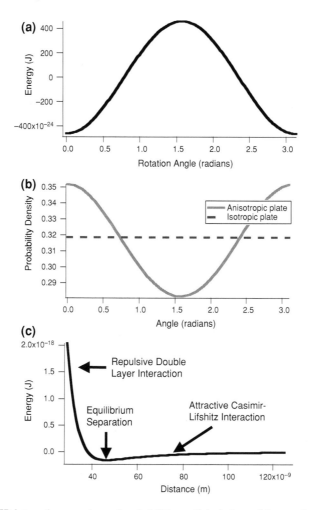

Fig. 8.17 QED interaction energies and probabilities. **a** Calculations of the angular dependence of the QED interaction energy between a 2 lm diameter LiNbO3 disk and a calcite plate. **b** Probability of detecting a rotation θ. **c** Energy as a function of separation between the disk and plate including contributions from the double layer interaction (dominant at close range), the Casimir–Lifshitz interaction (dominant at longer range) and gravity (negligible). Equilibrium is obtained around 50 nm

can be modified by changing the electrolyte concentration. Thus, the floatation height can be adjusted in this way. Figure 8.17c shows the approximate interaction energy following from the forces of (8.9), where we have chosen a double layer interaction leading to a levitation height of approximately 50 nm, with deviations of a few nm due to thermal energy $(k_B T)$, for a lithium niobate disk with radius $R = 1$ μm and thickness $h = 0.5$ μm in an aqueous solution above a calcite plate.

In order to track the motion of the disk above the plate, a video microscopy setup similar to the one described in [106] is used. The disk's motion is tracked and

Fig. 8.18 Schematic of the Brownian motion detection scheme showing data for a non-birefringent spherical particle. Light is recorded via a CCD and digitized to allow for tracking and determination of intensity fluctuations

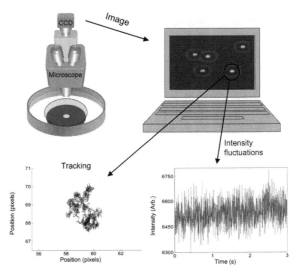

recorded as is the intensity of the transmitted light. The orientation of the disk is determined by placing it between a combination of polarizing optical components so that the intensity of the transmitted light can be related to the orientation of the optical axis. In order to determine the expected optical intensity at the output as a function of θ, the Jones matrix representation of the optical elements is used to determine the exiting E-field from which the intensity is calculated [107, 108]. For suitably chosen optical components (see Ref. [103]), the intensity is proportional to $[1 - \cos(2\theta)]$. From histogram of the intensities, we can determine the preferred angular orientation of the disk and hence the angular QED interaction energy and the torque.

Figure 8.18 shows the typical configuration for such experiments. The thermal fluctuations of the particles are recorded via a CCD camera attached to an upright microscope. The particles' centers can be determined and tracked by the method of [106] with a standard deviation of less than 1/10 pixel. Figure 8.18 shows both the tracking and intensity fluctuations recorded for a spherical non-birefringent particle. For non-birefringent particles, the intensity fluctuations are due to scattering by the particle as it undergoes Brownian motion. In order to study the QED torque, small birefringent disks should be used. Such disks have been fabricated using a combination of crystal ion slicing [109], mechanical polishing, and focused ion beam (FIB) sculpting as shown in Fig. 8.19. To date no experiments have been performed with birefringent particles; however, this detection scheme should be suitable for such QED torque experiments.

8.5 Future Directions

A number of other interesting QED phenomena await experimental investigation:

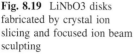

Fig. 8.19 LiNbO3 disks fabricated by crystal ion slicing and focused ion beam sculpting

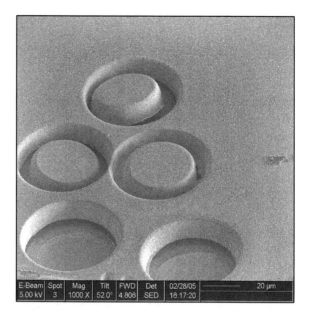

8.5.1 Phase Transitions and the Casimir Effect

Recently the influence of Casimir energy on the critical field of a superconducting film has been theoretically investigated, and it was shown that it might be possible to directly measure the variation of Casimir energy that accompanies the superconducting transition [110]. Another interesting experiment would be to use as one of the surfaces a vanadium oxide film. This material switches from insulator to metal above a temperature of ~ 60 C and such a transition with the attendant significant change in reflectivity is expected to lead to an observable increase of the Casimir force above that temperature. Recent measurements on another material (Ag-In-Sb-Te) that can undergo a phase transition has shown a modification of the Casimir force by up to 20% when samples prepared from either crystalline or amorphous phases were used [111].

8.5.2 Self Assembly and Sorting via the Casimir–Lifshitz Force

The ability to modify the Casimir–Lifshitz force opens the door to the possibility of engineering the potential energy landscape for particles based purely on their dielectric functions [84]. With the appropriate choice of fluid, repulsive forces will occur for asymmetric configurations (e.g. Au-SiO$_2$ across bromobenzene), while attractive forces will occur for symmetric configurations (e.g. Au-Au or SiO$_2$-SiO$_2$ across bromobenzene). By patterning a plate with these two different materials, one can study both non-additivity effects discussed above and the assembly and sorting of particles based solely on their dielectric functions. Similar sorting and aggregation effects have been observed in the thermodynamic Casimir effect, which is related to classical density fluctuations [112]. Other recent proposals include the ability to tune chemical reactions [113] and the self-assembly of colloidal scale devices [114, 115] based, at least partially, on manipulating the Casimir–Lifshitz forces.

8.5.3 Casimir Friction

There has been an interesting prediction that dissipative retarded van der Waals forces can arise between surfaces in relative motion due to the exchange of virtual photons which couple to acoustic phonons in the material [116]. Similar dissipative Casimir forces can arise between metals; here virtual photons would couple to particle-hole excitations in the metal [117]. This would lead to changes with position of the Q of suitable MEMS oscillators, such as the ones described in Sect. 8.2.2.

J. B. Pendry has considered another type of vacuum friction when two perfectly smooth featureless surfaces at T = 0, defined only by their respective dielectric functions, separated by a finite distance, move parallel to each other [118]. He found large frictional effects comparable to everyday frictional forces provided that the materials have resistivities of the order of 1 MΩ and that the surfaces are in close proximity. The friction depends solely on the reflection coefficients of the surfaces for electromagnetic waves, and its detailed behavior with shear velocity and separation is controlled by the dispersion of the reflectivity with frequency. There exists a potentially rich variety of vacuum friction effects, as discussed in a recent article [119]. See also the discussion in the Chap. 13 of Dalvit et al. and Henkel et al. Chap. 11 of this volume.

8.5.4 Dynamic Casimir Effect

It is also interesting to point out that the nonuniform relative acceleration of the metal and the sphere will lead, at least in principle, in the Casimir oscillator of Sect. 8.2.2 to an additional damping mechanism associated with the parametric down-conversion of vibrational quanta into pairs of photons, a QED effect associated with the nonlinear properties of vacuum. This phenomenon, which was investigated theoretically by Lambrecht, Jackel, and Reynaud in the context of a vibrating parallel plate capacitor [120], is an example of the so called dynamical Casimir effect, i.e. the nonthermal radiation emitted by uncharged metal or dielectric bodies in a state of nonuniform acceleration [121] (see also the Chap. 13

of Dalvit et al. of this volume). The extraction of photons from vacuum in a cavity vibrating at twice the fundamental frequency of the cavity can be viewed as a parametric "vacuum squeezing" phenomenon. Physically, photons are created as a result of the time dependent boundary conditions of cavity modes, which produce electromagnetic fields. The observation of this effect would require a very high cavity Q ($\sim 10^8$–10^9) typical of superconductive cavities and GHz oscillations frequencies [120]. Such frequencies have been achieved in NEMS [122].

It is worth pointing out that radiation can be extracted from QED fluctuations also from a beam of neutral molecules interacting with a grating. In this case coherent radiation can be generated as result of the time dependent modulation of the Casimir-Polder-van der Waals force between the molecules and the grating. Radiation in the far infrared region should be attainable with beam densities of 10^{17} cm^{-3} [123].

8.6 Conclusions

In conclusion following a comprehensive state-of-the-art overview of the Casimir effect from its original proposal, we have discussed our recent and ongoing research in this promising field.

Note: A review has recently appeared [124] discussing the physics of the Casimir effect in microstructured geometries.

Acknowledgments The authors would like to thank D. Iannuzzi, M Lisanti, L Spector, M B Romanowsky, N Geisse, K Parker, R M Osgood, R Roth, H Stone, Y Barash, V A Aksyuk, R N Kleinman, D J Bishop for their collaborations and R Guerra, R Onofrio, M Kardar, R L Jaffe, S G Johnson, J D Joannopoulos, L Levitov, V Parsegian, J. N. Israelachvili, E Tosatti, V. Pogrovski, M Scully, P W Milonni, W. Kohn, M. Cohen, A Lambrecht, F. Intravaia, S. Reynaud for helpful suggestions and discussions.

References

1. Milonni, P.W.: The Quantum Vacuum: An Introduction to Quantum Electrodynamics. Academic Press, San Diego (1993)
2. Casimir, H.B.G.: On the attraction between two perfectly conducting plates. Proc. K. Ned. Akad. Wet. **60**, 793–795 (1948)
3. Casimir, H.B.G.: Haphazard Reality. Half a Century of Science. Harper and Row, New York (1983)
4. Jaffe, R.L., Scardicchio, A.: Casimir effect and geometric optics. Phys. Rev. Lett. **92**, 070402 (2004)
5. Scardicchio, A., Jaffe, R.L.: Casimir effects: an optical approach I. Foundations and examples. Nuc. Phys. B **704**, 552–582 (2005)
6. Sparnaay, M.J.: Measurements of attractive forces between flat plates. Physica **24**, 751–764 (1958)

7. van Blokland, P.H.G.M., Overbeek, J.T.G.: van der Waals forces between objects covered with a chromium layer. J. Chem. Soc. Faraday Trans. **74**, 2637–2651 (1978)
8. Lamoreaux, S.K.: Demonstration of the Casimir force in the 0.6 to 6 μm range. Phys. Rev. Lett. **78**, 5–8 (1997)
9. Mohideen, U., Roy, A.: Precision Measurement of the Casimir force from 0.1 to 0.9 μm. Phys. Rev. Lett. **81**, 4549–4552 (1998)
10. Roy, A., Lin, C.-Y., Mohideen, U.: Improved Precision Measurement of the Casimir force. Phys. Rev. D **60**, 111101 (1999)
11. Harris, B.W., Chen, F., Mohideen, U.: Precision Measurement of the Casimir force using gold surfaces. Phys.Rev. A **62**, 052109 (2000)
12. Ederth, T.: Template-stripped gold surfaces with 0.4-nm rms roughness suitable for force measurements: Application to the Casimir force in the 20–100 nm range. Phys. Rev. A **62**, 062104 (2000)
13. Bressi, G., Carugno, G., Onofrio, R., Ruoso, G.: Measurement of the Casimir force between parallel metallic surfaces. Phys. Rev. Lett. **88**, 041804 (2002)
14. Decca, R.S., Lopez, D., Fischbach, E., Krause, D.E.: Measurement of the Casimir Force between dissimilar metals. Phys. Rev. Lett. **91**, 050402 (2003)
15. Decca, R.S., Lopez, D., Fischbach, E., Klimchitskaya, G.L., Krause, D.E., Mostepanenko, V.M.: Precise comparison of theory and new experiment for the Casimir force leads to stronger constraints on thermal quantum effects and long-range interactions. Ann. Phys. **318**, 37–80 (2005)
16. Chan, H.B., Aksyuk, V.A., Kleinman, R.N., Bishop, D.J., Capasso, F.: Quantum mechanical actuation of microelectromechanical systems by the Casimir force. Sci. **291**, 1941–1944 (2001)
17. Derjaguin, B.V., Abrikosova, I.I.: Direct measurement of the molecular attraction of solid bodies. Statement of the problem and measurement of the force by using negative feedback. Sov Phys. JETP **3**, 819–829 (1957)
18. Milonni, Peter.W., Shih, Mei.-Li.: Casimir Forces. Contemp. Phys. **33**, 313–322 (1992)
19. Spruch, L.: Long-range (Casimir) interactions. Sci. **272**, 1452–1455 (1996)
20. Parsegian, V.A.: Van der Waals forces: a Handbook for Biologists. Chemists, Engineers, and Physicists, Cambridge University Press, New York (2006)
21. Mostepanenko, V.M., Trunov, N.N.: The Casimir effect and its applications. Oxford University Press, Clarendon NY (1997)
22. Milton, K.A.: The Casimir effect: Physical manifestations of zero-point energy. World Scientific, singapore (2001)
23. Bordag, M., Mohideen, U., Mostepanenko, V.M.: New developments in the casimir effect. Phys. Rep. **353**, 1–205 (2001)
24. Martin, P.A., Buenzli, P.R.: The Casimir effect. Acta Phys. Polon. B **37**, 2503–2559 (2006)
25. Lambrecht, A.: The Casimir effect: a force from nothing. Physics World **15**, 29–32 (2002). (Sept. 2002)
26. Lamoreaux, S.K.: Resource Letter CF-1: Casimir Force. Am. J. Phys. **67**, 850–861 (1999)
27. For an extensive bibliography on the Casimir effect see: http://www.cfa.harvard.edu/~babb/casimir-bib.html.
28. Lifshitz, E.M.: The theory of molecular attractive forces between solids. Sov Phys. JETP **2**, 73–83 (1956)
29. Dzyaloshinskii, I.E., Lifshitz, E.M., Pitaevskii, L.P.: The general theory of van der Waals forces. Adv. Phys. **10**, 165–209 (1961)
30. Lambrecht, A., Reynaud, S.: Casimir force between metallic mirrors. Eur Phys. J. D **8**, 309–318 (2000)
31. Klimchitskaya, G.L., Mohideen, U., Mostepanenko, V.M.: Casimir and van der Waals forces between two plates or a sphere (lens) above a plate made of real metals. Phys. Rev. A **61**, 062107 (2000)
32. Palik, E.D. (ed.): Handbook of Optical Constants of Solids. Academic, New York (1998)

33. Palik, E.D. (ed.): Handbook of Optical Constants of Solids: II. Academic, New York (1991)
34. Ordal, M.A., Bell, R.J., Alexander Jr., R.W., Long, L.L., Querry, M.R.: Optical properties of fourteen metals in the infrared and far infrared: Al, Co, Cu, Au, Fe, Pb, Mo, Ni, Pd, Pt, Ag, Ti, V, and W. Appl. Opt. **24**, 4493–4499 (1985)
35. Maradudin, A.A., Mazur, P.: Effects of surface roughness on the van der Waals force between macroscopic bodies. Phys. Rev. B **22**, 1677–1686 (1980)
36. Neto, P.A.M., Lambrecht, A., Reynaud, S.: Roughness correction to the Casimir force: Beyond the proximity force approximation. Europhys. Lett. **69**, 924–930 (2005)
37. Mahanty, J., Ninham, B.W.: Dispersion Forces. Academic, London (1976)
38. Israelachvili, J.N.: Intermolecular and Surface Forces. Academic, London (1991)
39. Henkel, C., Joulain, K., Mulet, J.Ph., Greffet, J.-J.: Coupled surface polaritons and the Casimir force. Phys. Rev. A **69**, 023808 (2004)
40. Intravaia, F., Lambrecht, A.: Surface plasmon modes and the Casimir energy. Phys. Rev. Lett. **94**, 110404 (2005)
41. Intravaia, F.: *Effet Casimir et interaction entre plasmons de surface*, PhD Thesis, Université Paris, (2005)
42. Lamoreaux, S.K.: Comment on Precision Measurement of the Casimir Force from 0.1 to 0.9 μm. Phys. Rev. Lett. **83**, 3340 (1999)
43. Lamoreaux, S.K.: Calculation of the Casimir force between imperfectly conducting plates. Phys. Rev. A **59**, 3149–3153 (1999)
44. Iannuzzi, D., Gelfand, I., Lisanti, M., Capasso, F.: New Challenges and directions in Casimir force experiments. Proceedings of the Sixth Workshop on Quantum Field Theory Under the Influence of External Conditions, pp. 11-16. Rinton, Paramus, NJ (2004)
45. Pirozhenko, I., Lambrecht, A., Svetovoy, V.B.: Sample dependence of the Casimir force. New J. Phys. **8**, 238 (2006)
46. Buks, E., Roukes, M.L.: Metastability and the Casimir effect in micromechanical systems. Europhys. Lett. **54**(2), 220–226 (2001)
47. Chan, H.B., Aksyuk, V.A., Kleinman, R.N., Bishop, D.J., Capasso, F.: Nonlinear micromechanical Casimir oscillator. Phys. Rev. Lett. **87**, 211801 (2001)
48. Iannuzzi, D., Lisanti, M., Munday, J.N., Capasso, F.: The design of long range quantum electrodynamical forces and torques between macroscopic bodies. Solid State Commun. **135**, 618–626 (2005)
49. de Man, S., Heeck, K., Wijngaarden, R.J., Iannuzzi, D.: Halving the Casimir force with conductive oxides. Phys. Rev. Lett. **103**, 040402 (2009)
50. Lisanti, M., Iannuzzi, D., Capasso, F.: Observation of the skin-depth effect on the Casimir force between metallic surfaces. Proc. Natl. Acad. Sci. USA **102**, 11989–11992 (2005)
51. Golestanian, R., Kardar, M.: "Mechanical response of vacuum. Phys. Rev. Lett. **78**, 3421–3425 (1997)
52. Emig, T., Hanke, A., Kardar, M.: Probing the strong boundary shape dependence of the Casimir force. Phys. Rev. Lett. **87**, 260402 (2001)
53. Chan, H.B., Bao, Y., Zou, J., Cirelli, R.A., Klemens, F., Mansfield, W.M., Pai, C.S.: Measurement of the Casimir force between a gold sphere and a silicon surface with nanoscale trench arrays. Phys. Rev. Lett. **101**, 030401 (2008)
54. Parsegian, V.A., Weiss, G.H.: Dielectric anisotropy and the van der Waals interaction between bulk media. J. Adhes. **3**, 259–267 (1972)
55. Barash, Y.: On the moment of van der Waals forces between anisotropic bodies. Izv. Vyssh. Uchebn. Zaved. Radiofiz. **12**, 1637–1643 (1978)
56. Senturia, S.D.: Microsystem Design. Kluwer Academic, Dordrecht (2001)
57. Bishop, D.J., Giles, C.R., Austin, G.P.: The Lucent LambdaRouter MEMS: Technology of the future here today. IEEE Comun. Mag. **40**, 75–79 (2002)
58. Aksyuk, V.A., Pardo, F., Carr, D., Greywall, D., Chan, H.B., Simon, M.E., Gasparyan, A., Shea, H., Lifton, V., Bolle, C., Arney, S., Frahm, R., Paczkowski, M., Haueis, M., Ryf, R., Neilson, D.T., Kim, J., Giles, C.R., Bishop, D.: Beam-steering micromirrors for large optical cross-connects. J. Lightw. Technol. **21**, 634–642 (2003)

59. Serry, M., Walliser, D., Maclay, J.: The role of the Casimir effect in the static deflection and stiction of membrane strips in microelectromechanical systems (MEMS). J. Appl. Phys. **84**, 2501–2506 (1998)
60. De Los Santos, H.J.: Nanoelectromechanical quantum circuits and systems. Proc. IEEE **91**, 1907–1921 (2003)
61. Serry, F.M., Walliser, D., Maclay, G.J.: The anharmonic Casimir oscillator (ACO)-the Casimir effect in a model microelectromechanical system. J. Microelectromech. Syst. **4**, 193–205 (1995)
62. Landau, L.D., Lifshitz, E.M.: Mechanics. Pergamon, New York (1976)
63. Büscher, R., Emig, T.: Nonperturbative approach to Casimir interactions in periodic geometries. Phys. Rev. A **69**, 062101 (2004)
64. Klimchitskaya, G.L., Zanette, S.I., Caride, A.O.: Lateral projection as a possible explanation of the nontrivial boundary dependence of the Casimir force. Phys. Rev. A **63**, 014101 (2000)
65. Chen, F., Klimchitskaya, G.L., Mostepanenko, V.M., Mohideen, U.: Demonstration of the difference in the Casimir force for samples with different charge-carrier densities. Phys. Rev. Lett. **97**, 170402 (2006)
66. Iannuzzi, D., Lisanti, M., Capasso, F.: Effect of hydrogen-switchable mirrors on the Casimir force. Proc. Natl. Acad. Sci. USA **101**, 4019–4023 (2004)
67. Chu, W.K., Mayer, J.W., Nicolet, M.-A.: Backscattering Spectrometry. Academic Press, New York (1978)
68. Iannuzzi, D., Lisanti, M., Munday, J.N., Capasso, F.: Quantum fluctuations in the presence of thin metallic films and anisotropic materials. J. Phys. A: Math. Gen. **39**, 6445–6454 (2006)
69. Hough, D.B., White, L.R.: The Calcualtion of Hamaker constants from Lifshitz theory with applications to wetting phenomena. Adv. Colloid Interface Sci. **14**, 3–41 (1980)
70. Huiberts, J.N., Griessen, R., Rector, J.H., Wijngaarden, R.J., Dekker, J.P., de Groot, D.G., Koeman, N.J.: Yttrium and lanthanum hydride films with switchable optical properties. Nat. **380**, 231–234 (1996)
71. Richardson, T.J., Slack, J.L., Armitage, R.D., Kostecki, R., Farangis, B., Rubin, M.D.: Switchable mirrors based on nickel–magnesium films. Appl. Phys. Lett. **78**, 3047–3049 (2001)
72. de Man, S., Iannuzzi, D.: On the use of hydrogen switchable mirrors in Casimir force experiments. New J. Phys. **8**, 235 (2006)
73. de Man, S., Heeck, K., Wijngaarden, R.J., Iannuzzi, D.: Halving the Casimir force with conductive oxides. Phys. Rev. Lett. **103**, 040402 (2009)
74. Munday, J.N., Capasso, F.: Precision Measurement of the Casimir–Lifshitz force in a fluid. Phys. Rev. A **75**, 060102 (2007)
75. Munday, J.N., Capasso, F., Parsegian, V.A., Bezrukov, S.M.: Measurements of the Casimir–Lifshitz force in fluids: The effect of electrostatic forces and Debye the Casimir–Lifshitz force in fluids: The effect of electrostatic forces and Debye screening. Phys. Rev. A **78**, 032109 (2008)
76. Munday, J.N., Parsegian, V.A., Capasso, F.: Measured long-range repulsive Casimir–Lifshitz forces. Nat. **457**, 170–173 (2009)
77. Boyer, T.H.: 'Quantum electromagnetic zero-point energy of a conducting spherical shell and the Casimir model for a charged particle. Phys. Rev. **174**, 1764–1776 (1968)
78. Ambjorn, J., Wolfram, S.: Properties of the vacuum. I. mechanical and thermodynamic. Ann. Phys. **147**, 1–32 (1983)
79. Maclay, G.J.: Analysis of zero-point electromagnetic energy and Casimir forces in conducting rectangular cavities. Phys. Rev. A **61**, 052110 (2000)
80. Graham, N., Jaffe, R.L., Khemani, V., Quandt, M., Schroeder, O., Weigel, H.: The Dirichlet Casimir Problem. Nucl.Phys B **677**, 379–404 (2004)
81. Hertzberg, M.P., Jaffe, R.L., Kardar, M., Scardicchio, A.: Attractive Casimir forces in a closed geometry. Phys. Rev. Lett. **95**, 250402 (2005)

82. Tartaglino, U., Zykova-Timan, T., Ercolessi, F., Tosatti, E.: Melting and nonmelting of solid surfaces and nanosystems. Phys. Repts. **411**, 291–321 (2005)
83. Sabisky, E.S., Anderson, C.H.: Verification of the Lifshitz theory of the van der Waals potential using liquid-helium films. Phys. Rev. A **7**, 790–806 (1973)
84. Munday, J.N., Federico, C.: Repulsive Casimir and van der Waals forces: From measurements to future technologies. Inter. Journ. Mod. Phys. A **25**, 2252–2259 (2010)
85. Hutter, J.L., Bechhoefer, J.: Manipulation of van der Waals forces to improve image resolution in atomic-force microscopy. J. Appl. Phys. **73**, 4123 (1993)
86. Milling, A., Mulvaney, P., Larson, I.: Direct measurement of repulsive van der Waals interactions using an atomic force microscope. J. Col. Inter. Sci. **180**, 460 (1996)
87. Meurk, A., Luckham, P.F., Bergstrom, L.: Direct measurement of repulsive and attractive van der Waals forces between inorganic materials. Langmuir **13**, 3896 (1997)
88. Lee, S., Sigmund, W.M.: Repulsive van der Waals forces for silica and alumina. J. Col. Inter. Sci. **243**, 365 (2001)
89. Lee, S.W., Sigmund, W.M.: AFM study of repulsive van der Waals forces between teflon AF thin film and silica or alumina. Col. Surf. A **204**, 43 (2002)
90. Feiler, A., Plunkett, M.A., Rutland, M.W.: Superlubricity using repulsive van der Waals forces. Langmuir **24**, 2274 (2008)
91. Sader, J.E., et al.: Method for the calibration of atomic force microscope cantilevers. Rev. Sci. Instr **66**, 3789 (1995)
92. Cleveland, J.P., et al.: A nondestructive method for determining the spring constant of cantilevers for scanning force microscopy. Rev. Sci. Instr. **64**, 403 (1993)
93. Gibson, C.T., Watson, G.S., Myhra, S.: Scanning force microscopy—calibrative procedures for best practice. Scanning **19**, 564 (1997)
94. Craig, V.S.J., Neto, C.: In Situ calibration of colloid probe cantilevers in force microscopy: hydrodynamic drag on a sphere approaching a wall. Langmuir **17**, 6018 (2001)
95. Iannuzzi, D., Munday, J., Capasso, F.: Ultra-low static friction configuration. Patent submitted, (2005)
96. Ducker, W.A., Senden, T.J., Pashley, R.M.: Direct force measurements of colloidal forces using an atomic force microscope. Nat. **353**, 239–241 (1991)
97. van Enk, S.J.: Casimir torque between dielectrics. Phys. Rev. A **52**, 2569–2575 (1995)
98. Kenneth, O., Nussinov, S.: New polarized version of the Casimir effect is measurable. Phys. Rev. D **63**, 121701 (2001)
99. Shao, S.C.G., Tong, A.H., Luo, J.: Casimir torque between two birefringent plates. Phys. Rev. A **72**, 022102 (2005)
100. Torres-Guzmán, J.C., Mochán, W.L.: Casimir torque. J. Phys. A: Math. Gen. **39**, 6791–6798 (2006)
101. Rodrigues, R.B., Neto, P.A.M., Lambrecht, A., Reynaud, S.: Vacuum-induced torque between corrugated metallic plates. Europhys. Lett. **76**, 822 (2006)
102. Munday, J.N., Iannuzzi, D., Barash, Y., Capasso, F.: Torque on birefringent plates induced by quantum fluctuations. Phys. Rev. A **71**, 042102 (2005)
103. Munday, J.N., Iannuzzi, D., Capasso, F.: Quantum electrodynamical torques in the presence of brownian motion. New J. Phys. **8**, 244 (2006)
104. Prieve, D.C., Frej, N.A.: Total internal reflection microscopy: a quantitative tool for the measurement of colloidal forces. Langmuir **6**, 396–403 (1990)
105. Prieve, D.C.: Measurement of colloidal forces with TIRM. Adv. Colloid Interface Sci. **82**, 93–125 (1999)
106. Crocker, J.C., Grier, D.G.: Methods of digital video microscopy for colloidal studies. J. Colloid Interface Sci. **179**, 298–310 (1996)
107. Jones, R.C.: New calculus for the treatment of optical systems. I. Description and discussion of the calculus. J. Opt. A. **31**, 488–493 (1941)
108. Fowles, G.R.: Introduction to Modern Optics. Dover Publishing, New York (1968)

109. Levy, M., Osgood, R.M., Liu, R., Cross, L.E., Cargill III, G.S., Kumar, A., Bakhru, H.: Fabrication of single-crystal lithium niobate films by crystal ion slicing. Appl. Phys. Lett. **73**, 2293–2295 (1998)

110. Bimonte, G., Calloni, E., Esposito, G., Rosa, L.: Casimir energy and the superconducting phase transition. J. Phys. A: Math. Gen. **39**, 6161–6171 (2006)

111. Torricelli, G., van Zwol, P.J., Shpak, O., Binns, C., Palasantzas, G., Kooi, B.J., Svetovoy, V.B., Wuttig, M.: Switching Casimir forces with phase change materials. Phys. Rev. A **82**, 010101(R) (2010)

112. Soyka, F., et al.: Critical Casimir forces in colloidal suspensions on chemically patterned surfaces. Phys. Rev. Lett. **101**, 208301 (2008)

113. Sheehan, D.P.: Casimir chemistry. The J. Chem. Phys. **131**, 104706 (2009)

114. Cho, Y.K., et al.: Self-assembling colloidal-scale devices: selecting and using shortrange surface forces between conductive solids. Adv. Mater. **17**, 379 (2007)

115. Bishop, K.J.M., et al.: Nanoscale forces and their uses in self-assembly. Small **5**, 1600 (2009)

116. Levitov, L.S.: Van der Waals' friction. Europhys. Lett. **8**, 499–504 (1989)

117. Levitov L.S.: private communication

118. Pendry, J.B.: Shearing the vacuum-quantum friction. J. Phys.: Condens. Matter **9**, 10301–10320 (1997)

119. Kardar, M., Golestanian, R.: The friction of vacuum, and other fluctuation-induced forces. Rev Mod. Phys. **71**, 1233–1245 (1999)

120. Lambrecht, A., Jaekel, M., Reynaud, S.: Motion induced radiation from a vibrating cavity. Phys. Rev. Lett. **77**, 615–618 (1996)

121. Schwinger, J.: Casimir light: A glimpse. Proc. Natl. Acad. Sci. USA **90**, 958–959 (1993)

122. Huang, X.M.H., Feng, X.L., Zorman, C.A., Mehregany, M., Roukes, M.L.: VHF, UHF and microwave frequency nanomechanical resonators. New J. Phys. **7**, 247 (2005)

123. Belyanin, A., Kocharovsky, V., Kocharovsky, V., Capasso, F.: Coherent radiation from neutral molecules moving above a grating. Phys. Rev. Lett. **88**, 053602 (2002)

124. Rodriguez, A.W., Capasso, F., Johnson, S.G.: Nat. Photonics **5** 211 (2011)

Chapter 9
Casimir Force in Micro and Nano Electro Mechanical Systems

Ricardo Decca, Vladimir Aksyuk and Daniel López

Abstract The last 10 years have seen the emergence of micro and nano mechanical force sensors capable of measuring the Casimir interaction with great accuracy and precision. These measurements have proved fundamental to further develop the understanding of vacuum fluctuations in the presence of boundary conditions. These micromechanical sensors have also allowed to quantify the influence of materials properties, sample geometry and unwanted interactions over the measurement of the Casimir force. In this review we describe the benefits of using micro-mechanical sensors to detect the Casimir interaction, we summarize the most recent experimental results and we suggest potential optomechanical experiments that would allow measuring this force in regimes that are currently unreachable.

9.1 Introduction

During the last 60 years, there have been considerable studies trying to understand the forces acting between electrically neutral objects in vacuum, particularly, the van der Waals and Casimir forces. The experimental characterization and physical

R. Decca
Indiana University–Purdue University Indianapolis, 402 N. Blackford St.,
Bldg. LD154, Indianapolis, IN 46202, USA
e-mail:

V. Aksyuk
Center for Nanoscale Science and Technolog, National Institute of Standards
and Technology, 100 Bureau Dr., Gaithersburg, MD 20899, USA

D. López
Argonne National Laboratory, Center for Nanoscale Materials,
9700 South Cass Avenue, Lemont, IL 60439, USA

D. Dalvit et al. (eds.), *Casimir Physics*, Lecture Notes in Physics 834,
DOI: 10.1007/978-3-642-20288-9_9, © Springer-Verlag Berlin Heidelberg 2011

interpretation of these interactions is still generating discussions and stimulating the development of increasingly sophisticated experiments.

In the late 1940s, Hendrik Casimir [1] demonstrated theoretically that there is an *attractive* force between two electrically neutral, perfectly reflecting, and parallel conducting plates placed in vacuum. This attractive force is known as the Casimir force and is considered a quantum phenomenon since in classical electrodynamics the force acting between neutral planes is strictly zero. Casimir compared the quantum fluctuations of the electromagnetic field existing inside and outside these ideal parallel plates. The plates impose well-defined boundary conditions to the fluctuating electromagnetic modes existing between them and, as a consequence, the zero-point energy of this system is a function of the separation between plates (see Fig. 9.1). The difference between zero-point energy inside and outside the plates is responsible for the attractive force between plates. This force has the same origin as the van der Waals force but acts at larger separations between bodies and, as a consequence, relativistic retardation effects need to be considered.

According to Casimir's original calculation, the attractive force per unit area, i.e., the pressure between the plates, can be expressed as:

$$P(d) = -\frac{\pi^2 \hbar c}{240 \, d^4} \tag{9.1}$$

where d is the separation between plates, c is the speed of light and $\hbar$ is Planck's constant divided by 2π. The calculation of the Casimir's pressure for dielectric surfaces at finite temperature was obtained by Lifshitz [2] in 1956. Several excellent reviews describing the theoretical aspects of these calculations and alternative derivations are listed in the reference section [3–15].

The simple Casimir formulation of the pressure acting between two neutral metallic plates represented one of the first indications that the zero-point energy of the electromagnetic field could be experimentally detected. The physical reality of the Casimir effect was a very controversial subject when proposed for first time. In his biography [16], Casimir described the unsuccessful discussions he had with Wolfgang Pauli trying to convince him that this force could have observable effects. Since then, there have been so many experimental confirmations of this force that its effects are now routinely considered when studying objects at separations below 1 μm.

The first experiment intended to measure the Casimir force was performed in 1958 by Sparnaay [17] using parallel plates. While this experiment was not very successful due to the difficulties associated with moving parallel plates with high-precision, it provided the first indication that surface roughness needs to be reduced and surface charges must be removed. Blokland and Overbeek did the first convincing measurement of the Casimir force in 1978 [18]. By using a sphere in front of a metallic plate, to eliminate the problems associated with the parallelism of the plates, they measured the force with an experimental accuracy of 50%. The experiment done by S. Lamoreaux in 1997 [19] is considered the first high-

Fig. 9.1 Schematic representation of the photonic modes confined between two metallic surfaces. The Casimir force between these two metallic surfaces arises due to the dependence of the energy spectrum of the confined electromagnetic modes on the separation between the surfaces

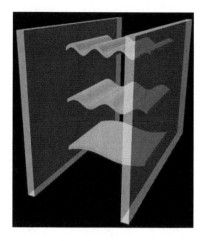

precision measurement of the Casimir force. He used a torsional pendulum in the sphere-plate configuration and obtained a 5% agreement between theory and experiment. Several variations of these experiments have been performed in the following years producing compelling evidence that the Casimir effect can be observed in various experimental conditions. A common feature of these experiments is that they involved *macroscopic* objects: They measured the Casimir force among objects having typical dimensions of several cm and for separations between bodies of the order of microns.

The first measurement of the Casimir force between *microscopic* objects separated by hundreds of nanometers or less, was performed by Mohideen in 1998 [20] using an atomic force microscope (AFM). In this experiment, a gold-coated 200 µm diameter sphere was attached to the tip of an AFM, which was used to measure the Casimir force between the sphere and a metalized plate. Similar experiments using AFM techniques are very popular today since they allow the measurement of this force at distances as short as 20 nm [21].

The use of micro-mechanical devices as a novel technique for characterization of this force was introduced by Ho Bun Chan and collaborators at Bell Laboratories in 2001 [22]. In this technique, a micro-mechanical torsional oscillator is used to detect the Casimir force induced by a metallic sphere approaching the oscillator. Furthermore, this experiment demonstrated that the Casimir force could be used to modify the mechanical state of microscopic devices introducing a novel mechanism for actuation at the micro- and nanoscale. Casimir force detectors based on micro-mechanical devices are currently the most sensitive devices to characterize this force [23].

Figure 9.2 summarizes the typical dimensions of the objects used in Casimir experiments performed in the last decades. The area labeled "torsion balance" represents the *macroscopic* experiments performed with objects having tens of cm in size. The use of MEMS (Micro Electro Mechanical Systems) and AFM (Atomic Force Microscopy) technology, what we call *microscopic* experiments, allowed

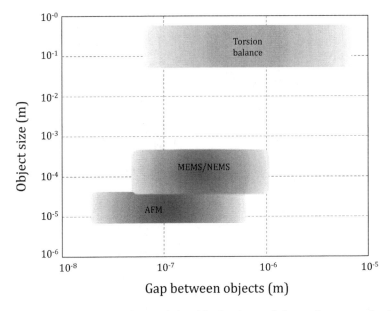

Fig. 9.2 Comparison of the characteristic object's size and interaction range (gap) of experiments performed in the last decades to measure the Casimir force between bodies

precise determination of the Casimir force down to 20 nm distances by involving objects with sizes below 1 mm.

In this review, we will describe the fundamentals of micro and nano electro mechanical devices (NEMS), we will explain the advantages they provide when used to detect the Casimir force and we will examine the most recent results obtained with this technology. We conclude this review with suggestions to improve the precision of micro/nano mechanical sensors to enable the investigation of the Casimir force in regimes that are currently not accessible.

9.2 Micro and Nano Electro Mechanical Systems

Mechanical devices with typical dimensions in the order of tens of microns, known as MEMS (Micro Electro Mechanical Systems), are already having a pervasive presence in science and technology [24]. They are widely employed as sensors and actuators due to their fast response time, enhanced sensitivity to external perturbations and the possibility of high-density integration of multiple elements into a single chip. By further reducing the size of these MEMS devices, we enter the world of NEMS (Nano Electro Mechanical Systems). In this size regime, the resonance frequency of these nano-devices becomes extremely large (up to GHz) and their mechanical quality factor remains very high ($Q \approx 10^6$). This

Fig. 9.3 Ultra dense array of NEMS mirrors for maskless lithography: (**a**) schematic representation of the array showing the mirror and spring layers and SEM images of the fabricated devices showing the mirrors array (**b**) and springs (**c**). Each mirror is 3 × 3 μm, the gap between them is around 100 nm. In (**c**), the spring's width and spacing is also ≈ 100 nm

combination implies exceptionally high force sensitivities, ultra-low power consumption and access to non-linear response with small actuation forces [25]. Furthermore, NEMS devices allow integration of even larger number of nano mechanical devices into extremely small areas.

As force detectors, MEMS and NEMS have been successfully used in a diversity of applications since they can routinely detect piconewtons (10^{-12} N) and, under special experimental conditions, they can detect forces as small as zeptonewtons (10^{-21} N). MEMS/NEMS based force sensors have been used to measure forces between individual biomolecules [26], to explore quantum effects in mechanical objects [27], to detect single spins by magnetic resonance force microscopy [28] and to study force fluctuations between closely spaced bodies [29]. As we will see in the following section, MEMS devices have also enabled the most precise measurement of the Casimir interaction between metallic objects in vacuum [23].

This long list of examples is also an indication of how vulnerable these devices are to local forces and to what extent local forces are to be considered in the design of MEMS/NEMS devices. In the particular case of the Casimir force, its effects become important when the distance between neutral objects is in the order of hundreds of nanometers. Fabrication of mechanical devices with features of this size is becoming common nowadays. Recently, high-density arrays of NEMS mirrors with critical dimensions of about 100 nm have been fabricated to modulate deep ultraviolet radiation (DUV) for maskless lithography applications [30]. These NEMS mirrors are separated by 100 nm gaps and they are supported at the center by 100 nm wide elastic springs providing the mechanical restoring forces (see Fig. 9.3). In the absence of electrostatic actuation, the Casimir force is the dominant interaction between these miniature objects and it needs to be taken into account in their design.

In the following section, we will describe the use of MEMS devices as force sensors for unambiguous detection of the Casimir interaction between metallic objects.

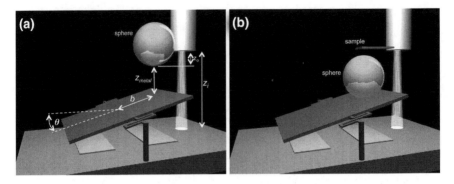

Fig. 9.4 Schematic of the two configurations used for the experimental setup. **a** Sphere attached to the optical fiber. All the relevant dimensions are included. z_{metal} is the separation between the bodies, z_o is the distance between the end of the fiber and the end of the sphere, b is the lever arm, θ is the angular deviation of the oscillator, and z_g (not shown in the graph) is the distance between the top of the oscillator and the reference plate. **b** Plate attached to the optical fiber assembly. All the dimensions have the same meaning, except for z_o which represents the distance between the end of the fiber and the bottom of the plate

9.3 Experimental Aspects on the Determination of the Casimir Interaction

A MEMS torsional oscillator is at the core of the experimental setup developed to measure the Casimir interaction between metallic bodies. The attractive force between two bodies can be measured by determining the changes induced in a MEMS oscillator due the Casimir interaction. These changes could be associated with either an induced displacement of the oscillator or a change of its natural resonance frequency due to the presence of the interaction between the two bodies. Beyond the requirement of a precise determination of the interaction itself, the separation between the two bodies also needs to be measured accurately and precisely. See also the Chap. 8 by Capasso et al. in this volume for additional discussions of MEMS based Casimir force measurements.

Our experimental setup has allowed us to obtain the most sensitive experimental determination to date of the Casimir interaction between similar and dissimilar metals. The current system consists of a MEMS torsional oscillator and a metal-coated sphere and is capable of extremely precise control of their relative position (see Fig. 9.4a). The MEMS oscillator and the sphere are independently coated with the materials under consideration. By approaching a coated sphere to one side of the coated torsional oscillator the attractive Casimir force induces a torque that rotates the MEMS device about the fixed supports. This rotation is detected by measuring the angular deflection of the plate as a function of the plate-sphere separation. Furthermore, these MEMS oscillators can be designed to concurrently have high resonance frequencies and large quality factor Q producing important improvements in sensitivity. This experimental setup can be operated in both static and dynamic regimes. In the static regime the sphere is maintained at a fixed vertical position and the Casimir force is measured directly. In the dynamic regime, the vertical separation between the sphere and the plate is changed harmonically with time, leading to an improvement of the sensitivity.

The MEMS is mounted onto a piezo-driven xyz stage which, in turn, is mounted on a micrometer controlled xy stage. This combination allows positioning the metal-coated sphere over the metal-coated MEMS plate. The separation z_i between the sphere and the substrate is controlled by the vertical axis of the xyz stage. A two color fiber interferometer-based closed-loop system is used to measure and control z_i.

Measurements of the Casimir interaction have been performed in two different configurations. In the first one, Fig. 9.4a, the polysilicon oscillator plate was coated with a thin adhesion layer (≈ 10 nm of either Cr or Ti) and subsequently a thick (≈ 200 nm) Au layer was evaporated. Similarly, the $R \approx 150$ mm sapphire sphere was coated with ≈ 10 nm Cr and ≈ 200 nm Au was thermally evaporated on it. The Au coating in both the plate and the sphere is thick enough to ensure that the Casimir interaction can be regarded as arising from solid Au bodies, which was checked by calculating the Casimir interaction between bodies for a multilayer system [31] and, more importantly, by measuring the interaction using a sphere with a thinner (≈ 180 nm) Au layer. No significant differences between both experimental runs were observed. The Au-coated sphere was glued with conductive silver epoxy to the sides of an Al-coated optical fiber that is part of an optical interferometer. In the second setup, Fig. 9.4b, the position of the sphere and the plate has been interchanged. This new configuration permits easier exchange of samples without modification of the fragile MEMS sensor.

When confronted with the measurement of small forces, the isolation of the detecting device from external vibrations is of supreme importance. Hence, using a MEMS torsional oscillator is preferable, since torsional oscillators are less sensitive to vibrations that couple with the motion of their center of mass. Further decoupling from external vibrations is achieved by mounting the rigid sample setup by soft springs to a vacuum chamber, which in turn is on top of a passive damping air table. The incorporation of magnetic damping, along all axes of motion, between the sample setup and the vacuum chamber reducing vibrations to a peak-to-peak amplitude $\Delta z_{pp} < 0.02$ nm for frequencies above 50 Hz. The small dimensions of the oscillator aids in improving its intrinsic quality factor and sensitivity [32]. The high quality Q of the oscillator, however, cannot be fully utilized in the presence of a dissipative medium. The effects of energy damping of the surrounding air are minimized by working in a vacuum. The vacuum is achieved by pumping the system down to 1.3×10^{-5} Pa (10^{-7} Torr) with an oil free diaphragm-turbomolecular pump system. While measurements are performed, pumping in the sample chamber is stopped and the pump is physically disconnected from the system. A low pressure (never higher than 0.3×10^{-3} Pa (10^{-5} Torr) is maintained by a chemical pump made of a cold (~ 77 K) activated carbon trap located inside the vacuum chamber.

In both experimental situations the optical fiber can be moved relatively to the oscillator assembly by means of a five-axis micrometer driven mechanical stage, and a xyz piezo-driven stage.

The separation dependent attractive force $F(z_{metal})$ between the sphere and the plate will cause the oscillator to rotate under the influence of the torque

$$\tau = bF(z_{metal}) = k_{torsion}\theta \qquad (9.2)$$

where $k_{torsion}$ is the torsional spring constant for the oscillator. Since the torsional angles are small, they are proportional to the change in capacitance between the underlying electrodes and the oscillator. Consequently,

$$\theta \propto \Delta C = C_{right} - C_{left} \qquad (9.3)$$

where C_{right} (C_{left}) is the capacitance between the right (left) electrode and the plate (Fig. 9.4). Hence the force between the two metallic surfaces separated by a distance z_{metal} is $F(z_{metal}) = k\Delta C$, where k is a proportionality constant that needs to be determined by calibration.

Alternatively, the force sensitivity of the oscillator can be enhanced by performing a dynamical measurement [22, 33, 34]. In this approach, the separation between the sphere and the MEMS oscillator plate is varied as $\Delta z = A\cos(\omega_{rest})$, where ω_{res} is the resonant angular frequency of the oscillator, and A is the amplitude of motion. The linearized solution for the oscillatory motion, valid for $A \ll z_{metal}$, yields [22, 33]

$$\omega_{res}^2 = \omega_0^2\left[1 - \frac{b^2}{I\omega_0^2}\frac{\partial F}{\partial z}\right], \qquad (9.4)$$

Where, for $Q \gg 1$, $\omega_o \approx \sqrt{k_{torsion}/I}$ is the natural resonance frequency of the oscillator, I is its moment of inertia. It has been shown [35] that there is an optimal value of A which is a function of the separation. If A is too small, then the error in the determination of ω_{res} increases due to thermal motion. If A is too large, then non-linearities can not be neglected. In general, A is selected to be between 2 nm and 5 nm to satisfy the aforementioned constrains. The resonance frequency can also be measured by recording the thermal vibration of the oscillator at temperature T, but it was found that driving the system with a sinusoidal signal and a phase-lock loop [33] provided a more stable signal.

Unlike the static regime where forces are measured, in the dynamic regime the force gradient $\partial F/\partial z$ is measured by observing the change in the resonant frequency as the sphere-plate separation changes. When F is given by the Casimir interaction, the gradient of the interaction within the applicability range of the proximity force, is found to be

$$-\frac{\partial F_c}{\partial z} = 2\pi R P_c(z_{metal}) \qquad (9.5)$$

where $P_C(z) = (F_C/S)$ is the force per unit area between two infinite metallic plates at the same separation z_{metal} as the sphere and the plate. In (5) F_C has been used to denote the Casimir interaction.

9.4 Calibrations

The characterization of the system and the determination of the calibration constants are performed by applying a known electrostatic force between the sphere and the MEMS plate, i.e., by applying a known potential difference, V_b, between them. In this case, the electrostatic force can be approximated by the force between a sphere and an infinite plate [36]:

$$F_e(z_{metal}, V) = -2\pi\varepsilon_0(V_b - V_0)^2 \sum_{n=1}^{\infty} \frac{\coth(u) - n\coth(nu)}{\sinh(nu)} \cong \Xi(z_{metal})(V_b - V_0)^2$$

$$(9.6)$$

$$\Xi(z_{metal}) = -2\pi\varepsilon_0 \sum_{m=0}^{7} B_m t^{m-1} \qquad (9.7)$$

In (9.6) and (9.7) ε_o is the permittivity of free space, V_o is a residual potential difference between the plate and the sphere, $u = 1 + t, t = z_{metal}/R$, and B_m are fitting coefficients. While the full expression (9.6) is exact, the series is slowly convergent, and it is easier to use the approximation developed in Refs. [37, 38]. The values of the B_m parameters are 0.5, -1.18260, 22.2375, -571.366, 9592.45, -90200.5, 383084, and -300357. Using these values, errors smaller than 1 part in 10^5 are obtained. In (9.6) it has been assumed that the contact potential V_o is independent of separation. If this is not the case a more involved analysis where the $V_o(z_{metal})$ dependence is taken into account would be needed [39]. See also the Chap. 7 by Lamoreaux in this volume for additional discussions of distance-dependent contact potentials.

To complete the electrostatic calibration (as well as to perform the measurement of the separation dependence of the Casimir interaction) it is necessary to determine z_{metal}. This variable can be determined precisely by using the following geometrical relationship (see Fig. 9.4a):

$$z_{metal} = z_i - z_0 - z_g - b\theta \qquad (9.8)$$

In (9.8), z_g is the distance between the top of the MEMS oscillator and the substrate. This distance is measured interferometrically with an error of $\delta z_g \approx 0.1$ nm. The distance b is measured optically ($\delta b \approx 2$ μm), θ is determined through the change in capacitance between the oscillator and the underlying electrodes ($\delta\theta \approx 10^{-7}$ rad) and z_i is measured with a two color interferometer where the light reflected at the end of the fiber is combined with the light reflected

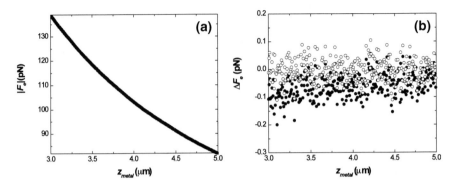

Fig. 9.5 a Absolute value of the electrostatic force as a function of separation determined using the procedure described in the text and (9.6). **b** Difference between the measured values and (9.6) for when either the best z_o (*open circles*) or $z_o^* = z_o + 1.5$ nm (*full circles*) are used in (9.6). Data shown was obtained for $(V_b - V_o) = 322.0$ mV

at the reference platform. The two color interferometer, which operates with a low coherence source (superluminescent diode, coherence length ≈ 20 µm) at 1310 nm and a stabilized laser at 1550 nm, is a fiber version of the one developed in Ref. [40]. The distance z_o is not known a priori and the electrostatic calibration is also used to determine it [41].

The electrostatic calibration is done at z_{metal} large enough such that the Casimir interaction does not have a measurable contribution. For a fixed $(V_b - V_o)$, z_i is measured. With the best estimate for z_o (optically measured with an error of ≈ 2 µm), an iterative method is then used. As a function of measured separations z_i, the change in capacitance between electrodes and the plate is found [34] and from here the corresponding values of θ are obtained. This is repeated for up to 150 different $(V_b - V_o)$. With the measured values of θ and the estimated value for z_o, a set of z_{metal} values is obtained from (8). Using

$$\theta = \frac{b}{k_{torsion}} \Xi(z_{metal} + \delta z_0)(V_b - V_0)^2 \tag{9.9}$$

$k_{torsion}/b$ and δz_o are used as fitting parameters. The improved value of z_o is used in (9.7) and the procedure is repeated until no further changes are obtained. The sensitivity of this approach is shown in Fig. 9.5. When all the errors are combined, it is found that z_{metal} can be measured to within $\delta z_{metal} = 0.6$ nm [35].

The electrostatic interaction is also used to obtain b^2/I. Typically for the first configuration (sphere on the fiber) $b^2/I \approx (1.2500 \pm 0.0005)$ mg^{-1}. When the sphere is attached to the oscillator the values of b^2/I are reduced by up to an order of magnitude (and vary significantly depending where the sphere is attached).

Once the electrostatic force has been used to calibrate the system, selecting $V_b = V_o$ makes the effect of the electrostatic interaction undetectable in our experiment. This is accomplished by applying a potential difference between the

Fig. 9.6 Magnitude of the derivative of the force dF/dV_{DC} as a function of V_{DC}. The plot was obtained at $z_{metal} = 3.5$ μm when $V_{DC} = V_o$. Data do not fall in a straight line due to the increase of the electrostatic force (and θ) when $|V_{DC} - V_o|$ increases

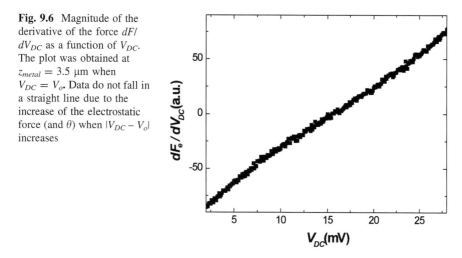

sphere and the plate $V_b = V_{DC} + \delta V\cos(\omega t)$, where the amplitude of the oscillatory component $\delta V \approx 1$ mV. The response of the oscillator is then proportional to $\partial F_e/\partial V_{DC}$, and V_o is obtained when the derivative of the force equals 0, as shown in Fig. 9.6.

9.5 Determination of the Casimir Interaction

Upon completion of the calibration procedure, the Casimir interaction can be determined. The electrical potential between the sphere and the plate is adjusted as to obtain a null F_e, $V_b = \langle V_o \rangle$, where $\langle V_o \rangle$ is the average potential for z_{metal} in the 160 nm to 5000 nm range, found as described in the previous section. The position of the fiber is then changed in ≈ 2 nm increments, as measured by the two color interferometer. The actual z_{metal} is obtained using (9.8) with previous determination of the corresponding θ. The resonance frequency of the oscillator is measured, and by means of (9.5), the equivalent Casimir pressure $P_C(z_{metal})$ is obtained. The procedure is repeated for many runs (where the measurements are performed at the same set of z_{metal} within the experimental error) and the average of the different runs is reported as $P_C(z_{metal})$. When taking into account the errors in the determination of ω_{res}, $\delta\omega_{res} \approx 5$ mHz, and R, $\delta R \approx 0.3$ μm as determined in a scanning electron microscope, the total error in P_C can be determined [35, 42]. Figure 9.7 shows the determination of P_C obtained by both experimental setups, with the sphere attached either to the fiber or to the sensor. Also in Fig. 9.7 the difference between both determinations is plotted. It is worth mentioning that these experiments were done with a separation of four years, in different vacuum chambers, using different experimental setups, and, more important, with Au deposited by different techniques. The data reported in Fig. 9.7 represent the most

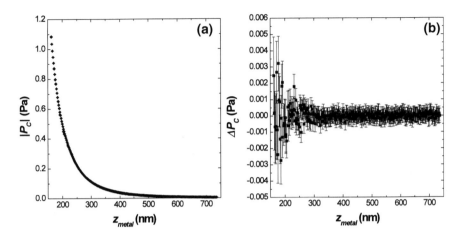

Fig. 9.7 a Absolute value of the measured Casimir pressure as a function of separation for the setup from Fig. 4a, sample from Ref. [42] *(open circles)*, and Fig 9.4b, sample electrodeposited on a Si single crystal, Ref. [47] *(closed circles)*. Both data sets are indistinguishable at this scale. **b** Difference between the data sets in **a**. The difference was obtained at the separations measured in the newest sample. The pressure at these separations for the older sample was determined by linear interpolation. Error bars represent the 95% confidence level in both the separation and pressure determinations

precise measurements of the Casimir interaction up to date. The error bars represent the 95% confidence level in both the separation and pressure determination.

The characterization of the frequency dependence of the dielectric function $\varepsilon(\omega)$ of the material is required in order to calculate the Casimir force between real metals. Figure 9.8 shows our experimental measurements of the real, ε', and imaginary ε'' parts of the dielectric function of the Au layer deposited on a Si single crystal. The measurements were performed between 196 nm and 820 nm. While there are differences between the values measured on our samples and the ones reported on standard reference books [43], it is important to notice that these differences are too small to produce any significant difference in [44] the calculation of the Casimir interaction [31]. See also the Chap. 10 by van Zwol et al. in this volume for additional discussions of characterization of optical properties of surfaces in Casimir force measurements.

In fact, the calculation of the Casimir pressure at finite temperatures for real samples is given by the Lifshitz formula [2, 45]

$$P_C(z_{metal}) = -\frac{k_B T}{\pi} \sum_{l=0}^{\infty}{}' \int_0^{\infty} k_\perp dk_\perp q_l \times \left\{ [r_\parallel^{-2}(\xi_l, k_\perp) e^{2q_l z} - 1]^{-1} + [r_\perp^{-2}(\xi_l, k_\perp) e^{2q_l z} - 1]^{-1} \right\}$$

$$(9.10)$$

where $k_\perp$ is the wave vector component in the plane of the plates, $q_l^2 = k_\perp^2 + \xi_l^2/c^2$, $\xi_l = (k_B T l) h^{-1}$ are the Matsubara frequencies, and $r_\parallel$ and $r_\perp$ are the

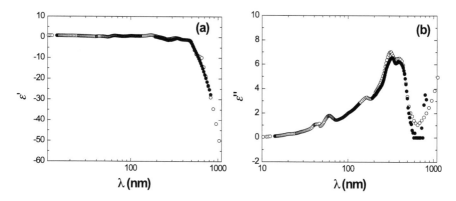

Fig. 9.8 *Filled circles show* (**a**) *ellipsometrically* measured values of ε' and (**b**) ε'' as a function of wavelength. Tabulated data from Ref. [43] are displayed as open circles

reflection coefficients for two independent polarization states computed for imaginary frequencies $\omega_l = i \, \xi_l$. The prime on the summation in (9.10) refers to the inclusion of a factor ½ for the term with $l = 0$.

As described in the Refs. [35, 42], the roughness of the sample also needs to be taken into account. By using the atomic force microscope image of the surfaces the fraction of the sample at different absolute separations are determined. The Casimir pressure between the two surfaces is obtained as the weighted average (weighted by the fraction of the sample at a given separation) of the Casimir pressure between samples of finite conductivity and at finite temperatures as given by (9.10). See also the Chap. 10 by van Zwol et al. in this volume for additional discussions of surface roughness in Casimir force measurements. When the dielectric function is used in (9.10), different results are obtained when the zero order term of the Matsubara series is computed using a Drude model or a plasma model. A detailed discussion of the comparisons can be found in Refs. [35, 42, 46]. Here it suffices to include the obtained results as a function of separation, as shown in Fig. 9.9. While the plasma model shows an excellent agreement with the data, no agreement within the experimental error is obtained when the Drude model is used. This remarkable result is still waiting for explanation, and has given rise to numerous problems and controversies in the interpretation of the data.

9.6 Current Discussions in the Precise Determination of the Casimir Force

As aforementioned, the discrepancies between experiments and the Drude model have resulted in numerous controversies. It is difficult to understand why while the low frequency transport of Au is very well described with a Drude model, the effects of dissipation on the conduction electrons are absent when performing

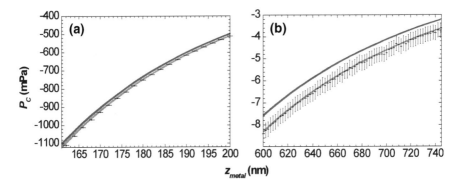

Fig. 9.9 Measured and calculated Casimir pressure as a function of separation, for the closest (**a**) and furthest separations (**b**). The crosses represent the 95% confidence levels in the measured values. The *dark gray* band is the calculation (with the error represented by the *thickness* of the band) using the *Drude* model. The light gray is the same when the plasma model is used, see Ref. [46]

Casimir pressure measurements. Among the arguments brought forward to explain these discrepancies, it has been hypothesized that differences in the Au layer could account for them. While this has not been completely ruled out, the data showed in Fig. 9.9 is a strong indication that this is not the case. Furthermore the effect of having a poor Au metallic coating would be to decrease the strength of the Casimir interaction, making the difference with the observed data more pronounced. In a recent experiment [47] we intended to provide an answer to this problem by measuring the Casimir pressure at different temperatures, ≈ 2, 4.2, 77 and 300 K. The idea here was to find out if as the temperature was reduced the measured Casimir pressure remained constant (thus supporting the plasma model) or changed (as it would be the result expected when dissipation is reduced). Unfortunately, while the average of $P_C(z_{metal})$ remains the same at all temperatures, the data shows a significantly larger amount of noise at low temperatures precluding the exclusion of either model. Other possibilities that have been mentioned is the existence of a systematic effect associated with either an improper electrostatic calibration, or the presence of patch potentials that provide an extra attractive interaction.

Additionally, there has been some controversy regarding the electrostatic calibration of the experimental setup. Particularly, the dependence of V_o between metallic layers has been significantly studied as a function of position, separation, and time. Differently from what other groups have found [39], in our samples V_o was observed to be independent of time, position or separation, as shown in Fig. 9.10, in agreement with what has been reported on Ref. [38]. Alternatively, there is an experimental report indicating that an electrostatic calibration free of the problems can be obtained even when V_o changes with separation [48]. While the results obtained by our group are in good agreement with theoretical expectations, the reasons behind the different observations in different configurations

Fig. 9.10 Measurement of
the residual potential V_o using
the method shown in Fig. 3 as
a function of (**a**) separation
(**b**) time, and (**c**) lateral
position

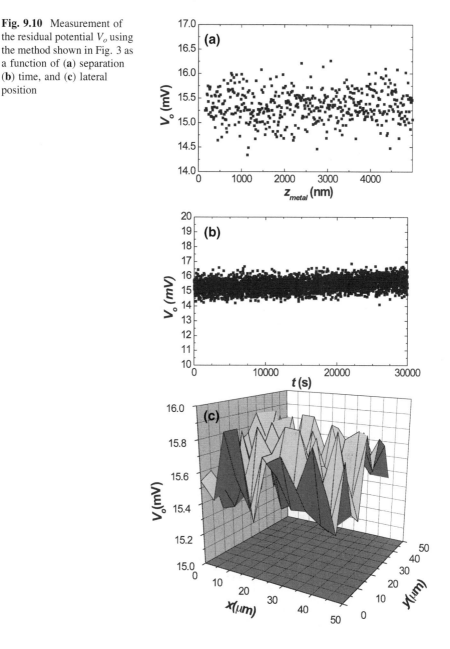

require further study. This dependence could be associated with patch potentials,
which would yield a separation dependence of V_o and a residual electrostatic force
that cannot be counterbalanced [49, 50]. In Ref. [35] it was calculated that the
effect of patch potentials would be undetectable if their extent were to
be ≈ 300 nm (estimated Au grain size in the samples). If, on the other hand, the

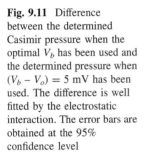

Fig. 9.11 Difference between the determined Casimir pressure when the optimal V_b has been used and the determined pressure when $(V_b - V_o) = 5$ mV has been used. The difference is well fitted by the electrostatic interaction. The error bars are obtained at the 95% confidence level

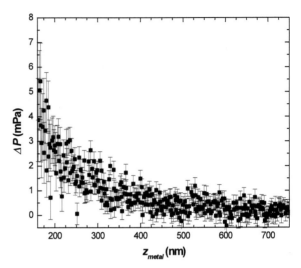

patches are very large, much larger than the effective interacting area, then their effect also would be cancelled by the effect of the applied V_b.

Finally, to shed more light on the effect of V_o on the interaction measurements, an experiment was performed where the applied V_b did not completely cancel V_o, leading to an effective "residual potential." The Casimir pressure was determined for this situation and when the optimal V_b was applied, and their difference plotted as a function of separation, as shown in Fig. 9.11. It is worth mentioning that firstly, a "residual potential" larger than the error in the average of V_o is needed to observe any effect in the interaction. Secondly, the interaction associated with the residual potential is well fitted by (9.6) with a leading $1/z_{metal}^2$ term at small separations. It appears from the totality of electrostatic measurements performed that the effects of residual potentials can be counterbalanced in the precision measurements of the Casimir interaction.

9.7 Future Directions: Improved Micromechanical Force Sensors

Investigations of the Casimir interaction stand to benefit considerably from the ongoing improvement in the precision micromechanical sensors and the associated position and force measurement techniques. The precision of the current generation of micromechanical sensors is significantly limited by the combination of the thermal noise of the mechanical sensor itself and the readout noise of the electrostatic or optical detector used in the MEMS or AFM based sensors respectively. In the case of MEMS sensors in particular, the Casimir force measurement precision has benefitted considerably from the stabilization of the measurement apparatus enabling very long acquisition times to improve the signal to noise ratio by signal averaging. It is however still desirable to improve the force and displacement measurement precision of these devices. This would enable the investigation of Casimir force in the regimes that were not previously easily accessible.

In the regime of large separation distance the Casimir force and its gradient are very weak and better force sensitivity would lead to an immediate improvement in that regime. In the regime of small separation distances the current limitation is the stability of the sensor against the destabilizing effect of the Casimir force gradient — negative "Casimir spring" —leading to inability to maintain controlled constant separation. This can be counteracted by oscillating the mechanical sensor with large amplitude and essentially sampling the Casimir force at small separations only over a short portion of the oscillation cycle. This however leads to stringent oscillation amplitude control and measurement requirement, as well as a nontrivial relationship between the Casimir potential, oscillation amplitude and the measured oscillation frequency shift. A more straightforward way to access this regime is to increase the stiffness of the mechanical sensor to maintain its stability. However combining the stiffer micromechanical sensor (lower mechanical gain) with the decreased oscillation amplitude needed to maintain a simple linear measurement leads to a significant reduction in the signal to noise that need to be compensated for.

Finally, there is a significant recent interest in measuring the forces acting on objects that are spatially finite and have micron or even submicron dimensions, in order to observe size and shape dependence of the Casimir forces as well as potentially spatially inhomogeneous electrostatic forces due to so-called "patch potentials". To realize such measurements, again, dramatic improvements in sensor precision are required.

In considering the force measurement by a mechanical sensor we need to essentially consider two transduction or "amplification" stages. The first one is mechanical, whereby a mechanical force is transduced to a displacement of a linear mechanical oscillator. It is characterized by stiffness (gain), effective mass or resonance frequency, and mechanical loss (with the corresponding thermal noise). In the second stage the mechanical displacement is transduced into an electrical signal, typically through either an electrostatic or an optical measurement. This stage can also be characterized by its gain and the input-referred noise. In an ideal case the gains are such that the noise of the first stage is dominant at all frequencies, but this is typically not the case.

In the first stage the mechanical loss essentially couples the oscillator to a thermal bath and introduces a mechanical thermal noise. The input-referred force noise spectrum of this Langevin force is white, independent of the frequency of the measurement. Consequently if the mechanical thermal noise of the transducer is the dominant noise source, the signal-to-noise (SNR) ratio of the measurement is also uniform and independent of frequency. The SNR is however inversely proportional to the square root of the loss, that is proportional to the square root of the mechanical quality factor Q. Note that when we measure the force at frequencies below the mechanical resonance frequency, the gain of the sensor is independent of Q, while the mechanical force and displacement noises decrease as $Q^{1/2}$. On the other hand, when we measure on resonance, the gain increases as Q, the force noise decreases as $Q^{1/2}$ and the corresponding displacement noise increases as $Q^{1/2}$. Thus the SNR improves with Q equally for off-resonance and on resonance measurement when thermal mechanical noise is dominant. When this is the case, SNR can only be improved by either increasing the force signal being measured, or by decreasing the equivalent temperature of the mechanical mode of the transducer during the measurement (see below).

In reality, however, there are technical and other noises, often referred to as 1/f noise, which can increase the noise floor at low frequencies above the thermal noise. In addition, in most practical situations the gain of the mechanical transducer off-resonance is too low, resulting in electrical or optical noise of the second stage dominating everywhere except the narrow window around the mechanical resonance frequency. To take advantage of the high mechanical gain and high SNR around the mechanical resonance, the input force should be applied at the appropriate resonance frequency. With a force that is constant in time but that is a strong function of the separation gap this is achieved by modulating the gap. While the gap can be modulated by an external actuator, more often this is achieved by exciting the mechanical vibration of the transducer itself by applying an external force to it in parallel with the force to be measured. For example this force could be an electrostatic force, or an inertial force applied by vibrating the whole transducer in space.

Typically the measured transducer position is used in a phase locked loop to apply the external excitation force exactly at 90-degree phase shift to the measured transducer displacement while maintaining the constant transducer vibration amplitude. This insures that the transducer always vibrates on resonance. In turn the interaction force now has a component that is AC modulated by the oscillating transducer gap. For a potential force that is only a function of the gap, this AC component is in phase with the mechanical motion and results in the shift in the resonance frequency of the transducer, which is then being detected. For a small oscillation amplitude the measurement is particularly easy to interpret, as the frequency shift is proportional to the gradient of the force of interest at a given separation, see (9.4). However it should be noted that the AC force component that is being generated and measured in this way is almost always smaller than the total DC force at that gap. The SNR of this measurement is proportional to the vibration amplitude for small amplitudes.

In both DC and AC measurements the gain of the mechanical sensor is inversely proportional to the sensor stiffness. However, making the sensors softer leads to earlier onset of instability for small separations. Moreover, as long as the sensor effective mass is limited by the need for extended sensor position readout areas, such as the case for electrostatic readout, decreasing the sensor stiffness leads to lowering the resonance frequency. While dynamic measurement bandwidth is not a concern where a DC force is measured and the averaging time is seconds or even minutes, staying above the low frequency technical noise and maintaining high Q of the sensor prevents further reduction of the stiffness. The issues of the measurement bandwidth indeed come to the foreground as one considers scanning probe sensors where the force is measured as a function of location.

We can thus conclude that the DC and off-resonance force measurement SNR is currently limited directly by the mechanical displacement readout, while for on resonance measurements with high Q transducers in vacuum thermal mechanical noise and the gap modulation amplitude determine the SNR at room temperature. Furthermore, decreasing the physical size of the position readout areas without compromising the readout precision would be required for more robust and higher bandwidth sensors.

While electrostatic readout has been widely used for MEMS sensors due to its relative simplicity of implementation in a MEMS transducer, it has significant limitations. It does not scale well with decreasing sensor size, as the capacitance derived signal is proportional to the area of the sensor. Even when the stray capacitance of the cables connecting the sensor is eliminated, the input capacitance of the readout transistor, together with the electronic Johnson noise, limits this readout scheme.

Optical readout, however, has been shown to achieve much lower mechanical displacement noise levels, while requiring the minimum interaction areas only of the order of the wavelength of light used. The fundamental noise limit is in this case imposed by the quantum optical shot noise, and is generally independent of temperature as the energy of a photon in the visible to near-IR range of the spectrum is much larger than k_BT, where k_B is the Boltzmann's constant.

To realize the full benefit of the optical readout scheme one needs to use an optical interferometer that has as high finesse as possible and is modulated as strongly as possible by the mechanical motion of the sensor. In one recent example of a comparably low finesse (≈ 20) cavity using a gold-coated micromechanical cantilever as one of the mirrors, the mechanical noise level of the order of 10^{-15} m/Hz$^{1/2}$ was achieved [51] with incident optical power of 1mW and the readout spot on the cantilever of only 3 μm in size. In another remarkable example using a high finesse cavity (≈ 30000), spot size of ≈ 60 μm and incident power of 1.5 mW the noise level of 4×10^{-19} m/Hz$^{1/2}$ was achieved [52]. In both cases the optical cavities were of the order 1 mm in size, external to the mechanical devices, and in the high finesse case the mechanical device was fairly large, 1 mm × 1 mm × 60 μm, and included a high reflectivity coating.

The next step in the transducer evolution is to integrate the high finesse optical interferometer on the same chip, optomechanically coupled to the mechanical transducer for optical readout (and even possibly excitation with an all-optical force). The optical resonators with optical Q of over 10^5 and as small as a few microns in size can be realized on chip via appropriate microfabrication processes. Planar structures such as photonic crystals and disk and ring resonators are some of the possible candidates, combining compactness and high mode confinement with excellent Qs and the ability to integrate with connecting waveguides as well as mechanical sensors.

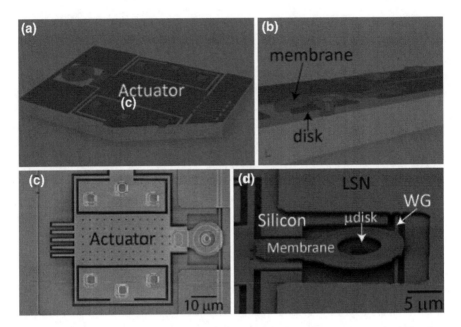

Fig. 9.12 Schematic (**a** and **b**) and Scanning Electron Microscope images (**c** and d) of an integrated opto-mechanical transducer. Membrane is microfabricated from low stress silicon nitride (LSN). Actuator, microdisk optical resonator (μdisk), and waveguide (WG) are single crystal silicon

Such an integrated device coupling a MEMS transducer with an optical interferometer has been recently realized [53–55]. The concept is shown schematically in Fig. 9.12. A high optical Q 10 μm diameter Si microdisk resonator is mechanically fixed to a substrate. The light can be coupled in and out of the resonator via a fixed microfabricated Si waveguide (WG) on a side of the resonator. A movable dielectric membrane (blue), made from low stress silicon nitride (LSN), is fabricated above the resonator. The membrane is attached to a MEMS transducer such as an electrostatic actuator and is capable of mechanical motion in the vertical direction. The optical mode in the microdisk is evanescently coupled to the membrane and as the membrane moves toward and away from the microdisk, never touching it mechanically, the motion significantly shifts the resonance frequency of the mode. While this is still work in progress, given the observed parameters of the current devices we estimate the shot noise limit of the mechanical motion readout to be below 10^{-15} m/Hz$^{1/2}$. With the evanescent field coupling approach the optical and mechanical devices are fabricated side by side and can be optimized essentially separately. No compromises are required such as integrating complicated and heavy coatings on micromechanical devices. Another advantage is the potential for completely fiber-pigtailed simplicity, without need for maintaining external optical alignment.

This type of device would in principle allow one to exploit various effects observed in cavity optomechanical systems [56]. One particular possibility is to excite the resonant vibration of the mechanical mode with an optical force by blue-detuning the optical excitation from resonance and use this as an alternative to the phase locked loop of the frequency sensing scheme described above. An even more exciting possibility is to use the position sensing for cooling the mechanical mode. This can be done either through feedback, or even directly by red-detuning the excitation light. For example, cooling factor of 60 from room temperature was achieved by using feedback approach [52]. While the cooling feedback is turned on, the effective mechanical Q is dramatically reduced, however as soon as the cooling is completed and the feedback is turned off, the Q is high and the thermal noise in sensor displacement is still low, while it takes the time of order Q/f_{res} for the mechanical mode to thermalize back to room temperature. If the cooling rate with the feedback turned on (in principle limited just by the opto-mechanical position sensing bandwidth and noise) can be made much faster than $1/Q$, the sensor can in principle be operated at the effective temperature much lower than room temperature.

Acknowledgments R. S. D. acknowledges support from the National Science Foundation (NSF) through grants Nos. CCF-0508239 and PHY-0701236, Los Alamos National Laboratories (LANL) support through contract No. 49423-001-07. The authors are grateful to the Defense Advanced Research Projects Agency (DARPA) grant No. 09-Y557.

References

1. Casimir, H.B.G.: On the attraction between two perfectly conducting plates. Proc. K. Ned. Akad. Wet. **60**, 793 (1948)
2. Lifshitz, E.M.: The theory of molecular attractive forces between solids. Sov. Phys. JETP **2**, 73 (1956)
3. Milonni, P.W.: The Quantum Vacuum: An introduction to Quantum Electrodynamics. Academic, San Diego, CA (1993)
4. Bordag, M., Klimchitskaya, G.L., Mohideen, U., Mostepanenko, V.M.: Advances in the Casimir effect. Oxford University Press Inc., New York (2009)
5. Milton, K.A.: The Casimir Effect: Physical Manifestations of Zero-point Energy. World Scientific, Singapore (2001)
6. Mostepanenko, V.M., Trunov, N.N.: The Casimir Effect and its Applications. Oxford University Press, Oxford (1997)
7. Parsegian, V.A.: Van der Waals Forces: a Handbook for Biologists, Chemists, Engineers and Physicist. Cambridge University Press, New York (2006)
8. Spruch, L.: Long range (Casimir) interactions. Science **272**, 145 (1996)
9. Daniel, K.: With apologies to Casimir. Physics Today, October 1990, p 9
10. Steve, L.: Casimir forces: still surprising after 60 years, Physics Today, February 2007
11. Jaffe, R.L., Scardicchio, A.: Casimir effect and geometric optics. Phys. Rev. Lett. **92**, 070402 (2004)
12. Scardicchio, A., Jaffe, R.L.: Casimir effect: An optical approach. Nucl. Phys. B. **704**, 552 (2005)
13. Special issue, Focus on Casimir forces, New J. Phys. 8, October (2006)

14. Klimchiskaya, G.L., Mohideen, U., Mostepanenko, V.M.: The Casimir force between real materials: experiment and theory. Rev. Mod. Phys. **81**, 1827 (2009)
15. Capasso, F., Munday, J.N., Iannuzzi, D., Chan, H.B.: Casimir forces and quantum electrodynamical torques: physics and nanomechanics, IEEE J. Selected. Topics Quantum Electron. **13**, 400 (2007)
16. Elizalde, E.: A remembrance of Hendrik Casimir in the 60th anniversary of his discovery, with some basic considerations on the Casimir effect, J. Phys. Conf. Series **161**, 012019 (2009)
17. Spaarnay, M.J.: Measurements of attractive forces between flat plates. Physica **24**, 751, (1958)
18. van Blokland, P.H.G.M., Overbeek, J.T.G.: van der Waals forces between objects covered with a chromium layer. J. Chem. Soc. Faraday Trans. **74**, 2637 (1978)
19. Lamoreaux, S.K.: Demonstration of the Casimir Force in the 0.6 to 6 μm Range. Phys. Rev. Lett. **78**, 5 (1997)
20. Mohideen, U., Roy, A.: Precision measurement of the Casimir Force from 0.1 to 0.9 μm. Phys. Rev. Lett. **81**, 4549 (1998)
21. Ederth, T.: Template-stripped gold surfaces with 0.4 nm rms roughness suitable for force measurements. Application to the Casimir force in the 20–100 nm range. Phys. Rev. A **62**, 62104 (2000)
22. Chan, H., Aksyuk, V.A., Kleiman, R.N., Bishop, D.J., Capasso, F.: Quantum mechanical actuation of microelectromechanical systems by the Casimir force, Science **291**, 1942 (2001)
23. Decca, R., López, D., Fischbach, E., Klimchitskaya, G., Krause, D., Mostepanenko, V.: Precise comparison of theory and new experiment for the Casimir Force leads to stronger constraints on thermal quantum effects and long-range interaction. Ann. Phys. **318**, 37 (2005)
24. Stephen S.: Microsystems Design. Springer, New York (2000)
25. Ekinci, K., Roukes, M.: Nanoelectromechanical systems. Rev. Sci. Instrum. **76**, 061101 (2005)
26. Bustamante, C., et al.: Mechanical processes in biochemistry. Annu. Rev. Biochem. **73**, 705 (2004)
27. Connell, A.D.O., et al.: Quantum ground state and single-phonon control of a mechanical resonator. Nature **464**, 697 (2010)
28. Schwab, K., Roukes, M.: Putting mechanics into quantum mechanics. Physics Today. 36. July (2005)
29. Stipe, B.C., et al.: Noncontact friction and force fluctuations between closely spaced bodies. Phys. Rev. Lett. **87**, 096801 (2001)
30. López, D. et al.: Two dimensional MEMS piston array for Deep UV optical Pattern generation; Proc. IEEE/LEOS International Conference on Optical MEMS and their applications, 148 (2006); V. Aksyuk et al., MEMS spatial light modulator for optical maskless lithography; Solid State Sensors, Actuators and Microsystems workshop, Hilton Hear 2006
31. Bordag, M., Mohideen, U., Mostepanenko, V.M.: New Developments in the Casimir Effect. Phys. Rep. **353**, 1 (2001)
32. Roukes, M.L.: Technical digest of the 2000 solid-state sensor and actuator workshop Hilton Head Island, South Carolina (2000)
33. Decca, R.S., López, D., Fischbach, E., Krause, D.E.: Measurement of the Casimir Force Between Dissimilar Metals. Phys. Rev. Lett. **91**, 050402 (2003)
34. López, D., Decca, R.S., Fischbach, E., Krause, D.E.: MEMS Based Force Sensors: Design and Applications. Bell Labs Tech. J. **10**, 61 (2005)
35. Decca, R., López, D., Fischbach, E., Klimchitskaya, G., Krause, D., Mostepanenko, V.: Precise comparison of theory and new experiment for the casimir force leads to stronger constraints on thermal quantum effects and long-range interaction. Ann. Phys. **318**, 37 (2005)
36. The expression for the force is obtained through the derivative of the capacitance between a sphere and a plane. This capacitance is obtained in Boyer, L., Houzé, F., Tonck, A., Loubet, J-L., Georges, J-M.: The influence of surface roughness on the capacitance between a sphere and a plane. J. Phys. D Appl. Phys. **27**, 1504 (1994)

37. Chen, F., Mohideen, U., Klimchitskaya, G.L., Mostepanenko, V.M.: Experimental test for the conductivity properties from the Casimir force between metal and semiconductor. Phys. Rev. A **74**, 022103 (2006)

38. Chiu, H.-C., Chang, C.-C., Castillo-Garza, R., Chen, F., Mohideen, U.: Experimental procedures for precision measurements of the Casimir force with an atomic force microscope. J. Phys. A: Math. Theor. **41**, 164022 (2008)

39. Kim, W. J., Brown-Hayes, M., Dalvit, D.A.R., Brownell, J.H., Onofrio, R.: Anomalies in electrostatic calibrations for the measurement of the Casimir force in a sphere-plane geometry. Phys. Rev. A **78**, 020101 (2008); (R); Reply to Comment on 'Anomalies in electrostatic calibrations for the measurement of the Casimir force in a sphere-plane geometry'. Phys. Rev. A **79**, 026102 (2009); Decca R.S., Fischbach, E., Klimchitskaya, G.L., Krause, D.E., López, D., Mohideen, U., Mostepanenko, V.M.: Comment on "Anomalies in electrostatic calibrations for the measurement of the Casimir force in a sphere-plane geometry". Phys. Rev. A **79**, 026101 (2009)

40. Yang, C., Ramachandra, R., Wax, A., Dasari, M., Feld, S.: 2p Ambiguity-Free Optical Distance Measurement with Subnanometer Precision with a Novel Phase-Crossing Low-Coherence Interferometer. Opt. Lett. **27**, 77 (2002)

41. Decca, R.S., López, D.: Measurement of the Casimir force using a microelectromechanical torsional oscillator: electrostatic calibration. Int. J. Mod. Phys. A **24**, 1748 (2009)

42. Decca, R.S., López, D., Fischbach, E., Klimchitskaya, G.L., Krause, D.E., Mostepanenko, V.M.: Novel constraints on light elementary particles and extra-dimensional physics from the casimir effect. Eur. Phys. J. C **51**, 963 (2007)

43. Palik, E.D.: editor, Handbook of Optical Constants of Solids. Academic Press, New York (1995)

44. Lambrecht, A., Reynaud, S.: Casimir Force Between Metallic Mirrors. Eur. Phys. J. D **8**, 309 (2000)

45. Lifshitz, E.M., Pitaevskii, L.P.: Statistical Physics. Pergamon Press, Oxford, Pt.II (1980)

46. Decca, R.S., López, D., Fischbach, E., Klimchitskaya, G.L., Krause, D.E., Mostepanenko, V.M.: Tests of new physics from precise measurements of the casimir pressure between two gold-coated plates. Phys. Rev. D. **75**, 077101 (2007)

47. Decca, R.S., López, D., Osquiguil, E.: New Results for the Casimir interaction: sample characterization and low temperature measurements. Proceedings of the quantum field theory under the influence of external conditions, Oklahoma to be published (2009)

48. de Man, S., et al.: No anomalous scaling in electrostatic calibrations for Casimir force measurements. Phys. Rev. A. **79**, 024102 (2009)

49. Speake, C., Trenkel, C.: Forces between Conducting Surfaces due to Spatial Variations of Surface Potential Phys. Rev. Lett. **90**, 160403 (2003)

50. Kim, W.J., Sushkov, O., Dalvit, D.A.R., Lamoreaux, S.K.: Measurement of the Short-Range Attractive Force between Ge Plates Using a Torsion Balance. Phys. Rev. Lett. **103**, 060401 (2009)

51. Hoogenboom, B.W., et al.: A Fabry–Perot interferometer for micrometer-sized cantilevers. Appl. Phys. Lett. **86**, 074101 (2005)

52. Arcizet, O., et al.: High-Sensitivity Optical Monitoring of a Micromechanical Resonator with a Quantum-Limited Optomechanical Sensor. Phys. Rev. Lett. **97**, 133601 (2006)

53. Miao, H., Srinivasan, K., Aksyuk, V.: Integrated MEMS tunable high quality factor optical cavity for optomechanical transduction. In: Conference on Lasers and Electro-Optics, OSA Technical Digest (CD) (Optical Society of America, paper CPDA10 (2010)

54. Srinivasan, K., Miao, H., Rakher, M.T., Davanco, M., Aksyuk, V.: Optomechanical Transduction of an Integrated Silicon Cantilever Probe Using a Microdisk Resonator. Nano Lett., **11**, 791–797 (2011)

55. Miao, H., Srinivasan, K., Rakher, M.T., Davanco, M., Aksyuk, V.: CAVITY OPTOMECHANICAL SENSORS, Digest Tech. Papers, Transducers'2011 Conference, Beijing, China, June 5–10 (2011)

56. Kippenberg, T.J., Vahala, K.J.: Cavity optomechanics: Back-action at the mesoscale. Science **29, 321**, 1172–1176 (2008)

Chapter 10
Characterization of Optical Properties and Surface Roughness Profiles: The Casimir Force Between Real Materials

P. J. van Zwol, V. B. Svetovoy and G. Palasantzas

Abstract The Lifshitz theory provides a method to calculate the Casimir force between two flat plates if the frequency dependent dielectric function of the plates is known. In reality any plate is rough and its optical properties are known only to some degree. For high precision experiments the plates must be carefully characterized otherwise the experimental result cannot be compared with the theory or with other experiments. In this chapter we explain why optical properties of interacting materials are important for the Casimir force, how they can be measured, and how one can calculate the force using these properties. The surface roughness can be characterized, for example, with the atomic force microscope images. We introduce the main characteristics of a rough surface that can be extracted from these images, and explain how one can use them to calculate the roughness correction to the force. At small separations this correction becomes large as our experiments show. Finally we discuss the distance upon contact separating two rough surfaces, and explain the importance of this parameter for determination of the absolute separation between bodies.

P. J. van Zwol (✉) · G. Palasantzas
Materials Innovation Institute and Zernike Institute for Advanced Materials,
University of Groningen, AG 9747, Groningen, The Netherlands
e-mail: petervanzwol@gmail.com

G. Palasantzas
e-mail: g.palasantzas@rug.nl

V. B. Svetovoy
MESA+ Institute for Nanotechnology, University of Twente, PO 217,
AE 7500 , Enschede, The Netherlands
e-mail: v.b.svetovoy@ewi.utwente.nl

D. Dalvit et al. (eds.), *Casimir Physics*, Lecture Notes in Physics 834,
DOI: 10.1007/978-3-642-20288-9_10, © Springer-Verlag Berlin Heidelberg 2011

10.1 Introduction

The Casimir force [1] between two perfectly reflecting metals does not depend on the material properties. This is a rather rough approximation as the Lifshitz theory demonstrates [2–4] (see Chap. 2 by Pitaevskii in this volume). In this theory material dependence of the force enters via the dielectric functions of the materials. Because the Casimir–Lifshitz force originates from fluctuations of the electromagnetic field, it is related to the absorption in the materials via the fluctuation dissipation theorem. The dissipation in the material at a frequency ω is proportional to the imaginary part of the dielectric function $\varepsilon(\omega) = \varepsilon'(\omega) + i\varepsilon''(\omega)$. Thus, to predict the force one has to know the dielectric properties of the materials.

In most of the experiments where the Casimir force was measured (see reviews [5, 6] and the Chap. 7 by Lamoreaux in this volume) the bodies were covered with conducting films but the optical properties of these films have never been measured. It is commonly accepted that these properties can be taken from tabulated data in handbooks [7, 8]. Moreover, for conducting materials one has to know also the Drude parameters, which are necessary to extrapolate the data to low frequencies [9]. This might still be a possible way to estimate the force, but it is unacceptable for calculations with controlled precision. The reason is very simple [10–13]: optical properties of deposited films depend on the method of preparation, and can differ substantially from sample to sample.

Analysis of existing optical data for Au [14] revealed appreciable variation of the force in dependence on the optical data used for calculations. Additionally, we measured our gold films using ellipsometry in a wide range of wavelengths $0.14 – 33\,\mu m$ [15], and found significant variation of optical properties for samples prepared at different conditions. Considerable dependence of the force on the precise dielectric functions of the involved materials was also stressed for the system solid–liquid–solid [16].

The Lifshitz formula can be applied to two parallel plates separated by a gap d. In reality each plate is rough and the formula cannot be applied directly. When the separation of the plates is much larger than their root–mean–square (rms) roughness w one can calculate correction to the force due to roughness using the perturbation theory. But even in this case the problem is far from trivial [17–19]. The roughness correction can be easily calculated only if one can apply the Proximity Force Approximation (PFA) [20]. Application of this approximation to the surface profile is justified when this profile changes slowly in comparison with the distance between bodies. The typical lateral size of a rough body is given by the correlation length ξ. Then the condition of applicability of PFA is $\xi \gg d$. This is very restrictive condition since, for example, for thermally evaporated metallic films the typical correlation length is $\xi \sim 50$ nm.

The roughness of the interacting bodies restricts the minimal separation d_0 between the bodies. This distance (distance upon contact) has a special significance for adhesion, which under dry conditions is mainly due to Casimir/van der

Waals forces across an extensive noncontact area [21]. It is important for micro (nano) electro mechanical systems (MEMS) because stiction due to adhesion is the major failure mode in MEMS [22]. Furthermore, the distance upon contact plays an important role in contact mechanics, is very significant for heat transfer, contact resistivity, lubrication, and sealing.

Naively one could estimate this distance as the sum of the rms roughnesses of body 1 and 2, $d_0 \approx w_1 + w_2$ [23], however, the actual minimal separation is considerably larger. This is because d_0 is determined by the highest asperities rather then those with the rms height. An empirical rule [24] for gold films gives $d_0 \approx 3.7 \times (w_1 + w_2)$ for the contact area of $\sim 1\mu m^2$. The actual value of d_0 is a function of the size of the contact area L. This is because for larger area the probability to find a very high peak on the surface is larger.

Scale dependence (dependence on the size L) is also important for the Casimir force in the noncontact regime. In this case there is an uncertainty in the separation $\delta d(L)$, which depends on the scale L. The reason for this uncertainty is the local variation of the zero levels, which defines the mathematical (average) surfaces of the bodies. This uncertainty depends on the roughness of interacting bodies and disappears in the limit $L \rightarrow \infty$.

In this paper we explain how one can collect the information about optical properties of the materials, which is necessary for the calculation of the Casimir–Lifshitz force. It is also discussed how the optical spectra of different materials manifest themselves in the force. We introduce the main characteristics of rough surfaces and discuss how they are related to the calculation of the roughness correction to the force. Scale dependence of the distance upon contact is discussed, and we explain significance of this dependence for the precise measurements of the force.

10.2 Optical Properties of Materials and the Casimir Force

Most Casimir force measurements were performed between metals [5, 6, 25–27] either e-beam evaporated or plasma sputtered on substrates. For such metallic films the grains are rather small in the order of tens of nanometers, and the amount of defects and voids is large [28]. The force measured between silicon single crystal and gold coated sphere [29] simplify situation only partly: the optical properties of the Si-crystal are well defined but properties of Au coating are not known well.

A detailed literature survey performed by Pirozhenko et al. [14] revealed significant scatter in the dielectric data of gold films collected by different groups. The measurement errors were not large and could not explain the data scattering. It was concluded that scattering of the data for gold films could lead to uncertainty in the calculated force up to 8% at separations around 100 nm. Most of the optical data for metals do not extend beyond the wavelength of 14 μm in the infrared range [30, 31]. Thus, it would be important to explore more the infrared regime and compare modern measured optical properties of samples used in force measurements with the old data. Moreover, mid and far infrared data are very important for the force prediction (see Sect. 10.2.2.2). This was accomplished by Svetovoy et al. [15] where ellipsometry from the far infrared (IR) to near ultraviolet (UV) was used over the wavelength range 140 nm–33 μm to obtain the frequency dependent dielectric functions for gold films prepared in different conditions. Analysis of different literature sources investigating the dielectric functions of a number of dielectrics such as silica and some liquids was performed by van Zwol et al. [16]. Situations where the data scattering can change even the qualitative behavior of the force (from attractive to repulsive) were indicated.

10.2.1 Dielectric Function in the Casimir Force

In this section we discuss how the dielectric functions of materials enter the Lifshitz theory and how these functions can be found experimentally.

10.2.1.1 The Force

Let us start the discussion from the Lifshitz formula [4] between two parallel plates separated by a gap d. It has the following form

$$F(T,d) = \frac{kT}{\pi} \sum_{n=0}^{\infty}{}' \int_0^{\infty} dq q \kappa_0 \sum_{v=s,p} \frac{r_1^v r_2^v e^{-2\kappa_0 d}}{1 - r_1^v r_2^v e^{-2\kappa_0 d}}, \qquad (10.1)$$

where "prime" at the sign of sum means that the $n = 0$ term must be taken with the weight 1/2, the wave vector in the gap is $\mathbf{K} = (\mathbf{q}, \kappa_0)$ with the z-component κ_0 defined below. The index "0" is related with the gap. Here $r_{1,2}^v$ are the reflection coefficients of the inner surfaces of the plates (index 1 or 2) for two different polarizations: $v = s$ or transverse electric (TE) polarization, and $v = p$ or transverse magnetic (TM) polarization. Specific of the Lifshitz formula in the form (10.1) is that it is defined for a discrete set of imaginary frequencies called the Matsubara frequencies

$$\omega_n = i\zeta_n = i\frac{2\pi kT}{\hbar}n, \quad n = 0, 1, 2, \ldots, \qquad (10.2)$$

where T is the temperature of the system and n is the summation index and k is the Boltzmann constant.

In practice the interacting bodies are some substrates covered with one or a few layers of working materials. If the top layer can be considered as a bulk layer then the reflection coefficients for body i are given by simple Fresnel formulas [32]:

$$r_i^s = \frac{\kappa_0 - \kappa_i}{\kappa_0 + \kappa_i}, \quad r_i^p = \frac{\varepsilon_i(i\zeta)\kappa_0 - \varepsilon_0(i\zeta)\kappa_i}{\varepsilon_i(i\zeta)\kappa_0 + \varepsilon_0(i\zeta)\kappa_i}, \tag{10.3}$$

where

$$\kappa_0 = \sqrt{\varepsilon_0(i\zeta)\frac{\zeta^2}{c^2} + q^2}, \quad \kappa_i = \sqrt{\varepsilon_i(i\zeta)\frac{\zeta^2}{c^2} + q^2}. \tag{10.4}$$

For multilayered bodies these formulas can be easily generalized (in relation with the dispersive forces see Ref. [33]). Only the reflection coefficients depend on the dielectric functions of the plate materials; the function $\varepsilon_0(i\zeta)$ of the gap material enters additionally in κ_0.

At small separations the thermal dependence of the force is very weak and in many cases can be neglected. Because important imaginary frequencies are around the so called characteristic frequency $\zeta_c = c/2d$, then the relative thermal correction can be estimated as $kT/\hbar\zeta_c$. For room temperature $T = 300$ K and separations smaller than 100 nm the correction will be smaller than 3%. If one can neglect this correction then in the Lifshitz formula ζ can be considered as a continuous variable and the sum in (10.1) is changed by the integral according to the rule:

$$\frac{kT}{\pi}\sum_{n=0}^{\infty}{}' \rightarrow \frac{\hbar}{2\pi^2}\int_0^{\infty} d\zeta. \tag{10.5}$$

The material function $\varepsilon(i\zeta)$ (we suppress the indexes for a while) cannot be measured directly but can be expressed via the observable function $\varepsilon''(\omega)$ with the help of the Kramers–Kronig (KK) relation [32]:

$$\varepsilon(i\zeta) = 1 + \frac{2}{\pi}\int_0^{\infty} d\omega \frac{\omega\varepsilon''(\omega)}{\omega^2 + \zeta^2}. \tag{10.6}$$

The knowledge of $\varepsilon(i\zeta)$ is of critical importance for the force calculations. Equation (10.6) demonstrates the main practical problem. To find the function $\varepsilon(i\zeta)$ for $\zeta \sim \zeta_c$ in general one has to know the physical function $\varepsilon''(\omega)$ in a wide range of real frequencies, which not necessarily coincides with $\omega \sim \zeta_c$. This property of the Casimir force was stressed in Ref. [12] and then was demonstrated experimentally [34–36]. It will be discussed below on specific examples.

10.2.1.2 The Optical Data

The optical properties of materials are described by two measurable quantities: the index of refraction $n(\lambda)$ and the extinction coefficient $k(\lambda)$, which both depend on the wavelength of the electromagnetic field λ. Combined together they define the complex index of refraction $\tilde{n}(\lambda) = n(\lambda) + ik(\lambda)$. The real part defines the phase velocity in a medium $v = c/n$ where c is the speed of light. The imaginary part tells us how much light is absorbed when it travels through the medium. The dielectric response of a material for the UV ($\hbar\omega > 5\,\text{eV}$), IR ($0.01--1\,\text{eV}$) and MicroWave (MW) or TeraHertz range ($10^{-4} - 10^{-2}\,\text{eV}$), is related to electronic polarization resonances, atomic polarization resonances (in case of metals this is a gas of quasi free electrons), and dipole relaxation, respectively.

The complex dielectric function $\varepsilon(\omega) = \varepsilon'(\omega) + i\varepsilon''(\omega)$ and the complex index of refraction are related as $\varepsilon(\omega) = \tilde{n}^2(\omega)$, which is equivalent to the following equations:

$$\varepsilon' = n^2 - k^2, \quad \varepsilon'' = 2nk. \tag{10.7}$$

In many cases only the absorbance is measured for a given material. In this case the refraction index can be found from the KK relation at real frequencies:

$$\varepsilon'(\omega) = 1 + \frac{2}{\pi} P \int_0^\infty dx \frac{x\varepsilon''(x)}{x^2 - \omega^2}, \tag{10.8}$$

where P means the principal part of the integral.

Kramers–Kronig relations originating from causality have a very general character. They are useful in dealing with experimental data, but one should be careful since in most cases dielectric data are available over limited frequency intervals. As a result specific assumptions must be made about the form of the dielectric data outside of measurement intervals, or the data should be combined with other (tabulated) experimental data before performing the KK integrals.

A powerful method to collect optical data simultaneously for both ε' and ε'' is ellipsomery. This is a non destructive technique where one measures an intensity ratio between incoming and reflected light and the change of the polarization state. Ellipsometry is less affected by intensity instabilities of the light source or atmospheric absorption. Because the ratio is measured no reference measurement is necessary. Another advantage is that both real and imaginary parts of the dielectric function can be extracted without the necessity to perform a Kramers–Kronig analysis. The ellipsometry measures two parameters Ψ and Δ, which can be related to the ratio of complex Fresnel reflection coefficients for p- and s-polarized light [37, 38]

$$\rho = \frac{r^p}{r^s} = \tan\Psi e^{i\Delta}, \tag{10.9}$$

where $r^{p,s}$ are the reflection coefficients of the investigated surface, and the angles Ψ and Δ are the raw data collected in a measurement as functions of the wavelength λ. When the films are completely opaque (bulk material), then the reflection coefficients are related to the dielectric function as follows

$$r_p = \frac{\langle \varepsilon \rangle \cos \vartheta - \sqrt{\langle \varepsilon \rangle - \sin^2 \vartheta}}{\langle \varepsilon \rangle \cos \vartheta + \sqrt{\langle \varepsilon \rangle - \sin^2 \vartheta}}, \quad r_s = \frac{\cos \vartheta - \sqrt{\langle \varepsilon \rangle - \sin^2 \vartheta}}{\cos \vartheta + \sqrt{\langle \varepsilon \rangle - \sin^2 \vartheta}}, \tag{10.10}$$

where ϑ is the angle of incidence and $\langle \varepsilon \rangle = \langle \varepsilon(\lambda) \rangle$ is the "pseudo" dielectric function of the films. The term "pseudo" is used here since the films may not be completely isotropic or uniform; they are rough, and may contain absorbed layers of different origin because they have been exposed to air. If this is the case then the dielectric function extracted from the raw data will be influenced by these factors. The dielectric function is connected with the ellipsometric parameter ρ for an isotropic and uniform solid as

$$\varepsilon = \sin^2 \vartheta \left[1 + \tan^2 \vartheta \left(\frac{1 - \rho}{1 + \rho} \right)^2 \right]. \tag{10.11}$$

As it was stated before the spectral range of our measured data is from 137 nm to 33 μm. Even longer wavelengths have to be explored to predict the force between metals without using the extrapolation. Ellipsometry in the terahertz range 0.1–8 THz (wavelengths 38–3000 μm) is difficult due to lack of intense sources in that range, and these systems are still in development [39]. Typically synchrotron radiation is used as a source deeming these measurements very expensive. Nonetheless for gold films it would be extremely interesting to have dielectric data in this regime.

Dielectric data obtained by ellipsometry or absorption measurements [40] in the far UV regime are also rare. The most obvious reason for this is that these measurements are expensive because high energy photons must be produced, again at synchrotrons [41]. Furthermore, ellipsometry in this range becomes complicated as polarizing materials become non transparent. For this range a few ellipsometry setups exist around the world covering the range 5–90 eV (wavelengths 12–200 nm) [41]. The vacuum UV (VUV) and extreme UV (XUV) parts may not be very important for metals but for low permittivity dielectrics such as all liquids, and, for example, silica or teflon, there is a major absorption band in the range 5–100 eV (see Fig. 10.6). It is precisely this band that dominates in the calculations of the Casimir force for these materials. It is very unfortunate that precisely for this band dielectric data are lacking for most substances except for a few well know cases as, for example, water.

10.2.2 Gold Films

In this section we discuss optical characterization of our gold films prepared in different conditions using ellipsometers. Then we discuss the dielectric function at imaginary frequencies for Au films and for metals in general stressing the

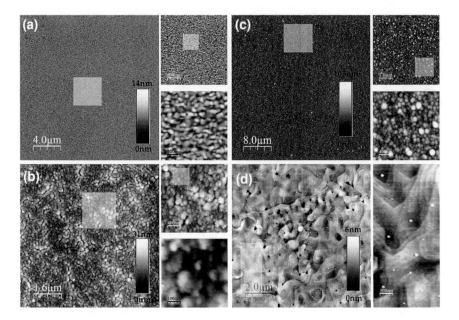

Fig. 10.1 Flattened roughness scans (up to 4000 × 4000 pixels) of gold surfaces where the highlighted areas are magnified. The scale bars can be applied only to the large images. **a** 100 nm Au on Si. **b** Au coated polysterene sphere (first plasma sputtered then 100 nm Au evaporated. **c** 1600 nm Au on Si. **d** very high quality 120 nm Au on mica, annealed for a few hours and slowly cooled down

importance of very low real frequencies for precise evaluation of $\varepsilon(i\zeta)$ at $\zeta \sim \zeta_c \sim 1\,\mathrm{eV}$ (separations around 100 nm). Finally we describe variation in the Casimir force if different samples would be used for the force measurements.

10.2.2.1 $\varepsilon(\omega)$ for Au Films

Let us have now a closer look at the dielectric functions of our gold films [15] used for the force measurements in Refs. [24, 42]. For optical characterization we have prepared five films by electron beam evaporation. Three of these films of different thicknesses 100, 200 and 400 nm were prepared within the same evaporation system on Si with 10 nm titanium sublayer and were not annealed. Different evaporation system was used to prepare two other films. These films were 120 nm thick. One film was deposited on mica and was extensively annealed. The other one was deposited on Si with chromium sublayer and was not annealed.

The AFM scans of the 100 nm film and the annealed 120 nm film on mica are shown in Fig. 10.1. In the same figure are shown also the gold covered sphere and 1600 nm film, which where not used in optical characterization. One can see that the annealed sample is atomically smooth over various length scales with atomic steps and terraces visible. Nevertheless, the local trenches of 5 nm deep are still present.

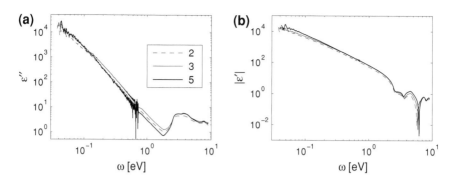

Fig. 10.2 a Measured ε'' as a function of frequency ω. **b** The same for $|\varepsilon'|$. For clearness the results are presented only for three samples 2, 3, and 5

Optical characterization of the films was performed by J. A. Woollam Co., Inc. [43]. Vacuum ultraviolet variable angle spectroscopic ellipsometer (VASE) was used in the spectral range 137–1698 nm. In the spectral range 1.9–32.8 μm the infrared variable angle spectroscopic ellipsometer (IR-VASE) was used for two incidence angles of 65 and 75°. The roughness and possible absorbed layer on the film surface can have some significance in the visible and ultraviolet ranges but not in the infrared, where the absorption on free electrons of metals is very large. Moreover, the effect of roughness is expected to be small since for all films the rms roughness is much smaller than the smallest wavelength 137 nm. Because the infrared domain is the most important for the Casimir force between metals, we will consider $\langle \varepsilon(\lambda) \rangle$ extracted from the raw data as a good approximation for the dielectric function of a given gold film.

Figure 10.2a shows the experimental results for $\varepsilon''(\omega)$ for three of five investigated samples. Around the interband transition (minimum of the the curves) the smallest absorption is observed for the sample 5 (annealed on mica) indicating the smallest number of defects in this sample [28]. On the contrary, this sample shows the largest $|\varepsilon'(\omega)|$ in the infrared as one can see in Fig. 10.2b. An important conclusion that can be drawn from our measurements is the sample dependence of the dielectric function. The sample dependence can be partly attributed to different volumes of voids in films as was proposed by Aspnes et al. [28]. The values of ε and their dispersion for different samples are in good correspondence with old measurements [30, 31]. The log-log scale is not very convenient for having an impression of this dependence. We present in Table 10.1 the values of ε for all five samples at chosen wavelengths $\lambda = 1, 5, 10 \, \mu m$. One can see that the real part of ε varies very significantly from sample to sample.

One could object that the real part of ε does not play role for $\varepsilon(i\zeta)$ and only variation of $\varepsilon''(\omega)$ from sample to sample is important. However, both $\varepsilon''(\omega)$ and $\varepsilon'(\omega)$ are important for precise determination of the Drude parameters. Let us discuss now how one can extract these parameters from the data.

All metals have finite conductivity. It means that at low frequencies $\omega \to 0$ the dielectric function behaves as $\varepsilon(\omega) \to 4\pi\sigma/\omega$, where σ is the material

Table 10.1 Dielectric function for different samples at fixed wavelengths $\lambda = 1, 5, 10\,\mu m$

Sample	$\lambda = 1\,\mu m$	$\lambda = 5\,\mu m$	$\lambda = 10\,\mu m$
1, 400 nm/Si	$-29.7 + i2.1$	$-805.9 + i185.4$	$-2605.1 + i1096.3$
2, 200 nm/Si	$-31.9 + i2.3$	$-855.9 + i195.8$	$-2778.6 + i1212.0$
3, 100 nm/Si	$-39.1 + i2.9$	$-1025.2 + i264.8$	$-3349.0 + i1574.8$
4, 120 nm/Si	$-43.8 + i2.6$	$-1166.9 + i213.9$	$-3957.2 + i1500.1$
5, 120 nm/mica	$-40.7 + i1.7$	$-1120.2 + i178.1$	$-4085.4 + i1440.3$

conductivity. It has to be stressed that this behavior is a direct consequence of the Ohm's law and, therefore, it has a fundamental character. Because the dielectric function has a pole at $\omega \to 0$, the low frequencies will give a considerable contribution to $\varepsilon(i\zeta)$ even if ζ is high (for example, in visible part of the spectrum) as one can see from (10.6).

Usually it is assumed that at low frequencies the dielectric functions of good metals follow the Drude model:

$$\varepsilon(\omega) = 1 - \frac{\omega_p^2}{\omega(\omega + i\gamma)}, \tag{10.12}$$

where ω_p is the plasma frequency and γ is the relaxation frequency of a given metal. When $\omega \to 0$ we reproduce the $1/\omega$ behavior with the conductivity $\sigma = \omega_p^2/4\pi\gamma$. For good metals such as Au, Ag, Cu, Al typical values of the Drude parameters are $\omega_p \sim 10^{16}$ rad/s and $\gamma \sim 10^{14}$ rad/s.

Separating real and imaginary parts in (10.12) one finds for ε' and ε''

$$\varepsilon'(\omega) = 1 - \frac{\omega_p^2}{\omega^2 + \gamma^2}, \quad \varepsilon''(\omega) = \frac{\omega_p^2 \gamma}{\omega(\omega^2 + \gamma^2)}. \tag{10.13}$$

These equations can be applied below the interband transition $\omega < 2.45\,eV$ ($\lambda > 0.5\,\mu m$) [44], but this transition is not sharp and one has to do analysis at lower frequencies. Practically (10.13) can be applied at wavelengths $\lambda > 2\,\mu m$ that coincides with the range of the infrared ellipsometer. The simplest way to find the Drude parameters is to fit the experimental data for ε' and ε'' with both equations (10.13). Alternatively to find the Drude parameters one can use the functions $n(\omega)$ and $k(\omega)$ and their Drude behavior, which follows from the relation $\varepsilon(\omega) = \tilde{n}^2(\omega)$. This approach uses actually the same data but weights noise differently.

Completely different but more complicated approach is based on the KK relations (10.8) (see Refs. [14, 15] for details). In this case one uses measured $\varepsilon''(\omega)$ extrapolated to low frequencies according to the Drude model and extrapolated to high frequencies as A/ω^3, where A is a constant. In this way we will get $\varepsilon''(\omega)$ at all frequencies. Using then (10.8) we can predict $\varepsilon'(\omega)$. Comparing the prediction with the measured function we can determine the Drude parameters. Similar procedure can be done for $n(\omega)$ and $k(\omega)$.

Table 10.2 Drude parameters γ, ω_p and roughness parameters, the correlation length ξ and rms roughness w, for all five measured samples

	1, 400 nm/Si	2, 200 nm/Si	3, 100 nm/Si	4, 120 nm/Si	5, 120 nm/mica
γ (meV)	40.5 ± 2.1	49.5 ± 4.4	49.0 ± 2.1	35.7 ± 5.1	37.1 ± 1.9
ω_p (eV)	6.82 ± 0.08	6.83 ± 0.15	7.84 ± 0.07	8.00 ± 0.16	8.38 ± 0.08
ξ (nm)	22	26	32	70	200
w (nm)	4.7	2.6	1.5	1.5	0.8

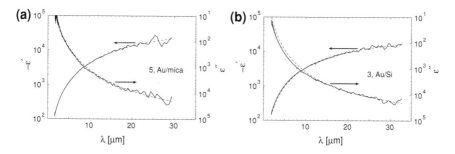

Fig. 10.3 The infrared data as functions of the wavelength λ for ε' and ε'' (*solid lines*) and the best Drude fits (*dashed lines*) for two gold films. **a** Shows the data for annealed sample 5 and **b** shows the same for unannealed sample 3

All methods for determination of the Drude parameters give reasonably close values of both parameters. We cannot give preference to any specific method. Instead, we average the values of the parameters determined by different methods, and define the rms error of this averaging as uncertainty in the parameter value. The averaged parameters and rms errors are given in the Table 10.2. We included in this table also the correlation lengths ξ and the rms roughness w for the sample roughness profiles. One can see the ω_p and ξ correlate with each other.[1] This correlation has sense because ξ describes the average size of the crystallites in the film; the larger the crystallites the smaller number of the defects has the film and, therefore, the larger value of the plasma frequency is realized.

One can see the quality of the Drude fit in Fig. 10.3 for samples 3 and 5. The fit is practically perfect for high quality sample 5, but there are some deviations for sample 3 at short wavelengths especially visible for ε''. More detailed analysis [15] revealed presence of a broad absorption peak of unknown nature around $\lambda = 10\,\mu\text{m}$. The magnitude of this absorption is the largest for poor quality samples 1 and 2.

---

[1] This correlation was not noted in [15] and stressed here for the first time.

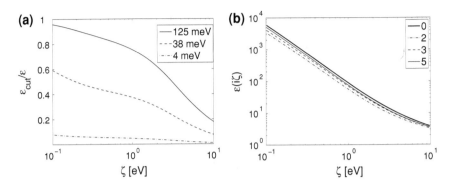

Fig. 10.4 **a** Relative contribution $\varepsilon_{cut}(i\zeta)/\varepsilon(i\zeta)$ to the dielectric function of gold at imaginary frequencies originating from the extrapolated region $\omega < \omega_{cut}$ (see explanations in the text). **b** The dielectric functions at imaginary frequencies for samples 2, 3, and 5; the thick curve marked as 0 corresponds to an "ideal sample" with the plasma frequency of single crystal

10.2.2.2 $\varepsilon(i\zeta)$ for Metals

Any method of optical characterization has a minimal accessible frequency ω_{cut} (cut-off frequency). At $\omega > \omega_{cut}$ one can measure the dielectric function but at lower frequencies $\omega < \omega_{cut}$ one has to make an assumption on the behavior of $\varepsilon(\omega)$, i. e. extrapolate to low frequencies. In KK relation (10.6) one can separate two intervals: $\omega < \omega_{cut}$, where $\varepsilon''(\omega)$ has to be extrapolated and $\omega > \omega_{cut}$, where $\varepsilon''(\omega)$ is measured. Then we can present $\varepsilon(i\zeta)$ as

$$\varepsilon(i\zeta) = 1 + \varepsilon_{cut}(i\zeta) + \varepsilon_{exper}(i\zeta),$$

$$\varepsilon_{cut}(i\zeta) = \frac{2}{\pi} \int_0^{\omega_{cut}} d\omega \frac{\omega\varepsilon''(\omega)}{\omega^2 + \zeta^2}, \quad \varepsilon_{exper}(i\zeta) = \frac{2}{\pi} \int_{\omega_{cut}}^{\infty} d\omega \frac{\omega\varepsilon''(\omega)}{\omega^2 + \zeta^2}. \qquad (10.14)$$

Of course, at very high frequencies we also do not know $\varepsilon''(\omega)$ but, for metals high frequencies are not very important. For this reason we include the high frequency contribution to $\varepsilon_{exper}(i\zeta)$.

We can estimate now the contribution of $\varepsilon_{cut}(i\zeta)$ to $\varepsilon(i\zeta)$. For that we assume the Drude behavior at $\omega < \omega_{cut}$ with the parameters $\omega_p = 9.0$ eV and $\gamma = 35$ meV [9]. At higher frequencies $\omega > \omega_{cut}$ we take the data from the handbook [7], for which the cut-off frequency is $\omega_{cut} = 0.125$ eV. These extrapolation and data were used for interpretation of most experiments, where the Casimir force was measured. In Fig. 10.4a the solid curve is the ratio $\varepsilon_{cut}(i\zeta)/\varepsilon(i\zeta)$ calculated with these data. One can see that at $\zeta = 1$ eV ($d \sim 100$ nm) the contribution to $\varepsilon(i\zeta)$ from the frequency range $\omega < \omega_{cut}$ is 75%. It means, for example, that if we change the Drude parameters, three fourths of $\varepsilon(i\zeta)$ will be sensitive to this change and only one forth will be defined by the measured optical data. Therefore, the extrapolation procedure becomes very important for reliable prediction of $\varepsilon(i\zeta)$.

The Drude parameters can vary from sample to sample due to different density of defects. The plasma frequency is related to the density of quasi-free electrons N as $\omega_p^2 = 4\pi Ne^2/m_e^*$, where for gold $m_e^* \approx m_e$ is the effective mass of electron. The value $\omega_p = 9.0\,\text{eV}$ is the maximal value of this parameter, which corresponds to N in a single crystal Au. In this way ω_p was estimated in Ref. [9]. All deposited films have smaller values of ω_p as one can see from Table 10.2 because the density of the films is smaller than that for the single crystal material. Precise values of the Drude parameters are extremely important for evaluation of $\varepsilon(i\zeta)$ and finally for calculation of the force.

The dielectric function $\varepsilon(i\zeta)$ becomes less dependent on the Drude parameters if the cut-off frequency is smaller. For example, our optical data [15] were collected up to minimal frequency $\omega_{cut} = 38\,\text{meV}$ that is about four times smaller than in the handbook data. The dashed curve in Fig. 10.4a shows the ratio $\varepsilon_{cut}/\varepsilon$ calculated for our sample 3 with the Drude parameters $\omega_p = 7.84\,\text{eV}$ and $\gamma = 49\,\text{meV}$. Now the relative contribution of $\varepsilon_{cut}(i\zeta)$ at $\zeta = 1\,\text{eV}$ is 37%. It is much smaller than for handbook data, but still dependence on the precise Drude parameters is important. Let us imagine now that we have been able to measure the dielectric response of the material for the same sample 3 from [15] to frequencies as low as $1\,\text{THz}$. In this case the cut-off frequency is $\omega_{cut} = 4\,\text{meV}$ and the relative contribution of the extrapolated region $\varepsilon_{cut}/\varepsilon$ is shown in Fig. 10.4a by the dash-dotted line. Now this contribution is only 5% at $\zeta = 1\,\text{eV}$.

The dielectric functions $\varepsilon_i(i\zeta)$, where $i = 1, 2, .., 5$ is the number of the sample, were calculated using the Drude parameters from Table 10.2. As a reference curve we use $\varepsilon_0(i\zeta)$, which was evaluated with the parameters $\omega_p = 9.0\,\text{eV}$ and $\gamma = 35\,\text{meV}$ in the range $\omega < 0.125\,\text{eV}$ and at higher frequencies the handbook data [7] were used. The results are shown in Fig. 10.4b. As was expected the maximal dielectric function is $\varepsilon_0(i\zeta)$, which corresponds in the Drude range to a perfect single crystal. Even for the best sample 5 (annealed film on mica) the dielectric function is 15% smaller than $\varepsilon_0(i\zeta)$ at $\zeta = 1\,\text{eV}$. For samples 1 and 2 the deviations are as large as 40%.

10.2.2.3 The Force Between Au Films

It is convenient to calculate not the force itself but the so called reduction factor η, which is defined as the ratio of the force to the Casimir force between ideal metals:

$$\eta(d) = \frac{F(d)}{F^c(d)}, \quad F^c(d) = -\frac{\pi^2 \hbar c}{240 d^4}. \tag{10.15}$$

We calculate the force between similar materials at $T = 0$ using the substitute (10.5) in (10.1). For convenience of the numerical procedure one can make an appropriate change of variables so that the reduction factor can be presented in the form

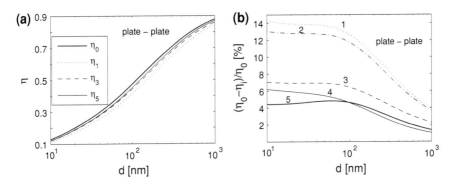

Fig. 10.5 a Reduction factor η as a function of the separation d for samples 1, 3, and 5. The thick line shows the reference result calculated with $\varepsilon_0(i\zeta)$. **b** Relative deviations of the reduction factors for different samples from the reference curve $\eta_0(d)$, which were evaluated using the handbook optical data [7] and the Drude parameters $\omega_p = 9\,\text{eV}$, $\omega_\tau = 35\,\text{meV}$

$$\eta(d) = \frac{15}{2\pi^4} \sum_{\mu=s,p} \int_0^1 dx \int_0^\infty \frac{dy\, y^3}{r_\mu^{-2} e^y - 1}, \qquad (10.16)$$

where the reflection coefficients as functions of x and y are defined as

$$r_s = \frac{1-s}{1+s}, \quad r_p = \frac{\varepsilon(i\zeta_c xy) - s}{\varepsilon(i\zeta_c xy) + s}, \qquad (10.17)$$

with

$$s = \sqrt{1 + x^2[\varepsilon(i\zeta_c xy) - 1]}, \quad \zeta_c = \frac{c}{2d}. \qquad (10.18)$$

The integral (10.16) was calculated numerically with different dielectric functions $\varepsilon_i(i\zeta)$. The results are presented in Fig. 10.5a for samples 1, 3, and 5. The reference curve (thick line) calculated with $\varepsilon_0(i\zeta)$ is also shown for comparison. It represents the reduction factor, which is typically used in the precise calculations of the Casimir force between gold surfaces. One can see that there is significant difference between this reference curve and those that correspond to actual gold films. To see the magnitude of the deviations from the reference curve, we plot in Fig. 10.5b the ratio $(\eta_0 - \eta_i)/\eta_0$ as a function of distance d for all five samples.

At small distances the deviations are more sensitive to the value of ω_p. At large distances the sample dependence becomes weaker and more sensitive to the value of ω_τ. For samples 1 and 2, which correspond to the 400 and 200 nm films deposited on Si substrates, the deviations are especially large. They are 12–14% at $d < 100$ nm and stay considerable even for the distances as large as 1 μm. Samples 3, 4, and 5 have smaller deviations from the reference case but even for these samples the deviations are as large as 5–7%.

10.2.3 Low Permittivity Dielectric Materials

The Lifshitz theory predicts [3] that the dispersive force can be changed from attractive to repulsive by immersing the interacting materials immersed in a liquid. Recently this prediction was confirmed experimentally [45] (see Chap. 8 by Capasso et al. in this volume). Repulsive forces arise when the dielectric function at imaginary frequencies in the liquid gap, $\varepsilon_0(i\zeta)$, is in between the functions of the interacting bodies 1 and 2: $\varepsilon_1(i\zeta) > \varepsilon_0(i\zeta) > \varepsilon_2(i\zeta)$. One can expect significant dependence on precise dielectric functions nearby the transition from attractive to repulsive force. This situation is exactly the case for the system silica–liquid–gold. In this section we present calculations for multiple liquids with various degrees of knowledge of the dielectric functions.

Liquids do not have grains or defects, but the density of a liquid is a function of temperature [46], and as a result the number of absorbers in the liquid varies with temperature. Furthermore, liquids can contain impurities like salt ions which can change the dielectric function (see discussion in Ref. [47]). Although for metals the dielectric function is very large in the IR regime, for liquids and glasses it is not the case. Consequently for low dielectric materials the UV and VUV dielectric data have a strongest effect on the forces.

For gold, silica, and water the dielectric functions are well known and measured by various groups. Let us consider first the interaction in the system gold–water–silica. We will use two sets of data for gold from the previous subsection (sample 1 and the "ideal sample"). Also two sets of data will be used for silica as obtained by different groups [48]. Finally, for water we will use the data of Segelstein compiled from different sources [49], and an 11-order oscillator model [50] that has been fit to different sets of data [51, 52]. All the dielectric data are collected in Fig. 10.6a. The corresponding functions at imaginary frequencies are shown in Fig. 10.6b. One can see considerable difference between solid and dashed curves corresponding to different sets of the data.

It has to be noted that $\varepsilon''(\omega)$ for water and silica are very close in a wide range of frequencies $5 \cdot 10^{-2} < \omega < 5 \cdot 10^2$ eV. As the result at imaginary frequencies $\varepsilon_{SiO_2}(i\zeta)$ and $\varepsilon_{H_2O}(i\zeta)$ differ on 30% or less in the range $10^{-2} < \zeta < 10^2$ eV, which is comparable with the magnitude of variation of $\varepsilon(i\zeta)$ due to data scattering. This similarity in the dielectric functions results in a strong dependence of the Casimir force in the system gold–water–silica on the used optical data. It is illustrated in Fig. 10.7a where the relative change of the force is shown. The spreading of the force data reaches a level of 60% for separation $d < 500$ nm. The effect can even be more clearly seen in Fig. 10.7b. The solid curve shows variation of the force in Au–water–Au system when different optical data for Au are used. In this case the relative change of the force is not very large. However, for the system silica–water–silica the use of different optical data for silica influence the force very significantly (dashed curve).

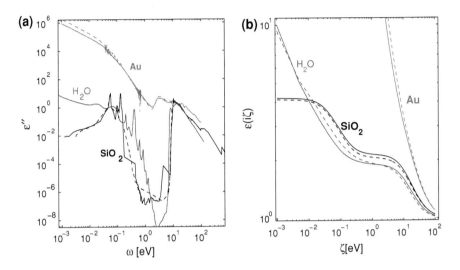

Fig. 10.6 **a** Dielectric data of the materials obtained from references in text. **b** Dielectric functions at imaginary frequencies. The *solid* and *dashed lines* for silica and gold are two different sets of optical data. For water the *solid line* is from the data in Ref. [49], and the *dashed line* is an 11-order oscillator model [50], which has been fitted to a different set of optical data

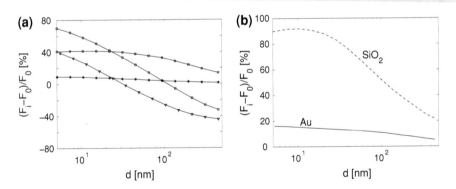

Fig. 10.7 **a** Variation of the force relative to $F_0(d)$ in the system gold–water–silica. *Circles, squares, triangles* and *stars* mark the curves which were calculated using different sets of the dielectric data. **b** Variation of the force for silica–water–silica (*dashed line*) with different sets of data for SiO_2 and for gold–water–gold (*solid line*) using different data for Au

We have to conclude that comparison of force measurements with prediction of the Lifshitz theory becomes reliable when the dielectric properties of the specific samples used in force measurement are measured over a wide range of frequencies.

In most of the papers where liquid gap between bodies is studied the dielectric function of the liquid is approximated using the oscillator models [53, 54]. For illustration purposes we mention that alcohols (and other liquid substances) can be

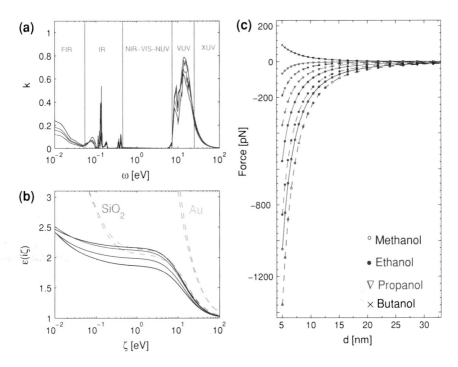

Fig. 10.8 Dielectric data at real **a** and at imaginary frequencies **b** for methanol, ethanol, propanol, and butanol, as described in the text. For comparison the dielectric data for silica and gold samples are also shown in (**b**). In **c** we show the forces for the gold–alcohol–silica, for two different sets of measured dielectric data for silica. In the case of butanol the force is attractive for one set, and repulsive for another one

described, for example, by a three oscillator model for the dielectric function $\varepsilon(i\zeta)$ at imaginary frequencies [53]

$$\varepsilon(i\zeta) = 1 + \frac{\varepsilon_0 - \varepsilon_{IR}}{1 + \zeta/\omega_{MW}} + \frac{\varepsilon_{IR} - n_0^2}{1 + (\zeta/\omega_{IR})^2} + \frac{n_0^2 - 1}{1 + (\zeta/\omega_{UV})^2}. \qquad (10.19)$$

Here n_0 is the refractive index in the visible range, ε_0 is the static dielectric constant, and ε_{IR} is the dielectric constant where MW relaxation ends and IR begins. The parameters ω_{MW}, ω_{IR}, and ω_{UV} are the characteristic frequencies of MW, IR, and UV absorption, respectively. It has to be stressed that oscillator models should be used with caution, because some of them are of poor quality [16].

For ethanol rather detailed information on the dielectric function exists, but even in this case variation in dielectric data was found [16, 40]. An interesting fact for higher alcohols is that the absorption in the UV range increases when increasing the alkane chain. In Fig. 10.8 we show the dielectric data for the first four alcohols.

The VUV data were taken from [55]. These measurements were done in the gas phase, but they can be converted to the liquid case by considering the number of absorbers in gas and liquid. The near UV data was taken from [56]. For the XUV we have only data for ethanol and propanol [57]. For methanol and butanol we used cubic extrapolation, $\varepsilon'' \sim 1/\omega^3$, which is in very good agreement with the cases of ethanol and propanol. In the near IR (NIR) to visible (VIS) ranges the extinction coefficient k of ethanol, and other alcohols, is very low and can be taken to be zero, $k = 0$, which is qualitatively consistent with the fact that all alcohols are transparent in the visible range. The IR data can be found in Ref. [58]. The far IR (FIR) data are known only for methanol [59]. For the other alcohols we take the similar functional behavior as for methanol but with different parameters and extrapolate the data to far IR in this way.

If one has to estimate the dielectric functions for some alcohols, first of all one has to have measured data in the range of major IR peaks and even more importantly the UV absorption must be carefully measured. Thus the dielectric functions at imaginary frequencies should be reasonably accurate to within the scatter of the data as it was found for ethanol [40].

With the optical data for alcohols we calculated the forces in the system gold–alcohol–silica. The forces are attractive and become weaker for methanol, ethanol and propanol. For butanol they are extremely weak, but still either repulsive or attractive. Caution is required in the analysis of optical properties in liquids since in general the KK consistency has to be applied properly in order to correct for variation of the dielectric properties observed in between different measuring setups. Effectively the force for gold–butanol–silica is screened to within the scatter of the forces related to sample dependence of the optical properties of silica. Measurements between gold and glass with simple alcohols were performed, but experimental uncertainty, and double layer forces prevented the measurement of this effect [60].

10.3 Influence of Surface Roughness on the Casimir–Lifshitz Force

The Lifshitz formula (10.1) does not take into account inevitable roughness of the interacting bodies. When rms roughness of the bodies is much smaller than the separation, then the roughness influence on the force can be calculated using the perturbation theory [17–19]. However, when the separation becomes comparable with the roughness the perturbation theory cannot be applied. It was demonstrated experimentally [24] that in this regime the force deviates significantly from any theoretical prediction. The problem of short separations between rough bodies is one of the unresolved problems. In this section we give introduction into interaction of two rough plates and a sphere and a plate.

10.3.1 Main Characteristics of a Rough Surface

Suppose there is a rough plate which surface profile can be described by the function $h(x, y)$, where x and y are the in-plane coordinates. An approximation for this function is provided for example, by an AFM scan of the surface. It gives the height h_{ij} at the pixel position $x_i = \Delta \cdot i$ and $y_j = \Delta \cdot j$, where $i, j = 1, 2, \ldots, N$ and Δ is the pixel size that is related to the scan size as $L = \Delta \cdot N$. We can define the mean plane of the rough plate as the averaged value of the function $h(x, y)$: $\bar{h} = A^{-1} \int dx dy h(x, y)$, where A is the area of the plate. This definition assumes that the plate is infinite. In reality we have to deal with a scan of finite size, for which the mean plane is at

$$h_{av} = \frac{1}{N^2} \sum_{i,j} h(x_i, y_j). \tag{10.20}$$

The difference $\bar{h} - h_{av}$, although small, is not zero and is a random function of the scan position on the plate. This difference becomes larger the smaller the scan size is. Keeping in mind this point, which can be important in some situations (see below), we can consider (10.20) as an approximate definition of the mean plane position.

An important characteristic of a rough surface is the rms roughness w, given by

$$w = \frac{1}{N^2} \sum_{i,j} \left[h(x_i, y_j) - h_{av} \right]^2, \tag{10.21}$$

that can be interpreted as the interface thickness. More detailed information on the rough surface can be extracted from the height-difference correlation function defined for the infinite surface as

$$g(R) = \frac{1}{A} \int dx dy \left[h(\mathbf{r} + \mathbf{R}) - h(\mathbf{r}) \right]^2, \tag{10.22}$$

where $\mathbf{r} = (x, y)$ and $\mathbf{R} = \mathbf{r}' - \mathbf{r}$.

A wide variety of surfaces, as for example, deposited thin films far from equilibrium, exhibit the so called self-affine roughness which is characterized besides the rms roughness amplitude w by the lateral correlation length ξ (indicating the average lateral feature size), and the roughness exponent $0 < H < 1$ [61–63]. Small values of $H \sim 0$ correspond to jagged surfaces, while large values of $H \sim 1$ to a smooth hill valley morphology. For the self-affine rough surfaces $g(R)$ scales as

$$g(R) = \begin{cases} R^{2H}, & R \ll \xi, \\ 2w^2, & R \gg \xi. \end{cases} \tag{10.23}$$

The parameters w, ξ and H can be determined from the measured height-difference correlation function $g(R)$. This function can be extracted approximately from the AFM scans of the surface.

To find the roughness correction to the force one has to know (see below) the spectral density $\sigma(k)$ of the height-height correlation function $C(R)$. The latter is related to $g(R)$ as $g(R) = 2w^2 - C(R)$. An analytic form of the spectral density for a self-affine surface is given by [64, 65]

$$\sigma(k) = \frac{CHw^2\xi^2}{\left(1 + k^2\xi^2\right)^{1+H}}, \quad C = \frac{2}{1 - \left(1 + k_c^2\xi^2\right)^{-H}}. \tag{10.24}$$

Here C is a normalization constant [63–65] and $k_c = 2\pi/L_c$ is the cutoff wavenumber.

10.3.2 Roughness Correction

While the separation between two surfaces is large in comparison with the rms roughness, $d \gg w$, one can use the perturbation approach to calculate the roughness correction to the Casimir force. This correction was calculated first using the proximity force approximation [66]. This approximation assumes that the surface profile varied slowly in comparison with the distance between the bodies. The lateral size of a rough profile is given by the correlation length ξ, therefore, PFA can be applied if $\xi \gg d$. This condition is very restrictive since typical values of ξ for deposited metals films (grain size) are in the range 20–100 nm. In most of the experimental situations the condition $\xi \gg d$ is broken. More general theory [17–19] for the roughness correction can be applied at $\xi \leq d$ and treats the correction perturbatively within the scattering formalism (see the Chap. 4 by Lambrecht et al. in this volume). Here we discuss application of this theory to realistic rough surfaces and describe situations, for which one has to go beyond the perturbation theory to find agreement with experiments.

10.3.2.1 Application of the Perturbation Theory

Let us consider two parallel rough plates. A plate surface can be described by the roughness profile $h_i(x, y)$ ($i = 1, 2$ for plate 1 or 2) as shown in Fig. 10.9a. The averaged value over large area is assumed to be zero $\langle h_i(x, y) \rangle = 0$. Then the local distance between the plates is

$$d(x, y) = d - h_1(x, y) - h_2(x, y). \tag{10.25}$$

This distance depends on the combined rough profile $h(x, y) = h_1(x, y) + h_2(x, y)$. As explained in Sect. 10.3.3 the interaction of two rough plates is equivalent to the interaction of a smooth plate and a rough plate with the roughness given by the combined profile $h(x, y)$ (see Fig. 10.9b).

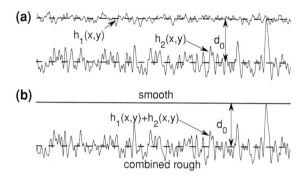

Fig. 10.9 Contact of two rough surfaces. **a** Two rough plates in contact. Roughness of each plate, $h_i(x, y)$, is counted from the mean plane shown by the dashed line. The distance between bodies is the distance between these mean planes. **b** The interaction between two rough plates is equivalent to the interaction between a smooth plate and a rough plate with the roughness given by the combined profile $h(x, y)$. The distance upon contact, d_0 has well defined meaning in this case. See Sect. 10.3.3 and Ref. [73] for details

Let us assume further that the interaction energy per unit area of two flat plates is $E_{pp}(d)$. If the rms roughness of the combined profile $h(x, y)$ is small, $w \ll d$, but the correlation length is large, $\xi \gg d$, we can present the interaction between rough plates as

$$E_{pp}^{rough} = \langle E_{pp}(d(x, y)) \rangle \approx E_{pp}(d) + \frac{E_{pp}''}{2}\langle h^2 \rangle, \qquad (10.26)$$

where $\langle h^2 \rangle = w^2 = w_1^2 + w_2^2$. Equation (10.27) defines the PFA roughness correction $\delta E_{pp} = E_{pp}'' w^2 / 2$. This correction was used in all early studies to estimate the roughness effect.

It was noted [17] that in most experimental configurations the condition $\xi \gg d$ is broken and PFA cannot be applied. In Refs. [18, 19] was developed a theory, which is not restricted by the condition $\xi \gg d$. Within this theory the roughness correction is expressed via the spectral density of the rough surface $\sigma(k)$ as

$$\delta E_{pp}(d) = \int \frac{d^2k}{(2\pi)^2} G(k, d)\sigma(k), \qquad (10.27)$$

where $G(k, d)$ is a roughness response function derived in [19]. The PFA result (10.27) is recovered from here in the limit of small wavenumbers $k \to 0$ when $G(k, d) \to E_{pp}''(d)/2$. The roughness power spectrum is normalized by the condition $\int d^2k \sigma(k)/(2\pi)^2 = w^2$. The spectrum itself can be obtained from AFM scans and in the case of self-affine rough surfaces is given by (10.24).

Let us enumerate the conditions at which (10.27) is valid. (i) The lateral dimensions of the roughness ξ must be much smaller than the system size L, $\xi \ll L$. This is usually the case in experiments. (ii) The rms roughness w must be

Table 10.3 The parameters characterizing the sphere-film systems (all in nm) for $R = 50 \, \mu m$. The first three rows were determined from combined images. The last row for d_0^{el} gives the values of d_0 determined electrostatically. The errors are indicared in brackets

	100 nm	200 nm	400 nm	800 nm	1600 nm
w	3.8	4.2	6.0	7.5	10.1
ξ	26.1(3.8)	28.8(3.7)	34.4(4.7)	30.6(2.4)	42.0(5.5)
d_0^{im}	12.8(2.2)	15.9(2.7)	24.5(4.8)	31.3(5.4)	55.7(9.3)
d_0^{el}	17.7(1.1)	20.2(1.2)	23.0(0.9)	34.5(1.7)	50.8(1.3)

small compared to the separation distance, $w \ll d$. This condition means that roughness is treated as perturbative effect. (iii) The lateral roughness dimensions must be much larger than the vertical dimensions, $w \ll \xi$ [19]. The last two assumptions are not always satisfied in the experiment.

In the plate-plate configuration the force per unit area can be calculated as the derivative of $E_{pp}^{rough}(d)$. For the sphere-plate configuration, which is used in most of the experiments, the force is calculated with the help of PFA as $F_{sp}(d) = 2\pi R E_{pp}^{rough}(d)$. In contrast with the roughness correction the latter relation is justified for $d \ll R$, which holds true for most of the experimental configurations. We use the sphere-plate configuration to illustrate the roughness effect. The deposited gold films can be considered as self-affine. For all calculations reported here we are using our smoothest spheres with the parameters $w = 1.8 \, nm$, $\xi = 22 \, nm$, and $H = 0.9$. We alter only the plate roughness since it is easy to prepare and replace during experiments. We use the optical data for gold films described previously. It was found that the PFA limit is quickly recovered for increasing correlation length. Deviations from PFA prediction for real films were found to be about 1–5% in the range $d = 50$–$200 \, nm$.

Therefore, for real rough surfaces the scattering theory gives a few percent correction to the force compared to the PFA. This difference is difficult to measure. However, at small separations both PFA and perturbation theory fail since the rms roughness becomes comparable in size to the separation distance. It would be interesting to calculate the roughness effect when d is comparable with w. At the moment there is no a theoretical approach to estimate the effect except a direct numerical analysis similar to that used in [67]. It will therefore be interesting to do a full numerical analysis for films with high local slopes instead of using perturbation theory. On the other hand it is experimentally possible to go to sufficiently small distances as it will be discussed below.

10.3.2.2 Experimental Evidence of Large Roughness Effect

The Casimir forces between a $100 \, \mu m$ gold coated sphere and substrates covered with Au of different thicknesses from 100 nm to 1600 nm were measured in [24]. Different layers of Au resulted to different roughnesses and different correlation lengths, which are collected in Table 10.3. The roughness exponent was constant

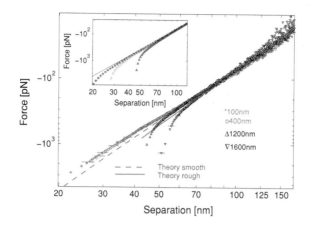

Fig. 10.10 Casimir forces measured for the films of different roughness. The roughness effect manifests itself as a strong change in scaling at smaller separations, where the forces become much stronger. Errors in separation are shown for some points by the horizontal bars. The theoretical curves are shown for the 100 nm (*smooth*) and 1600 nm (*rough*) films. The inset shows the forces calculated by integrating over the roughness scans using PFA (see text)

$H = 0.9 \pm 0.05$ in agreement with former growth studies of thin films [62, 64, 65]. The sphere was attached to a cantilever with a spring constant of 0.2 N/m. The calibration procedure is described in [24].

The force results are shown in Fig. 10.10. Our measurements were restricted to separations below 200 nm where the Casimir force is large enough compared to the approximately linear signal from laser light surface backscattering [24, 26]. The small separation limit or contact point is restricted by the jump to a contact (~ 5 nm) [68–70] and surface roughness. Note that an error of 1.0 nm in absolute distance leads to errors in the forces of up to 20% at close separations as for example $d \sim 10$ nm [24]. Thus we cannot detect the scattering effects described above. What we do see is the failure of the perturbation theory, for the roughest films, for which the roughness strongly increases the force. These deviations are quite large, resulting in much stronger forces at the small separations <70 nm. At larger separations, 70–130 nm (within our measurement range), where the roughness influence is negligible, the usual $1/d^{2.5}$ scaling of the force observed also in other experiments with gold is recovered and agreement with the theory is restored. For the smoother films deviations from theory below 40 nm can be explained with the error in the distance.

Qualitatively the roughness effect can be reproduced by calculating the force between small areas on the surfaces separated by the local distance $d(x, y)$. One can call this procedure the non-perturbative PFA approach. Although it is qualitative, it can be used to obtain an estimate of the force at close proximity (2 nm above the point upon contact), where the roughness has an enormous influence on the Casimir force (see inset in Fig. 10.10). This explains the jump to contact only partially, since approximately 5 nm above the point of contact, the capillary force

Fig. 10.11 Capillary forces
in air (relative humidity
2–60%) for a smooth gold
coated sphere, $w_{sph} \sim 1$ nm,
measured with a stiff
cantilever, $k = 4$ N/m, and
different rough films. The
inset shows a fully wetted
sphere (*upper dashed line*),
and a roughness asperity
wetted sphere (*lower dashed
line*)

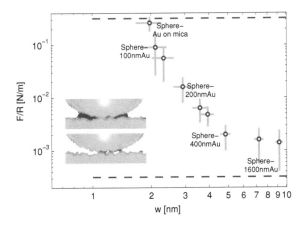

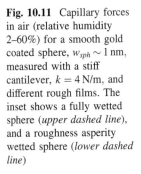

will act as well. This force appears due to absorbed water and Kelvin condensation
[22, 71] resulting in water bridges formation between bodies. In the limit of fully
wetted surface (see Fig. 10.11) the capillary force is given by $F_{cap} \approx 4\pi\gamma R \cos\vartheta$
(upper dashed line), while for a single asperity (of size ξ) wetting the minimum
capillary force is $F_{cap} \approx 4\pi\gamma\xi \cos\vartheta$ (lower dashed line). Here γ is the surface
tension of liquid, and ϑ is the contact angle [68, 70].

At this point we have to compare the Casimir adhesion between rough films
with adhesion by capillary forces [69, 70]. While Casimir forces may lead to
stiction between movable parts, once the surfaces are in contact capillary forces
(being present in air between hydrophilic surfaces into close proximity) are much
stronger. The roughness effect on capillary adhesion is also much stronger as
shown in Fig. 10.11. Note that the Casimir force for a $R = 50$ μm sphere is in the
order of 10 nN at 10 nm separations. Capillary forces between a mica substrate
and the same sphere are as large as 10 μN deeming contact measurements with soft
cantilevers (spring constant < 1 N/m) even impossible since the retraction range
is outside of that of most piezo z-ranges. Measurements of the capillary forces with
a smooth sphere used for the Casimir force measurement are shown in Fig. 10.11.
One can see that when roughness increases a few times the force decreases by
more than a factor of 100. This can be related to full sphere wetting and asperity
wetting. The size of the sphere $R = 50$ μm is 1000 times larger than that of an
asperity $\xi \sim 50$ nm. Multiple asperity capillary bridge formation is likely to happen
in the rough regime giving increasing forces.

Furthermore, formation of capillary bridges means that under ambient conditions gold absorbs water, and as a result it is covered with an ultra thin water layer.
The experiments [68, 70] suggest that the thickness of this layer is in the nanometer range, 1–2 nm. The natural questions one could ask is how thick the water
layer is, and what is the influence of this water layer on the dispersive force [72]?
At short separations, $d < 20$ nm, these questions become of crucial importance
because they place doubts in our understanding of the dispersive forces when
experiments under ambient conditions are compared with predictions of the

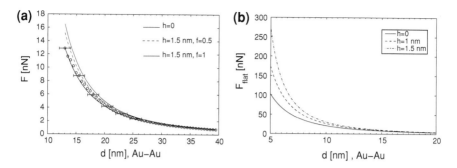

Fig. 10.12 a Experimental data for the force vs distance (*circles*) down to 13 nm separations, and the theoretical prediction without water layer (*lower solid curve*). Errors in the absolute separation are shown for some points by the bars. The *upper solid curve* is the prediction for continuous water layer of thickness $h = 1.5$ nm. The *dashed curve* corresponds to the same water layer with 50% of voids ($f = 0.5$). **b** Theoretical calculations using the Lifsthiz theory for flat surfaces and continuous water layer films with thicknesses $h = 0, 1, 1.5$ nm for small separations

Lifshitz theory. Figure 10.12 shows the Casimir force measured at short distances together with theoretical calculations made for gold with or without a water layer on top. The errors are shown to arise mainly from the experimental uncertainty in determining the separation upon contact d_0 due to nanoscale surface roughness. We can conclude that the experiment can exclude the water layer thicker than 1.5 nm. Figure 10.12b shows that the effect of water becomes very significant at separations below 10 nm, which were not accessible in our measurements due to jump-into-contact. We presented only the forces between flat surfaces because at these small separations there is no a reliable way to estimate the roughness correction.

10.3.3 Distance upon Contact

The absolute distance separating two bodies is a parameter of principal importance for the determination of dispersive forces. It becomes difficult to determine when the separation gap approaches nanometer dimensions. This complication originates from the presence of surface roughness, which manifests itself on the same scale. In fact, when the bodies are brought into gentle contact they are still separated by some distance d_0, which we call the distance upon contact due to surface roughness. This distance has a special significance for weak adhesion, which is mainly due to van der Waals forces across an extensive noncontact area [21]. It is important for MEMS and NEMS as unremovable reason for stiction [22]. In the modern precise measurements of the dispersive forces d_0 is the main source of errors (see reviews [5, 6]). This parameter is typically determined using electrostatic calibration. The distance upon contact is usually considerably larger than the rms roughness because it is defined by the highest asperities. It is important to clear understand the origin of d_0, its dependence on the lateral size L of involved surfaces, and possible uncertainties in its value [73]. These are the questions we address in this subsection.

10.3.3.1 Plate-Plate Contact

Two plates separated by the distance d and having roughness profiles $h_i(x,y)$ are locally separated by the distance $d(x,y)$ given by (10.25) (see Fig. 10.9). Indeed, the averaged local distance has to give d, $\langle d(x,y) \rangle = d$. We can define the distance upon contact d_0 as the largest distance $d = d_0$, for which $d(x,y)$ becomes zero.

It is well known from contact mechanics [74] that the contact of two elastic rough plates is equivalent to the contact of a rough hard plate and an elastic flat plate with an effective Young's modulus E and a Poisson ratio v. Here we analyze the contact in the limit of zero load when both bodies can be considered as hard. This limit is realized when only weak adhesion is possible, for which the dispersive forces are responsible. Strong adhesion due to chemical bonding or due to capillary forces is not considered here. This is not a principal restriction, but the case of strong adhesion has to be analyzed separately. Equation (10.25) shows that the profile of the effective rough body is given by

$$h(x,y) = h_1(x,y) + h_2(x,y). \tag{10.28}$$

The latter means that $h(x,y)$ is given by the combined image of the surfaces facing each other. If topography of the surfaces was determined with AFM, we have to take the sum of these two images and the combined image will have the size of the smallest image.

To determine d_0 we collected [73] high resolution megascans (size $40 \times 40\,\mu m^2$, lateral resolution 4096×4096 pixels) for gold films of different thicknesses described before. The maximal area, which we have been able to scan on the sphere, was $8 \times 8\,\mu m^2$ (2048×2048 pixels). The images of 100 nm film, sphere, and 1600 nm film are shown in Fig. 10.1 a–c, respectively. Combining two images and calculating from them the maximal peak height we can find d_0 for a given size of the combined image. Of course, taking the images of the same size every time we will get different value of d_0 and averaging over a large number of images we will find the averaged d_0 and possible rms deviations. This is quite obvious. What is less obvious is that if we take images of different size and will do the same procedure the result for d_0 will be different.

Let L_0 be the size of the combined image. Then, in order to obtain information on the scale $L = L_0/2^n$, we divide this image on 2^n subimages. For each subimage we find the highest point of the profile (local d_0), and average all these values. This procedure gives us $d_0(L)$ and the corresponding statistical error. Megascans are very convenient for this purpose otherwise one has to collect many scans in different locations. For the 100 nm film above the 400 nm film the result of this

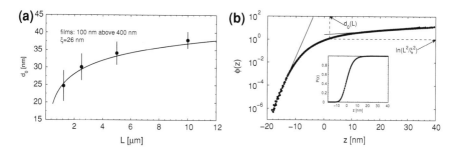

Fig. 10.13 a Distance upon contact as a function of the length scale. Dots with the error bars are the values calculated from the megascans. The *solid curve* is the theoretical expectation according to (10.30). Note that d_0 is considerably larger than w. **b** Statistics of the surface roughness. Four $10 \times 10\,\mu m^2$ images were used. The main graph shows the "phase" as a function of z. The *solid lines* show the best fits at large positive and negative z. The *dashed lines* demonstrate the solution of (10.30). The inset shows the cumulative distribution $P(z)$

procedure is shown in Fig. 10.13a. We took the maximum area to be $10 \times 10\,\mu m^2$. The figure clearly demonstrates the dependence of d_0 on the scale L although the errors appear to be significant. The dependence of the rms roughness w on the length scale L is absent in accordance with the expectations, while only the error bars increase when L is decreasing.

To understand the dependence $d_0(L)$ let us assume that the size L of the area of nominal contact is large in comparison with the correlation length, $L \gg \xi$. It means that this area can be divided into a large number $N^2 = L^2/\xi^2$ of cells. The height of each cell (asperity) can be considered as a random variable h. The probability to find h smaller than some value z can be presented in a general form

$$P(z) = 1 - e^{-\phi(z)}, \tag{10.29}$$

where the "phase" $\phi(z)$ is a nonnegative and nondecreasing function of z. Note that (10.29) is just a convenient way to represent the data: instead of cumulative distributions $P(z)$ we are using the phase $\phi(z)$, which is not a so sharp function of z.

For a given asperity the probability to find its height above d_0 is $1 - P(d_0)$, then within the area of nominal contact one asperity will be higher than d_0 if

$$e^{-\phi(d_0)}\left(L^2/\xi^2\right) = 1 \quad \text{or} \quad \phi(d_0) = \ln\left(L^2/\xi^2\right). \tag{10.30}$$

This condition can be considered as an equation for the asperity height because due to a sharp exponential behavior the height is approximately equal to d_0. To solve (10.30) we have to know the function $\phi(z)$, which can be found from the roughness profile.

The cumulative distribution $P(z)$ can be extracted from combined images by counting pixels with the height below z. Then the "phase" is calculated as $\phi(z) = -\ln(1 - P)$. The results are presented in Fig. 10.13b. The procedure of solving (10.30) is shown schematically in Fig. 10.13b by dashed lines, and the solution itself is the curve in Fig. 10.13a. It has to be mentioned that the normal distribution fails to describe the data at large z. Other known distributions cannot satisfactorily describe the data for all z. Asymptotically at large $|z|$ the data can be reasonably well fit with the generalized extreme value Gumbel distributions (solid lines in Fig. 10.13b) [75]:

$$\ln \phi(z) = \begin{cases} -\alpha z, \ z \to -\infty \\ \beta z, \ z \to \infty \end{cases} \tag{10.31}$$

The observed dependence $d_0(L)$ can be understood intuitively. The probability to have one high asperity is exponentially small but the number of asperities increases with the area of nominal contact. Therefore, the larger the contact area, the higher probability to find a high feature within this area. Scale dependence of d_0 shows that smaller areas getting into contact will be bound more strongly than larger areas because upon the contact they will be separated by smaller distances. This fact is important for weak adhesion analysis.

10.3.3.2 Sphere-Plate Contact

Most of the Casimir force experiments measure the force in the sphere-plate configuration to avoid the problem with the plates parallelism. Let us consider how the scale dependence of d_0 manifests itself in this case. Assuming that the sphere is large, $R \gg d$, the local distance is

$$d(x,y) = d + (x^2 + y^2)/2R - h(x,y), \tag{10.32}$$

where $h(x,y)$ is the combined topography of the sphere and the plate. As in the plate-plate case d_0 is the maximal d, for which the local distance becomes zero. This definition gives

$$d_0 = \max_{x,y}\left[h(x,y) - (x^2 + y^2)/2R\right]. \tag{10.33}$$

Now d_0 is a function of the sphere radius R, but, of course, one can define the length scale L_R corresponding to this radius R.

As input data in (10.33) we used the combined images of the sphere and different plates. The origin $(x = 0, y = 0)$ was chosen randomly in different positions and then d_0 was calculated according to (10.33). We averaged d_0 found in 80 different locations to get the values of d_0^{im}, which are collected in Table 10.3. The same values can be determined theoretically using $d_0(L)$ found between two plates (see (10.30) and Ref. [73]).

The values of d_0^{im} for rougher films are in agreement with those found from the electrostatic calibration. However, for smoother films (100 and 200 nm) d_0^{im} and d_0^{el} do not agree with each other. This is most likely to be attributed to the roughness on the sphere, which varies considerably locally. For example, between those 80 d_0^{im} found in different locations 5% are in agreement with d_0^{el} found from the electrostatic calibration [73]. This is illustrated by the fact that when the roughness of the

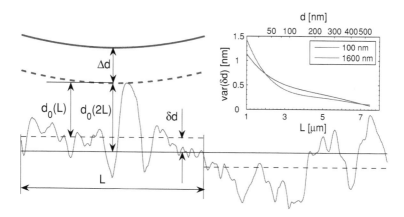

Fig. 10.14 Schematic explanation of additional uncertainty δd in d_0 (see text). The sphere in two positions is shown by the *dashed* (contact) and *solid blue curves*. The inset shows the variance of δd as a function of the scale L or separation d

plate dominates the discrepancy between d_0^{im} and d_0^{el} disappears. Note that the standard deviations for d_0^{im} are larger than that for d_0^{el}. The standard deviations in d_0^{im} originate from place to place variations of d_0^{im}. In the case of electrostatic determination of d_0 statistical variation of d_0 from place to place is not included in the errors of d_0^{el}. This explains why the errors in d_0^{el} are smaller.

Consider the experimental situation when the dispersive force is measured in the sphere-plate configuration. The system under consideration is equivalent to a smooth sphere above a combined rough profile $h(x, y)$. The position of the average plane depends on the area of averaging L^2 especially for small scales L. The profile shown in Fig. 10.14 demonstrates different mean values in the left and right segments shown by the dashed black lines. Both of these values deviate from the middle line for the scale $2L$ (solid black line). The true average plane is defined for $L \to \infty$.

Position of the average plane define the absolute separation of the bodies. It has to be stressed that the electrostatic and Casimir interactions "see" different areas on the plate. This is due to different dependence on d and quite often the electrostatic calibration is performed at larger separations than measurement of the Casimir force. The size L of the interaction area is determined by the relation $L^2 = \alpha \pi R d$, where $\alpha = 2$ for the electrostatic and $\alpha = 2/3$ for pure Casimir force. Therefore, these two interactions can "see" different positions of the average planes. It introduces an additional uncertainty δd in the absolute separation [73]. For a fixed L this uncertainty is a random variable distributed roughly normally around $\delta d = 0$. However, it has to be stressed that δd manifests itself not as a statistical error but rather as a kind of a systematic error. This is because at a given lateral position of the sphere this uncertainty takes a fixed value. The variance of δd is defined by the roughness statistics. It was calculated from the images and shown as inset in Fig. 10.14. One has to remember that with a probability of 30% the value of δd can be larger than that shown in Fig. 10.14.

10.4 Conclusions

In this chapter we considered the Casimir force between realistic materials containing defects, which influence the optical properties of interacting materials, and having surface roughness, which contributes to the force.

It was demonstrated that the gold films prepared in different conditions have dielectric functions, which differ from sample to sample, and this difference cannot be ignored in the calculation of the Casimir force aimed at precision better than 10%. The main conclusion is that for metals one has to measure the dielectric function of used materials in a wide range of frequencies, where far and mid IR are especially important. Precise knowledge of the dielectric functions is also important for low permittivity dielectric materials. In this case significant sensitivity of the force to the dielectric functions is realized nearby the attractive-to-repulsive transition in solid–liquid–solid systems.

The roughness correction to the Casimir force can be reliably calculated if rms roughness w is small in comparison with the separation, $w \ll d$, when one can apply the perturbation theory. In the experiments at short separations this condition can be violated. The current situation with the theory is that there is no direct method to calculate the force between rough bodies when $d \sim w$ except using rather complicated numerical calculations.

We gave also a detailed analysis of the distance upon contact, d_0, which is an important parameter in Casimir physics. Analysis of AFM scans demonstrated that d_0 is always a few times larger than the rms roughness. Moreover, d_0 is a function of the size L of the nominal area of contact. This dependence is important for weak adhesion, which is due to van der Waals forces across an extensive noncontact area. Uncertainty in d_0 is the main source of errors in the Casimir force measurements. We demonstrated here that there is an additional indefiniteness in d_0, which cannot be excluded by the electrostatic calibration. It becomes very important for small areas of interaction. Also, this indefiniteness has to be taken into account if one compares two independent experiments.

Acknowledgements The research was carried out under Project No. MC3.05242 in the framework of the Strategic Research Programme of the Materials Innovation Institute M2i (the former Netherlands Institute for Metals Research NIMR) The authors benefited from exchange of ideas by the ESF Research Network CASIMIR.

References

1. Casimir, H.B.G.: On the attraction between two perfectly conducting plates. Proc. K. Ned. Akad. Wet. **51**, 793–795 (1948)
2. Lifshitz, E.M. (1955) Zh. Eksp. Teor. Fiz. **29**, 894 [Soviet Phys. JETP 2, 73 (1956)]
3. Dzyaloshinskii, I.E., Lifshitz, E.M., Pitaevskii, L.P.: General theory of van der Waals' forces. Usp. Fiz. Nauk **73**, 381–421 (1961) [Soviet Phys. Usp. 4, 153–176 (1961)]
4. Lifshitz, E.M., Pitaevskii, L.P.: Statistical Physics, Part 2. Pergamon, Oxford (1980)

5. Lamoreaux, S.K.: The Casimir force: background, experiments, and applications. Rep. Prog. Phys. **68**, 201–236 (2005)
6. Capasso, F., Munday, J.N., Iannuzzi, D., Chan, H.B.: Casimir Forces and Quantum Electrodynamical Torques: Physics and Nanomechanics. IEEE J. Sel. Top. Quantum Electron. **13**, 400–414 (2007)
7. Palik, E.D.: Handbook of Optical Constants of Solids. Academic Press, New York (1995)
8. Weaver,J.H., Krafka, C., Lynch, D.W., Koch, E.E.:(1981) Optical properties of metals, Part II, Physics Data No. 18-2. Fachinformationszentrum Energie, Physik, Mathematik, Karsruhe
9. Lambrecht, A., Reynaud, S.: Casimir force between metallic mirrors. Eur. Phys. J. D **8**, 309–318 (2000)
10. Lamoreaux, S.K.: Calculation of the Casimir force between imperfectly conducting plates. Phys. Rev. A **59**, R3149–R3153 (1999)
11. Lamoreaux, S.K.: Comment on "Precision Measurement of the Casimir Force from 0.1 to 0.9 μm". Phys. Rev. Lett. **83**, 3340 (1999)
12. Svetovoy, V.B., Lokhanin, M.V.: Do the precise measurements of the Casimir force agree with the expectations?. Mod. Phys. Lett. A **15**, 1013–1021 (2000)
13. Svetovoy, V.B., Lokhanin, M.V.: Precise calculation of the Casimir force between gold surfaces. Mod. Phys. Lett. A **15**, 1437–1444 (2000)
14. Pirozhenko, I., Lambrecht, A., Svetovoy, V.B.: Sample dependence of the Casimir force. New J. Phys. **8**, 238 (2006)
15. Svetovoy, V.B., van Zwol, P.J., Palasantzas, G., DeHosson, J.Th.M.: Optical properties of gold films and the Casimir force. Phys. Rev. B **77**, 035439 (2008)
16. van Zwol, P.J., Palasantzas, G., De Hosson, J.Th.M.: Influence of dielectric properties on van der Waals/Casimir forces in solid-liquid systems. Phys. Rev. **79**, 195428 (2009)
17. Genet, C., Lambrecht, A., Maia Neto, P., Reynaud, S.: The Casimir force between rough metallic plates. Europhys. Lett. **62**, 484–490 (2003)
18. Maia Neto, P., Lambrecht, A., Reynaud, S.: Roughness correction to the Casimir force: Beyond the Proximity Force Approximation. Europhys. Lett. **69**, 924–930 (2005)
19. Maia Neto, P., Lambrecht, A., Reynaud, S.: Casimir effect with rough metallic mirrors. Phys. Rev. A **72**, 012115 (2005)
20. Derjaguin, B.V., Abrikosova, I.I., Lifshitz, E.M.: Direct measurement of molecular attraction between solids separated by a narrow gap. Q. Rev. Chem. Soc. **10**, 295–329 (1956)
21. DelRio, F.W., de Boer, M.P., Knapp, J.A., Reedy, E.D. Jr, Clews, P.J., Dunn, M.L.: The role of van derWaals forces in adhesion of micromachined surfaces. Nat. Mater. **4**, 629–634 (2005)
22. Maboudian, R., Howe, R.T.: Critical review: Adhesion in surface micromechanical structures. J. Vac. Sci. Technol. B **15**, 1 (1997)
23. Houston, M.R., Howe, R.T., Maboudiana, R.: Effect of hydrogen termination on the work of adhesion between rough polycrystalline silicon surfaces. J. Appl. Phys. **81**, 3474–3483 (1997)
24. van Zwol, P.J., Palasantzas, G., De Hosson, J.Th.M.: Influence of random roughness on the Casimir force at small separations. Phys. Rev. B **77**, 075412 (2008)
25. Lamoreaux, S.K.: Demonstration of the Casimir force in the 0.6−6 μm range. Phys. Rev. Lett. **78**, 5–8 (1997)
26. Harris, B.W. et al.: Precision measurement of the Casimir force using gold surfaces. Phys. Rev. A. **62**, 052109 (2000)
27. Decca, R.S. et al.: Tests of new physics from precise measurements of the Casimir pressure between two gold-coated plates. Phys. Rev. D **75**, 077101 (2007)
28. Aspnes, D.E., Kinsbron, E., Bacon, D.D.: Optical properties of Au: Sample effects. Phys. Rev B **21**, 3290–3299 (1980)
29. Chen, F., Mohideen, U., Klimchitskaya, G.L., Mostepanenko, V.M.: Investigation of the Casimir force between metal and semiconductor test bodies. Phys. Rev. A **72**, 020101 (2005)
30. Dold, B., Mecke, R.: Optische Eigenschaften von Edelmetallen, Übergangsmetallen und deren Legierungen im Infratot. Optik **22**, 435–446 (1965)

31. Motulevich, G.P., Shubin, A. A.: Influence of Fermi surface shape in gold on the optical constants and hall effect. Zh. Eksp.Teor. Fiz. **47**, 840 (1964) [Soviet Phys. JETP 20, 560–564 (1965)]

32. Landau, L.D., Lifshitz, E.M.: Electrodynamics of Continuous Media. Pergamon Press, Oxford (1963)

33. Zhou, F., Spruch, L.: van der Waals and retardation (Casimir) interactions of an electron or an atom with multilayered walls. Phys. Rev. A **52**, 297–310 (1995)

34. Iannuzzi, D., Lisanti, M., Capasso, F.: Effect of hydrogen-switchable mirrors on the Casimir force. Proc. Natl. Acad. Sci. U.S.A. **101**, 4019–4023 (2004)

35. Lisanti, M., Iannuzzi, D., Capasso, F.: Observation of the skin-depth effect on the Casimir force between metallic surfaces. Proc. Natl. Acad. Sci. U.S.A. **102**, 11989–11992 (2005)

36. de Man, S., Heeck, K., Wijngaarden, R.J., Iannuzzi, D.: Halving the Casimir force with Conductive Oxides. Phys. Rev. Lett. **103**, 040402 (2009)

37. Azzam, R.M.A., Bashara, N.M.: Ellipsometry and Polarized Light. North Holland, Amsterdam (1987)

38. Tompkins, H.G., McGahan, W.A.: Spectroscopic Ellipsometry and Reflectometry. Wiley, New York (1999)

39. Hofmann, T. et al.: Terahertz Ellipsometry Using Electron-Beam Based Sources. Mater. Res. Soc. Symp. Proc. **1108**, A08–04 (2009)

40. Feng, R., Brion, C.E.: Absolute photoabsorption cross-sections (oscillator strengths) for ethanol (5–200 eV). Chem. Phys. **282**, 419–427 (2002)

41. Tompkins, H.G., Irene, E.A.: Handbook of ellipsometry William Andrew (2005)

42. van Zwol, P.J., Palasantzas, G., van de Schootbrugge, M., De Hosson, J.Th.M.: Measurement of dispersive forces between evaporated metal surfaces in the range below 100 nm. Appl. Phys. Lett. **92**, 054101 (2008)

43. http://www.JAWoollam.com

44. Thèye, M.-L.: Investigation of the Optical Properties of Au by Means of Thin Semitransparent Films. Phys. Rev. B **2**, 3060–3078 (1970)

45. Munday, J.N., Capasso, F., Parsegian, A.: Measured long-range repulsive CasimirLifshitz forces. Nature **457**, 170–173 (2009)

46. Dagastine, R.R., Prieve, D.C., White, L.R.: The Dielectric Function for Water and Its Application to van der Waals Forces. J. Colloid Interface Sci. **231**, 351–358 (2000)

47. Munday, J.N., Capasso, F., Parsegian, A., Bezrukov, S.M.: Measurements of the Casimir-Lifshitz force in fluids: The effect of electrostatic forces and Debye screening. Phys. Rev. A **78**, 032109 (2008)

48. Kitamura, R. et al.: Optical constants of silica glass from extreme ultraviolet to far infrared at near room temperature. Appl. Optics **46**, 8118 (2007)

49. Segelstein, D.J.: The complex refractive index of water. PhD thesis, University of Missouri, Kansas City, USA (1981)

50. Nguyen, A.V.: Improved approximation of water dielectric permittivity for calculation of hamaker constants. J. Colloid Interface Sci. **229**, 648–651 (2000)

51. Parsegian, V.A., Weiss, G.H.: Spectroscopic parameters for computation of van der Waals forces. J. Colloid Interface Sci. **81**, 285–289 (1981)

52. Roth, C.M., Lenhoff, A.M.: Improved parametric representation of water dielectric data for Lifshitz theory calculations. J. Colloid Interface Sci. **179**, 637–639 (1996)

53. van Oss, C.J., Chaudhury, M.K., Good, R.J.: Interfacial Lifshitz-van der Waals and polar interactions in macroscopic systems. Chem. Rev. **88**, 927–941 (1988)

54. Milling, A., Mulvaney, P., Larson, I.: Direct measurement of repulsive van der Waals interactions using an atomic force microscope. J. Colloid Interface Sci. **180**, 460465 (1996)

55. Ogawa, M., Cook, G.R.: Absorption Coefficients of Methyl, Ethyl, N-Propyl and N-Butyl Alcohols. J. Chem. Phys **28**, 747–748 (1957)

56. Salahub, D.R., Sandorfy, C.: The far-ultraviolet spectra of some simple alcohols and fluoroalcohols. Chem. Phys. Lett. **8**, 71–74 (1971)

57. Koizumi, H. et al.: VUV-optical oscillator strength distributions of C_2H_6O and C_3H_8O isomers. J. Chem. Phys. **85**, 4276–4279 (1986)
58. Sethna, P.P., Williams, D.: Optical constants of alcohols in the infrared. J. Phys. Chem. **83**, 405–409 (1979)
59. Bertie, J.E., Zhang, S.L., Keefe, C.D.: Measurement and use of absolute infrared absorption intensities of neat liquids. Vib. Spectro. **8**, 215–229 (1995)
60. van Zwol, P.J., Palasantzas, G., DeHosson, J.Th.M.: Weak dispersion forces between glass and gold macroscopic surfaces in alcohols. Phys. Rev. E. **79**, 041605 (2009)
61. Meakin, P.: The growth of rough surfaces and interfaces. Phys. Rep. **235**, 189–289 (1993)
62. Palasantzas, G., Krim, J.: Experimental Observation of Self-Affine Scaling and Kinetic Roughening at Sub-Micron Lengthscales. Int. J. Mod. Phys. B, **9**, 599–632 (1995)
63. Zhao, Y., Wang, G.-C., Lu, T.-M.: Characterization of Amorphous and Crystalline Rough Surfaces-principles and Applications. Academic Press, New York (2001)
64. Palasantzas, G.: Power spectrum and surface width of self-affine fractal surfaces via the k-correlation model. Phys. Rev. B **48**, 14472–14478 (1993)
65. Palasantzas, G.: Phys. Rev. B **49**, 5785 (1994)
66. Klimchitskaya, G.L., Pavlov, Yu.V.: The correction to the Casimir forces for configurations used in experiments: the spherical lens above the plane and two crossed cylinders. Int. J. Mod. Phys. A **11**, 3723–3742 (1996)
67. Rodriguez, A., Ibanescu, M., Iannuzzi, D., Capasso, F., Joannopoulos, J.D., Johnson, S.G.: Computation and Visualization of Casimir Forces in Arbitrary Geometries: Nonmonotonic Lateral-Wall Forces and the Failure of Proximity-Force Approximations. Phys. Rev. Lett. **99**, 080401 (2007)
68. van Zwol, P.J., Palasantzas, G., De Hosson, J.Th.M.: Influence of roughness on capillary forces between hydrophilic sur-faces. Phys. Rev. E **78**, 031606 (2008)
69. DelRio, F.W., Dunn, M.L., Phinney, L.M., Bourdon, C.J.: Rough surface adhesion in the presence of capillary condensation. Appl. Phys. Lett. **90**, 163104 (2007)
70. van Zwol, P.J., Palasantzas, G., De Hosson, J.Th.M.: Influence of random roughness on the adhesion between metal surfaces due to capillary condensation. Appl. Phys. Lett. **91**, 101905 (2007)
71. Israelachvili, J.: Intermolecular and Surface Forces. Academic, New York (1992)
72. Palasantzas, G., Svetovoy, V.B., van Zwol, P.J.: Influence of water adsorbed on gold on van der Waals/Casimir forces. Phys. Rev. B **79**, 235434 (2009)
73. van Zwol, P.J., Svetovoy, V.B., Palasantzas, G.: Distance upon contact: Determination from roughness profile. Phys. Rev. B **80**, 235401 (2009)
74. Greenwood, J.A., Williamson, J.B.P.: Contact of nominally flat surfaces. Proc. R. Soc. A **295**, 300–319 (1966)
75. Coles, S.: An Introduction to Statistical Modelling of Extreme Values. Springer, Berlin (2001)

Chapter 11
Fluctuation-Induced Forces Between Atoms and Surfaces: The Casimir–Polder Interaction

Francesco Intravaia, Carsten Henkel and Mauro Antezza

Abstract Electromagnetic fluctuation-induced forces between atoms and surfaces are generally known as Casimir–Polder interactions. The exact knowledge of these forces is rapidly becoming important in modern experimental set-ups and for technological applications. Recent theoretical and experimental investigations have shown that such an interaction is tunable in strength and sign, opening new perspectives to investigate aspects of quantum field theory and condensed-matter physics. In this chapter we review the theory of fluctuation-induced interactions between atoms and a surface, paying particular attention to the physical characterization of the system. We also survey some recent developments concerning the role of temperature, situations out of thermal equilibrium, and measurements involving ultra-cold atoms.

11.1 Introduction

In the last decade remarkable progress in trapping and manipulating atoms has opened a wide horizon to new and challenging experimental set-ups. Precision tests of both quantum mechanics and quantum electrodynamics have become

F. Intravaia (✉)
Theoretical Division, MS B213,
Los Alamos National Laboratory, Los Alamos, NM 87545, USA
e-mail: francesco_intravaia@lanl.gov

C. Henkel
Institut für Physik und Astronomie, Universität Potsdam, 14476, Potsdam, Germany
e-mail: carsten.henkel@physik.uni-potsdam.de

M. Antezza
Laboratoire Kastler Brossel, Ecole Normale Suprieure, CNRS and UPMC,
24 rue Lhomond, 75231, Paris, France
e-mail: mauro.antezza@lkb.ens.fr

D. Dalvit et al. (eds.), *Casimir Physics*, Lecture Notes in Physics 834,
DOI: 10.1007/978-3-642-20288-9_11, © Springer-Verlag Berlin Heidelberg 2011

possible through the capacity of addressing single trapped particles [1, 2] and of cooling ultracold gases down to Bose–Einstein condensation [3–5]. This stunning progress is also very profitable to other fundamental areas of physics and to technology. For example, ultracold gases have been suggested as probes in interesting experimental proposals aiming at very accurate tests of the gravity law [6–8], looking for extra forces predicted by different grand-unified theories [9] (see also the Chap. 3 by Milton in this volume for detailed discussions on the interplay between Casimir energy and gravity). Technologically speaking, one paradigmatic example of this new frontier is provided by atom chips [10, 11]. In these tiny devices, a cloud of atoms (typically alkalis like Sodium, Rubidium or Cesium) is magnetically or optically trapped above a patterned surface, reaching relatively short distances between a few microns to hundreds of microns [12–14]. The micro-machined surface patterns form a system of conducting wires, which are used to control the atomic cloud by shaping electromagnetic trapping fields (also super-conducting wires have been demonstrated [15–17]).

At a fundamental level, all these systems have in common to be strongly influenced by all kinds of atom–surface interactions. A particular category are fluctuation-induced forces, of which the most prominent representative is the van der Waals interaction [18]. These forces usually derive from a potential with a characteristic power-law dependence

$$\text{van der Waals limit:} \quad V = V_{\text{vdW}} \propto \frac{1}{L^n}, \tag{11.1}$$

where L is the distance between the objects (two atoms or a surface and a atom) and the exponent depends upon physical parameters and geometry of the system ($n = 3$ for an atom and a thick plate, see Fig. 11.1). Historically speaking, the existence of this kind of interaction was postulated long before it was experi-mentally possible to address single atoms [20]. The first quantum-mechanical theory was formulated by F. London in the 1930s using the idea that the quantum mechanical uncertainty of electrons in atoms can be translated into fluctuating electric dipole moments [21]. London found that two atoms attract each other following (11.1) with an exponent $n = 6$. London's theory was extensively applied in studying colloidal suspensions [22] which provided confirmations of its validity but also showed its limitations.

The next step was taken by H.B.G. Casimir and his student D. Polder [23] who applied the framework of quantum electrodynamics, including the concept of vacuum (field) fluctuations. They generalized the London–van der Waals formula by relaxing the electrostatic approximation, in other words, including the effect of retardation. The main success of Casimir–Polder theory was to provide an explanation for the change in the power law exponent observed in some experi-ments [22]. Indeed for distances larger than a characteristic length scale λ_0 of the system, the effect of retardation can no longer be neglected, and this leads to

$$\text{Casimir–Polder limit } L \gg \lambda_0: \quad V = V_{\text{CP}} \propto \frac{\lambda_0}{L^{n+1}}, \tag{11.2}$$

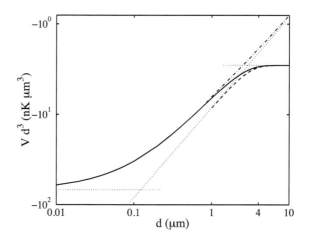

Fig. 11.1 Atom–surface potential [free energy of interaction $V(d)$] versus distance d between a ^{87}Rb atom and a SiO_2 (*sapphire*) substrate, multiplied by d^3. The potential is calculated using the theory of Dzyaloshinskii, Lifshitz, and Pitaevskii (see see chapter VIII of [19]). Note the logarithmic scale and the sign. The figure, adapted from Fig. 3 of [4], shows the potential at $T = 300$ K (*solid line*), at $T = 0$ K (*dash-dotted line*), and the three asymptotic behaviors (*dotted lines*): van der Waals–London $\propto -1/d^3$, Casimir–Polder ($\propto -1/d^4$), and Lifshitz ($\propto -T/d^3$)

and therefore to the L^{-7} dependence typical of the Casimir–Polder interaction between two atoms. The scale λ_0 is in this case the wavelength of the main atomic absorption lines, which is in the visible to near infrared for typical alkali atoms (of the order of a few hundreds of nanometers).

The estimates (11.1, 11.2) apply at $T = 0$ when only quantum fluctuations play a role. If the temperature is nonzero, another length scale comes into play, the thermal or Wien wavelength

$$\lambda_T = \frac{\hbar c}{k_B T}, \tag{11.3}$$

which roughly corresponds to the wave length where the thermal radiation spectrum peaks. Calculations of the atom–surface interaction using thermal quantum field theory have been pioneered by Dzyaloshinskii, Lifshitz, and Pitaevskii [19, 24]. They were able to recover the van der Waals and the Casimir–Polder potentials as limit behavior of a more general expression and to confirm, quite surprisingly, that at distances $L \ll \lambda_T$ the interactions are typically dominated by quantum fluctuations, the main reason being that their spectrum is much wider than that of the thermal field (which is constrained by the Bose–Einstein distribution [25]). At distances $L \gg \lambda_T$ they show that the potential shows again a crossover from the Casimir–Polder to the Lifshitz asymptote:

$$\text{Lifshitz limit } L \gg \lambda_T : \quad V = V_L \propto \frac{\lambda_0}{\lambda_T L^n} \sim \frac{k_B T}{\hbar \omega_0} V_{vdW}, \tag{11.4}$$

where ω_0 is the (angular) frequency corresponding to λ_0. This potential that scales with temperature is actually a free energy of interaction and is also known as the Keesom potential between polar molecules: there, the dipoles are rotating freely under the influence of thermal fluctuations [26]. In this case (Rydberg atoms provide another example), the particle resonances overlap with the thermal spectrum, and the Casimir–Polder regime is actually absent. Equation (11.4) predicts an apparent enhancement, at nonzero temperature, of the fluctuation-induced interaction, relative to the Casimir-Polder limit.. This does not necessarily happen, however, because the molecular polarizabilities are also temperature-dependent [26–28].

In Fig.11.1 above, we considered the case of an alkali atom whose peak absorption wavelength λ_0 is much shorter than the Wien wavelength. The Lifshitz tail is then much smaller than the van der Waals potential. Note that λ_T is of the order of a few micrometers at room temperature, comparable to the smallest atom–surface distances achieved so far in atom chips. The crossover between the Casimir–Polder and the Lifshitz regimes can thus be explored in these set-ups. We discuss corresponding experiments in Sect. 11.5.

In the following sections, we start with a derivation of the interaction between an atom and a general electromagnetic environment (Sect. 11.2). We will refer to it using the term "atom–surface interaction" or "Casimir–Polder interaction". This term is also of common use in the literature to stress the fluctuation-induced nature of the interaction, although the term "Casimir–Polder" more correctly indicates the potential in the retarded limit (see (11.2)). The result will be valid within a second-order perturbation theory and can be easily adapted to specific geometries. We provide some details on a planar surface (Sect. 11.3). Situations out of global thermal equilibrium are discussed in Sect. 11.4, dealing with forces on ultracold atoms in a general radiation environment (the temperature of the surface and that of the surrounding environment are not necessarily the same), and with radiative friction. The final Sect. 11.5 sketches experiments with atomic beams and ultra-cold samples.

11.2 Understanding Atom–Surface Interactions

The interaction between atoms and surfaces plays a fundamental role in many fields of physics, chemistry and technology (see also the Chap. 12 of DeKieviet et al. in this volume for detailed discussions on modern experiments on atom–surface Casimir physics). From a quantum-mechanical point of view, fluctuation-induced forces are not surprising and almost a natural consequence of the initial assumptions. Indeed the existence of fluctuations, even at zero temperature, is one of the most remarkable predictions of the theory. Each observable corresponding to a physically measurable quantity can be zero on average but its variance will always be nonzero if the system is not in one of its eigenstates. When two quantum systems interact, the dynamics of the fluctuations becomes richer: each subsystem

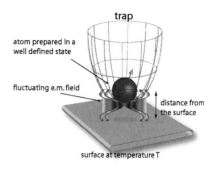

Fig. 11.2 A schematic representation of the atom in the trap near a surface

trap

atom prepared in a well defined state

fluctuating e.m. field

distance from the surface

surface at temperature T

experiences, in addition to its own fluctuations, an external, fluctuating force. This becomes particularly clear in the case of a polarizable particle (atom or nanosphere) interacting with the vacuum electromagnetic field. In vacuum, the electromagnetic field fluctuates not only by itself, but also because there are fluctuating sources for it, like the electric dipole moment of the particle. At the same time, the particle's dipole is not only fluctuating on its own, but is also responding to the fluctuations of the electromagnetic field [29, 30]. As a result, when two atoms are brought nearby, they interact through their fluctuations mediated by the electromagnetic field. Similarly when a particle is in proximity of a macroscopic object, electric currents fluctuating inside the object and the fluctuations of the particle lead to a distance-dependent force. This second case is complicated by the fact of dealing with a macroscopic object and its quantum-mechanical description. However, if the medium responds linearly to an electromagnetic perturbation, the fluctuation-dissipation theorem [31] provides a connection between the field's autocorrelation function and its macroscopic response (or Green function) (Fig. 11.2).

We will use the previous considerations as a starting point for the derivation of the Casimir–Polder interaction between a surface and an atom or also a nanoparticle. We will follow Refs. [27, 32–34]. Although this is not the unique approach [23, 24, 35–42], it provides a physically transparent way to reach our final result.

11.2.1 Energy of a Polarizable Particle in an Electromagnetic Field

When a polarizable particle is introduced in an electric field, the change in energy of the system is given by [43]

$$\mathscr{F} = -\frac{\langle \mathbf{d}(t) \cdot \mathbf{E}(\mathbf{r}_0, t) \rangle}{2}, \tag{11.5}$$

where, since we are working in the Heisenberg representation, all the operators are time dependent. From the thermodynamic point of view the previous quantity is a *free energy* and gives the amount of work that can be extracted from the system by moving the particle: in our (thermodynamic) convention a negative free energy will correspond to an attractive interaction (binding energy).

The expectation value $\langle \cdots \rangle$ is taken over the (initial) state of the non-coupled system; $\mathbf{d}$ is the (electric or magnetic) dipole operator and $\mathbf{E}$ the corresponding (electric or magnetic) field operator, evaluated at the dipole position $\mathbf{r}_0$. We are implicitly assuming that the size of the particle is small enough to locally probe the electromagnetic field (dipole approximation). The factor $1/2$ in (11.5) arises from the fact that we are considering the energy of a linearly polarizable particle in an external field, rather than a permanent dipole [43]. Note that the choice of a particular ordering does not seem to be necessary at this stage since the dipole operator and the electric field operator commute. The symmetric order proves, however, to be particularly useful if we want to attach a physical meaning to each single contribution to the energy [29, 30], see (11.15).

The Hamiltonian of the coupled system can be in general written as $H = H_0 + V$, where H_0 is the sum of the Hamiltonians of the two isolated subsystems and V describes the interaction between them. Starting from this, the equation of motion for an operator $A(t)$ can be written in the following integral form (Heisenberg picture)

$$A(t) = A^{\text{free}}(t) + \frac{i}{\hbar} \int_0^t d\tau \, e^{\frac{i}{\hbar}H_0(t-\tau)}[V, A(\tau)] \, e^{\frac{i}{\hbar}H_0(t-\tau)}, \qquad (11.6)$$

where $t = 0$ was chosen as initial time and the superscript free indicates that the operator evolves with respect to the Hamiltonian of the uncoupled system (H_0), i.e.

$$A^{\text{free}}(t) = e^{\frac{i}{\hbar}H_0 t} A e^{-\frac{i}{\hbar}H_0 t}. \qquad (11.7)$$

Now, within first-order perturbation theory, (11.6) can be solved by replacing the operator $A(\tau)$ under the integral by its corresponding free evolution. If the interaction Hamiltonian is bilinear like in the case of the electric dipole interaction, $V = -\mathbf{d} \cdot \mathbf{E}(\mathbf{r}_0)$, we get the following approximate expression for the dipole operator:

$$\mathbf{d}(t) \approx \mathbf{d}^{\text{free}}(t) + \mathbf{d}^{\text{ind}}(t), \qquad (11.8a)$$

$$\mathbf{d}^{\text{ind}}(t) = \int_{-\infty}^{\infty} d\tau \left(\frac{i}{\hbar} [\mathbf{d}^{\text{free}}(t), \mathbf{d}^{\text{free}}(t-\tau)]\theta(\tau) \right) \cdot \mathbf{E}^{\text{free}}(\mathbf{r}_0, t-\tau), \qquad (11.8b)$$

and similarly for the field operator:

$$\mathbf{E}(\mathbf{r}, t) \approx \mathbf{E}^{\text{free}}(\mathbf{r}, t) + \mathbf{E}^{\text{ind}}(\mathbf{r}_0, t), \qquad (11.9a)$$

$$\mathbf{E}^{\text{ind}}(\mathbf{r}_0, t) = \int_{-\infty}^{\infty} d\tau \left(\frac{i}{\hbar} [\mathbf{E}^{\text{free}}(\mathbf{r}, t), \mathbf{E}^{\text{free}}(\mathbf{r}_0, t-\tau)]\theta(\tau) \right) \cdot \mathbf{d}^{\text{free}}(t-\tau). \qquad (11.9b)$$

The τ-integral runs effectively over $\tau \geq 0$ (note the step function $\theta(\tau)$) because in (11.6), only times $\tau > 0$ after the initial time are relevant (causality). In addition, we have set the upper limit to $\tau = \infty$ assuming that there exists a transient time τ_c after which the system behaviour becomes stationary. This time can be estimated from the system operators in (11.8b, 11.9b): the commutators are either c-number functions that die out for time arguments that differ by more than τ_c, or taking the expectation value, one gets subsystem correlation functions with τ_c as correlation time.

We have therefore that, within the first order perturbation theory, the dipole in addition to its unperturbed evolution ($\mathbf{d}^{\text{free}}(t)$) "responds" linearly ($\mathbf{d}^{\text{ind}}(t)$) to an external perturbation (in this case the electromagnetic field). The same also happens to the electromagnetic field where now the dipole is the external source of perturbation. The term in parenthesis under the integrals (11.8b, 11.9b), when evaluated over a particular state, is called *susceptibility* and contains the detailed physical information about the linear response of the system to the perturbation [29, 30]. In the particular case of a dipole the Fourier transform of the susceptibility tensor is the polarizability

$$\int_{-\infty}^{\infty} dt \frac{i}{\hbar} \langle a | [\mathbf{d}^{\text{free}}(t), \mathbf{d}^{\text{free}}(0)] | a \rangle \theta(t) e^{i\omega t} = \overset{\leftrightarrow}{\alpha}^a(\omega), \tag{11.10}$$

where we have taken the expectation value for a given quantum state $|a\rangle$. In the time domain (see (11.8b)), the atomic susceptibility links the hermitean dipole operator to a hermitean field operator; hence it must be a real function. The polarizability, being its Fourier transform, therefore satisfies

$$[\overset{\leftrightarrow}{\alpha}^a(\omega)]^* = \overset{\leftrightarrow}{\alpha}^a(-\omega^*). \tag{11.11}$$

In addition, because of causality, (11.10) implies that $\overset{\leftrightarrow}{\alpha}^a(\omega)$ must be analytical in the upper-half of the complex ω-plane.

Similar conclusions hold for the electromagnetic field. If we assume that the dynamics of the field and the surrounding matter (other than the atom) can be completely described in terms of bosonic operators [39, 44, 45], the result of the commutator in (11.9b) is a c-number and the susceptibility does not depend on the state of the radiation. The linearity of the Maxwell equations then ensures that the result of (11.9a,b) for a point-like dipole is correct to all orders. A simple identification leads to the following expression:

$$\int_{-\infty}^{\infty} \frac{i}{\hbar} [\mathbf{E}^{\text{free}}(\mathbf{r}, t), \mathbf{E}^{\text{free}}(\mathbf{r}_0, 0)] \theta(t) e^{i\omega t} dt = \overset{\leftrightarrow}{G}(\mathbf{r}, \mathbf{r}_0, \omega), \tag{11.12}$$

where $\overset{\leftrightarrow}{G}$ is the electric field Green tensor, solution to the macroscopic Maxwell equation

$$-\nabla_{\mathbf{r}} \times \nabla_{\mathbf{r}} \times \overset{\leftrightarrow}{G}(\mathbf{r}, \mathbf{r}_0, \omega) + \frac{\omega^2}{c^2} \varepsilon(\mathbf{r}, \omega) \overset{\leftrightarrow}{G}(\mathbf{r}, \mathbf{r}_0, \omega) = -\frac{\omega^2}{\varepsilon_0 c^2} \delta(\mathbf{r} - \mathbf{r}_0) \overset{\leftrightarrow}{I}, \tag{11.13}$$

where $\overleftrightarrow{I}$ is the identity tensor and $\varepsilon(\mathbf{r}, \omega)$ is the local dielectric function (here supposed to be a scalar for simplicity) of the matter surrounding the dipole. In conclusion, in frequency space the induced quantities can be described in terms of the retarded response functions [43]

$$
\begin{aligned}
\mathbf{d}^{\text{ind}}(\omega) &= \overleftrightarrow{\alpha}(\omega) \cdot \mathbf{E}^{\text{free}}(\mathbf{r}_0, \omega), \\
\mathbf{E}^{\text{ind}}(\mathbf{r}, \omega) &= \overleftrightarrow{G}(\mathbf{r}, \mathbf{r}_0, \omega) \cdot \mathbf{d}^{\text{free}}(\omega),
\end{aligned}
\tag{11.14}
$$

where the frequency dependence and causality allow for a temporal delay. This is slightly schematic because the polarizability tensor is defined only when the average is taken.

Expressions (11.8a,b) and (11.9a,b) formalize the considerations made at the beginning of this section: both the dipole moment $\mathbf{d}(t)$ and the field $\mathbf{E}(t)$ can be split into two parts, the (free) fluctuating part describes the free intrinsic fluctuation, while the induced part arises in perturbation theory from the dipole coupling [25]. Equation (11.5) becomes

$$
\mathscr{F} = -\frac{\langle \mathbf{d}^{\text{ind}}(t); \mathbf{E}^{\text{free}}(\mathbf{r}_0, t) \rangle}{2} - \frac{\langle \mathbf{E}^{\text{ind}}(\mathbf{r}_0, t); \mathbf{d}^{\text{free}}(t) \rangle}{2}.
\tag{11.15}
$$

We assumed a factorized initial state in which each free evolution operator is zero on average and where the correlations between the fluctuating parts are entirely encoded in the linear response functions, setting the correlation between the freely fluctuating components to zero. This assumption would break down at higher orders of perturbation theory. Note that while in (11.5) the total dipole and field operators (Heisenberg picture) commute at equal times, this is no longer true for their 'ind' and 'free' constituents in (11.15). The choice of the symmetric order (indicated by the semicolon) allows one to see each term of the previous expression as the result of the quantum expectation value of a Hermitian operator and therefore to attach to it a physical meaning [29, 30, 32]. The first term on the right hand side of (11.15) can be seen as the contribution to the Casimir–Polder energy coming from the fluctuations of the vacuum field; the second will be called the self-reaction term since it arises from the interaction of the dipole with the field generated by the dipole itself.

11.2.2 Equilibrium Fluctuations

Consider now a configuration at global thermal equilibrium, i.e. when both the dipole and the field are in a thermal state at temperature T. In this case we can apply the *fluctuation-dissipation theorem* [31]. This milestone of the linear response theory connects the correlation of a generic observable of a system in thermal equilibrium at temperature T with the imaginary part of the linear susceptibility which characterizes the response to a weak perturbation. In our case the theorem holds separately for the dipole and the field and we have [46–48]

$$\langle E_i^{\text{free}}(\mathbf{r}, \omega) E_j^{\text{free}}(\mathbf{r}, \omega') \rangle_T = 2\pi\hbar\delta(\omega + \omega') \coth\left(\frac{\hbar\omega}{2k_BT}\right) \text{Im}\left[G_{ij}(\mathbf{r}, \mathbf{r}, \omega)\right], \quad (11.16)$$

$$\langle d_i^{\text{free}}(\omega) d_j^{\text{free}}(\omega') \rangle_T = 2\pi\hbar\delta(\omega + \omega') \coth\left(\frac{\hbar\omega}{2k_BT}\right) \text{Im}\left[\alpha_{ij}^T(\omega)\right], \quad (11.17)$$

where the symbol $\langle \cdots \rangle_T$ define the quantum and the thermal average and, $\overset{\leftrightarrow}{\alpha}^T(\omega)$ defines the atomic polarizability operator evaluated at temperature T, according to (11.10) and (11.25).

$$\langle a^\dagger a + aa^\dagger \rangle_T = 1 + 2N(\omega) = \coth\left(\frac{\hbar\omega}{2k_BT}\right), \quad (11.18)$$

where $N(\omega)$ is the Bose–Einstein distribution. Note the asymptotic limits $(\omega > 0)$

$$\coth\left(\frac{\hbar\omega}{2k_BT}\right) \rightarrow \begin{cases} 1 & T \ll \hbar\omega/k_B \\ \frac{2k_BT}{\hbar\omega} & T \gg \hbar\omega/k_B \end{cases} \quad (11.19)$$

in the "quantum" (low-temperature) and "classical" (high-temperature) limits.

The expression given in (11.16) can be directly reconnected with the currents fluctuating inside the media surrounding the dipole. For these currents Rytov's theory [49] predicts a correlation similar to (11.17) where the role of the polarizability is now played by the dielectric function [46–48] (see Sect. 11.4). This picture also lends itself to a natural generalization where the bodies are assumed to be in "local thermal equilibrium" (see Sect. 11.4.4).

Note that the field correlations are needed at the same position $\mathbf{r}_0$. The Green function is, however, divergent in this limit due to its free-space contribution

$$\overset{\leftrightarrow}{G}(\mathbf{r}, \mathbf{r}_0, \omega) = \underbrace{\overset{\leftrightarrow}{G}_0(\mathbf{r}, \mathbf{r}_0, \omega)}_{\text{freespace}} + \underbrace{\overset{\leftrightarrow}{\mathscr{G}}(\mathbf{r}, \mathbf{r}_0, \omega)}_{\text{scattered}}. \quad (11.20)$$

The corresponding part of the free energy provides the Lamb shift of the internal levels of the dipole immersed in the electromagnetic field [37]. This contribution is position-independent and does not contain any information about the interaction between the bodies and the dipole. Therefore it can be safely "hidden" in the (renormalized) energy levels of the atom. The physical information about the interaction is indeed contained only in the *scattered* part of the Green function [36]. If the body happens to be a plane surface, it follows from symmetry that the result can only depend on the dipole-surface distance L and we can set $\overset{\leftrightarrow}{\mathscr{G}}(\mathbf{r}_0; \mathbf{r}_0, \omega) \equiv \overset{\leftrightarrow}{\mathscr{G}}(L, \omega)$.

Combining (11.15–11.17), we finally obtain that the free energy of a polarizable particle at nonzero temperature T has the following general form (Einstein summation convention)

$$\mathscr{F}(L,T) = -\frac{\hbar}{2\pi}\int\limits_{0}^{\infty}\mathrm{d}\omega\,\mathrm{coth}\!\left(\frac{\hbar\omega}{2k_BT}\right)\mathrm{Im}\left[\alpha_{ij}^T(\omega)\mathscr{G}_{ji}(L,\omega)\right]. \tag{11.21}$$

We have used the reality condition (11.11), implying that the imaginary part of both polarizability and Green tensors are odd in ω. Equation (11.21) coincides with the expression of the atom–surface interaction derived by many authors [23, 24, 27, 32–42]. It is often expressed in an equivalent form using the analyticity of $\overleftrightarrow{\alpha}^T(\omega)$ and $\overleftrightarrow{\mathscr{G}}(L,\omega)$ in the upper half of the complex frequency plane. Performing a (Wick) rotation to the imaginary frequency axis yields the so-called Matsubara expansion [24, 50]

$$\mathscr{F}(L,T) = -k_BT\sum_{n=0}^{\infty}{}'\alpha_{ij}^T(\mathrm{i}\xi_n)\mathscr{G}_{ji}(L,\mathrm{i}\xi_n), \tag{11.22}$$

where the Matsubara frequencies $\xi_n = 2\pi n k_B T/\hbar$ arise from the poles of the hyperbolic cotangent, and the prime on the sum indicates that the $n=0$ term comes with a coefficient 1/2. Both $\overleftrightarrow{\alpha}^T(\mathrm{i}\xi)$ and $\overleftrightarrow{\mathscr{G}}(L,\mathrm{i}\xi)$ are real expressions for $\xi > 0$ because of (11.10).

These considerations conclude our general analysis of the Casimir–Polder interaction. In the following section we will analyze the particle response function appearing in the previous formulation, namely the atomic polarizability, and mention also the case of a nanoparticle.

11.2.3 Polarizability Tensor

The previous results can be used for the interaction of a surface with atoms, molecules, particles or in general any (small) object that can be described with good approximation in terms of a electric-dipole polarizability tensor. Here we are going to review the polarizability of an atom and of a nanoparticle.

11.2.3.1 Atoms

The polarizability tensor is determined by the transition dipole matrix elements and the resonance frequencies. For an arbitrary atomic state $|a\rangle$ it can be written as

$$\alpha_{ij}^a(\omega) = \sum_b \frac{d_i^{ab}d_j^{ba}}{\hbar}\frac{2\omega_{ba}}{\omega_{ba}^2 - (\omega + \mathrm{i}0^+)^2}, \tag{11.23}$$

where d_i^{ba} is the matrix element between the states $|b\rangle$ and $|a\rangle$ of the i-th component of the electric dipole operator and $\omega_{ba} = (E_b - E_a)/\hbar$ the corresponding transition frequency. The introduction of an infinitesimal imaginary part shifts the poles of the expression into the lower part of the complex frequency plane $(\pm\omega_{ba} - i0^+)$, which is mathematically equivalent to the causality requirement. The tensorial form of the previous expression allows to take into account a possible anisotropic response of the atom to an electric field. A simplification can be obtained averaging over the different levels and directions so that the polarizability tensor becomes $\alpha_{ij}^a = \delta_{ij}\alpha_{iso}^a$ with the scalar function

$$\alpha_{iso}^a(\omega) = \sum_b \frac{|\mathbf{d}^{ba}|^2}{3\hbar} \frac{2\omega_{ba}}{\omega_{ba}^2 - (\omega + i0^+)^2}. \tag{11.24}$$

The polarizability is exactly isotropic when several excited sublevels that are degenerate in energy are summed over, like the $np_{x,y,z}$ orbitals of the hydrogen-like series. When the atom is in thermal equilibrium, we have to sum the polarizability over the states $|a\rangle$ with a Boltzmann weight:

$$\alpha_{ij}^T(\omega) = \sum_a \frac{e^{-E_a/k_BT}}{Z}\alpha_{ij}^a(\omega), \tag{11.25}$$

where Z is the partition function. In the limit $T \to 0$, we recover the polarizability for a ground state atom. For a single pair of levels $|a\rangle$ and $|b\rangle$, this leads to the following relation between the state-specific and the thermalized polarizabities:

$$\alpha_{ij}^T(\omega) \approx \alpha_{ij}^a(\omega)\tanh\frac{\hbar\omega_{ba}}{2k_BT}. \tag{11.26}$$

This is mainly meant to illustrate the temperature dependence, otherwise it is a quite crude approximation. The reason is that the coupling to other levels makes the polarizabilities α_{ij}^a and α_{ij}^b differ quite a lot. Electronically excited states are much more polarizable due to their larger electron orbitals.

11.2.3.2 Nanospheres

Let us consider now the case where the atom is replaced by a nanosphere [34, 51, 52]. When the sphere radius R is smaller than the penetration depth and the radiation wavelength, we can neglect higher order multipoles in the Mie expansion [53] and consider only the electric and magnetic dipole (the sphere is globally neutral).

In this long-wavelength limit, the Clausius–Mossotti relation [43, 54] provides the electric polarizability

$$\alpha_{sph}(\omega) = 4\pi\varepsilon_0 R^3 \frac{\varepsilon(\omega) - 1}{\varepsilon(\omega) + 2}, \tag{11.27}$$

where $\varepsilon(\omega)$ is the (scalar) dielectric function of the sphere's material. The nanosphere has also a magnetic polarizability that arises because a time-dependent magnetic field induces circulating currents (Foucault currents) [43]. This leads to a diamagnetic response [55]

$$\beta_{\text{sph}}(\omega) = \frac{2\pi}{15\mu_0} \left(\frac{R\omega}{c}\right)^2 [\varepsilon(\omega) - 1]R^3.$$
(11.28)

Both polarizabilities are isotropic (scalars). For a metallic sphere, the electric polarizability goes to a positive constant at zero frequency, while the magnetic one vanishes there and has a negative real part at low frequencies (diamagnetism).

For a qualitative comparison to an atom, one can estimate the oscillator strength [25], defined by the integral over the imaginary part of the polarizability. For the atom we have

$$\int_0^\infty d\omega \, \text{Im} \, \alpha_{\text{at}}(\omega) \sim \frac{\pi(ea_0)^2}{\hbar} \quad \text{and} \quad \int_0^\infty d\omega \, \text{Im} \, \beta_{\text{at}}(\omega) \sim \frac{\pi\mu_B^2}{\hbar},$$
(11.29)

where the Bohr radius a_0 and the Bohr magneton μ_B give the overall scaling of the transition dipole moments. The following dimensionless ratio allows a comparison between the two:

$$\frac{(ea_0)^2/\varepsilon_0}{\mu_B^2\mu_0} \sim \frac{1}{\alpha_{\text{fs}}^2},$$
(11.30)

where $\alpha_{\text{fs}} = e^2/(4\pi\varepsilon_0\hbar c) \approx 1/137$ is the fine structure constant. The electric oscillator strength clearly dominates in the atom.

Let us compare to a metallic nanosphere (gold is often used in experiments) and assume a Drude model (11.51) for the dielectric function. In terms of the volume $V = 4\pi R^3/3$, we get an electric oscillator strength

$$\int_0^\infty \text{Im} \, \alpha_{\text{sph}}(\omega)d\omega = \frac{3}{2}\varepsilon_0 \frac{\omega_p}{\sqrt{3}}V + \mathcal{O}\left(\frac{\gamma}{\omega_p}\right).$$
(11.31)

where $\omega_p/\sqrt{3}$ is the resonance frequency of the particle plasmon mode (the pole of $\alpha_{\text{sph}}(\omega)$, (11.27)). This is much larger than for an atom if the nanoparticle radius satisfies $a_0 \ll R \ll \lambda_p$, i.e., a few nanometers. The magnetic oscillator strength can be estimated as

$$\int_0^{\omega_p} \text{Im} \, \beta_{\text{sph}}(\omega)d\omega = \frac{2\pi}{3\mu_0}\gamma \log\left(\frac{\omega_p}{\gamma}\right)\left(\frac{R}{\lambda_p}\right)^2 V.$$
(11.32)

where we took ω_p as a cutoff frequency to make the integral convergent (at higher frequencies, (11.28) does not apply any more). We have used the plasma wavelength $\lambda_p = 2\pi c/\omega_p(\sim 100\,\text{nm}$ for gold). Similar to an atom, the nanoparticle response is dominantly electric, but the ratio of oscillator strengths can be tuned via the material parameters and the sphere size. The magnetic contribution to the particle–surface interaction is interesting because it features a quite different temperature dependence, see Ref. [27].

11.2.4 Non-perturbative Level Shift

In the previous section we saw that the main ingredient to derive the Casimir–Polder interaction between a particle and an object is the ability to solve for the dynamics of the joint system particle+electromagnetic field. Previously we limited ourself to a solution at the first order in the perturbation, implicitly motivated by the difficulty to solve *exactly* the dynamics of a multi-level atomic system coupled to a continuum of bosonic degrees of freedom (e.m. field). Things are different if we consider the linear coupling between two bosonic systems, i.e. if we describe the particle as a quantum harmonic oscillator. The linearity of the coupled system allows for an exact solution of its dynamics and even if the harmonic oscillator may be in some cases only a poor description of an atom [56], it is a good representation of a nanoparticle (the resonance frequency being the particle plasmon frequency). Generally, this approach gives a first qualitative indication for the physics of the interaction [57–61].

The main idea we follow in this section is based upon a generalization of the "remarkable formula" of Ford, Lewis and O'Connell [62, 63] (see also [57–61]). According to this formula, the free energy of a one-dimensional oscillator immersed in black body radiation is

$$\mathscr{F}_{\mathrm{FLOC}}(T) = \frac{1}{\pi} \int_0^\infty d\omega \, f(\omega, T) \mathrm{Im}[\partial_\omega \ln \alpha(\omega)], \qquad (11.33)$$

where $f(\omega, T)$ is the free energy per mode,

$$f(\omega, T) = k_B T \log\left[2 \sinh\left(\frac{\hbar\omega}{2k_B T}\right)\right], \qquad (11.34)$$

and $\alpha(\omega)$ is the (generalized) susceptibility of the oscillator derived from (11.39) below. More precisely, $\mathscr{F}_{\mathrm{FLOC}}(T)$ gives the difference between two free energies: the oscillator coupled to the radiation field and in equlibrium with it, on the one hand, and solely the radiation field, on the other. Equation (11.33) is "remarkable" because the only system-relevant information needed here is the susceptibility function.

In three dimensions, the polarizability becomes a tensor

$$\mathbf{d}(\omega) = \overleftrightarrow{\alpha}(\omega) \cdot \mathbf{E}(\omega), \qquad (11.35)$$

where $\mathbf{E}(\omega)$ is the external electric field. In the case considered by Ford, Lewis, and O'Connell, there was no need to include a spatial dependence because of the homogeneity and isotropy of the black body field. We are going to consider this symmetry to be broken by the presence of some scattering object. As a consequence, the generalized susceptibility tensor becomes position-dependent and anisotropic: $\overleftrightarrow{\alpha}(\omega, \mathbf{r}_0)$. The spatial dependence is connected with the scattered part of the Green function and leads both to a position-dependent frequency renormalization and a damping rate.

In order to get the expression of $\overleftrightarrow{\alpha}(\omega, \mathbf{r}_0)$, let us consider for simplicity the equation of motion of an isotropic oscillator with charge q interacting with the e.m. field near some scattering body (that is described by a dielectric constant). In frequency space, the (nonrelativistic) dynamics of the oscillator is described by

$$m\left[-\omega^2 \mathbf{d}(\omega) + \omega_0^2 \mathbf{d}(\omega)\right] = q^2 \mathbf{E}(\mathbf{r}_0, \omega), \tag{11.36}$$

where we have neglected the coupling with the magnetic field (first order in $\dot{d}/c$). For the field we have

$$\nabla \times \nabla \times \mathbf{E}(\mathbf{r}, \omega) - \frac{\omega^2}{c^2}\varepsilon(\omega, \mathbf{r})\mathbf{E}(\mathbf{r}, \omega) = i\omega\mu_0 \mathbf{j}(\mathbf{r}, \omega), \tag{11.37}$$

where the source current is $\mathbf{j}(\mathbf{r}, \omega) = -i\omega\mathbf{d}(\omega)\delta(\mathbf{r} - \mathbf{r}_0)$.

The formal exact solution for the operator $\mathbf{E}$ can be given in term of the (electric) Green tensor:

$$\mathbf{E}(\mathbf{r}, \omega) = \mathbf{E}^{\text{free}}(\mathbf{r}, \omega) + \overleftrightarrow{G}(\mathbf{r}, \mathbf{r}_0, \omega) \cdot \mathbf{d}(\omega), \tag{11.38}$$

where the Green tensor is the solution of (11.13) given above. The field $\mathbf{E}^{\text{free}}(\mathbf{r}, \omega)$ is the electromagnetic field we would have without the oscillator and it is connected with the intrinsic fluctuations of the polarization field, or equivalently, of the currents in the body. Physically (11.38) states that the total electromagnetic field is given by the field present near the scattering object plus the field generated by the dipole. Introducing (11.38) in (11.36) we get

$$-m(\omega^2 - \omega_0^2)\mathbf{d}(\omega) - q^2\overleftrightarrow{G}(\mathbf{r}_0; \mathbf{r}_0, \omega) \cdot \mathbf{d}(\omega) = q^2 \mathbf{E}^{\text{free}}(\mathbf{r}_0, \omega). \tag{11.39}$$

The Green function $\overleftrightarrow{G}(\mathbf{r}; \mathbf{r}_0, \omega)$ solves Eq. (11.37) which describes an electromagnetic scattering problem and therefore, it decomposes naturally into a free-space field $\overleftrightarrow{G}_0$ (as if the source dipole were isolated in vacuum), and the field scattered by the body, $\overleftrightarrow{\mathscr{G}}$. This is at the basis of the splitting in (11.20) discussed above. The free-space part $\overleftrightarrow{G}_0(\mathbf{r}; \mathbf{r}_0, \omega)$ is a scalar in the coincidence limit because of the isotropy of space: schematically, part of the divergence $(\text{Re}[G_0])$ can be reabsorbed into mass renormalization, $m\omega_0^2 \mapsto m\tilde{\omega}_0^2$, and part $(\text{Im}[G_0])$ gives rise to dissipation (damping rate $\gamma(\omega)$[62,63]). Therefore (11.39) can be rewritten as

$$\left(-\omega^2 - i\gamma(\omega)\omega + \tilde{\omega}_0^2 - \frac{q^2}{m}\overleftrightarrow{\mathcal{G}}(\mathbf{r}_0;\mathbf{r}_0,\omega)\right)\cdot \mathbf{d}(\omega) = \frac{q^2}{m}\mathbf{E}^{\text{free}}(\mathbf{r}_0,\omega). \quad (11.40)$$

The free electromagnetic field plays here the role of an external force and therefore the generalized (or "dressed") polarizability tensor is given by

$$\overleftrightarrow{\alpha}(\omega,\mathbf{r}_0) = \alpha_v(\omega)\left(1 - \alpha_v(\omega)\overleftrightarrow{\mathcal{G}}(\mathbf{r}_0;\mathbf{r}_0,\omega)\right)^{-1}, \quad (11.41)$$

where we have defined

$$\alpha_v(\omega) = \frac{q^2}{m}\left(-\omega^2 - i\gamma(\omega) + \tilde{\omega}_0^2\right)^{-1}. \quad (11.42)$$

If Ford, Lewis and O'Connell's result is generalized to a three-dimensional oscillator, a trace operation appears before the logarithm in (11.33). Using the identity $\text{tr}\log\overleftrightarrow{a} = \log\det\overleftrightarrow{a}$, one gets

$$\begin{aligned}
\mathscr{F}_{\text{FLOC}}(T)\mathbf{r}_0 : \mathscr{F}_{\text{FLOC}}(\mathbf{r}_0,T) &= \frac{1}{\pi}\int_0^\infty d\omega\, f(\omega,T)\text{Im}\left[\partial_\omega \ln\det\overleftrightarrow{\alpha}(\omega,\mathbf{r}_0)\right] \\
&= \frac{1}{\pi}\int_0^\infty d\omega\, f(\omega,T)\text{Im}[\partial_\omega \ln\alpha_v(\omega)] \\
&\quad - \frac{1}{\pi}\int_0^\infty d\omega\, f(\omega,T)\text{Im}\left[\partial_\omega \ln\det\left(1 - \alpha_v(\omega)\overleftrightarrow{\mathcal{G}}(\mathbf{r}_0;\mathbf{r}_0,\omega)\right)\right].
\end{aligned}$$

$$(11.43)$$

The first term is distance-independent and coincides with the free energy of an isolated oscillator in the electromagnetic vacuum. It can be interpreted as a free-space Lamb shift. The second part of (11.43) is distance-dependent and therefore gives rise to the Casimir–Polder interaction. In the case of a surface, with the help of a partial integration, we finally get

$$\mathscr{F}(L,T) = \frac{\hbar}{2\pi}\int_0^\infty d\omega \coth\left(\frac{\hbar\omega}{2k_BT}\right)\text{Im}\left[\ln\det\left(1 - \alpha_v(\omega)\overleftrightarrow{\mathcal{G}}(L,\omega)\right)\right]. \quad (11.44)$$

The previous result can be easily generalized to the case of an anisotropic oscillator by just replacing the vacuum polarizability with the respective tensor.

The usual expression (11.21) for the Casimir Polder free energy is recovered by assuming a weak atom-field interaction. Expanding the logarithm to first order we get

$$\mathscr{F}(L,T) = -\frac{\hbar}{2\pi}\text{Tr}\int_0^\infty d\omega \coth\left(\frac{\hbar\omega}{2k_BT}\right)\text{Im}\left[\alpha_v(\omega)\overleftrightarrow{\mathcal{G}}(L,\omega)\right]. \quad (11.45)$$

From a scattering point of view, this approximation is equivalent to neglecting the multiple reflections of the electromagnetic field between oscillator and surface. At short distance to the surface, these reflections become relevant; the next-order correction to the van der Waals interaction arising from (11.44) is discussed in Sect. 12.3.4 of the chapter by DeKieviet et al. in this volume.

Note that although very similar, (11.21) and (11.45) are not identical. (11.21), applied to an oscillator atom, would have featured the bare polarizability

$$\alpha(\omega) = \frac{q^2/m}{\omega_0^2 - (\omega + i0^+)^2},\tag{11.46}$$

where the infinitesimal imaginary part $i0^+$ ensures causality. Equation (11.45) involves, on the contrary, the renormalized or vacuum-dressed polarizability which is causal by default. In other words, it contains a summation over an infinite subclass of terms in the perturbation series.

Finally, a general remark that connects with the scattering interpretation of dispersion forces (see Chap. 4 by Lambrecht et al. and Chap. 5 by Rahi et al. in this volume for detailed discussions on the calculation of the Casimir effect within the framework of the scattering theory): within the theory of two linearly coupled linear systems, the susceptibilities involved in the description of the equilibrium Casimir–Polder interaction are the *isolated and dressed* ones (isolated scatters). This means that, within a linear response theory, or equivalently up to the first order in the perturbation theory, the susceptibilities are not modified by the presence of the other scatters but only dressed by the electromagnetic field. In our case, this means that $\gamma(\omega)$ or $\tilde{\omega}$ in (11.44) or (11.45) do not depend on $\mathbf{r}_0$.

11.3 Atoms and a Planar Surface

Let us consider for definiteness the Casimir–Polder potential near a planar surface, with a distance L between the atom and surface. The Green function is in this case explicitly known and is given in the following subsection.

11.3.1 Behaviour of the Green function

In the case of a planar surface, the electromagnetic Green tensor can be calculated analytically, and we present here some of its main features. In our description we let the atom (source dipole) be on the positive z-axis at a distance L from the medium that occupies the half space below the xy-plane.

11.3.1.1 Reflection Coefficients and Material Response

The electric Green tensor $\overset{\leftrightarrow}{\mathscr{G}}(\mathbf{r}, \mathbf{r}_0, \omega)$ is needed for coincident positions $\mathbf{r} = \mathbf{r}_0$; by symmetry it is diagonal and invariant under rotations in the xy-plane [36, 46, 64, 65]:

$$
\overset{\leftrightarrow}{\mathscr{G}}(L, \omega)
$$

$$
= \frac{1}{8\pi\varepsilon_0} \int\limits_0^\infty kdk \, \kappa \left[\left(r^{\mathrm{TM}}(\omega, k) + \frac{\omega^2}{c^2\kappa^2} r^{\mathrm{TE}}(\omega, k) \right) [\hat{\mathbf{x}}\hat{\mathbf{x}} + \hat{\mathbf{y}}\hat{\mathbf{y}}] + 2\frac{k^2}{\kappa^2} r^{\mathrm{TM}}(\omega, k)\hat{\mathbf{z}}\hat{\mathbf{z}} \right] e^{-2\kappa L},
$$

$$(11.47)$$

where ε_0 is the vacuum permittivity, $k = |\mathbf{k}|$ is the modulus of the in-plane wave vector, and $\hat{\mathbf{x}}\hat{\mathbf{x}}$, $\hat{\mathbf{y}}\hat{\mathbf{y}}$, $\hat{\mathbf{z}}\hat{\mathbf{z}}$ are the cartesian dyadic products. We consider here a local and isotropic medium, excluding the regime of the anomalous skin effect [66]. The Fresnel formulae then give the following reflection coefficients in the TE- and TM-polarization (also known as s- and p-polarization) [43]:

$$
r^{\mathrm{TE}}(\omega, k) = \frac{\mu(\omega)\kappa - \kappa_m}{\mu(\omega)\kappa + \kappa_m}, \quad r^{\mathrm{TM}}(\omega, k) = \frac{\varepsilon(\omega)\kappa - \kappa_m}{\varepsilon(\omega)\kappa + \kappa_m}, \tag{11.48}
$$

where $\epsilon(\omega), \mu(\omega)$ are the permittivity and permeability of the medium. κ, κ_m are the propagation constants in vacuum and in the medium, respectively:

$$
\kappa = \sqrt{k^2 - \frac{\omega^2}{c^2}}, \quad \kappa_m = \sqrt{k^2 - \varepsilon(\omega)\mu(\omega)\frac{\omega^2}{c^2}}. \tag{11.49}
$$

The square roots are defined so that $\mathrm{Im}\,\kappa, \mathrm{Im}\,\kappa_m \leq 0$ and $\mathrm{Re}\,\kappa, \mathrm{Re}\,\kappa_m \geq 0$. In particular κ is either real or pure imaginary. The corresponding frequencies and wave vectors define two regions in the (ω, k) plane [54]: *Evanescent region* $\omega < ck$: the electromagnetic field propagates only parallel to the interface and decays exponentially ($\kappa > 0$) in the orthogonal direction. *Propagating region* $\omega > ck$: the electromagnetic field also propagates ($\mathrm{Re}\,\kappa = 0$) in the orthogonal direction. Note that the magnetic Green tensor $\overset{\leftrightarrow}{\mathscr{H}}$ can be obtained from the electric one by swapping the reflection coefficients [67]:

$$
\varepsilon_0 \overset{\leftrightarrow}{\mathscr{G}} \equiv \frac{1}{\mu_0} \overset{\leftrightarrow}{\mathscr{H}}(r^{\mathrm{TE}} \leftrightarrow r^{\mathrm{TM}}). \tag{11.50}
$$

All information about the optical properties of the surface is encoded in the response functions $\varepsilon(\omega)$ and $\mu(\omega)$. For the sake of simplicity, we focus in the following on a nonmagnetic, metallic medium ($\mu(\omega) = 1$) and use the Drude model [43]:

$$
\varepsilon(\omega) = 1 - \frac{\omega_p^2}{\omega(\omega + i\gamma)}, \tag{11.51}
$$

where ω_p is the plasma frequency (usually for metals in the UV regime). The dissipation rate γ takes account of all dissipative phenomena (impurities, electron-phonon scattering, etc.) in the metal [68] and generally $\gamma/\omega_p \ll 1$ ($\sim 10^{-3}$ for gold).

11.3.1.2 Distance Dependence of the Green Tensor

The Drude model includes Ohmic dissipation in a very characteristic way, through the parameter γ. This affects the physical length scales of the system (see Ref. [69] for a review). In our case the relevant ones are the photon wavelength in vacuum λ_ω and the skin depth in the medium δ_ω. While the first is simply given by

$$\lambda_\omega = \frac{2\pi c}{\omega},\tag{11.52}$$

the second is defined in terms of the low frequency behavior of the dielectric function

$$\frac{1}{\delta_\omega} = \frac{\omega}{c}\,\mathrm{Im}\sqrt{\varepsilon(\omega)} \approx \sqrt{\frac{\omega}{2D}}\quad(\text{for }\omega \ll \gamma),\tag{11.53}$$

where $D = \gamma c^2/\omega_p^2$ is the diffusion coefficient for the magnetic field in a medium with Ohmic damping [43]. The skin depth gives a measure of the penetration of the electromagnetic field in the medium ($\sim 0.79\,\mu\mathrm{m}$ at 10 GHz for gold). If we have $\delta_\omega \ll \lambda_\omega$, the dependence of the Green function on L is quite different in the following three domains: (i) the *sub-skin-depth region*, $L \ll \delta_\omega$, (ii) the *non-retarded region*, $\delta_\omega \ll L \ll \lambda_\omega$, (iii) the *retarded region*: $\lambda_\omega \ll L$. In zones (i) and (ii), retardation can be neglected (van-der-Waals zone), while in zone (iii), it leads to a different power law (Casimir–Polder zone) for the atom–surface interaction.

In the three regimes, different approximations for the reflection coefficients that appear in (11.47) can be made. In the *sub-skin-depth zone* [67], we have $k \gg 1/\delta_\omega \gg 1/\lambda_\omega$ and

$$r^{\mathrm{TE}}(\omega, k) \approx [\varepsilon(\omega) - 1]\frac{\omega^2}{4c^2k^2},$$

$$r^{\mathrm{TM}}(\omega, k) \approx \frac{\varepsilon(\omega) - 1}{\varepsilon(\omega) + 1}\left[1 + \frac{\varepsilon(\omega)}{\varepsilon(\omega) + 1}\frac{\omega^2}{c^2k^2}\right].\tag{11.54}$$

At intermediate distances in the *non-retarded zone*, the wave vector is $1/\lambda_\omega \ll k \ll 1/\delta_\omega$, hence

$$r^{\mathrm{TE}}(\omega, k) \approx -1 + i\frac{2}{\sqrt{\varepsilon(\omega)}}\frac{ck}{\omega},$$

$$r^{\mathrm{TM}}(\omega, k) \approx 1 + i\frac{2}{\sqrt{\varepsilon(\omega)}}\frac{\omega}{ck}.\tag{11.55}$$

Table 11.1 Magnetic and electric Green tensors at a planar surface

	Sub-skin depth	Non-retarded	Retarded
$\mathscr{G}_{xx}$	$\frac{1}{32\pi\varepsilon_0 L^3}\left(1-\frac{2}{\varepsilon(\omega)}\right)$		$\frac{1}{32\pi\varepsilon_0 L^3}\left(1-\frac{2}{\varepsilon(\omega)}\right)\left(1-\mathrm{i}\frac{4\pi L}{\lambda_\omega}-\frac{1}{2}\left[\frac{4\pi L}{\lambda_\omega}\right]^2\right)e^{4\pi\mathrm{i}L/\lambda_\omega}$
$\mathscr{H}_{xx}$	$\frac{\mathrm{i}\mu_0}{32\pi\delta_\omega^2 L}$	$-\frac{\mu_0}{32\pi L^3}$	$-\frac{\mu_0}{32\pi L^3}\left(1-\mathrm{i}\frac{4\pi L}{\lambda_\omega}-\frac{1}{2}\left[\frac{4\pi L}{\lambda_\omega}\right]^2\right)e^{4\pi\mathrm{i}L/\lambda_\omega}$

In this case the other elements have the asymptotes $\mathscr{H}_{yy}=\mathscr{H}_{xx}$, $\mathscr{H}_{zz}=2\mathscr{H}_{xx}$, and similarly for $\mathscr{G}_{ii}$. The off-diagonal elements vanish. The expressions are for metals where $|\varepsilon(\omega)|\gg 1$.

Finally, in the *retarded zone* we can consider $k\ll 1/\lambda_\omega\ll 1/\delta_\omega$, so that

$$r^{\mathrm{TE}}(\omega,k)\approx -1+\frac{2}{\sqrt{\varepsilon(\omega)}},$$

$$r^{\mathrm{TM}}(\omega,k)\approx 1-\frac{2}{\sqrt{\varepsilon(\omega)}}. \qquad (11.56)$$

Note that the first terms in (11.55, 11.56) correspond to a perfectly reflecting medium (formally, $\varepsilon \to \infty$).

The asymptotics of the Green tensor that correlate to these distance regimes are obtained by performing the k-integration in (11.47) with the above approximations for the reflection coefficients. The leading-order results are collected in Table 11.1. One notes that the zz-component is larger by a factor 2 compared to the xx-and yy-components. This difference between the normal and parallel dipoles can be understood by the method of images [43].

The magnetic response for a normally conducting metal in the sub-skin-depth regime is purely imaginary and scales linearly with the frequency ω: the reflected magnetic field is generated by induction. A significant response to low-frequency magnetic fields appears for superconductors because of the Meissner–Ochsenfeld effect [70]. In contrast, the electric response is strong for all conductors because surface charges screen the electric field efficiently.

The imaginary part of the trace of the Green tensor determines the local mode density (per frequency) for the electric or magnetic fields [71]. These can be compared directly after multiplying by ε_0 (or $1/\mu_0$), respectively. As is discussed in Refs. [69, 71], in the sub-skin-depth regime near a metallic surface, the field fluctuations are mainly of magnetic nature. This can be traced back to the efficient screening by surface charges connected with electric fields. Magnetic fields, however, cross the surface much more easily as surface currents are absent (except for superconductors). This reveals, to the vacuum outside the metal, the thermally excited currents within the bulk.

11.3.2 Asymptotic Power Laws

To begin with, let us assume that the particle and the field are both at zero temperature. The Matsubara series in (11.22) can be replaced by an integral over imaginary frequencies:

$$\mathcal{F}(L) = -\frac{\hbar}{2\pi} \int_0^\infty d\xi \sum_j \alpha_{jj}^g(i\xi) \mathcal{G}_{jj}(L, i\xi), \tag{11.57}$$

where we have used the fact that the Green tensor is diagonal. Alternatively one can get the previous result by taking the limit $T \to 0$ of (11.21) and performing a Wick rotation on the imaginary axis. One of the main advantages of this representation is that all functions in (11.57) are real. For electric dipole coupling, one has $\alpha_{ii}^g(i\xi), \mathcal{G}_{ii}(L, i\xi) > 0$, and we can conclude that the (11.57) is a binding energy and corresponds to an attractive force (see the Chap. 8 by Capasso et al. in this volume for detailed discussion on repulsive fluctuation-induced forces in liquids).

Along the imaginary axis, the Green tensor is dominated by an exponential $e^{-2\xi L/c}$, see (11.47). This exponential suppresses large values of ξ and the main contribution to the integral comes from the region $\xi < c/(2L)$. If this value is smaller than the characteristic frequency, say Ω_e, of the atom or of the nanoparticle (the lowest transition in (11.23)), the polarizability can be approximated by its static value. Assuming an isotropic polarizability we get the Casimir–Polder asymptote ($\lambda_e = c/(2\Omega_e)$)

$$L \gg \lambda_e: \quad \mathcal{F}(L)_{\text{CP}} \approx -\frac{3\hbar c \alpha_{\text{iso}}^g(0)}{2^5 \pi^2 \varepsilon_0 L^4}, \tag{11.58}$$

which is the well known expression for the atom–surface Casimir–Polder interaction [23]. At short distance the polarizability limits the relevant frequency range to $\xi \lesssim \Omega_e$. Therefore for $L \ll \lambda_e$ we can replace Green tensor by its short distance approximation (see Table 11.1) where it becomes independent of ξ. We recover then the van der Waals asymptote

$$L \ll \lambda_e: \quad \mathcal{F}(L)_{\text{vdW}} \approx -\frac{\hbar}{2^4 \pi^2 \varepsilon_0 L^3} \int_0^\infty d\xi \alpha_{\text{iso}}^g(i\xi). \tag{11.59}$$

Similar expressions hold for the interaction due to a fluctuating magnetic dipole, the behaviour becoming more complicated when the distance becomes comparable to a characteristic skin depth (see (11.53) and Ref. [27]).

If we write the Matsubara frequencies as $\xi_n = 2\pi n c/\lambda_T$ ($n = 0, 1, 2, \ldots$), the temperature may be low enough so that the limit $\lambda_T \gg L$ holds. Then all Matsubara frequencies are relevant, and if they are dense enough ($\lambda_T \gg \lambda_e$), the effect of temperature is negligible. The series in (11.22) is then well approximated by the integral in (11.57). In the opposite (high-temperature) limit, one has $\lambda_T \ll L$ so that the exponential behavior of the Green tensor limits the series in (11.22) to its first term recovering the Lifshitz asymptote

$$\mathcal{F}(L, T) \approx -\frac{k_B T \alpha_{\text{iso}}^g(0)}{16\pi \varepsilon_0 L^3}. \tag{11.60}$$

We still have an attractive force. Note, however, that this attraction is mainly due to the classical part of the radiation, as the same result would be obtained with a polarizable object immersed into the thermal field.

11.4 Beyond Equilibrium

11.4.1 Overview

The theory presented so far has mainly considered the atom, the field, and the surface to be in a state of global thermal equilibrium, characterized by the same temperature T. When one moves away from these conditions, the atom–surface interaction assumes novel features like metastable or unstable states, driven steady states with a nonzero energy flux etc. We review some of these aspects here, since they have also appeared in recent experiments (Sect. 11.5). On the theoretical side, there are a few controversial issues that are currently under investigation [72–74].

We start with atoms prepared in non-thermal states: ground or excited states that decay by emission or absorption of photons, and with atoms in motion where frictional forces appear. We then consider field–surface configurations out of global equilibrium like a surface surrounded by a vacuum chamber at different temperature.

11.4.2 Atoms in a Given State and Field in Thermal Equilibrium

The generalization of the Casimir–Polder potential to an atom in a definite state $|a\rangle$ can be found, for example, in Wylie and Sipe [36], (4.3, 4.4). The fluctuation–dissipation theorem for the dipole, (11.17), does not apply, but perturbation theory is still possible, with the result (summation over repeated indices i, j)

$$\mathcal{F}(L, T) = -k_B T \sum_{n=0}^{\infty} {}' \alpha_{ij}^a(i\xi_n)\mathcal{G}_{ji}(L, i\xi_n) + \sum_b N(\omega_{ba})d_i^{ab}d_j^{ba}\mathrm{Re}[\mathcal{G}_{ji}(L, \omega_{ba})],$$

(11.61)

where $\overset{\leftrightarrow}{\alpha}^a$ is the state-dependent polarizability [36, 75]. The dipole matrix elements are written $d_i^{ab} = \langle a|d_i|b\rangle$. The thermal occupation of photon modes (Bose–Einstein distribution) is

$$N(\omega) = \frac{1}{e^{\hbar\omega/k_B T} - 1} = -1 - N(-\omega).$$

(11.62)

Note the second term in (11.61) that is absent in thermal equilibrium. It involves the absorption and (stimulated) emission of photons on transitions $a \to b$ to other quantum states, and the thermal occupation number $N(\omega_{ba})$ evaluated at the Bohr frequency $\hbar\omega_{ba} = E_b - E_a$. For this reason, it can be called *resonant* part. The first term that was also present in equilibrium now features the state-dependent polarizability tensor $\overleftrightarrow{\alpha}^a(i\xi_n)$. This is the *non-resonant* part of the interaction.

For the alkali atoms in their ground state $|a\rangle = |g\rangle$, the Bohr frequencies E_{bg} are all positive (visible and near-infrared range) and much larger than typical laboratory temperatures (equivalent to the THz range), hence the thermal occupation numbers $N(\omega_{ba})$ are negligibly small. By the same token, the ground-state polarizability is essentially the same as in thermal equilibrium $\overleftrightarrow{\alpha}^T(i\xi_n)$ because the thermal occupation of the excited states would come with an exponentially small Boltzmann weight. The atom–surface interaction is then indistinguishable from its global equilibrium form and dominated by the non-resonant part.

With suitable laser fields, one can perform the spectroscopy of atom–surface interaction of excited states $|a\rangle = |e\rangle$. It is also possible to prepare excited states by shining a resonant laser pulse on the atom. In front of a surface, the second term in (11.61) then plays a dominant role: the transition to the ground state where a real photon is emitted is accompanied by an energy shift proportional to $\text{Re}[\overleftrightarrow{\mathscr{G}}(L, -\omega_{eg})]$. This resonant contribution can be understood in terms of the radiation reaction of a classical dipole oscillator [36, 76]: one would get the same result by asking for the frequency shift of an oscillating electric dipole in front of a surface—a simple interpretation in terms of an image dipole is possible at short distances (where the k-dependence of the reflection coefficients (11.48) can be neglected). This term is essentially independent of temperature if the transition energy E_{eg} is above $k_B T$.

A more familiar effect for the excited state is spontaneous decay, an example for a non-stationary situation one may encounter out of thermal equilibrium. We can interpret the resonant atom–surface interaction as the 'reactive counterpart' to this dissipative process. Indeed, the spontaneous decay rate is modified relative to its value in free space by the presence of the surface. This can also be calculated in classical terms, leading to a modification that involves the imaginary part of the Green tensor $\overleftrightarrow{G}(L, \omega_{eg})$. (The free-space contribution $\overleftrightarrow{G}_0$ has a finite imaginary part.) In fact, both the decay rate and the interaction potential can be calculated from a complex quantity, formally equivalent to an "effective self-energy", of which (11.61) is the real part in the lowest non-vanishing order of perturbation theory.

What happens if the Bohr frequencies $\hbar\omega_{ba}$ become comparable to $k_B T$? This applies, for example, to optically active vibrational transitions and to atoms in highly excited states (Rydberg atoms) where the energy levels are closely spaced. It is obvious from (11.61) that the resonant term is subject to cancellations among "up" ($E_b > E_a$) and "down" transitions ($E_{b'} < E_a$) with nearly degenerate Bohr frequencies: the occupation numbers $N(\omega_{ba})$ and $N(\omega_{b'a})$ differ in sign, while $\text{Re}[\overleftrightarrow{\mathscr{G}}(L, \omega)]$ is even in ω. To leading order in the high-temperature limit, the resonant term becomes

$$\mathscr{F}^{\mathrm{res}}(L,T) \approx \frac{k_B T}{\hbar} \left[\sum_{b > a} \frac{\mathrm{Re}\,\mathscr{G}^{ba}(L,\omega_{ba})}{\omega_{ba}} - \sum_{b' < a} \frac{\mathrm{Re}\,\mathscr{G}^{b'a}(L,\omega_{ab'})}{\omega_{ab'}} \right], \qquad (11.63)$$

$$\mathscr{G}^{ba}(L,\omega) = d_i^{ab} d_j^{ba} \mathscr{G}_{ij}(L,\omega), \qquad (11.64)$$

where the notation $b > a$ and $b < a$ means summing over states with energies E_b above or below E_a. This is proportional to the anharmonicity of the atomic level spectrum around E_a. It vanishes exactly for a harmonic oscillator and reduces significantly the coefficient linear in temperature in weakly anharmonic regions of the atomic spectrum [28].

11.4.3 Moving Atoms

An atom that moves in a radiation field can be subject to a frictional force, as pointed out by Einstein in his seminal 1917 paper on the blackbody spectrum [77]. This force originates from the aberration and the Doppler shift between the field the atom "sees" in its co-moving frame, and the "laboratory frame". (The latter frame is actually defined in terms of the thermal distribution function of the radiation field that is not Lorentz-invariant. Only the field's vacuum state in free space is Lorentz-invariant.) In addition, electric and magnetic fields mix under a Lorentz transformation so that a moving electric dipole also carries a magnetic moment proportional to $\mathbf{d} \times \mathbf{v}$ where $\mathbf{v}$ is the (center-of-mass) velocity of the dipole (the Röntgen current discussed in Refs. [78–81]).

11.4.3.1 Black Body Friction

The free-space friction force $\mathbf{f}(\mathbf{v}, T)$ is given by [82, 83]:

$$\mathbf{f}(\mathbf{v}, T) = -\mathbf{v} \frac{\hbar^2/k_B T}{12\pi^2 \varepsilon_0 c^5} \int_0^\infty d\omega \frac{\omega^5 \mathrm{Im}\,\alpha(\omega)}{\sinh^2(\hbar\omega/2k_B T)}, \qquad (11.65)$$

where $\alpha(\omega)$ is the polarizability of the atom (in its electronic ground state) and the approximation of slow motion (first order in $\mathbf{v}/c$) has been made. For atomic transitions in the visible range, this force is exponentially suppressed by the Boltzmann factor $\sim e^{-\hbar\omega_{eg}/k_B T}$ that is winning against the prefactor $1/T$ in (11.65). The physics behind this effect is the same as in Doppler cooling in two counter-propagating laser beams: the friction arises from the frequency shift in the frame co-moving with the atom that breaks the efficiency of absorbing photons with counter- and co-propagating momenta. Einstein derived the Maxwell–Boltzmann distribution for the atomic velocities by balancing this radiative friction with the momentum recoil in randomly distributed directions as the absorbed photons are re-emitted, which leads to Brownian motion in velocity space [77, 84]. Conversely, assuming thermal equilibrium and the validity of the Einstein relation between momentum diffusion and friction, one can calculate the (linear) friction tensor $\overleftrightarrow{\Gamma}$ in $\mathbf{f}(\mathbf{v}) = -\overleftrightarrow{\Gamma}\mathbf{v}$ from the correlation function of the force operator [83, 85]:

$$\overleftrightarrow{\Gamma} = \frac{1}{k_B T} \int\limits_{-\infty}^{+\infty} d\tau \langle \mathbf{F}(t+\tau); \mathbf{F}(t) \rangle, \tag{11.66}$$

where $\mathbf{F}(t)$ is the force operator in the Heisenberg picture, the operator product is symmetrized (as in Sect. 11.2.1), and the average $\langle \cdots \rangle$ is taken at (global) thermal equilibrium. One recognizes in (11.66) the zero-frequency component of the force correlation spectrum.

The motion of atoms in the radiation field plays a key role for laser cooling of ultracold gases. Although a discussion of laser-induced forces is beyond the scope of this chapter, the basic principles can be illustrated by moving away from global equilibrium and assigning temperatures T_A, T_F to atom and field. An ultracold gas, immediately after switching off the lasers, would correspond to T_A in the nano-Kelvin range, while $T_F = 300\,\text{K}$ is a good assumption for the fields in a non-cryogenic laboratory apparatus. Dedkov and Kyasov calculated the separate contributions from fluctuations of the atomic dipole and the field, respectively. We follow here Ref. [86]. Qualitatively speaking, the fluctuating dipole experiences a force when it emits a photon; this force is nonzero and depends on velocity, even after averaging over all emission directions, because the emission is isotropic only in the rest frame of the atom. The absorption of photons from the fluctuating field is accompanied by photon recoil, and here isotropy is broken because the Doppler shift brings certain directions closer to the resonance frequency. The same principle is behind the so-called Doppler cooling in two counterpropagating beams. The sum of the two contributions takes the form (adapted from (29) of Ref. [86])

$$\hat{\mathbf{v}} \cdot \mathbf{f}(\mathbf{v}) = -\frac{\hbar}{4\pi\varepsilon_0 c \gamma_\mathbf{v}} \int \frac{d^3k}{\pi^2} (\hat{\mathbf{k}} \cdot \hat{\mathbf{v}})(\omega')^2 \text{Im}\,\alpha(\omega')(N(\omega, T_F) - N(\omega', T_A)) \tag{11.67}$$

$$\omega' = \gamma_\mathbf{v}(\omega + \mathbf{k} \cdot \mathbf{v}), \tag{11.68}$$

where $\gamma_\mathbf{v} = (1 - \mathbf{v}^2/c^2)^{-1/2}$ is the relativistic Lorentz factor, and $\hat{\mathbf{v}}$, $\hat{\mathbf{k}}$ are unit vectors along the atom's velocity and the photon momentum. The photon frequency in the "blackbody frame" (field temperature T_F) is $\omega = c|\mathbf{k}|$, and $N(\omega, T)$ is the Bose–Einstein distribution for a mode of energy quantum $\hbar\omega$ at temperature T. The term with $N(\omega, T_F)$ gives the force due to absorption of thermal photons, while $N(\omega', T_A)$ gives the force due to dipole fluctuations. The absorbed power has to be calculated in the atom's rest frame: the energy $\hbar\omega'$ times the photon number provides the electric field energy density in this frame, and the absorption spectrum $\text{Im}[\omega'\alpha(\omega')]$ must be evaluated at the Doppler-shifted frequency ω'. This shifted spectrum also appears in the fluctuation-dissipation theorem (11.17) now applied locally in the atom's rest frame, and determines the dipole fluctuations.

At equilibrium and in the non-relativistic limit, the difference between the Bose–Einstein distributions can be expanded to give

$$N(\omega, T_F) - N(\omega', T_A) \approx -(\mathbf{k} \cdot \mathbf{v})\partial_\omega N(\omega, T) = (\mathbf{k} \cdot \mathbf{v})\frac{\hbar/k_B T}{4\sinh^2(\hbar\omega/2k_B T)},$$

$$(11.69)$$

and performing the angular integration, one recovers (11.65). As another example, let us consider a ground-state atom ($T_A = 0$) moving in a "hot" field $T_F > 0$ with a small velocity. We can then put $N(\omega', T_A) = 0$ since for free space photons, $\omega' > 0$ (the positive-frequency part of the light cone in the $(\omega, \mathbf{k})$ space is a Lorentz-invariant set). Expanding in the Doppler shift, performing the angular integration and making a partial integration, one arrives at

$$\mathbf{f}(\mathbf{v}) \approx \mathbf{v}\frac{\hbar}{3\pi^2\varepsilon_0 c^5} \int\limits_0^\infty d\omega\, \omega^2 \text{Im}\,\alpha^g(\omega)\partial_\omega[\omega^3 N(\omega, T_F)]. \qquad (11.70)$$

Note that the function $\omega^3 N(\omega, T_F)$ has a positive (negative) slope for $\omega \lesssim k_B T_F/\hbar$ ($\omega \gtrsim k_B T_F/\hbar$), respectively. The velocity-dependent force thus accelerates the particle, $\mathbf{v} \cdot \mathbf{f} > 0$, if its absorption spectrum has a stronger weight at sub-thermal frequencies. This may happen for vibrational transitions in molecules and illustrates the unusual features that can happen in non-equilibrium situations. Drawing again the analogy to laser cooling, the radiative acceleration corresponds to the "anti-cooling" set-up where the laser beams have a frequency $\omega > \omega_{eg}$ ("blue detuning"). Indeed, we have just found that the peak of the thermal spectrum occurs on the blue side of the atomic absorption lines. As a rough estimate, consider the hydrogen atom (mass M) with transition dipoles of the order $|\mathbf{d}_{ag}| \sim ea_0$ (Bohr radius). If the atomic resonances are beyond the peak of the thermal spectrum, one gets from (11.70) a frictional damping time of the order of

$$\frac{1}{\Gamma_{bb}(T)} = \frac{Mv}{|\mathbf{f}(\mathbf{v})|} \sim \frac{10^{21}\text{s}}{(T/300\,\text{K})^5}, \qquad (11.71)$$

which is longer than the age of the Universe. For estimates including resonant absorption where faster frictional damping may occur, see Ref. [82].

11.4.3.2 Radiative Friction Above a Surface

Near a surface, the fluctuations of the radiation field are distinct from free space, and are encoded in the surface-dependent Green function $\overleftrightarrow{\mathscr{G}}(L, \omega)$, see (11.16, 11.20). In addition, one has to take into account that the available photon momenta differ, since also evanescent waves appear whose k-vectors have components larger than ω/c. All these properties can be expressed in terms of the electromagnetic Green tensor, assuming the field in thermal equilibrium. Let us consider for simplicity an atom with an isotropic polarizability tensor $\alpha_{ij} = \alpha\delta_{ij}$, moving at a non-relativistic velocity $\mathbf{v}$. From (118) in Ref. [87], one then gets a friction force

$$\hat{\mathbf{v}} \cdot \mathbf{f}(\mathbf{v}) = -\frac{2\hbar}{\pi} \int_0^\infty d\omega (-\partial_\omega N(\omega, T)) \text{Im}\alpha(\omega)(\hat{\mathbf{v}} \cdot \nabla)(\mathbf{v} \cdot \nabla') \text{tr Im} \overleftrightarrow{G}(\mathbf{r}, \mathbf{r}', \omega),$$

(11.72)

where the spatial derivatives are taken with respect to the two position variables of the Green tensor, and $\mathbf{r} = \mathbf{r}' = \mathbf{r}_A(t)$ is taken afterwards. This expression neglects terms of higher order in α that appear in the self-consistent polarizability (11.41); it can also be found from (25, 26) in Zurita Sanchez et al. [83].

Note that friction is proportional to the local density of field states, encoded in the imaginary part of the (electric) Green tensor $\overleftrightarrow{G}$. If the motion is parallel to a plane surface, the result only depends on the distance L and is independent of time. The friction force is comparable in magnitude to the free-space result (11.67) if the distance L is comparable or larger than the relevant wavelengths c/ω: the derivatives in (11.72) are then of the order $(\hat{\mathbf{v}} \cdot \nabla)(\mathbf{v} \cdot \nabla') \sim |\mathbf{v}|(\omega/c)^2$. At sub-wavelength distances, the non-retarded approximation for the Green tensor can be applied (see Table 11.1), and the previous expression becomes of the order of $|\mathbf{v}|/L^2$. The remaining integral is then similar to the temperature-dependent part of the atom–surface interaction discussed in Sect. 11.4.4.2 below. An order of magnitude estimate can be found in the non-retarded regime, using the approximate Green tensor of Table 11.1. For a metallic surface, the conductivity σ within the thermal spectrum is a relevant parameter; a typical value is $\sigma/\varepsilon_0 \sim 10^{16}\,\text{s}^{-1}$. Taking the atomic polarizability of hydrogen as before, one gets a slight increase relative to the blackbody friction rate (11.71):

$$\Gamma(L, T) \sim \Gamma_{\text{bb}}(T)\frac{\lambda_T^4}{L^4}\frac{c\varepsilon_0}{L\sigma} \sim 10^{-20}\,\text{s}^{-1}\frac{(T/300\,\text{K})^2}{(L/1\,\mu\text{m})^5},$$

(11.73)

but the effect is still negligibly small, even at distances in the nm range.

An expression that differs from (11.72) has been found by Scheel and Buhmann [81] who calculated the radiation force on a moving atom to first order in the velocity, and at zero temperature. Their analysis provides a splitting into resonant and non-resonant terms, similar to (11.61). For the ground state, the friction force is purely non-resonant and contains a contribution from the photonic mode density, similar to (11.72), and one from the Röntgen interaction that appears by evaluating the electric field in the frame co-moving with the atom. Another non-resonant friction force appears due to a velocity-dependent shift in the atomic resonance frequency, but it vanishes for ground-state atoms and for the motion parallel to a planar surface. The remaining friction force becomes in the non-retarded limit and for a Drude metal

$$\mathbf{f}(\mathbf{v}) \approx -\frac{\mathbf{v}}{16\pi\varepsilon_0 L^5} \sum_{a>g} \frac{|\mathbf{d}^{ag}|^2 \omega_s \Gamma_a}{(\omega_{ag}+\omega_s)^3}, \tag{11.74}$$

where the sum $a > g$ is over excited states, the relevant dipole matrix elements are $|\mathbf{d}^{ag}|^2 = \sum_i |d_i^{ag}|^2$, Γ_a is the radiative width of the excited state (which also depends on L), and $\omega_s = \omega_p/\sqrt{2}$ the surface plasmon resonance in the non retarded limit. The fluctuations of the electromagnetic field are calculated here without taking into account the "back-action" of the atom onto the medium (see Ref. [85] and the discussion below). In order of magnitude, atomic and plasmon resonances are often comparable, $\omega_{ag} \sim \omega_s$. Taking hydrogen dipole elements, one thus gets a friction rate

$$\frac{|\mathbf{f}(\mathbf{v})|}{M v} \approx \frac{\alpha_{\rm fs}^2}{3} \frac{\hbar\omega_s a_0^4}{M L^5 c} \sim \frac{10^{-16}\,{\rm s}^{-1}}{(L/1\mu{\rm m})^5}. \tag{11.75}$$

This is significantly larger than (11.73) and does not depend on temperature.

We briefly mention that the behaviour of friction forces in the limit of zero temperature ("quantum friction") has been the subject of discussion that is still continuing (see also the Chap. 13 by Dalvit et al. in this volume for further discussion on quantum friction). An early result of Teodorovich on the friction force between two plates, linear in $\mathbf{v}$ with a nonzero coefficient as $T \rightarrow 0$ [88], has been challenged by Harris and Schaich [85]. They point out that a charge or current fluctuation on one metallic plate can only dissipate by exciting electron-hole pairs in the other plate, but the cross-section for this process vanishes like T^2. This argument does not hold, however, for Ohmic damping arising from impurity scattering. In addition, Ref. [85] points out that the fluctuations of the atomic dipole should be calculated with a polarizability that takes into account the presence of the surface. This self-consistent polarizability has been discussed in Sect. 11.2.4 and reduces the friction force, in particular at short (non-retarded) distances. Carrying out the calculation for a metallic surface and in the non-retarded regime, Harris and Schaich find the scaling

$$\mathbf{f}(\mathbf{v}) \approx -\mathbf{v}\frac{\hbar\alpha_{\rm fs}^2}{L^{10}}\left(\frac{\alpha(0)c}{4\pi\varepsilon_0\omega_s}\right)^2, \tag{11.76}$$

where $\alpha_{\rm fs}$ is the fine structure constant, $\alpha(0)$ is the static polarizability of the atom, and again ω_s the surface plasmon frequency. Note the different scaling with distance L compared to (11.74). Let us consider again $L = 100\,{\rm nm} - 1\,\mu{\rm m}$ for the ease of comparison, although shorter distances are required to get into the non-retarded regime. Taking the hydrogen polarizability and a surface plasmon at $\omega_s \sim 10^{16}\,{\rm s}^{-1}$, (11.76) gives a friction rate of the order $10^{-24}\,{\rm s}^{-1}(L/1\,\mu{\rm m})^{-10}$. This is about eight (three) orders of magnitude smaller that the estimate (11.75) at a distance $L = 1\,\mu{\rm m}$ (100 nm), respectively.

11.4.4 Non-equilibrium Field

A radiation field that is not in thermal equilibrium is a quite natural concept since under many circumstances, an observer is seeing radiation where the Poynting vector is non-zero (broken isotropy) and where the frequency spectrum is not given by the (observer's) temperature. The modelling of these fields can be done at various levels of accuracy: "radiative transfer" is a well-known example from astrophysics and from illumination engineering—this theory can be understood as a kinetic theory for a "photon gas". It is, in its simplest form, not a wave theory and therefore not applicable to the small length scales (micrometer and below) where atom–surface interactions are relevant. "Fluctuation electrodynamics" is a statistical description based on wave optics, developed by the school of S. M. Rytov [19, 89], and similar to optical coherence theory developed by E. Wolf and co-workers [25]. The main idea is that the radiation field is generated by sources whose spectrum is related to the local temperature and the material parameters of the radiating bodies. The field is calculated by solving the macroscopic Maxwell equations, where it is assumed that the matter response can be treated with linear response theory (medium permittivity or dielectric function $\varepsilon(\mathbf{x}, \omega)$ and permeability $\mu(\mathbf{x}, \omega)$). This framework has been used to describe the quantized electromagnetic field, as discussed by Knöll and Welsch and their co-workers [90], by the group of Barnett [91], see also the review paper [92]. Another application is radiative heat transfer and its enhancement between bodies that are closer than the thermal (Wien) wavelength, as reviewed in [87, 93]. The non-equilibrium heat flux between two bodies at different temperatures is naturally calculated from the expectation value of the Poynting vector.

In this section we review the atom–surface interaction in the out-of-equilibrium configuration similar to the one studied in [94]: the atom is close to a substrate hold at temperature T_S, the whole being enclosed in a "cell" with walls (called "environment") at temperature T_E. In the following we will only consider the electric atom–surface interaction and we will use the zero temperature atomic electric polarizability. In fact, the electric dipole transitions are mainly in the visible range and their equivalent in temperature (10^{3-4} K) are not achieved in the experiment. Therefore the atom does not participate in the thermal exchange and can be considered in its ground state.

11.4.4.1 Fluctuation Electrodynamics and Radiative Forces

A very simple non-equilibrium situation occurs when an atom is located near a "heated body" whose temperature is larger than its "surroundings" (see Fig. 11.3). As mentioned above, it is quite obvious that the Poynting vector of the radiation field does not vanish: there is radiative heat flux from the body into the surrounding space. This flux is accompanied by a radiative force on the atom that depends on the atomic absorption spectrum, but also on the angular distribution of

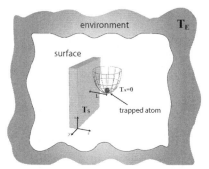

Fig. 11.3 Sketch of an atom–surface system with the field being out of thermal equilibrium. T_S is the temperature of the substrate and T_E is the temperature of the walls of the cell surrounding the atom–substrate system. If $T_S > T_E$, there is a nonzero radiative heat flux from the surface into the surrounding environment

radiated and re-scattered photons. The atom–photon interaction, in this case, does not derive from the gradient of a potential. The basic concept is that of the radiation force **F**; it is given by [43]

$$\mathbf{F}(\mathbf{r}) = \langle d_i \nabla E_i(\mathbf{r}) \rangle + \langle \mu_i \nabla B_i(\mathbf{r}) \rangle + \text{higher multipoles}, \tag{11.77}$$

where we have written out only the contributions from the electric and magnetic dipole moments and the atom is assumed at rest at position **r**. (The generalization to a moving atom leads to the velocity-dependent forces discussed in Sect. 11.4.3.) As a general rule, the electric dipole is the dominant contribution for atoms whose absorption lines are in the visible range. (11.77) can be derived by averaging the Coulomb–Lorentz force over the charge and current distribution in the atom, assuming that the atomic size is small compared to the scale of variation (wavelength) of the electromagnetic field. The average $\langle \cdots \rangle$ is taken with respect to the quantum state of atom and field, and operator products are taken in symmetrized form.

The radiation force (11.77) can be evaluated with the scheme outlined in Sect. 11.2.1 where the operators **d** and **E(r)** are split into "freely fluctuating" and "induced" parts. Carrying this through for the contribution of field fluctuations, leads to an expression of the form

$$\langle d_i^{\text{ind}} \nabla E_i^{\text{free}}(\mathbf{r}) \rangle = \int \frac{d\omega \, d\omega'}{2\pi \, 2\pi} \alpha_{ij}(\omega) \langle E_j^{\text{free}}(\mathbf{r}, \omega) \nabla E_i^{\text{free}}(\mathbf{r}, \omega') \rangle, \tag{11.78}$$

where the spatial gradient of a field autocorrelation function appears. In a non-equilibrium situation, the fluctuation-dissipation theorem of (11.16) cannot be applied, and this field correlation must be calculated in a different way. In a similar way, one gets

$$\langle d_i^{\text{free}} \nabla E_i^{\text{ind}}(\mathbf{r}) \rangle = \int \frac{d\omega \, d\omega'}{2\pi \, 2\pi} \nabla_1 G_{ij}(\mathbf{r}, \mathbf{r}, \omega') \langle d_i^{\text{free}}(\omega) d_j^{\text{free}}(\omega') \rangle, \tag{11.79}$$

where the gradient $\nabla_1 G_{ij}$ is evaluated with respect to the first position coordinate of the Green function. This term requires some regularization because of the divergent Green function at coincident positions. The correlation function of the atomic dipole can be calculated in its stationary state which could be a thermal equilibrium state or not, as discussed in Sect. 11.4.2. In the case of an ultracold atomic gas, it is clear that the atom can be at an effectively much lower temperature compared to the macroscopic bodies nearby. This is consistent with the perturbation theory behind the operator splitting into fluctuating and induced parts. In global equilibrium, when both fluctuation spectra are given by the fluctuation-dissipation theorem (11.16, 11.17), it can be seen easily that the force reduces to the gradient of the equilibrium interaction potential (11.21).

Within Rytov's fluctuation electrodynamics, the fluctuating field is given in terms of its sources and the macroscopic Green function. Generalizing (11.14), one gets

$$E_i(\mathbf{r}, \omega) = \int d^3r' G_{ij}(\mathbf{r}, \mathbf{r}', \omega) P_j(\mathbf{r}', \omega) + \text{magnetization sources}, \qquad (11.80)$$

where we have not written down the contribution from the magnetization field that can be found in Ref. [39]. The polarization density $P_j(\mathbf{r}', \omega)$ describes the excitations of the material (dipole moment per unit volume). If the material is locally stationary, the polarization operator averages to zero and its correlations $\langle P_i(\mathbf{r}, \omega) P_j(\mathbf{r}', \omega') \rangle$ determine the field spectrum

$$\langle E_i(\mathbf{r}, \omega) E_j(\mathbf{r}', \omega') \rangle = \int d^3x d^3x' G_{ik}(\mathbf{r}, \mathbf{x}, \omega) G_{jl}(\mathbf{r}', \mathbf{x}', \omega') \langle P_k(\mathbf{x}, \omega) P_l(\mathbf{x}', \omega') \rangle,$$

$$(11.81)$$

Making the key assumption of local thermal equilibrium at the temperature $T(\mathbf{r})$, the source correlations are given by the *local version* of the fluctuation-dissipation theorem [89]:

$$\langle P_i(\mathbf{r}, \omega) P_j(\mathbf{r}', \omega') \rangle = 2\pi\hbar\delta(\omega + \omega')\delta_{ij}\delta(\mathbf{r} - \mathbf{r}') \coth\left(\frac{\hbar\omega}{2k_B T(\mathbf{r})}\right) \text{Im}\left[\varepsilon_0\varepsilon(\mathbf{r}, \omega)\right],$$

$$(11.82)$$

where $\varepsilon(\mathbf{r}, \omega)$ is the (dimensionless) dielectric function of the source medium, giving the polarization response to a local electric field. The assumption of a local and isotropic (scalar) dielectric function explains the occurrence of the terms $\delta_{ij}\delta(\mathbf{r} - \mathbf{r}')$; this would not apply to ballistic semiconductors, for example, and to media with spatial dispersion, in general. The local temperature distribution $T(\mathbf{r})$ should in that case be smooth on the length scale associated with spatial dispersion (Fermi wavelength, screening length, mean free path). Similar expressions for random sources are known as "quasi-homogeneous sources" and are studied in the theory of partially coherent fields [25, Sect. 5.2].

If we define a polarization spectrum by

$$S_P(\mathbf{r}, \omega) = \hbar \coth\left(\frac{\hbar\omega}{2k_B T(\mathbf{r})}\right) \text{Im}\left[\varepsilon_0 \varepsilon(\mathbf{r}, \omega)\right], \tag{11.83}$$

the field correlation function (11.81) becomes

$$\langle E_i(\mathbf{r}, \omega) E_j(\mathbf{r}', \omega') \rangle = 2\pi\delta(\omega + \omega') \int d^3x\, G_{ik}^*(\mathbf{r}, \mathbf{x}, \omega') G_{jk}(\mathbf{r}', \mathbf{x}, \omega') S_P(\mathbf{x}, \omega'), \tag{11.84}$$

where (11.11) has been applied to the Green function. This expression was named "fluctuation-dissipation theorem of the second kind" by Eckhardt [47] who analyzed carefully its limits of applicability to non-equilibrium situations. Note that even if the sources are spatially decorrelated (the $\delta(\mathbf{r} - \mathbf{r}')$ in (11.82)), the propagation of the field creates spatial coherence, similar to the lab class diffraction experiment with a coherence slit. The spatial coherence properties of the field determine the order of magnitude of the field gradient that is relevant for the radiation force in (11.78).

Let us focus in the following on the correction to the atom–surface force due to the thermal radiation created by a "hot body". We assume that the atom is in its ground state and evaluate the dipole fluctuation spectrum in (11.79) at an atomic temperature $T_A = 0$. To identify the non-equilibrium part of the force, it is useful to split the field correlation spectrum in (11.78) into its zero-temperature part and a thermal contribution, using $\coth(\hbar\omega/2k_B T) = \text{sign}(\omega)[1 + 2N(|\omega|, T)]$ with the Bose–Einstein distribution $N(\omega, T)$. The Rytov currents are constructed in such a way that at zero temperature, the fluctuation-dissipation theorem (11.16) for the field is satisfied. This can be achieved by allowing formally for a non-zero imaginary part $\text{Im}\,\varepsilon(\mathbf{r}, \omega) > 0$ everywhere in space [48, 90], or by combining the radiation of sources located inside a given body and located at infinity [92, 95]. The two terms arising from dipole and field fluctuations then combine into a single expression where one recognizes the gradient of the Casimir–Polder potential Equation (11.21). This is discussed in detail in Refs. [33, 96]. The remaining part of the atom–surface force that depends on the body temperature is discussed now.

11.4.4.2 Radiation Force Near a Hot Body

Let us assume that the body has a homogeneous temperature $T(\mathbf{x}) = T_S$ and a spatially constant dielectric function. Using (11.84) and subtracting the $T = 0$ limit, the spectrum of the non-equilibrium radiation (subscript 'neq') can then be expressed by the quantity

$$\langle E_i(\mathbf{r}, \omega) E_j(\mathbf{r}', \omega') \rangle_{\text{neq}} = 2\pi\delta(\omega + \omega') N(|\omega|, T_S) \hbar S_{ij}(\mathbf{r}, \mathbf{r}', \omega), \tag{11.85}$$

$$S_{ij}(\mathbf{r}, \mathbf{r}', \omega) = \mathrm{Im}\left[\varepsilon_0 \varepsilon(|\omega|)\right] \int_S d^3x \, G_{ik}^*(\mathbf{r}, \mathbf{x}, \omega) G_{jk}(\mathbf{r}', \mathbf{x}, \omega), \qquad (11.86)$$

where the space integral is over the volume of the body. The tensor $S_{ij}(\mathbf{r}, \mathbf{r}', \omega)$ captures the material composition of the body and its geometry relative to the observation points.

Referring to the force due to field fluctuations in (11.78), let us assume for simplicity that the atomic polarizability is isotropic, $\alpha_{ij}(\omega) = \delta_{ij}\alpha(\omega)$. We combine the integrand over positive and negative frequencies to isolate dispersive and absorptive contributions ($\omega > 0$)

$$\begin{aligned}
&\alpha^*(\omega)\nabla_2 S_{ii}(\mathbf{r}, \mathbf{r}, \omega) + (\omega \mapsto -\omega) \\
&= \mathrm{Re}[\alpha(\omega)]\nabla_{\mathbf{r}}[S_{ii}(\mathbf{r}, \mathbf{r}, \omega)] + 2\mathrm{Im}[\alpha(\omega)]\mathrm{Im}[\nabla_2 S_{ii}(\mathbf{r}, \mathbf{r}, \omega)],
\end{aligned} \qquad (11.87)$$

where ∇_2 is the gradient with respect to the second coordinate of S_{ij}, while $\nabla_{\mathbf{r}}$ differentiates both coordinates. This form highlights that the non-equilibrium force separates in two [97, 98] contributions that are familiar in laser cooling [99, 100]: a *dipole force* equal to the gradient of the electric energy density (proportional to $S_{ii}(\mathbf{r}, \mathbf{r}, \omega) \geq 0$). This is proportional to the real part of α and can be interpreted as the polarization energy of the atom in the thermal radiation field. The second term in (11.87) gives rise to *radiation pressure*, it is generally[1] proportional to the atomic absorption spectrum and the phase gradient of the field. The phase gradient can be identified with the local momentum of the emitted photons. By inspection of (11.86) for a planar surface, one indeed confirms that the radiation pressure force pushes the atom away from the thermal source. An illustration is given in Fig. 11.4 for a nanoparticle above a surface, both made from semiconductor. The dielectric function $\varepsilon(\omega)$ is of Lorentz–Drude form and uses parameters for SiC (see Ref. [33]). The arrows mark the resonance frequencies of transverse bulk phonon polaritons ω_T and of the phonon polariton modes of surface (ω_1) and particle (ω_2).

The radiation pressure force is quite difficult to observe with atomic transitions in the visible range because the peaks of the absorption spectrum are multiplied by the exponentially small Bose–Einstein factor $N(\omega_{ag}, T_S)$, even if the body temperature reaches the melting point. Alternative settings suggest polar molecules or Rydberg atoms [28] with lower transition energies. In addition, some experiments are only sensitive to force gradients (see Sect. 11.5), and it can be shown that the radiation pressure above a planar surface (homogeneous temperature, infinite lateral size) does not change with distance.

---

[1] Note that when the dressed polarizability is used instead of the bare one, the polarizability has an imaginary part even in the absence of absorption (see discussion at the end of Sect. 11.2.4). This is equivalent to include the effects of the "radiative reaction" in the dynamic of the dipole [97, 101, 102] as required by the conservation of energy and the optical theorem.

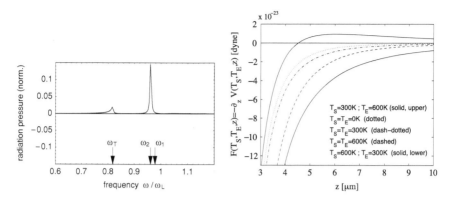

Fig. 11.4 *Left* spectrum of thermal radiation pressure force exerted on a small spherical particle above a planar substrate (positive = repulsive, does not depend on distance). The arrows mark the substrate and particle resonances at $\mathrm{Re}\,\varepsilon^{-1}(\omega_T) = 0$ (bulk phonon polariton) and $\mathrm{Re}\,\varepsilon(\omega_{1,2}) = -1, -2$ (surface and particle phonon polariton). The force spectrum is given by $\hbar N(\omega, T_S)$ times the second term in (11.87), and normalized to $(16/3)\hbar k_L(k_L a)^3 N(\omega, T_S)$ where a is the particle radius and $k_L = \omega_L/c$ the wavenumber of the longitudinal bulk polariton $(\mathrm{Re}\,\varepsilon(\omega_L) = 0)$. *Right* theoretical calculation of the atom–surface force, in and out of thermal equilibrium, taken from Ref. [94], Fig. 11.2. The atom is Rubidium 87 in its electronic ground state, the surface is made of sapphire (SiO_2). Note the variation with temperature(s) of the non-equilibrium force, both in magnitude and in sign. A negative sign corresponds to an attractive interaction

For the evaluation of the dipole force, the same argument related to the field temperature can be applied so that the atomic polarizability in (11.87) is well approximated by its static value, $\mathrm{Re}\,\alpha(\omega) \approx \alpha(0)$. We thus find

$$\mathbf{F}_{\mathrm{neq}}^{\mathrm{dip}}(\mathbf{r}, T_S, T_E = 0) = -\nabla U_{\mathrm{neq}}^{\mathrm{dip}}(\mathbf{r}, T_S, T_E = 0), \tag{11.88}$$

$$U_{\mathrm{neq}}^{\mathrm{dip}}(\mathbf{r}, T_S, T_E = 0) = -\alpha(0) \int\limits_0^\infty \frac{d\omega}{2\pi} \hbar N(\omega, T_S) S_{ii}(\mathbf{r}, \mathbf{r}, \omega). \tag{11.89}$$

11.4.4.3 General Non-Equilibrium Configuration and Asymptotic Behaviours

In the general case both T_S and T_E can be different from zero [94]. The total force will be the sum of $\mathbf{F}_{\mathrm{neq}}^{\mathrm{dip}}(\mathbf{r}, T_S, T_E = 0)$ given in the previous expression and of

$$\mathbf{F}_{\mathrm{neq}}^{\mathrm{dip}}(\mathbf{r}, T_S = 0, T_E) = \mathbf{F}_{\mathrm{eq}}^{\mathrm{dip}}(\mathbf{r}, T_E) - \mathbf{F}_{\mathrm{neq}}^{\mathrm{dip}}(\mathbf{r}, T_E, T_S = 0), \tag{11.90}$$

that is the difference between thermal force at equilibrium at the temperature T_E and the force in (11.88) and (11.89) with T_S and T_E swapped. An illustration of the resulting force is given in Fig. 11.4 (right) for a planar surface. A large-distance asymptote of the non-equilibrium interaction can be derived in the form [94, 103, 104]

$$L \gg \frac{\lambda_T}{\sqrt{\varepsilon(0) - 1}} : \qquad U_{\text{neq}}^{\text{dip}}(L, T_S, T_E) \approx -\frac{\pi}{12}\alpha(0)\frac{\varepsilon(0) + 1}{\sqrt{\varepsilon(0) - 1}}\frac{k_B^2(T_S^2 - T_E^2)}{\hbar c L^2},$$

(11.91)

where $\varepsilon(0) < \infty$ is the static dielectric constant. The previous expression is valid for dielectric substrates, see Ref. [94] for Drude metals. For $T_S = T_E$, this formula vanishes, and one ends with the "global equilibrium" result of (11.61).

Expression (11.91) shows that the configuration out of thermal equilibrium presents new features with respect to the equilibrium force. Indeed the force scales as the difference of the square of the temperatures and can be attractive or repulsive. For $T_E > T_S$ the force changes sign, going from attractive at small distance to repulsive at large distance (i.e. featuring a unstable equilibrium position in between), and it decays slower than the equilibrium configuration ($\propto L^{-4}$), leading therefore to a stronger force. This new feature was important for first measurement of the thermal component of the surface-atom surface (see next section). Moreover, when a gas of atoms is placed in front of the surface, the non-equilibrium interaction can lead to interesting non-additive effects [96, 105].

11.5 Measurements of the Atom–Surface Force with Cold Atoms

11.5.1 Overview

We do not discuss here the regime of distances comparable to the atomic scale where atomic beams are diffracted and reveal the crystallography of the atomic structure of the surface. We consider also that the atoms are kept away even from being physisorbed in the van der Waals well (a few nanometers above the surface). One is then limited to distances above approximately one micron (otherwise the attractive forces are difficult to balance by other means), but can take advantage of the techniques of laser cooling and micromanipulation and use even chemically very reactive atoms like the alkalis.

The first attempts to measure atom–surface interactions in this context go back to the sixties, using atomic beam set-ups. In the last 20 years, technological improvements have achieved the sensitivity required to detect with good accuracy and precision tiny forces. As examples, we mention the exquisite control over atomic beams provided by laser cooling [106, 107], spin echo techniques that reveal the quantum reflection of metastable noble gas beams [108, 109] (see also the Chap. 12 by DeKieviet et al. for detailed discussions on this topic), or the trapping of an ultracold laser-cooled gas in atom chip devices [5, 110]. In this section we will briefly review some of the experiments which exploited cold atoms in order to investigate the Casimir–Polder force.

By using different techniques, it has been possible to measure the atom–surface interaction (atomic level shift, potential, force, or force gradient, depending on the cases) both at small $(0.1 \, \mu m < L < \lambda_{opt} \approx 0.5 \, \mu m)$ and large $(L > 1 \, \mu m)$ distances. Because the interaction rapidly drops as the atom-surface separation becomes larger, the small-distance (van der Waals–London, (11.59)) regime at $L < \lambda_{opt}$ is somewhat easier to detect. Recall that in this limit, only the vacuum fluctuation of electromagnetic field are relevant, and retardation can be ignored. More recent experiments explored the weaker interaction in the Casimir–Polder regime (11.58), $\lambda_{opt} < L < \lambda_T$, where retardation effects are relevant, but thermal fluctuations still negligible. Also the Lifshitz–Keesom regime at $L > \lambda_T$ has been explored, where thermal fluctuations are dominant [see (11.60)]. The theory of Dzyaloshinskii, Lifshitz, and Pitaevski (DLP, [19,Chap.VIII]) encompasses the three regimes with crossovers that are illustrated in Fig. 11.1 for a Rubidium atom and a room temperature sapphire surface.

11.5.2 From van der Waals to Casimir–Polder: Equilibrium

Typically, experiments have been performed at room temperature, and at thermal equilibrium, and used several techniques to measure the interaction, usually of mechanical nature.

The van der Waals–London regime has been first explored by its effect on the deflection of an atomic beam passing close to a substrate [111–114]. Such kind of experiments were almost qualitative, and hardly in agreement with the theory. Subsequently, more accurate measurements of the atom–surface interaction in this regime have been done by using dielectric surfaces "coated" with an evanescent laser wave that repels the atoms (atom mirrors, [115]), by atom diffraction from transmission gratings [116, 117], by quantum reflection [108, 109], and by spectroscopic studies [106, 118].

The Casimir–Polder regime, where vacuum fluctuations of the electromagnetic field and the finite speed of light are relevant, was first studied experimentally in [107].[2] Here the force has been measured through an atomic beam deflection technique, which consists in letting an atomic beam (Na atoms in their ground state) pass across in a cavity made by two walls (gold plates), as one can see in Fig. 11.5. The atoms of the beam are drawn by the Casimir–Polder force to the walls, whose intensity depends on the atomic position within the cavity. Part of the atoms are deflected during their path and stick to the cavity walls without reaching

---

[2] Reprinted figures with permission from C. I. Sukenik, M. G. Boshier, D. Cho, V. Sandoghdar, and E. A. Hinds, Phys. Rev.Lett. **70**, 560 (1993). Copyright (1993) by the American Physical Society.

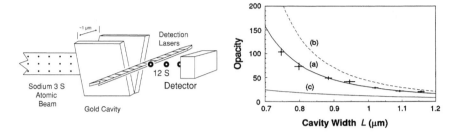

Fig. 11.5 (*left*) Scheme of the experiment of Sukenik and al, taken from Ref. [107]. An atomic beam enters a micron-sized gold cavity, and the flux of atoms emerging the cavity is detected and related to the atom–surface potential inside the cavity. (*right*) Measurement of the atom–surface interaction in the Casimir–Polder regime, in the experiment of [107], taken from the same paper. The opacity is proportional to the number of atoms which do not exit from the cavity, and is related to the atom–surface potential. The solid lines are the theoretical prediction based on: full DLP potential (**a**), van der Waals–London (*short-distance*) potential (**b**), and no atom–surface potential (**c**)

the end of the cavity. Only few atoms will pass the whole cavity, and their flux is measured and related to the atom–surface interaction in the cavity. Such a measurement is shown in Fig. 11.5, where the theoretical curves are based on atomic trajectories in the atom–surface potential that is assumed to be either of van der Waals-London or of Casimir–Polder form. The data are clearly consistent with the CP interaction, hence retardation already plays a role for typical distances in the range of 500 nm.

Subsequent measurements of the Casimir–Polder force have been done, among others, using the phenomenon of quantum reflection of ultra-cold atomic beams from a solid surface [108, 109]. In the experiment, a collimated atomic beam of metastable Neon atoms impinges on a surface (made of Silicon or some glass) at a glancing angle (very small velocity normal to the surface). In this regime, the de Broglie wave of the incident atoms must adapt its wavelength to the distance-dependent potential, and fails to do so because the potential changes too rapidly on the scale of the atomic wavelength. This failure forces the wave to be reflected, a quantum effect that would not occur in an otherwise attractive potential. In the limit of zero normal velocity (infinite wavelength), the reflection probability must reach 100%. The variation with velocity depends on the shape of the atom–surface potential and reveals retardation effects [119, 120]. In Fig. 11.6 is shown the measurement of quantum reflection performed by Shimizu [108].[3] In this case, the accuracy was not high enough to distinguish reliably between theoretical predictions. More recent data are shown in Fig. 12.5 of the chapter by DeKieviet et al. in this volume.

---

[3] Reprinted figure with permission from F. Shimizu, Phys. Rev. Lett. **86**, 987 (2001). Copyright (2001) by the American Physical Society.

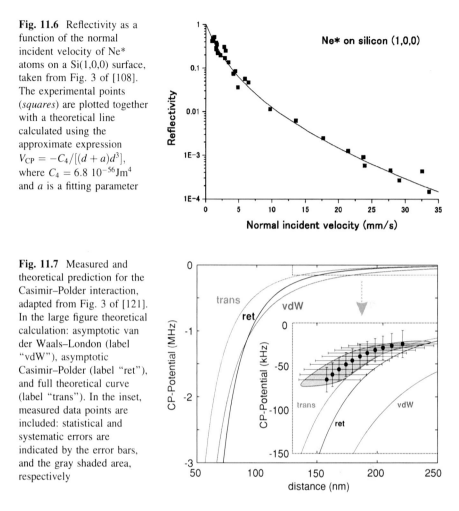

Fig. 11.6 Reflectivity as a function of the normal incident velocity of Ne* atoms on a Si(1,0,0) surface, taken from Fig. 3 of [108]. The experimental points (*squares*) are plotted together with a theoretical line calculated using the approximate expression $V_{CP} = -C_4/[(d+a)d^3]$, where $C_4 = 6.8 \ 10^{-56} \mathrm{Jm}^4$ and a is a fitting parameter

Fig. 11.7 Measured and theoretical prediction for the Casimir–Polder interaction, adapted from Fig. 3 of [121]. In the large figure theoretical calculation: asymptotic van der Waals–London (label "vdW"), asymptotic Casimir–Polder (label "ret"), and full theoretical curve (label "trans"). In the inset, measured data points are included: statistical and systematic errors are indicated by the error bars, and the gray shaded area, respectively

The crossover between the van der Waals and Casimir–Polder regimes has been recently measured by the group of C. Zimmermann and S. Slama [121], using the reflection of a cloud of ultracold atoms at an evanescent wave atomic mirror. This experiment improves previous data obtained by the A. Aspect group [115] into the crossover region. The data are shown in Fig. 11.7 where "vdW" and "ret" label the asymptotes van der Waals and Casimir–Polder potentials, respectively.[4] The full calculation (DLP theory) is labelled "trans" and show some deviation from both asymptotes in the crossover region. The data (shown with error bars) are clearly favoring the full (DPL) theory.

---

[4] Reprinted figure with permission from H. Bender, P.W. Courteille, C. Marzok, C. Zimmermann, and S. Slama, Phys. Rev. Lett. **104**, 083201 (2010). Copyright (2010) by the American Physical Society.

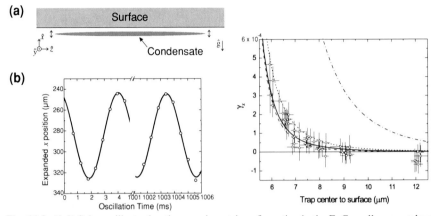

Fig. 11.8 (*left*) Scheme illustrating the experimental configuration in the E. Cornell group, taken from Fig. 1 of [122]. Typical arrangement of the condensate close to the surface. The cloud is formed by a few hundred thousand Rubidium 87 atoms, its axial length is ∼ 100 μm. The surface is made of fused silica. The coordinate axes and the direction of gravity are indicated (**a**). Typical data showing the center-of-mass (dipole) oscillation (*x*-direction normal to the surface). This is obtained after holding the BEC near the surface and then shifting it rapidly away from the surface; the "expanded position" is proportional to the velocity component v_x. (*right*) Measured and theoretical frequency of the BEC center-of-mass motion, relative to the nominal trap frequency ω_{trap} and normalized as $\gamma(x) = (\omega_{cm} - \omega_{trap})/\omega_{trap}$. Each data point represents a single measurement, with both statistical and systematic errors. The mean oscillation amplitude is ≈ 2.06μm, and the typical size of the BEC (Thomas–Fermi radius) in the oscillation direction is ≈ 2.40μm. Theory lines, calculated using theory from [124], consider the full atom- surface potential: $T = 0$ K (*dashed line*), $T = 300$ K (*solid line*) and $T = 600$ K (*dotted line*). No adjustable parameters have been used. The result of the van der Waals–London potential has been added (*dash-dotted line*) (**b**)

11.5.3 The Experiments of the E. Cornell Group

11.5.3.1 Lifshitz Regime

The atom–surface interaction in the Lifshitz regime has been explored in E. Cornell's group [5, 122]. Here a quantum degenerate gas in the Bose–Einstein condensed phase has been used as local sensor to measure the atom–surface interaction, similar to the work in V. Vuletić's group where smaller distances were involved [110]. The Cornell experiments use a Bose–Einstein condensate (BEC) of a few 10^{5} ^{87}Rb atoms that are harmonically trapped at a frequency ω_{trap}. The trap is moved towards the surface of a sapphire substrate, as illustrated in Fig. 11.8.

Center-of-mass (dipole) oscillations of the trapped gas are then excited in the direction normal to the surface. In absence of atom–surface interaction, the frequency of the center-of-mass oscillation would correspond to the frequency of the trap: $\omega_{cm} = \omega_{trap}$. Close to the substrate, the atom–surface potential changes the effective trap frequency, shifting the center-of-mass frequency by a quantity $\gamma =$

$(\omega_{cm} - \omega_{trap})/\omega_{trap}$. The value of γ is related to the atom–surface force [4] and for small oscillation amplitudes we have:

$$\omega_{cm}^2 = \omega_{trap}^2 + \frac{1}{m} \int_{-R_z}^{R_z} dz\, n_0^z(z)\, \partial_z^2 \mathscr{F}(z), \tag{11.92}$$

where m is the atomic mass, $\mathscr{F}(z)$ is the atom–surface free energy in (11.21) (z is the direction normal to the surface), and $n_0^z(z)$ is the xy-integrated unperturbed atom density profile [4] that takes into account the finite size of the gas cloud. In the Thomas–Fermi approximation

$$n_0^z(z) = \frac{15}{16R_z}[1 - \left(\frac{z}{R_z}\right)^2]^2, \tag{11.93}$$

where R_z (typically of few microns) is the cloud radius along z, which depends on the chemical potential. In the comparison with the experiment, non-linear effects due to large oscillation amplitudes [4] may become relevant [122]. The experiment of Ref. [122] was performed at room temperature and succeeded in measuring the atom–surface interaction for the first time up to distances $L \approx 7\,\mu m$ (see Fig. 11.8). Although the relative frequency shift in (11.89) is only $\sim 10^{-4}$, the damping of this dipole oscillator is so weak that its phase can be measured even after hundreds of periods, (see Fig. 11.8 (left)). The same technique has been recently proposed to test the interaction between an atom and a non-planar surface [124, 125].

11.5.3.2 Temperature Dependence and Non-Equilibrium Force

The experiment of Ref. [122] did not reach the accuracy to discriminate between the theoretical predictions at $T = 0\,K$ and the $T = 300\,K$, and a clear evidence of thermal effects was still missing. In this experiment there was no room to increase the temperature of the surface: at high temperatures some atoms thermally desorb from the walls of the cell, the vacuum in the cell degrades, resulting in the impossibility to produce a BEC. To overcome this experimental limitation a new configuration was studied, where only the surface temperature was increased: the quality of the vacuum was not affected because of the relatively small size of the substrate. The non-equilibrium theory of atom–surface interactions in this system was developed in Refs. [94, 96, 104, 105, 126], as outlined in Sect. 11.4. It predicts new qualitative and quantitative effects with respect to global equilibrium that are illustrated in Fig. 11.4 (right). The experimental measurement has been achieved in 2007 [5] and remains up to now the only one that has detected thermal effects of the electromagnetic dispersion interactions in this range of distances. A sketch of the experimental apparatus is given in Fig. 11.9, the experimental results in Fig. 11.10.

Fig. 11.9 Scheme of the experiment of Ref. [5] (from which the figure is taken), where atom–surface interactions out of thermal equilibrium have been measured

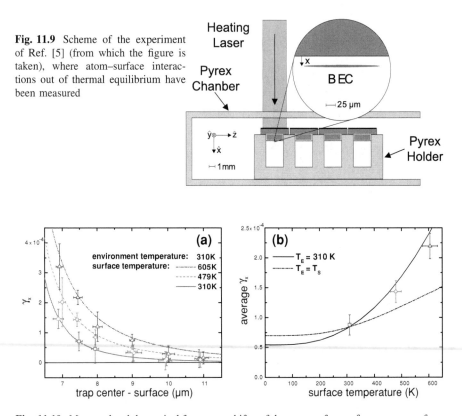

Fig. 11.10 Measured and theoretical frequency shift γ of the center-of-mass frequency ω_{cm} for a trapped atomic BEC (^{87}Rb atoms) close to surface (fused silica), in a system in and out of thermal equilibrium. The figure is taken from Fig. 4 of [5]. **a** The figure shows three sets of data and accompanying theoretical curves with no adjustable parameters for various substrate temperatures. Data are shown for different substrate temperatures: $T_S = 310$ K (*squares*), $T_S = 479$ K (*circles*), and $T_S = 605$ K (*triangles*). The environment temperature is maintained at $T_E = 310$ K. The error bars represent the total uncertainty (statistical and systematic) of the measurement. **b** Average values γ_x over the trap center–surface separations of 7.0, 7.5, and 8.0 μm, plotted versus substrate temperature. It is evident a clear increase in strength of the atom–surface interaction for elevated temperatures. The solid theory curve represents the non-equilibrium effect, while the dash–dotted theory curve represents the case of equal temperatures

11.5.3.3 Outlook

Precision measurements of the atom–surface interaction may shed light on the ongoing discussion about the temperature dependence of dispersion interactions with media that show absorption, like any conducting medium. It has been pointed out by Klimchitskaya and Mostepanenko [127] that if the small, but nonzero conductivity of the glass surface in the Cornell experiment [5, 122] had been taken into account, the Lifshitz–Keesom tail would involve an infinite static dielectric function, and hence deviate from a dielectric medium where $1 < \varepsilon(0) < \infty$. This theoretical prediction would also be inconsistent with the data. This issue is related to similar problems that arise in the macroscopic Casimir interaction, see other chapters in this volume and Refs. [128–130] for reviews. In the atom–surface case, Pitaevskii has pointed out that a smooth crossover from a metal to a dielectric is obtained within a non-local description that takes into account electric screening in the surface (wave-vector-dependent dielectric function $\varepsilon(\mathbf{k}, \omega)$) [131] which Ref. [127] did not include.

New interesting experimental proposals have been presented in order to measure the atom–surface force with higher accuracy, essentially based on interferometric techniques. All of them deal with atoms trapped in a periodic lattice made by laser beams ("optical lattice" [132, 133]), and placed close to a substrate. Gradients in the potential across the lattice can be detected with coherent superposition states of atoms delocalized over adjacent lattice sites [7]. These gradients also induce Bloch oscillations through the reciprocal space of the lattice: if $\hbar q$ is the width of the Brillouin zone, the period τ_B of the Bloch oscillations is [6]

$$\frac{1}{\tau_B} = \frac{\overline{-\partial_L U}}{\hbar q}. \tag{11.94}$$

where the average (overbar) is over the cloud size in the lattice. The atom–surface interaction would, in fact, shift the Bloch period by a relative amount of $10^{-4} \ldots 10^{-3}$ if the main force is gravity and the atoms are at a distance $L \approx 5\,\mu\text{m}$ [105]. Distance-dependent shifts in atomic clock frequencies have also been proposed [134]. They arise from the differential energy shift of the two atomic states which are related to the difference in polarizabilities. Finally, a corrugated surface produces a periodic Casimir–Polder potential that manifests itself by a band gap in the dispersion relation of the elementary excitations of the BEC (Bogoliubov modes). The spectrum of these modes is characterized by a dynamic structure factor that can be detected by the Bragg scattering of a pair of laser beams [135].

11.6 Conclusion

In this chapter, we have outlined a quantum theory of fluctuations, focussing on "discrete systems" like atoms or nanoparticles and on a quantized field. As atoms interact with the electromagnetic field, the fluctuations of the two entities get correlated, and this becomes manifest in forces, energy shifts, and damping rates. We have presented a formalism where these radiative interactions are described by response functions (polarizabilities, Green functions), and thermodynamics enters via the fluctuation-dissipation theorem, a powerful generalization of the Einstein relation familiar from Brownian motion. These interactions have been explored recently over an extended range of distances between atoms and macroscopic bodies, using experimental techniques from laser cooling and ultracold atom manipulation. Atom-surface interactions are setting now non-trivial limitations for the stability of micro-traps and nano-devices, as their dimensions are reduced. Non-equilibrium situations ("cold atoms near a hot body") are quite common in these settings, and we have outlined in this chapter the corresponding fluctuation theory based on local thermodynamic equilibrium. For atom–surface interactions, this theory is relatively simple to formulate since to a good approximation, atoms are "probe particles" that do not affect the quantum state of the field (and of the macroscopic surroundings). Still, subtle issues have emerged in controversies, both long standing and more recent, that are related to the choice of placing the "cut" between "system" (dynamically responding) and "bath" (in thermodynamic equilibrium). We anticipate that experimental and theoretical advances in the near future will help to resolve these issues within the challenging realm of open quantum systems.

Acknowledgments The authors would like to thank H. Haakh for fruitful discussions and comments, and R. Behunin for a critical reading. F.I. and C.H. acknowledge support from the *Deutsche Forschungsgemeinschaft*, the Humboldt foundation, and the European Science Foundation.

References

1. Diedrich, F., Bergquist, J.C., Itano, M.W., Wineland, D.J.: Laser cooling to the zero-point energy of motion. Phys. Rev. Lett. **62**, 403 (1989)
2. Meekhof, D.M., Monroe, C., King, B.E., Itano, W.M., Wineland, D.J: Generation of nonclassical motional states of a trapped ion. Phys. Rev. Lett. **76**, 1796 (1996)
3. Pitaevskii, L.P., Stringari, S.: Bose–Einstein Condensation, 116 of International Series of Monographs on Physics. Clarendon Press, Oxford (2003)
4. Antezza, M., Pitaevskii, L.P., Stringari, S.: Effect of the Casimir–Polder force on the collective oscillations of a trapped Bose–Einstein condensate. Phys. Rev. A. **70**, 053619 (2004)
5. Obrecht, J.M., Wild, R.J., Antezza, M., Pitaevskii, L.P., Stringari, S., Cornell, E.A.: Measurement of the temperature dependence of the Casimir–Polder force. Phys. Rev. Lett. **98**, 063201 (2007)
6. Carusotto, I., Pitaevskii, L., Stringari, S., Modugno, G., Inguscio, M.: Sensitive measurement of forces at micron scale using bloch oscillations of ultracold atoms. Phys. Rev. Lett. **95**, 093202 (2005)
7. Wolf, P., Lemonde, P., Lambrecht, A., Bize, S., Landragin, A., Clairon, A.: From optical lattice clocks to the measurement of forces in the Casimir regime. Phys. Rev. A. **75**, 063608 (2007)
8. Sorrentino, F., Alberti, A., Ferrari, G., Ivanov, V.V., Poli, N., Schioppo, M., Tino, G.M.: Quantum sensor for atom–surface interactions below 10 μm. Phys. Rev. A. **79**, 013409 (2009)
9. Onofrio, R.: Casimir forces and non-Newtonian gravitation. New J. Phys. **8**, 237 (2006)
10. Folman, R., Krüger, P., Schmiedmayer, J., Denschlag, J.H., Henkel, C.: Microscopic atom optics: from wires to an atom chip. Adv. At. Mol. Opt. Phys. **48**, 263 (2002)
11. Fortágh, J., Zimmermann, C.: Magnetic microtraps for ultracold atoms. Rev. Mod. Phys. **79**, 235 (2007)
12. Reichel, J., Hänsel, W., Hänsch, T.W.: Atomic micromanipulation with magnetic surface traps. Phys. Rev. Lett. **83**, 3398 (1999)

13. Folman, R., Krüger, P., Cassettari, D., Hessmo, B., Maier, T., Schmiedmayer, J.: Controlling cold atoms using nanofabricated surfaces: atom chips. Phys. Rev. Lett. **84**, 4749 (2000)
14. Ott, H., Fortagh, J., Schlotterbeck, G., Grossmann, A., Zimmermann, C.: Bose–Einstein condensation in a surface microtrap. Phys. Rev. Lett. **87**, 230401 (2001)
15. Roux, C., Emmert, A., Lupascu, A., Nirrengarten, T., Nogues, G., Brune, M., Raimond, J.-M., Haroche, S.: Bose–Einstein condensation on a superconducting atom chip. Europhys. Lett. **81**(56004), 6 (2008)
16. Cano, D., Kasch, B., Hattermann, H., Kleiner, R., Zimmermann, C., Koelle, D., Fortágh, J.: Meissner effect in superconducting microtraps. Phys. Rev. Lett. **101**, 183006 (2008)
17. Müller, T., Zhang, B., Fermani, R., Chan, K.S., Wang, Z.W., Zhang C.B., Lim M.J., Dumke, R.: Trapping of ultra-cold atoms with the magnetic field of vortices in a thin-film superconducting micro-structure. New J. Phys. **12**, 043016 (2010)
18. Parsegian, V.A.: van der Waals Forces—A Handbook for Biologists, Chemists, Engineers, and Physicists. Cambridge University Press, New York (2006)
19. Lifshitz, E.M., Pitaevskii, L.P.: Statistical Physics (Part 2), 9 of Landau and Lifshitz Course of Theoretical Physics. 2nd ed. Pergamon, Oxford (1980)
20. Margenau, H.: van der Waals forces. Rev. Mod. Phys. **11**, 1 (1939)
21. London, F.: Zur Theorie und Systematik der Molekularkräfte. Z. Physik. **63**, 245 (1930)
22. Verweey, E., Overbeek, J.: Theory of the Stability of Lyophobic Colloids. Elsevier Science Publishers, Amsterdam (1948)
23. Casimir, H.B., Polder, D.: The Influence of retardation on the London-van der Waals forces. Phys. Rev. **73**, 360 (1948)
24. Dzyaloshinskii, I.E., Lifshitz, E.M., Pitaevskii, L.P.: General theory of van der Waals' forces. Sov. Physics Usp. **4**, 153 (1961)
25. Mandel, L., Wolf, E.: Optical Coherence and Quantum Optics. Cambridge University Press, Cambridge (1995)
26. Linder, B.: van der Waals interaction potential between polar molecules. Pair Potential and (Nonadditive) Triple Potential. J. Chem. Phys. **40**, 2003 (1964)
27. Haakh, H., Intravaia, F., Henkel, C., Spagnolo, S., Passante, R., Power, B., Sols, F.: Temperature dependence of the magnetic Casimir–Polder interaction. Phys. Rev. A **80**, 062905 (2009)
28. Ellingsen, S.A., Buhmann, S.Y., Scheel, S.: Temperature-invariant Casimir–Polder forces despite large thermal photon numbers. Phys. Rev. Lett. **104**, 223003 (2010)
29. Dalibard, J., Dupont-Roc, J., Cohen-Tannoudji, C.: Vacuum fluctuations and radiation reaction: identification of their respective contributions. J. Physique (France) **43**, 1617 (1982)
30. Dalibard, J., Dupont-Roc, J., Cohen-Tannoudji, C.: Dynamics of a small system coupled to a reservoir: reservoir fluctuations and self-reaction. J. Physique (France) **45**, 637 (1984)
31. Callen, H.B., Welton, T.A.: Irreversibility and generalized noise. Phys. Rev. **83**, 34 (1951)
32. Meschede, D., Jhe, W., Hinds, E.A.: Radiative properties of atoms near a conducting plane: an old problem in a new light. Phys. Rev. A **41**, 1587 (1990)
33. Henkel. C., Joulain. K., Mulet.J.-P., Greffet.J.-J.: Radiation forces on small particles in thermal near fields. J. Opt. A: Pure Appl. Opt. **4**, S109 (2002), special issue "Electromagnetic Optics" (EOS Topical Meeting, Paris 26–30 August 2001)
34. Novotny, L., Henkel, C.: van der Waals versus optical interaction between metal nanoparticles. Opt. Lett. **33**, 1029 (2008)
35. McLachlan, A.D.: Retarded dispersion forces between molecules. Proc. R. Soc. London A. **271**, 387 (1963)
36. Wylie, J.M., Sipe, J.E.: Quantum electrodynamics near an interface. II. Phys. Rev. A. **32**, 2030 (1985)
37. Milonni, P.W.: The Quantum Vacuum. Academic Press Inc., San Diego (1994)
38. Sols, F., Flores, F.: Dynamic interactions between a charge or an atom and a metal surface. Solid State Commun. **42**, 687 (1982)

39. Buhmann, S.Y., Welsch, D.-G.: Dispersion forces in macroscopic quantum electrodynamics. Progr. Quantum Electr. **31**, 51 (2007)
40. Bezerra, V.B., Klimchitskaya, G.L., Mostepanenko, V.M., Romero, C.: Lifshitz theory of atom-wall interaction with applications to quantum reflection. Phys. Rev. A. **78**, 042901 (2008)
41. Bimonte, G.: Bohr-van Leeuwen theorem and the thermal Casimir effect for conductors. Phys. Rev. A. **79**, 042107 (2009)
42. Skagerstam, B.-S., Rekdal, P.K., Vaskinn, A.H.: Theory of Casimir–Polder forces. Phys. Rev. A. **80**, 022902 (2009)
43. Jackson, J.D.: Classical Electrodynamics. John Wiley and Sons Inc, New York, Berlin u.a., (1975) XIX, 938 S
44. Huttner, B., Barnett, S.M.: Dispersion and loss in a Hopfield dielectric. Europhys. Lett. **18**, 487 (1992)
45. Barnett, S.M., Huttner, B., Loudon, R., Matloob, R.: Decay of excited atoms in absorbing dielectrics. J. Phys. B: Atom. Mol. Opt. Phys. **29**, 3763 (1996)
46. Agarwal, G.S.: Quantum electrodynamics in the presence of dielectrics and conductors. I. electromagnetic-field response functions and black-body fluctuations in finite geometries. Phys. Rev. A. **11**, 230 (1975)
47. Eckhardt, W.: First and second fluctuation-dissipation-theorem in electromagnetic fluctuation theory. Opt. Commun. **41**, 305 (1982)
48. Eckhardt, W.: Macroscopic theory of electromagnetic fluctuations and stationary radiative heat transfer. Phys. Rev. A. **29**, 1991 (1984)
49. Rytov, S.M.: Theory of Electrical Fluctuations and Thermal Radiation. Publishing House, Academy of Sciences USSR (1953)
50. Matsubara, T.: A New approach to quantum statistical mechanics. Prog. Theor. Phys. **14**, 351 (1955)
51. Håkanson, U., Agio, M., Kühn, S., Rogobete, L., Kalkbrenner, T., Sandoghdar, V.: Coupling of plasmonic nanoparticles to their environments in the context of van der Waals—Casimir interactions. Phys. Rev. B. **77**, 155408 (2008)
52. Klimov, V., Lambrecht, A.: Plasmonic nature of van der Waals forces between nanoparticles. Plasmonics. **4**, 31 (2009)
53. Mie, G.: Beiträge zur Optik trüber Medien, speziell kolloidaler Metallösungen. Ann. Phys. **330**, 377 (1908)
54. Born, M., Wolf, E.: Principles of Optics. seventh ed. Cambridge University Press, Cambridge. (1999)
55. Feinberg, G., Sucher, J.: General form of the Van der Waals interaction: a model-independent approach. Phys. Rev. A. **2**, 2395 (1970)
56. Babb, J.F.: Long-range atom–surface interactions for cold atoms. J. Phys. Conf. Series. **19**, 1 (2005)
57. Renne, M.J.: Microscopic theory of retarded van der Waals forces between macroscopic dielectric bodies. Physica. **56**, 125 (1971)
58. Renne, M.J.: Retarded van der Waals interaction in a system of harmonic oscillators. Physica **53**, 193 (1971)
59. Davies, B.: Calculation of van der Waals forces from dispersion relations. Chem. Phys. Lett. **16**, 388 (1972)
60. Mahanty, J., Ninham, B.W.: Dispersion forces between oscillators: a semi-classical treatment. J. Phys. A: Gen. Phys. **5**, 1447 (1972)
61. Mahanty, J., Ninham, B.W.: Boundary effects on the dispersion force between oscillators. J. Phys. A: Math. Nucl. Gen. **6**, 1140 (1973)
62. Ford, G.W., Lewis, J.T., O'Connell, R.F.: Quantum oscillator in a blackbody radiation field. Phys. Rev. Lett. **55**, 2273 (1985)
63. Ford, L.H.: Spectrum of the Casimir effect. Phys. Rev. D **38**, 528 (1988)
64. Agarwal, G.S.: Quantum electrodynamics in the presence of dielectrics and conductors. I. Electromagnetic-field response functions and black-body fluctuations in finite geometries. Phys. Rev. A **11**, 230 (1975)

65. Wylie, J.M., Sipe, J.E.: Quantum electrodynamics near an interface. Phys. Rev. A **30**, 1185 (1984)
66. Sondheimer, E.H.: The mean free path of electrons in metals. Adv. Phys. **1**, 1 (1952)
67. Henkel, C., Pötting, S., Wilkens, M.: Loss and heating of particles in small and noisy traps. Appl. Phys. B **69**, 379 (1999)
68. Kittel, C.: Introduction to Solid State Physics. 7th ed. John Wiley and Son Inc., New York (1996)
69. Henkel, C.: Magnetostatic field noise near metallic surfaces. Eur. Phys. J. D **35**, 59 (2005), topical issue on Atom chips: manipulating atoms and molecules with microfabricated structures
70. Schrieffer, J.R.: (1999) Theory of Superconductivity (Perseus Books, Reading MA)
71. Joulain, K., Carminati, R., Mulet, J.-P., Greffet, J.-J.: Definition and measurement of the local density of electromagnetic states close to an interface. Phys. Rev. B **68**, 245405 (2003)
72. Philbin, T.G., Leonhardt, U.: No quantum friction between uniformly moving plates. New J. Phys. **11**(033035), 18 (2009)
73. Pendry, J.B.: Quantum friction—fact or fiction? New J. Phys. **12**, 033028 (2010)
74. Leonhardt, U.: Comment on quantum friction—fact or fiction? New J. Phys. **12**, 068001 (2010)
75. McLachlan, A.D.: Retarded dispersion forces between molecules. Proc. R. Soc. London A **271**, 387 (1963)
76. Gorza, M.P., Ducloy, M.: Van der Waals interactions between atoms and dispersive surfaces at finite temperature. Eur. Phys. J. D **40**, 343 (2006)
77. Einstein, A.: Zur Quantentheorie der Strahlung. Physik. Zeitschr. **18**, 121 (1917)
78. Wilkens, M.: Spurious velocity dependence of free-space spontaneous emission. Phys. Rev. A **47**, 671 (1993)
79. Wilkens, M.: Significance of Röntgen current in quantum optics: spontaneous emission of moving atoms. Phys. Rev. A **49**, 570 (1994)
80. Wilkens, M.: Quantum phase of a moving dipole. Phys. Rev. Lett. **72**, 5 (1994)
81. Scheel, S., Buhmann, S.Y.: Casimir–Polder forces on moving atoms. Phys. Rev. A **80**, 042902 (2009)
82. Mkrtchian, V., Parsegian, V.A., Podgornik, R., Saslow, W.M.: Universal Thermal Radiation Drag on Neutral Objects. Phys. Rev. Lett. **91**, 220801 (2003)
83. Zurita-Sánchez, J.R., Greffet, J.-J., Novotny, L.: Friction forces arising from fluctuating thermal fields. Phys. Rev. A **69**, 022902 (2004)
84. Kleppner, D.: Rereading Einstein on Radiation. Physics Today (Feb 2005) p. 30
85. Schaich, W.L., Harris, J.: Dynamic corrections to van der Waals potentials. J. Phys. F: Metal Phys. **11**, 65 (1981)
86. Dedkov, G.V., Kyasov, A.A.: Tangential force and heating rate of a neutral relativistic particle mediated by equilibrium background radiation. Nucl. Instr. Meth. Phys. Res. B **268**, 599 (2010)
87. Volokitin, A.I., Persson, B.N.J.: Near-field radiative heat transfer and noncontact friction. Rev. Mod. Phys. **79**, 1291 (2007)
88. Teodorovich, E.V.: On the Contribution of Macroscopic van der Waals Interactions to frictional force. Proc. R. Soc. London A **362**, 71 (1978)
89. Rytov, S.M., Kravtsov, Y.A., Tatarskii, V.I.: Elements of Random Fields, 3 of Principles of Statistical Radiophysics. (Springer, Berlin) (1989)
90. Knöll, L., Scheel, S., Welsch, D.-G.: In: Peřina, J. (ed) Coherence and statistics of photons and atoms, John Wiley & Sons. Inc., New York (2001)
91. Barnett. S.M.: (1997) . In: Quantum fluctuations. Les Houches, Session LXIII, 1995, edited by S. Reynaud, E. Giacobino, and J. Zinn-Justin (Elsevier, Amsterdam) 137–179
92. Henry, C.H., Kazarinov, R.F.: Quantum noise in photonics. Rev. Mod. Phys. **68**, 801 (1996)
93. Joulain, K., Mulet, J.-P., Marquier, F., Carminati, R., Greffet, J.-J.: Surface electromagnetic waves thermally excited: Radiative heat transfer, coherence properties and Casimir forces revisited in the near field. Surf. Sci. Rep. **57**, 59 (2005)

94. Antezza, M., Pitaevskii, L.P., Stringari, S.: New asymptotic behaviour of the surface-atom force out of thermal equilibrium. Phys. Rev. Lett. **95**, 113202 (2005)
95. Savasta, S., Stefano, O.D., Girlanda, R.: Light quantization for arbitrary scattering systems. Phys. Rev. A **65**, 043801 (2002)
96. Antezza, M., Pitaevskii, L.P., Stringari, S., Svetovoy, V.B.: Casimir–Lifshitz force out of thermal equilibrium. Phys. Rev. A **77**, 022901 (2008)
97. Chaumet, P.C., Nieto-Vesperinas, M.: Time-averaged total force on a dipolar sphere in an electromagnetic field. Opt. Lett. **25**, 1065 (2000)
98. Arias-González, J.R., Nieto-Vesperinas, M.: Optical forces on small particles: attractive and repulsive nature and plasmon-resonance conditions. J. Opt. Soc. Am. A **20**, 1201 (2003)
99. Dalibard, J., Cohen-Tannoudji, C.: Dressed-atom approach to atomic motion in laser light: the dipole force revisited. J. Opt. Soc. Am. B **2**, 1707 (1985)
100. Cohen-Tannoudji, C., Dupont-Roc, J., Grynberg, G.: Atom–photon interactions. Wiley, New York (1998)
101. Draine, B.: The discrete-dipole approximation and its application to interstellar graphite grains. Astrophys. J. **333**, 848 (1988)
102. Milonni, P.W., Boyd, R.W.: Momentum of Light in a Dielectric Medium. Adv. Opt. Photon. **2**, 519 (2010)
103. Antezza, M.: Surface–atom force out of thermal equilibrium and its effect on ultra-cold atoms force out of thermal equilibrium and its effect on ultra-cold atoms. J. Phys. A: Math. Theor. **39**, 6117 (2006)
104. Pitaevskii, L.P.: Long-distance behaviour of the surface-atom Casimir–Polder force out of thermal equilibrium. J. Phys. A: Math. Gen. **39**, 6665 (2006)
105. Antezza, M., Pitaevskii, L.P., Stringari, S., Svetovoy, V.B.: Casimir–Lifshitz force out of thermal equilibrium and asymptotic non-additivity. Phys. Rev. Lett. **97**, 223203 (2006)
106. Sandoghdar, V., Sukenik, C.I., Hinds, E.A., Haroche, S.: Direct measurement of the van der Waals interaction between an atom and its images in a micron-sized cavity. Phys. Rev. Lett. **68**, 3432 (1992)
107. Sukenik, C.I., Boshier, M.G., Cho, D., Sandoghdar, V., Hinds, E.A.: Measurement of the Casimir–Polder force. Phys. Rev. Lett. **70**, 560 (1993)
108. Shimizu, F.: Specular reflection of very slow metastable neon atoms from a solid surface. Phys. Rev. Lett. **86**, 987 (2001)
109. Druzhinina, V., DeKieviet, M.: Experimental Observation of Quantum Reflection far from Threshold. Phys. Rev. Lett. **91**, 193202 (2003)
110. Lin, Y.-J., Teper, I., Chin, C., Vuletić, V.: Impact of Casimir–Polder potential and Johnson noise on Bose–Einstein condensate stability near surfaces. Phys. Rev. Lett. **92**, 050404 (2004)
111. Raskin, D., Kusch, P.: Interaction between a Neutral Atomic or Molecular Beam and a Conducting Surface. Phys. Rev. **179**, 712 (1969)
112. Shih, A., Raskin, D., Kusch, P.:Investigation of the interaction potential between a neutral molecule and a conducting surface. Phys. Rev. A **9**, 652 (1974)
113. Shih, A., Parsegian, V.A.: Van der Waals forces between heavy alkali atoms and gold surfaces: Comparison of measured and predicted values. Phys. Rev. A **12**, 835 (1975)
114. Anderson, A., Haroche, S., Hinds, E.A., Jhe, W., Meschede, D., Moi, L.: Reflection of thermal Cs atoms grazing a polished glass surface. Phys. Rev. A **34**, 3513 (1986)
115. Landragin, A., Courtois, J.-Y., Labeyrie, G., Vansteenkiste, N., Westbrook C.I., Aspect, A.: Measurement of the van der Waals force in an atomic mirror. Phys. Rev. Lett. **77**, 1464 (1996)
116. Grisenti, R.E., Schöllkopf, W., Toennies, J.P., Hegerfeldt ,G.C., Köhler, T.: Determination of atom–surface van der Waals potentials from transmission-grating diffraction intensities. Phys. Rev. Lett. **83**, 1755 (1999)
117. Brühl, R., Fouquet, P., Grisenti R.E., Toennies, J.P., Hegerfeldt, G.C., Kohler, T., Stoll, M., Walter, C.: The van der Waals potential between metastable atoms and solid surfaces: novel diffraction experiments vs. theory. Europhys. Lett. **59**, 357 (2002)

118. Fichet, M., Dutier, G., Yarovitsky, A., Todorov, P., Hamdi, I., Maurin, I., Saltiel, S., Sarkisyan, D., Gorza, M.-P., Bloch, D. Ducloy, M.: Exploring the van der Waals atom-surface attraction in the nanometric range. Europhys. Lett. **77**, 54001 (2007)
119. Côté, R., Segev, B., Raizen, M.G.: Retardation effects on quantum reflection from an evanescent-wave atomic mirror. Phys. Rev. A **58**, 3999 (1998)
120. Friedrich, H., Jacoby, G., Meister, C.G.: Quantum reflection by Casimir–van der Waals potential tails. Phys. Rev. A **65**, 032902 (2002)
121. Bender, H., Courteille, P.W., Marzok, C., Zimmermann, C., Slama, S.: Direct measurement of intermediate-range Casimir–Polder potentials. Phys. Rev. Lett. **104**, 083201 (2010)
122. Harber, D.M., Obrecht, J.M., McGuirk, J.M., Cornell, E.A.: Measurement of the Casimir–Polder force through center-of-mass oscillations of a Bose–Einstein condensate. Phys. Rev. A **72**, 033610 (2005)
123. Antezza, M., Pitaevskii, L.P., Stringari, S.: Effect of the Casimir–Polder force on the collective oscillations of a trapped Bose–Einstein condensate. Phys. Rev. A **70**, 053619 (2004)
124. Dalvit, D.A.R., Lamoreaux, S.K.: Contribution of drifting carriers to the Casimir–Lifshitz and Casimir–Polder interactions with semiconductor materials. Phys. Rev. Lett. **101**, 163203 (2008)
125. Messina, R., Dalvit, D.A.R., Neto, P.A.M., Lambrecht, A., Reynaud, S.: Dispersive interactions between atoms and nonplanar surfaces. Phys. Rev. A **80**, 022119 (2009)
126. M. Antezza, Ph.D. thesis, University of Trento, 2006
127. Klimchitskaya, G.L., Mostepanenko, V.M.: Conductivity of dielectric and thermal atom—wall interaction. J. Phys. A. Math. Gen. **41**, 312002(F) (2008)
128. Brown-Hayes, M., Brownell, J.H., Dalvit, D.A.R., Kim, W.J., Lambrecht, A., Lombardo, F.C., Mazzitelli, F.D., Middleman, S.M., Nesvizhevsky, V.V., Onofrio, R., Reynaud, S.: Thermal and dissipative effects in Casimir physics. J. Phys. A. Math. Gen. **39**, 6195 (2006)
129. Klimchitskaya, G.L., Mohideen, U., Mostepanenko, V.M.: The Casimir force between real materials: experiment and theory. Rev. Mod. Phys. **81**, 1827 (2009)
130. Milton, K.A.: Recent developments in the Casimir effect. J. Phys. Conf. Series **161**, 012001 (2009)
131. Pitaevskii, L.P.: Thermal Lifshitz Force between Atom and Conductor with Small Density of Carriers. Phys. Rev. Lett. **101**, 163202 (2008)
132. Grynberg, G., Robilliard, C.: Cold atoms in dissipative optical lattices. Phys. Rep. **355**, 335 (2001)
133. Bloch, I., Dalibard, J., Zwerger, W.: Many-body physics with ultracold gases. Rev. Mod. Phys. **80**, 885 (2008)
134. Derevianko, A., Obreshkov, B., Dzuba, V.A.: Mapping Out Atom-Wall Interaction with Atomic Clocks. Phys. Rev. Lett. **103**, 133201 (2009)
135. Moreno, G.A., Dalvit, D.A.R., Calzetta, E.: Bragg spectroscopy for measuring Casimir–Polder interactions with Bose–Einstein condensates above corrugated surfaces. New J. Phys. **12**, 033009 (2010)

Chapter 12
Modern Experiments on Atom-Surface Casimir Physics

Maarten DeKieviet, Ulrich D. Jentschura and Grzegorz Łach

Abstract In this chapter we review past and current experimental approaches to measuring the long-range interaction between atoms and surfaces, the so-called Casimir-Polder force. These experiments demonstrate the importance of going beyond the perfect conductor approximation and stipulate the relevance of the Dzyaloshinskii-Lifshitz-Pitaevskii theory. We discuss recent generalizations of that theory, that include higher multipole polarizabilities, and present a list of additional effects, that may become important in future Casimir-Polder experiments. Among the latter, we see great potential for spectroscopic techniques, atom interferometry, and the manipulation of ultra-cold quantum matter (e.g. BEC) near surfaces. We address approaches based on quantum reflection and discuss the atomic beam spin-echo experiment as a particular example. Finally, some of the advantages of Casimir-Polder techniques in comparison to Casimir force measurements between macroscopic bodies are presented.

12.1 Introduction

In this chapter we will be dealing with experimental aspects of the Casimir-Polder force. (See Chap. 11 by Intravaia et al. in this volume for additional discussions about the theoretical aspects of the Casimir-Polder force.) Although no strict criterion can be implied to distinguish between the Casimir and the Casimir-Polder

M. DeKieviet (✉) · G. Łach
Physikalisches Institut der Universität Heidelberg, Philosophenweg 12,
69120, Heidelberg, Germany
e-mail: maarten.dekieviet@physi.uni-heidelberg.de

U. D. Jentschura
Department of Physics, Missouri University of Science and Technology,
Rolla, MO 65409-0640, USA

D. Dalvit et al. (eds.), *Casimir Physics*, Lecture Notes in Physics 834,
DOI: 10.1007/978-3-642-20288-9_12, © Springer-Verlag Berlin Heidelberg 2011

effect, it is common use to describe the interaction between two macroscopic, polarizable (neutral and nonmagnetic) bodies due to the exchange of virtual photons as Casimir interactions. In contrast, Casimir-Polder effects should involve at least one microscopic, polarizable body, typically an atom or molecule. Although the transition from Casimir to Casimir-Polder physics is a continuous one,[1] the distinction goes back to the seminal 1948 paper by Casimir and Polder [1]. Herein, the authors discuss the influence of retardation on the London-van der Waals forces. In particular, they derive the long-distance behavior of the quantum electrodynamic interaction between an atom and a perfectly conducting surface and that between two atoms. The works by Casimir and Casimir and Polder address the fact that the mutually induced polarization between two neighboring objects may be delayed as a consequence of the finiteness of the velocity of light. The Casimir and the Casimir-Polder forces could thus semiclassically be termed long-range retarded dispersion van der Waals forces. Indisputably the two 1948 papers by Casimir and Casimir and Polder mark the beginning of a whole new branch of research addressing fundamental questions about quantum field theory in general and the structure of the vacuum in particular. The concepts developed in 1948 are now being used in order to describe a rich field of physics, and have been supplemented by a variety of methods to address also practically important areas of applications. Especially with the rise of nanotechnology, there is a growing need for a quantitative understanding of this interaction and experimental tests are indispensable.

12.2 The History of Casimir-Polder Experiments

Before we turn to modern aspects of Casimir-Polder physics it is useful to review some of the developments that led us here. The correct explanation for the non-retarded dispersive van der Waals interaction between two neutral, but polarizable bodies was possible only after quantum mechanics was properly established. Using a perturbative approach, London showed in 1930 [2] for the first time that the above mentioned interaction energy is approximately given by

$$\mathscr{V}_{London}(z) \approx -\frac{3\hbar\omega_0\alpha_1^A\alpha_1^B}{2^6\pi^2\varepsilon_0^2 z^6} \tag{12.1}$$

where α_1^A and α_1^B are the static dipole polarizabilities of atoms A and B, respectively. ω_0 is the dominant electronic transition frequency and z is the distance between the objects. Experiments on colloidal suspensions in the 1940s by Verwey and Overbeek [3] showed that London's interaction was not correct for large distances. Motivated by this disagreement, Casimir and Polder were the first in

---

[1] It is in fact a useful and instructive exercise to derive the latter from the former by simple dilution.

1948 to consider retardation effects on the van der Waals forces. They showed that in the retarded regime the van der Waals interaction potential between two identical atoms is given by

$$\mathcal{V}_{retard}(z) = -\frac{23\hbar c \alpha_1^A \alpha_1^B}{2^6 \pi^3 \varepsilon_0^2 z^7}.$$ (12.2)

The reason is that, at larger separations, the time needed to exchange information on the momentary dipolar states between the two objects may become comparable to or larger than the typical oscillation period of the fluctuating dipole. The length scale at which this happens can be expressed in terms of a reduced wavelength $\hat{\lambda}$ of the lowest allowed atomic transition that relates to the transition frequency ω_0 (or wavelength λ) and the speed of light c

$$z \geq \frac{c}{\omega_0} = \frac{\lambda}{2\pi} = \hat{\lambda}.$$ (12.3)

The onset of retardation is thus a property of the system. In their paper, Casimir and Polder also showed that the retarded van der Waals interaction potential between an atom and a perfectly conducting wall falls at large distances as $1/z^4$, in contrast to the result obtained in the short-distance regime (non-retarded regime), where it is proportional to $1/z^3$. They derived a complete interaction potential valid also for intermediate distances. By the use of the dynamic polarizability of the atom $\alpha(i\omega)$ the Casimir and Polder result can be rewritten as

$$\mathcal{V}^{\infty}(z) = -\frac{\hbar}{(4\pi)^2 \varepsilon_0 z^3} \int_0^{\infty} d\omega \alpha_1(i\omega)\left[1 + 2\frac{\omega z}{c} + 2\left(\frac{\omega z}{c}\right)^2\right]e^{-2\omega z/c}.$$ (12.4)

Its short-range limit, equivalent to the nonrelativistic approximation ($c \to \infty$) reproduces the van der Waals result for the atom-surface interaction energy

$$\mathcal{V}^{\infty}(z) \stackrel{z \to 0}{=} -\frac{\hbar}{(4\pi)^2 \varepsilon_0 z^3} \int_0^{\infty} d\omega \alpha_1(i\omega).$$ (12.5)

The long-distance limit for the perfect conductor case is especially important and has become the signature for the Casimir-Polder force

$$\mathcal{V}^{\infty}(z) \stackrel{z \to \infty}{=} -\frac{3\hbar c \alpha(0)}{2^5 \pi^2 \varepsilon_0 z^4}.$$ (12.6)

In the 1960s, Lifshitz and collaborators developed a general theory of van der Waals forces [4] and extended the result of Casimir and Polder to arbitrary solids. (See also the Chap. 2 by Pitaevskii in this volume for additional discussions about the Lifshitz theory.) Their continuum theory is valid for both dielectrics and semiconductors, and for conductors, as long as their electromagnetic properties can be described by a local $\varepsilon(\omega)$. The result for a perfect conductor is recovered by taking the limit of infinite permittivity: $\varepsilon(\omega) \to \infty$. Lifshitz [4] computed the interaction energy between an atom and a realistic material, described by its frequency dependent permittivity evaluated for imaginary frequencies $\varepsilon(i\omega)$. His result gives the dominant contribution for large atom-surface distances, and reads

$$\mathscr{V}(z) = -\frac{2\hbar}{(4\pi)^2 \varepsilon_0 c^3} \int_0^\infty d\omega \omega^3 \alpha_1(i\omega) \int_1^\infty d\xi e^{-2\xi\omega z/c} H(\xi, \varepsilon(i\omega)), \qquad (12.7)$$

where the $H(\xi, \varepsilon)$ function is defined as

$$H(\xi, \varepsilon) = (1 - 2\xi^2)\frac{\sqrt{\xi^2 + \varepsilon - 1} - \varepsilon\xi}{\sqrt{\xi^2 + \varepsilon - 1} + \varepsilon\xi} + \frac{\sqrt{\xi^2 + \varepsilon - 1} - \xi}{\sqrt{\xi^2 + \varepsilon - 1} + \xi}. \qquad (12.8)$$

The Lifshitz formula (12.7) reproduces the Casimir and Polder result (12.4) for perfect conductors. Both in the short- and the long-range limits, the Lifshitz formula simplifies considerably. For $z \to 0$ the interaction potential behaves as

$$\mathscr{V}(z) \overset{z\to 0}{\sim} -\frac{\hbar}{(4\pi)^2 \varepsilon_0 z^3} \int_0^\infty d\omega \alpha_1(i\omega)\frac{\varepsilon(i\omega) - 1}{\varepsilon(i\omega) + 1} \equiv -\frac{C_3}{z^3}. \qquad (12.9)$$

In the long-distance limit the interaction potential for a generic $\varepsilon(\omega)$ is

$$\mathscr{V}(z) \overset{z\to\infty}{\sim} -\frac{3\hbar c\alpha_1(0)}{2(4\pi)^2 \varepsilon_0 z^4}\frac{\varepsilon(0) - 1}{\varepsilon(0) + 1} \equiv -\frac{C_4}{z^4}. \qquad (12.10)$$

Under some special conditions the Casimir and Polder force can also be inferred (both theoretically and experimentally) from the interaction energy of two macroscopic bodies, if we consider one of the media sufficiently dilute [5]. These conditions were fulfilled in the experiments by Sabisky and Anderson [6]. In this sense this 1972 experiment can be viewed as the first experimental verification of the Casimir-Polder force. In their beautiful cryogenic experiments Subisky and Anderson measured the thickness of helium films adsorbed on cleaved SrF$_2$ surfaces in thermal equilibrium at 1.4 K as a function of hyperpressure. The authors get exceptionally good agreement between the film thickness measured and the one calculated using the Lifshitz formula. The agreement is even more remarkable since the method strongly depends on the assumption that the non-additivity of the van der Waals forces does not play a major role. In this case that assumption works so well because of the extraordinary weakness of the interaction between helium atoms. The authors note that the best results were obtained on atomically flat regions of the substrate and that the influence of surface roughness in these experiments was certainly significant.

Only two years later Shih and Parsegian [7] published an atomic beam deflection experiment to measure van der Waals forces between heavy alkali atoms and gold surfaces. With these precision measurements the authors were able for the first time to pin down the distance dependency of the van der Waals force in

Fig. 12.1 The Cs beam profile, normalized to the full beam intensity, measured as a function of the deflection distance S, into the geometric shadow of the gold surface. The derived interaction potential is consistent with the van der Waals form $1/z^3$ and has a strength which is best described within the macroscopic continuum theory of Lifshitz. (Reproduced with permission from [7].)

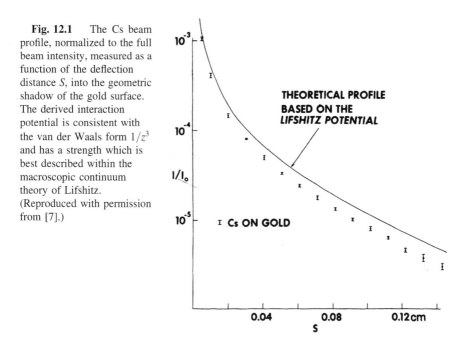

the non-retarded regime to be proportional to $1/z^3$. The authors include very accurate calculations of the interaction potential performed using the Lifshitz formula. They used *ab-initio* computed (on the Hartree-Fock level) atomic polarizabilities and included the finite conductivity of the gold substrate. The resulting strength coefficient C_3 of the van der Waals interaction is not a constant but varies with separation (the retardation effect). Unfortunately, the measurements (see Fig. 12.1) were not precise enough in the region of interest to reveal these retardation effects as a deviation from the $1/z^3$ behavior. It is worth noting that although these calculations were the best available at the time the theoretical values they predict are systematically 60% larger than the values for the interaction observed in experiments. The authors suggest surface roughness and contamination as possible sources for the discrepancy between calculation and experiment. Still, this seminal paper is the first definite confirmation of the validity of the Dzyaloshinskii-Lifshitz-Pitaevskii [8] formalism for the interaction between an atom and a surface, and establishes QED vacuum fluctuations as the common basis for van der Waals and Casimir-Polder forces.

It took almost two more decades before a deviation from the $1/z^3$ behavior for the van der Waals potential was experimentally resolved. In Ed Hinds's group [9] a beam of ground state sodium atoms was passed through a micron-sized cavity. This geometry does not exactly correspond to the Casimir-Polder atom single plate arrangement, but instead the atoms travel slowly through a cavity made of plates that include a small angle (V-shape). The cavity width can be varied by moving the source vertically. As a function of plate separation the transmission loss due to the long-range interaction with the cavity walls was measured and compared to Monte-Carlo simulations. At large plate distances the purely geometric losses dominate, but below a critical width Casimir-Polder losses become significant.[2] Hinds et al. could clearly show a modification of the ground state Lamb shift for the sodium atoms within the cavity, scaling as $1/z^4$. Furthermore, the numerical coefficient of the $1/z^4$ interaction was found to be in agreement with theory. Unfortunately the data do not show a smooth transition to the expected van der Waals behavior at short distances. The experiment relies heavily on the fact that any atoms not traveling through the center of the cavity are accelerated towards one of the two plates, stick there and are thus removed from the beam. This accumulation and its influence on the spectral properties of the plates were not taken into account.

With the advent of methods for laser cooling and manipulating atoms it became possible to use light in order to control ultracold atom-wall collisions. Aspect et al. [10] were the first to use laser cooled and trapped atoms to investigate the Casimir-Polder potential. ^{87}Rb atoms, caught at a temperature of 10 μK, were released from a magneto-optical trap into the gravitational potential. After 15 mm of free fall, the ultracold beam of ^{87}Rb atoms impinged at normal incidence on an atomic mirror. This mirror consists of a prism that is irradiated from the back by laser light. This laser light forms an evanescent wave extending along the face of the dielectric, and as it is slightly detuned from the atomic resonance frequency of the ^{87}Rb atoms it can provide a controlled repulsive potential for the atoms. Within this light potential the equilibrium between a repulsive light force and the attractive surface interaction is used to establish a potential barrier that can be well controlled by changing the light field. The reflectivity for incident atoms is measured as a function of barrier height. Landragin's data seem to favor the inclusion of retardation effects in the atom-wall interaction potential.

In 2001 Shimizu succeeded in reflecting very slow metastable neon atoms specularly from a solid surface[3] [11]. Shimizu prepared an ultracold beam of neon atoms by trapping the metastables in a magneto-optical trap and then releasing them into the gravitational potential. After tens of centimeters of free fall, the metastables impinged under grazing incidence on the substrate. By varying the angle of incidence he could change the normal incident velocity of the metastable neon atoms between 0 and 35 mm/s. New in his approach was that the quantum reflectivity, that is the reflection from the attractive part of the interaction potential only, depends heavily on the perpendicular impinging energy of the particle. Plotting the reflectivity of metastable neon atoms versus the normal incident velocity of the beam the author was able to uniquely identify that the attractive

---

[2] It is not entirely clear how much of the nontrivial geometrical effects were taken into account in the data analysis.

[3] In early experiments of quantum deflecting H atoms, liquid helium was used as a target. Collectiveness of He atoms within the fluid was ignored, giving information only on the H–He interatomic potential. It is not quite clear how many of the exotic properties of this superfluid have an effect on the interaction potential with this macroscopic quantum object.

interaction was of the Casimir-Polder sort, that is having a $1/z^4$ dependency. A simple combined van der Waals Casimir-Polder potential was used to fit the data, but the measurements were not accurate enough to allow for a quantitative comparison with theory. The surfaces Shimizu used were rough on a scale much larger than the de Broglie wavelength of the atoms. This and the fact that the Ne atoms were in a metastable state guarantees that none of the atoms is specularly reflected from the repulsive core potential at short distances.[4]

In 2003 our group published the first experimental observation of quantum reflection of ground state atoms from a solid surface [12]. Using the extreme sensitivity of the earlier developed atomic beam spin echo method [13] we monitored ultracold collisions of ^{3}He atoms from the rough surface of a quartz single crystal. The roughness of the substrate guaranteed that the reflection could only originate from the attractive part of the potential. The quantum reflectivity data confirmed the theory by Friedrich et al. [14] on the asymptotic behavior of the quantum reflection probability at incident energies far from the threshold $E_i \rightarrow 0$.[5] The data confirm this high energy asymptotic behavior which is determined by the van der Waals interaction and show a gradual transition towards the retarded regime starting as early as 30 Å above the surface. From the data the gas-solid interaction potential is deduced quantitatively covering both regions. Using a simple approximation for the inhomogeneous attractive branch of the helium-quartz interaction potential

$$\mathscr{V}(z) = -\frac{C_4}{z^3(z+\hat{\lambda})} \qquad (12.11)$$

we find excellent quantitative agreement with the quantum reflection experimental data for the potential coefficient $C_4 = 23.6$ eV Å^3 and the reduced atomic transition wavelength $\hat{\lambda} = 93$Å. The experiment establishes quantitative evidence for the $1/z^4$ Casimir-Polder attraction and the transition towards the $1/z^3$ van der Waals regime. Surface roughness was taken into account explicitly, based on an ex-situ atomic force microscopic measurement. It is worthwhile stressing that both the atom and the substrate are in their ground state and no lasers are involved. It thus represents the situation of a single atom interacting with a well defined, extended, dielectric body through the virtual photons of the quantum fluctuating vacuum exclusively quite accurately.

The experimental data were also investigated with respect to the inhomogeneity of the potential model used in (12.11). The analysis shows that neither of the two homogeneous parts, the van der Waals and the Casimir-Polder branches, can describe the data without the inclusion of the other. Even though at the distances explored in this experiment the energies are dominated by the van der Waals interaction, there is a definite need to include the Casimir-Polder term explicitly.

---

[4] Mind that metastables may exchange real photons with the substrate.

[5] For a review on WKB waves far from the semiclassical limit see [15] and references therein.

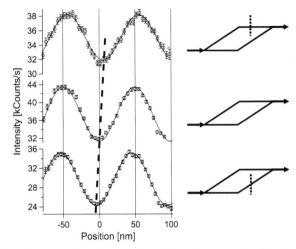

Fig. 12.2 Interference pattern observed for sodium atoms passing the atom interferometer without a grating (*middle*), or when the grating is inserted into path α (*upper*) or path β (*lower*). The dashed line illustrates the measured phase shift of 0.3 rad. From the measured induced phase shift as a function of atomic velocity, the authors obtain a van der Waals coefficient $C_3 = 3$ meV nm^3. (Reproduced with permission from [16].)

This stipulates that the onset of retardation is not only smooth, but extends to well within the van der Waals regime.

Around the same period, Perreault and Cronin [16] used an atom interferometer, in which a well collimated beam of sodium atoms illuminates a silicon nitride grating with a period of 100 nm. During passage through the grating slots atoms acquire a phase shift due to the van der Waals interaction with the grating walls. As a result the relative intensities of the matter-wave diffraction peaks deviate from those expected for a purely absorbing grating. A complex transmission function was developed to explain the observed diffraction envelopes. By fitting a modified Fresnel optical theory to the experimental data the authors obtain a van der Waals coefficient $C_3 = 2.7 \pm 0.8$ meV nm^3 for the interaction between atomic sodium and a silicon nitride surface. A few years later, the experiment was refined by going to a 50 nm wide cavity. The magnitude of the measured phase shift (see Fig. 12.2) caused by atom-surface interaction is in agreement with that predicted by quantum electrodynamics for a non-retarded van der Waals interaction.

Steady progress in laser cooling and quantum optics and the application of evaporative cooling at the turn of the century led to the first successful creation of a macroscopic quantum system. Nowadays Bose-Einstein condensates are produced routinely in dozens of labs all over the world. Their behavior and stability in the vicinity of a solid wall has been investigated for both fundamental and technical reasons. The first such experiment focusing on the Casimir-Polder interaction was performed in the group of Ketterle [17] using a Bose-Einstein condensate of ^{23}Na atoms. Since the BEC is much colder than the thermal ensemble in a MOT (magnetic-optical trap), like the one Shimizu used, the authors observed quantum reflectivity for incident angles up to normal. Another major difference is that the silicon substrate used in this experiment is approached while the atoms are still trapped in a weak gravito-magnetic trap. The measured reflection probability as a function of incident normal velocity was compared with a numerical simulation done for sodium atoms interacting with a conducting wall. The comparison shows qualitative agreement. Even though the authors simulate the interaction between the sodium atoms and the semiconductive surface through a C_4 coefficient which corresponds to a conducting surface, the calculated reflection probability is still systematically higher than the measured data points. The experiment confirms the $1/z^4$ behavior but the range of velocities the authors could explore is not large enough to investigate the region closer to the surface where the potential has $1/z^3$ dependence. The authors refer to earlier work by Cornell et al. [18] (see also Chap. 11 by Intravaia et al. in this volume for additional discussions) and Vuletic [19], who discuss serious problems that condensate-based experiments near surfaces may experience due to the pollution of even only a small number of atoms at the surface. Spurious electric and magnetic fields caused by the adsorbates may lead to large local anomalies in the interaction potential and may severely limit sensitive Casimir-Polder force measurements.[6] A few years later, a refined version of the experiment was published and interpreted as a measurement of the temperature dependence of the Casimir-Polder force [20]. In the analysis, however, the temperature dependence of the coverage and the IR properties of the adsorbates were not included. In a recent experiment by Zimmermann et al. [21], which is comparable to the experiment in [10], an unresolved inconsistency in the quantitative comparison between Casimir-Polder theory and experiment is still present.

It is fair to say that in present day Casimir-Polder physics the quantitative comparison between experiments and theory is limited by the lack of detailed knowledge of the exact experimental parameters like surface roughness and cleanliness, identity of the adsorbates and their influence on the overall atom-surface interaction potential. Consequently, modern experiments on atom-surface Casimir physics should not only aim for higher precision, but simultaneously for a better characterization and control of these experimental conditions.

Before discussing the different experimental approaches in this context, it may be useful to identify the physics that could possibly be revealed in future Casimir-Polder studies. In the following section we summarize some of the current theoretical issues.

---

[6] In [17] Ketterle makes the following observation: "Surfaces are traditionally considered enemies of cold atoms: Laser cooling and atom optics have developed thanks to magnetic and optical traps that confine atoms with non-material walls in ultrahigh vacuum environments designed to prevent contact with the surfaces. Paradoxically, it turns out that in the extreme quantum limit of nano Kelvin matter waves, a surface at room temperature might become a useful device to manipulate atoms."

12.3 The Atom-Surface Interaction

12.3.1 Practical Application of the Lifshitz Formula

The properties of the solid enter the formula (12.7) in the form of its frequency dependent permittivity. Although it has been demonstrated that ε can in some cases be computed *ab-initio* [22], the accuracy and reliability of the obtained results are uncertain at present. Possible difficulties include the contribution of vibrational excitations within the solid to the atom-surface interaction energy [23]. This term can be relatively large, like in the case of glass or α–quartz (10%) and has so far been neglected in the full *ab-initio* approach.

Some of the solids for which the contemporary experiments are performed are transition metals, for which both the relativistic effects and the fact that metals are open shell systems make accurate theoretical predictions extremely difficult. Another drawback of the presently available theoretical results for solids is the neglect of temperature effects. As a result the theoretically computed permittivity as a function of frequency typically exhibits more structure than the ones derived from experimental optical data. This makes the use of published optical data a useful alternative to theoretical computation of the dielectric properties of solids (e.g. [24] and references therein). The procedure, however, is complicated by the fact that experimental data for the complex permittivity exist only for real frequencies and its values along the imaginary axis have to be reconstructed using the Kramers-Kroenig relation:

$$\varepsilon(i\omega) = 1 + \frac{2}{\pi} \int_0^\infty d\xi \frac{\xi \operatorname{Im}\varepsilon(\xi)}{\omega^2 + \xi^2}. \tag{12.12}$$

Apart from the frequency dependent permittivity of the solid the computation of the atom-surface interaction energy requires the dynamic dipole (or multipole) polarizability of the atom as an input. In the past a single resonance model for the dynamic polarizability was used:

$$\alpha_1(i\omega) = \frac{\alpha_1(0)}{1 + \omega^2/\omega_1^2}, \tag{12.13}$$

where ω_1 is the lowest allowed transition frequency. This simplification may lead to an error in the short range interaction energy as large as 40% (for the case of He) when compared to results that use more accurate polarizabilities. Slight improvement is obtained by using a few-resonance approximation, in which a few lowest lying excited states are included in the spectral expansion:

$$\alpha_1(i\omega) = \sum_i \frac{f_{i0}^{(1)}}{\omega_i^2 + \omega^2}, \tag{12.14}$$

where ω_i denote the allowed transition frequencies, and $f_{i0}^{(1)}$ (see for example [23]) are the corresponding oscillator strengths. Even when all discrete transition frequencies are included, this approximate approach still leads to a 20% error (for ground state He) as it neglects the significant contribution coming from excitations to the continuous spectrum. We therefore recommend using *ab-initio* computed dynamic atomic polarizabilities whenever possible. Dynamic polarizabilities of the noble gas, alkali and alkaline atoms evaluated for imaginary frequencies have recently been published [25].

12.3.2 Limitations of the Lifshitz Theory

The Lifshitz formula describes the interaction of an arbitrary atom with any surface and is a good starting point for the comparison between atom-surface theory and experiment. It should be realized, however, that it suffers from a number of drawbacks that may become important in future Casimir-Polder experiments. As pointed out earlier, formula (12.7) becomes invalid for atom-surface distances comparable to the charge radius of the atom (of the order of Å), due to exchange effects. At slightly larger distances, of the order of a few Å (for ground state atoms, but as large as a few nm for metastable helium atoms), the Lifshitz approach, being based on second order perturbation theory, breaks down due to higher order effects. At even larger distances (a few nm for He) its validity is reduced further, when quadrupole and higher multipole polarizabilities of the atom are to be included. These effects can be theoretically computed and are considered in the next sections. These, and other corrections to the Lifshitz formula, that cannot easily be assigned a definite length scale will be considered in the following sections. Their list includes:

- Corrections from higher orders of perturbation theory
- Contributions from higher multipole polarizabilities of the atom
- Temperature effects
- Relativistic and radiative corrections
- Effects of magnetic susceptibilities of both the atom and the solid
- Corrections from the nonplanar geometry and imperfections of the surface

From the experimental point of view some of these effects are already in reach with current methods. Others, like the relativistic and radiative corrections, or the effects of nonlocal response of the surface, may only become visible in future generations of experiments.

12.3.3 Higher Orders of Perturbation Theory

The applicability of the Lifshitz formula is limited to distances z larger than $\sqrt[3]{\alpha_1(0)/(4\pi\varepsilon_0)}$, at which higher orders of perturbation theory become important.

The next nonzero contribution to the atom-surface interaction energy comes from the fourth order in atom-field coupling, i.e. second order in atomic dipole polarizability. This has been calculated by Marvin and Toigo [26] and reads

$$\mathscr{V}_{1,1}(z) = -\frac{\hbar}{2^9 \pi^2 \varepsilon_0 z^6} \int_0^\infty d\omega \alpha_1^2(i\omega) \left[\frac{\varepsilon(i\omega) - 1}{\varepsilon(i\omega) + 1} \right]^2 e^{-2\omega z/c} P_{1,1}(\omega z/c), \quad (12.15)$$

where the polynomial $P_{1,1}(x)$ is defined as

$$P_{1,1}(x) = 3 + 12x + 16x^2 + 8x^3 + 4x^2. \quad (12.16)$$

The $\{1,1\}$ index indicates that (12.15) depends on the square of the dipole polarizability of the atom. It is only the first term of the series of corrections and the ones depending on products of multipole polarizabilities followed by $\mathscr{V}_{1,2}$, $\mathscr{V}_{1,3}$, $\mathscr{V}_{2,2}$, $\mathscr{V}_{1,1,1}$, etc., all of which are at present not calculated.

It is worth noting that together with the above result the same authors present an incorrect formula for the first order term, given here in (12.7). Their leading order result [26] including retardation is in disagreement with [4, 23, 27]. Its short-range, nonrelativistic limit is, however, in agreement with other works. The short-range limit of (12.15) is

$$\mathscr{V}_{1,1}(z) = -\frac{\hbar}{2^9 \pi^2 \varepsilon_0 z^6} \int_0^\infty d\omega \alpha_1^2(i\omega) \left[\frac{\varepsilon(i\omega) - 1}{\varepsilon(i\omega) + 1} \right]^2 \quad (12.17)$$

and is consistent with the nonrelativistic result of McLachlan [28].

12.3.4 Effect of Multipole Polarizabilities

Another group of effects limiting the use of (12.7) are those resulting from the higher multipole polarizabilities of the atom, like quadrupole and octupole ones. The corrections from quadrupole polarizability of the atom become important when z is comparable to $\sqrt{\alpha_2(0)/\alpha_1(0)}$, where α_1 and α_2 denote the dipole and quadrupole polarizabilities of the atom.

The effects of the quadrupolar polarizability of the atom on interaction potential $\mathscr{V}_2(R)$ were first calculated in the nonrelativistic case by Zaremba and Hutson [29]. In general the second order interaction energy can be written as a multipole expansion

$$\mathscr{V}(z) = \mathscr{V}_1(z) + \mathscr{V}_2(z) + \mathscr{V}_3(z) + \cdots \quad (12.18)$$

where each of the $\mathscr{V}_n(z)$ terms comes from the 2^n-pole atomic polarizability coupled to the fluctuating electromagnetic field. The first term $\mathscr{V}_1(z)$ is the previously considered Lifshitz contribution, while the next ones constitute corrections, which in the limit of small distances behave as C_{2n+1}/z^{2n+1}.

For short distances, higher terms in the multipole expansion used to derive (12.7) become important. The generalization of the derivation of the Lifshitz formula to include higher multipoles has been presented in [23], and gives contributions to the interaction energy from an arbitrary 2^n-pole polarizability of the atom:

$$\mathcal{V}_n(z) = -\frac{\hbar}{8\pi^2\varepsilon_0 c^{2n+1}} \int_0^\infty d\omega\, \omega^{2n+1} \alpha_n(i\omega) \int_1^\infty d\xi\, e^{-2\xi\omega z/c} P_n(\xi) H(\xi, \varepsilon(i\omega)),$$

(12.19)

where $\alpha_n(\omega)$ is the 2^n-pole polarizability and $P_n(\xi)$ are polynomials which for $n = 1, 2, 3, 4$ (dipole, quadrupole, octupole and hexadecupole components) explicitly read

$$P_1(\xi) = 1,$$ (12.20a)

$$P_2(\xi) = \frac{1}{6}(2\xi^2 - 1),$$ (12.20b)

$$P_3(\xi) = \frac{1}{90}(4\xi^4 - 4\xi^2 + 1),$$ (12.20c)

$$P_4(\xi) = \frac{1}{90}(8\xi^6 - 12\xi^4 + 6\xi^2 - 1).$$ (12.20d)

In the limit of small distances or, equivalently, in the nonrelativistic limit ($c \to \infty$) (12.19) leads to a surprisingly simple result:

$$\mathcal{V}_n(z) \stackrel{z\to 0}{=} -\frac{\hbar}{(4\pi)^2\varepsilon_0 z^{2n+1}} \int_0^\infty d\omega\, \alpha_n(i\omega) \frac{\varepsilon(i\omega) - 1}{\varepsilon(i\omega) + 1}.$$ (12.21)

The $n = 2$ case reproduces the short-range asymptotic result of Zaremba and Hutson [29]. The derivation of the long-distance limit of the multipole contributions lead to

$$\mathcal{V}_n(z) \stackrel{z\to\infty}{=} -\frac{\hbar c \alpha_n(0)}{(4\pi)^2\varepsilon_0 z^{2n+2}} D_n \frac{\varepsilon(0) - 1}{\varepsilon(0) + 1},$$ (12.22)

where the D_n constants are: $D_1 = 3/8$, $D_2 = 25/12$, $D_3 = 301/120$, $D_4 = 1593/560$.

The quadrupole and octupole dynamic atomic polarizabilities of helium atoms (in the ground and metastable states) together with accurate (at 10^{-8}) and simple global fits have been published [30]. Simulations based on potentials computed using these multipole polarizabilities have shown that, in the quantum reflection experiments of the type reported in [12], the quadrupolar contribution could be seen at the few percent level.

12.3.5 Effects of Non-zero Temperature

Modification of the atom-surface interaction energy due to non-zero temperature
has two distinct origins. The permittivity of the solid, entering the formulae
(12.7)–(12.19), is itself temperature dependent. More importantly there is also an
explicit temperature dependence resulting from thermal fluctuations of the electromagnetic field, calculated first by Dzyaloshinskii et al. [8]. The complete formula in a form closely resembling (12.7) reads [31, 32]

$$\mathscr{V}_1(z;T) = \frac{\hbar k_B T}{4\pi\varepsilon_0} \sum_{i=0}^{\infty} \left(1 - \frac{1}{2}\delta_{0i}\right) \alpha_1(i\omega_i) \int_1^{\infty} d\xi e^{-2\xi\omega_i/c} H(\xi, \varepsilon(i\omega_i)), \qquad (12.23)$$

where k_B is the Boltzmann constant, $\omega_i = 2\pi i k_B T z/\hbar$ are the Matsubara
frequencies and $\delta_{0i} = 1$ for $i = 0$, and is equal to 0 otherwise. The above result
relies on the assumption that the temperature is small enough to neglect thermal
excitations of the atom. This assumption is true as long as $\hbar\omega_1 \gg k_B T$, which is
fulfilled for most ground state atoms at any reasonable temperature (i.e. below the
melting point of the experimentally considered solids), but does not have to be true
for experiments with metastable excited states.

Just as for the $T = 0$ case, in the short-range limit formula (12.23) can be
approximated by

$$\mathscr{V}_1(z;T) \overset{z\to 0}{=} -\frac{C_3(T)}{z^3}, \qquad (12.24)$$

where

$$C_3(T) = \frac{\hbar k_B T}{16\pi\varepsilon_0} \left[\alpha(0) \frac{\varepsilon(0) - 1}{\varepsilon(0) + 1} + 2 \sum_{i=1}^{\infty} \alpha(i\omega_i) \frac{\varepsilon(i\omega_i) - 1}{\varepsilon(i\omega_i) + 1} \right]. \qquad (12.25)$$

In the low temperature limit the above result approaches the $T = 0$ one (12.9).
It has already been discovered by Lifshitz that, in comparison to the zero-temperature case, the long-distance behavior of (12.23) is qualitatively different. For
distances much greater than the thermal length scale introduced in Matsubara
frequencies

$$z \gg \lambda_T \equiv \frac{\hbar c}{k_B T}, \qquad (12.26)$$

the atom-surface interaction energy behaves as:

$$\mathscr{V}(z) \overset{z\to\infty}{\sim} -\frac{k_B T \alpha_1(0)}{4(4\pi)\varepsilon_0 z^3} \frac{\varepsilon(0) - 1}{\varepsilon(0) + 1}. \qquad (12.27)$$

It is worth noting that the coefficient at the long distance $1/z^3$ asymptotics given by
(12.27) vanishes for $T \to 0$, but the range of distances where it is valid, given by

(12.26), also widens. The distance scale of the temperature effects λ_T is usually orders of magnitude larger than the retardation distance scale λ and for experimentally considered cases the $1/z^4$ asymptotics given by (12.6) is valid for a wide range of distances. At room temperature λ_T is on the order of 100 μm (infrared wavelength), while λ is on the order of 100 nm for most ground state atoms (ultraviolet wavelength). For conductors, due to the singularity of $\varepsilon(\omega)$ at zero frequency, (12.27) takes a universal form independent of any characteristics of the solid. It remains a subject of current research whether or not this result requires modification. This peculiarity is especially pronounced for conductors with a very low concentration of carriers, or with slowly moving carriers (for example ions) [33].

12.3.6 Relativistic and Radiative Corrections

The theory of relativistic corrections to the interaction between closed shell atoms was recently developed by Pachucki [34]. Earlier studies of the interplay between relativistic and radiative corrections had been done by Meath and Hirschfelder [35, 36]. The general result is that the relativistic corrections of the lowest order (α^2) can be grouped into three categories, where the ones dominant at large interatomic distances (orbit-orbit, or Breit interaction) are already incorporated in the Casimir-Polder formula. Corrections can be classified as relativistic corrections to dynamic atomic polarizabilities, or as coupling between electric polarizability of one atom and the magnetic susceptibility of the other [34]. The relativistic and QED corrections to atomic polarizabilities are, at least for the well studied case of He, negligible at the level of precision considered here [37], and so are contributions from atomic magnetic susceptibility. On the other hand the contribution to the Casimir-Polder force resulting from the magnetic susceptibility of the solid can lead to measurable effects in the case of ferromagnetic materials [38].

12.3.7 Effects of Nonplanar Geometry and Nonuniformity

As described in the introductory chapter, many of the original experiments investigating the atom-surface interactions have been performed for geometries different from the planar one. The cases of an atom interacting with macroscopic spheres and cylinders (or wires) have been partly solved [39–41]. For more complicated geometries corrections can be estimated, at least for dielectric materials, using the pairwise summation method [5]. When the radius of curvature of the surface is much larger than the relevant range of atom-surface distances (which is frequently the case) the resulting geometric corrections are negligible. For all other cases, a first rigorous theoretical study on the inclusion of uniaxial corrugations has been performed for a scalar field by Gies et al. [42]. Unfortunately the extension of these results to electromagnetic fields is nontrivial. An important exception results from the fact that even if the experiments are meant to be performed on flat surfaces, the nontrivial geometry enters in the form of corrections induced by corrugations or imperfections of the surface. In the limit of atom-surface distance much larger than the surface corrugations the roughness of the surface can be taken into account as a modification of its reflection coefficients [43].

The case when neither of the conditions mentioned above is fulfilled, which means that the curvature radius of the surface is neither very small nor very large in comparison to the atom-surface distance, and the surface is a conductor, is an open problem.

Modern Casimir-Polder Experiments

Modern experiments on atom-surface Casimir physics should not only be more sensitive and precise as a technique, but also be able to characterize and control the system under investigation in a quantitative manner. The first requirement is obvious; the second, however, no less important. Progress in this field will critically depend on the applicability of direct comparisons between theory and experiment and thus on the accuracy of the system parameters used on both sides. Let us relate this to the different type of modern Casimir-Polder experiments according to the approach they apply.

- Spectroscopic measurements: It is common knowledge in experimental physics that the best way for improving precision is to design a signature that can be measured as a frequency. In this respect the (quantum) optics experiments of the type mentioned above naturally promise a huge potential for Casimir-Polder physics. The atom-wall interaction leads to a distance-dependent shift of the atomic energy levels which may be detected by resonant spectroscopic techniques. An important and necessary requirement for this technique to work is that the energy shifts involved for the excited atoms behave very differently from those of the ground state species. The equilibrium between a repulsive light force and the attractive surface interaction can be used to establish a potential barrier. The barrier height can be controlled by changing the light field and its influence is then probed through the reflectivity for incident atoms [10]. For systems involving excited [44, 45] or Rydberg atoms [46, 47] additional resonance effects come into play, which may in fact turn the Casimir-Polder interaction itself into a repulsion (see for example [48]). Early experiments on the direct spectroscopic measurement of van der Waals forces have been successful up to a level of several tens of a percent [44, 46]. To date, it seems feasible to explore the progress in spectroscopic precision and to investigate the complex QED effects taking place during the atom-wall collisions an order of magnitude more accurately.
- Bose-Einstein condensates (BECs): From the manipulation of a BEC in front of a wall, information on the single atom-surface interaction can be obtained,

since all particles within this macroscopic quantum system are in the same state. As the trap-minimum is moved closer towards the surface, the Casimir-Polder forces change the curvature of the trapping potential which leads to a relative change of the center of mass oscillation frequency. Cornell's group measured this frequency shift for BEC atoms inside a chip-based trap [18, 49]. Refined measurements by the same group at the 10^{-4} level were interpreted by Antezza et al. in terms of equilibrium and nonequilibrium thermal corrections to the Casimir-Polder force [50]. The interaction strength was determined at a ten percent level. In a recent proposal it was argued that the precision in this type of BEC experiments may be improved by an order of magnitude through the use of atomic clocks [51].

In both types of quantum optics experiments above it remains unclear how the state of the surface and its deposits can be quantitatively measured and included in theory. The same argument also holds for the deflection experiments by Hinds et al. that rely on the fact that atoms are removed from the beam through sticking. Absorbed atoms contaminate the substrate and modify the interaction potential at a level that need not be constant, but may vary with the experimental conditions, like time, light intensity and temperature. Even at sub-monolayer coverage, adsorbates (in particular metallic ones) modify the dielectric properties of the system. This is in fact one of the reasons why IR spectroscopy is such a sensitive and well established tool in surface science (see for example [52]). Benedek et al. [53] recently reviewed spectral consequences of vibrations of alkali metal overlayers on metal surfaces, that depend very sensitively on atomic mass, adsorption position, coverage and substrate orientation. Pucci et al. [54] have recently found a strong enhancement of vibration signals by coupling an antenna to surface phonon polaritons, using (far) IR surface spectroscopy. In particular the investigation of temperature effects in the Casimir-Polder physics may be strongly influenced by such spectral features. Inversely, one could include (far) IR spectroscopy as a tool to quantify the state of the substrate in future BEC experiments.

• Atom interference: Another very useful step for improving experimental precision is shifting from intensity measurements to phase sensitive quantities. The experiments mentioned in Sect. 2 by Aspect [55] and Cronin [16] are examples along this line. At the heart of both techniques is an atom interferometer, which measures phase shifts in the de Broglie waves of the atoms. These first experiments to detect surface-induced phase shifts were not accurate enough to test the power law of the potential. Recently, however, by using the Toulouse interferometer significant improvement could be reported [56]. The interaction strength between an atom and the silicon nitride nano-grating was now determined with a precision of 6%. Since the setup was not equipped with any surface analytical tools, the authors used a trick to reduce the effect of surface contamination discussed above. By taking ratios of interaction strength coefficients obtained for different atoms, they compared the van der Waals systems at a level of better than 3%. It must be realized, however, that taking the ratio between two C_3 values is good only to first order approximation: different atomic species adsorb differently onto the same surface, with distinguished temperature behavior and spectral features. In comparing the two, the dielectric property changes of the wall need not be exactly the same. Still, this method is a definite improvement and could be useful for other techniques as well.

• Quantum reflection: In 2002 Shimizu et al. reported on a giant quantum reflection of neon atoms from a ridged silicon surface [57]. The enhancement of reflectivity with respect to that from flat silicon surfaces was initially explained as resulting from the reduced effective matter density in the outermost region of the structured sample. At comparable distances, the interaction potential would consequently be weaker for the latter and that would enhance the so called "Bad Land" condition for quantum reflectivity at a given value of z. Later it was established that the giant reflectivity was not so much a consequence of a modification in the Casimir-Polder potential, but rather due to the collision process itself.[7,8] The explanation was modified, interpreting the process of grazing incidence over the ridged surface in terms of a Fresnel diffraction [58]. This explanation, however, ignores the modification of the Casimir-Polder potential due to the periodic structure. This prompted us to experimentally investigate the underlying question of including geometric effects in this QED phenomenon more systematically. Extensive studies on the quantum reflectivity of ground state He atoms from nano-structured substrates of different shape and material were performed. The results have not been published yet, but have already initiated great activity in theory to develop a method for embedding these structures in the scattering formalism. In addition, our experiments show that the non-additivity of the Casimir-Polder potential leads to a strong azimuthal dependence of the measured reflectivity. The data allow for a phenomenological power law description of the interaction, which has a different exponent along different azimuthal directions.

12.4.1 The Heidelberg Approach

Thermal atom beam scattering was founded in the early 20th century by Stern and Estermann [59, 60] to verify the wave nature of matter. With the advent of high quality supersonic jets, the method of diffracting a beam of atoms from a crystalline surface developed in the 80's into a powerful tool in surface science.

---

[7] The need for a more sophisticated explanation was supported by independent observations in Ketterle's laboratory. In experiments at normal incidence dedicated to extremely low density materials quantum reflection was never observed.

[8] Recently, similar experiments were performed in Gerhard Meijer's group, scattering He atoms from a periodic micro-structure. It could not be substantiated that the process observed is in fact quantum reflection.

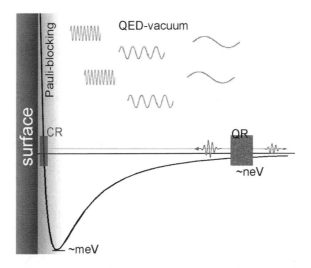

Fig. 12.3 Schematic representation of the full interaction potential between a neutral atom and the surface of a solid. At large distances, there is an attraction due to the Casimir-Polder and van der Waals effects. This region (marked QR) can be sampled by measuring the quantum reflectivity of an atomic beam impinging at grazing incidence. Close to the surface the interaction is dominated by an exponential repulsion, due to the Pauli blocking. The fraction of atoms scattering 'classically' from this wall can be used to extract in-situ information on the quality and the state of the surface on a sub-atomic level, like periodicity, corrugation, cleanliness and dynamics. Note that the typical energy ranges for these two processes differ by more than six orders of magnitude

"Classical" thermal atom scattering[9] is governed predominantly by the repulsive part of the full atom-surface interaction potential (see Fig. 12.3). This repulsion is due to Pauli exclusion during the partial overlap of the electron clouds of the atom and the solid at very short distances. The full interaction includes of course both the repulsive and attractive branches and shows a well of minimum potential energy.

The angular distribution of the specularly reflected atoms reveals information on the crystallinity and cleanliness of the exposed surface.[10] If the de Broglie wavelength of the atoms is on the order of the lattice constant of the crystal, Bragg diffraction may occur. The location of the diffraction peaks gives the periodicity of the surface structure, whereas their intensity distribution contains information on the corrugation height. With these signatures many detailed studies on clean and adsorbate covered substrates were performed. Within the impressive spectrum of

---

[9] The quotation marks merely indicate that scattering from a repulsive wall can of course be described according to the laws of classical physics. It should be clearly distinguished from quantum reflection, for which there is no classical analogue.

[10] Note that in the quantum mechanical formulation of atoms scattering from a repulsive wall, there always is a finite probability for the completely elastic channel.

achievements of this technique there are not only the crystallographic studies with sub-Angstrom resolution, but also a whole range of results on two dimensional dynamics, like surface phonon modes and charge density waves (to get a flavor of these accomplishments, see for example [61, 62] and references therein).

Thermal atom scattering is predominantly sensitive to potential features between the wall and the van der Waals range. Only limited information on the long distance Casimir-Polder contribution can be extracted from the scattering patterns. Physisorption and chemisorption processes, for example, are generally governed by the location and shape of the potential well near its energy minimum. The stronger bound states involved have their classical turning points generally near to the position of its minimum and normally also lie well within the non-retarded regime. They have been discussed to investigate quadrupolar and non-local effects in the physisorption of rare-gas atoms on metal surfaces [63].

Bound state resonance phenomena in atom-surface scattering [61, 62], however, are determined by the upper bound energy levels within the full interaction potential of 3. The outermost turning points of these higher levels are located at distances at which retardation effects may start playing a role. Analyzing all available data for helium, atomic and molecular hydrogen, however, Vidali et al. 64 established a surprising universality, indicating that these bound states lie still within the van der Waals regime. Choosing the mathematical shape of the repulsive wall to be exponential with distance $\sim \exp(-z/\sigma)$ and fixing the tail to be $\sim 1/z^3$ only, Vidali calculated the Bohr-Sommerfeld quantization condition for all bound state levels and plotted these with the experimental data (see Fig. 12.4). He found that independent of their type (metals, semiconductors and insulators) all measured systems fall onto the same curve. This impressive result suggests, that the adsorption potential energy functional form nearly universally takes the form of an exponential repulsion and a van der Waals attraction.

It remains worthwhile, however, to search for special cases in which the weakest bound state lies but barely underneath the dissociation limit. The outermost turning points may then be located at distances at which retardation effects are indeed important.[11] For such systems, bound state resonance experiments could indeed be a valuable tool in modern Casimir-Polder experimental physics, although maybe not a generic one.

About a decade ago, we designed a machine for performing atomic beam spin echo measurements on surfaces. Details on the experimental setup are given elsewhere [13]. In this apparatus, the nuclear magnetic moments of a flux of ^{3}He atoms are manipulated so as to obtain detailed information about (changes in) the velocity distribution of the beam before and after scattering from the surface. The ^{3}He beam is very slow (100 m/s $<v<$ 200 m/s), since the atomic beam source is cooled down to 1.1–4.2 K. The ^{3}He kinetic energy thus amounts to

---

[11] A famous example from atomic physics: the weak interaction between two neutral He atoms just happens to accommodate a single bound level at -150 neV only. As a consequence, the bond length in the He dimer is close to 45 Å!

Fig. 12.4 Plot from [64] illustrating the experimental correlation between energies of atom-surface bound states in units of potential well depth $|\varepsilon_n| = -E_n/D$ and $J(\varepsilon_n) = (n + \frac{1}{2})\hbar\pi/(C_3^{1/3}D^{1/6}(2m)^{1/2})$ (reproduced with permission). The solid line is the prediction given by a simple model potential $V(z) = 3/(u-3)e^{-uz/a} - 1/(z+a)^3$ with constant u and a

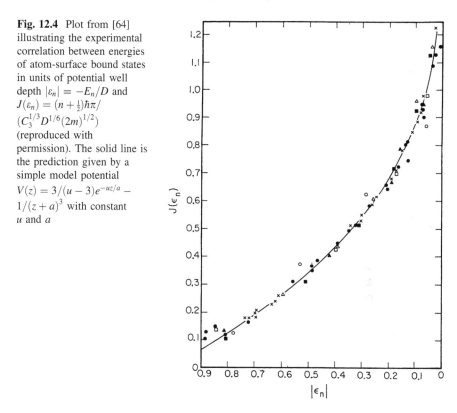

200–800 μeV, corresponding to a de Broglie wavelength λ_{dB} of 10–3 Å, or a wavevector $2\pi/\lambda_{dB}$ of 0.5–1.5 Å^{-1}. The He atoms are exclusively surface sensitive and have shown to probe corrugations at the surface down to some 0.01 Å [65]. The ^{3}He Atomic Beam Spin Echo method has been demonstrated to be a particularly sensitive and precise tool for characterizing the quality, structure and dynamics of clean, as well as adsorbate covered surfaces. As an example we summarize some of our results obtained with single crystal gold. In elastic ^{3}He diffraction experiments on the surface of single crystal Au(111), the dimensions of the reconstructed unit cell could be extracted to be $(p \times \sqrt{3})$, with $p = 21.5 \pm 0.5$ being expressed in the bulk gold nearest neighbor distance $a = 2.885$ Å. On top of this substrate, adsorbate molecules (coronene) were deposited and the structure at monolayer coverage was determined to be a commensurate (4×4) superstructure with respect to the unreconstructed gold surface. In addition, we studied diffusion of the diluted coronene molecules at submonolayer coverages. Their 2-D dynamics was shown to exhibit continuous and non-continuous (jump) 1-D diffusion. The activation barrier to this diffusion was inferred from an Arrhenius analysis of its temperature dependency. We investigated inelastic contributions to the scattering process and confirmed the existence and dispersion of anomalous phonon modes that are associated with the reconstruction.

Table 12.1 Problems limiting the accuracy of classical Casimir force measurements and corresponding solutions in the ^{3}He Atomic Beam Spin Echo technique

Casimir	versus	Casimir-Polder
Characterization of the surface quality		Surface science (typical <0.01 monolayer)
Quantization of surface roughness		Atom diffraction (single crystal)
Calibration of probe-surface distance		Atomic resolution
(plate-plate, sphere-plate)		(~ 0.01 Å)
Control over probe		Atomic beam
(sphere quality, coating, roughness)		($\sim 10^{19}$ He atoms s^{-1} sr^{-1})
Geometry		Atom size $\ll$ relevant length scale
(pl-pl: parallellism, sph-pl: proximity approx.)		(1 Å vs. 1 μm)
Spurious electrostatic effects		Neutral single atoms
Imperfect conductor		Any system (metal, semi-conductor, insulator)
Resolution		Atom interferometry

With the "classical" reflection governed by the repulsive and van der Waals regime of the potential, we have at our disposal an in-situ analytical tool from surface science. On the other hand, we can investigate He atoms being quantum reflected from the retarded regime and study Casimir-Polder physics. The advantages of such a powerful combination are summarized in the following Table 12.1.

In order to illustrate that this unique combination of tools is in fact at our disposal contemporaneously, we present here experimental data obtained for single crystal gold [66]. The measurements were performed much in the same way as our quantum reflection data described in Sect. 12.2 for the case of single crystal quartz were obtained [12]. The gold surface, however, is atomically flat[12] and also shows ^{3}He reflectivity from the repulsive wall (see Fig. 12.3).

Although both the "classical" and the quantum contributions consist of specularly reflected beams, the two can be distinguished through their line shape in the angular distribution. For each incident wave vector a detector angle scan was made and the two intensities evaluated. In this manner, the dots in upper set for the "classical" and the dots in lower set for the quantum reflectivities in Fig. 12.5 were obtained. The first curve in upper set is but a guide-to-the-eye for the classical reflection results. The first middle curve in lower set is an exact calculation based on the simple potential model used in 12.11, using the strength parameter for a perfect conductor. These quantum reflection data were also investigated with respect to the necessity for inhomogeneity of the model in 12.11. The additional curves in lower set in Fig. 12.5 clearly demonstrate that neither a strict van der Waals potential nor a pure Casimir-Polder attraction describe the data accurately. The dashed red line in lower set accounts for the homogeneous Casimir-Polder contribution (12.6), again using the interaction strength for a perfect conductor.

---

[12] In fact it is the exact same sample that was used in the surface science experiments reviewed earlier in this section.

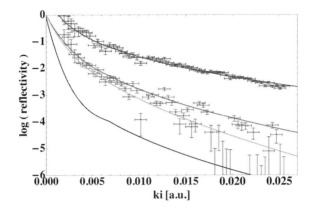

Fig. 12.5 Plot of the reflectivity of ^{3}He atoms from the single crystal surface of Au(111). The upper set of data points (including their statistical errors) represents "classical" reflection from the repulsive wall (see text). The lower set of data points and error bars corresponds to the simultaneously measured quantum reflectivity from the attractive tail of the interaction potential. Results for the corresponding simulation are shown in the lower three solid curves: the top one is based on the full potential derived from (12.19); the two lower curves are based on the homogeneous potentials for a perfect conductor, the Casimir-Polder result (12.6) (middle) and van der Waals result (12.5) (bottom), respectively. Finally, the bottom curve represents the homogeneous van der Waals result (12.5) for a perfect conductor. (Further details will be presented in [66].)

The corresponding homogeneous van der Waals interaction would give a quantum reflectivity, which is negligibly small. The second middle curve in lower set represents the quantum reflectivity based on a van der Waals interaction only, using the more realistic coefficient of $C_3 = 0.25$ eV Å^3. From the figure it is clear that at the distances explored in this experiment the energies are dominated by the Casimir-Polder interaction. Still, there is a definite need to include the van der Waals term explicitly. This demonstrates that with the Heidelberg approach it is now possible to resolve features in the potential at a few-percent level.

In summary, with the renaissance in atomic physics a great variety of new experimental techniques have been steadily improving the accuracy in Casimir-Polder measurements to well below the ten percent level. Current trends and proposals promise to be pushing this in the near future to below the one percent barrier. At that level of precision new contributions, like those explicitly dependent on temperature and higher order effects (additional multipoles, etc), are within reach. This will be important for our understanding of vacuum fluctuations. Tremendous theoretical progress has been made to include non-trivial boundary conditions to their spectrum. The feasibility of a direct comparison with measured data requires a quantitative experimental characterization of these conditions as well.

Acknowledgements G.L. would like to acknowledge his support from the Deutsche Forschungsgemeinschaft (DFG, contract Je285/5–1). U.J. acknowledges support from the National Science Foundation (Grant PHY–8555454) and from the National Institute of Standards and Technology (precision measurement grant).

References

1. Casimir, H.B.G., Polder, D.: The influence of retardation on the London-van der waals forces. Phys. Rev. **73**(4), 360–372 (1948)
2. London, F.: Zur Theorie und Systematik der Molekularkräfte. Z. Phys. **63**(3–4), 245 (1930)
3. Verwey. E.J.W., Overbeek. J.T.G., van Nes. K.: (1947) J Phys. Colloid Chem. 51:631
4. Lifshitz. E., Pitaevskii. L.P.: (1956) Landau and Lifshitz course of theoretical physics: Statistical physics part 2. J. Exp.Theor. Phys. **2**:(73)
5. Parsegian, V.A.: Van der Waals Forces; A Handbook for Biologists, Chemists, Engineers, and Physicists. Cambridge University Press, New York (2006)
6. Sabisky, E.S., Anderson, C.H: Verification of the Lifshitz theory of the van der Waals potential using liquid-helium films. Phys. Rev. A **7**(2), 790–806 (1973)
7. Shih, A., Parsegian, V.A.: Van der Waals forces between heavy alkali atoms and gold surfaces: Comparison of measured and predicted values. Phys. Rev. A **12**(3), 835–841 (1975)
8. Dzyaloshinskii, I.E., Lifshitz, E.M., Pitaevskii, L.P: General theory of van der Waals forces. Adv. Phys. **29**, 165 (1961)
9. Sukenik, C.I., Boshier, M.G., Cho, D., Sandoghdar, V., Hinds, E.A.: Measurement of the Casimir-Polder force. Phys. Rev. Lett. **70**(5), 560–563 (1993)
10. Landragin, A., Courtois, J.Y., Labeyrie, G., Vansteenkiste, N., Westbrook, C.I., Aspect, A.: Measurement of the van der Waals force in an atomic mirror. Phys. Rev. Lett. **77**(8), 1464–1467 (1996)
11. Shimizu, F.: Specular reflection of very slow metastable neon atoms from a solid surface. Phys. Rev. Lett. **86**(6), 987–990 (2001)
12. Druzhinina, V., DeKieviet, M.: Experimental observation of quantum reflection far from threshold. Phys. Rev. Lett. **91**(19), 193,202 (2003)
13. DeKieviet, M., Dubbers, D., Schmidt, C., Scholz, D., Spinola, U.:^3He Spin echo: New atomic beam technique for probing phenomena in the neV range. Phys. Rev. Lett. **75**(10), 1919–1922 (1995)
14. Friedrich, H., Jacoby, G., Meister, C.G.: Quantum reflection by Casimir–van der Waals potential tails. Phys. Rev. A **65**(3), 032,902 (2002)
15. Friedrich, H., Trost, J.: Working with WKB waves far from the semiclassical limit. Phys. Rep. **397**(6), 359–449 (2004)
16. Perreault, J.D., Cronin, A.D.: Observation of atom wave phase shifts induced by van der Waals atom-surface interactions. Phys. Rev. Lett. **95**(13), 133,201–133,204 (2005)
17. Pasquini, T.A., Shin, Y., Sanner, C., Saba, M., Schirotzek, A., Pritchard, D.E., Ketterle, W.: Quantum reflection from a solid surface at normal incidence. Phys. Rev. Lett. **93**(22), 223,201–223,204 (2004)
18. McGuirk, J.M., Harber, D.M., Obrecht, J.M., Cornell, E.A.: Alkali-metal adsorbate polarization on conducting and insulating surfaces probed with Bose-Einstein condensates. Phys. Rev. A **69**(6), 062,905–062,910 (2004)
19. Lin, Y.J., Teper, I., Chin, C., Vuletić, V.: Impact of the Casimir-Polder potential and Johnson noise on Bose-Einstein condensate stability near surfaces. Phys. Rev. Lett. **92**(5), 050–404 (2004)
20. Obrecht, J.M., Wild, R.J., Antezza, M., Pitaevskii, L.P., Stringari, S., Cornell, E.A: Measurement of the temperature dependence of the Casimir-Polder force. Phys. Rev. Lett. **98**(6), 063,201–063,204 (2007)
21. Bender, H., Courteille, P.W., Marzok, C., Zimmermann, C., Slama, S.: Direct measurement of intermediate-range Casimir-Polder potentials. Phys. Rev. Lett. **104**(8), 083,201–083,204 (2010)
22. Saniz, R., Barbiellini, B., Platzman, P.M., Freeman, A.J.: Physisorption of positronium on quartz surfaces. Phys. Rev. Lett. **99**(9), 096,101–096,104 (2007)
23. Łach, G., DeKieviet, M., Jentschura, U.D: Multipole effects in atom-surface interactions: A theoretical study with an application to He–α-quartz. Phys. Rev. A **81**(5), 052–507 (2010)

24. Palik, E.D.: Handbook of Optical Constants of Solids. Academic Press, San Diego (1985)
25. Derevianko, A., Porsev, S.G., Babb, J.F.: Electric dipole polarizabilities at imaginary frequencies for the alkali-metal, alkaline-earth, and inert gas atoms. arXiv:0902.3929v1 p.16 (2009)
26. Marvin, A.M., Toigo, F.: Van der Waals interaction between a point particle and a metallic surface. II. Applications. Phys. Rev. A **25**(2), 803–815 (1982)
27. Milonni, P.W.: The Quantum Vacuum: An Introduction to Quantum Electrodynamics. Academic Press, San Diego (1994)
28. McLachlan, A.D.: Van der Waals forces between an atom and a surface. Mol. Phys. Int. J. Int. Chem. Phys. **7**, 381 (1964)
29. Hutson, J.M., Fowler, P., Zaremba E.: Quadrupolar contributions to the atom-surface Van der Waals interaction. Surf. Sci. **175**, 775–781 (1986)
30. Lach, G., DeKieviet, M.F.M., Jentschura, U.D.: Noble gas, alkali and alkaline atoms interacting with a gold surface. Int. J. Mod. Phys. A **25**(11),2337–2344 (2010)
31. Babb, J.F., Klimchitskaya, G.L., Mostepanenko, V.M.: Casimir-Polder interaction between an atom and a cavity wall under the influence of real conditions. Phys. Rev. A **70**(4), 042,901–042,912 (2004)
32. Mostepanenko, V.M., Trutnov, N.N.: The Casimir Effect and Its Applications. Clarendon Press, Oxford (1997)
33. Pitaevskii, L.P.: Thermal Lifshitz force between an atom and a conductor with a small density of carriers. Phys. Rev. Lett. **101**(16), 163–202 (2008)
34. Pachucki, K.: Relativistic corrections to the long-range interaction between closed-shell atoms. Phys. Rev. A **72**(6), 062–706 (2005)
35. Meath, W.J., Hirschfelder, J.O.: Long-range (retarded) intermolecular forces. J. Chem. Phys. **44**(9), 3210–3215 (1966)
36. Meath, W.J., Hirschfelder, J.O: Relativistic intermolecular forces, moderately long range. J. Chem. Phys. **44**(9), 3197–3209 (1966)
37. Łach, G., Jeziorski, B., Szalewicz, K.: Radiative corrections to the polarizability of helium. Phys. Rev. Lett. **92**(23), 233001 (2004)
38. Bimonte, G., Klimchitskaya, G.L., Mostepanenko, V.M.: Impact of magnetic properties on atom-wall interactions. Phys. Rev. A **79**(4), 042,906–043,912 (2009)
39. Marvin, A.M., Toigo, F.: Van der Waals interaction between a point particle and a metallic surface. I. Theory. Phys. Rev. A **25**(2), 782–802 (1982)
40. Buhmann, S.Y., Dung, H.T., Welsch, D.G.: The van der Waals energy of atomic systems near absorbing and dispersing bodies. J. Opt. B: Quantum and Semiclassical Opt. **6**(3), S127 (2004)
41. Klimchitskaya, G.L., Blagov, E.V., Mostepanenko, V.M.: Casimir-Polder interaction between an atom and a cylinder with application to nanosystems. J. Phys. A Math. Gen. **39**(21), 6481 (2006)
42. Döbrich B., DeKieviet, M., Gies, H.: Scalar Casimir-Polder forces for uniaxial corrugations. Phys. Rev. D **78**(12), 25022 (2008)
43. Celli, V., Marvin, A., Toigo, F.: Light scattering from rough surfaces. Phys. Rev. B **11**(4), 1779–1786 (1975)
44. Failache, H., Saltiel, S., Fichet, M., Bloch, D., Ducloy, M.: Resonant van der Waals repulsion between excited Cs atoms and sapphire surface. Phys. Rev. Lett. **83**(26), 5467–5470 (1999)
45. Gorza, M.P., Saltiel, S., Failache, H., Ducloy, M.: Quantum theory of van der Waals interactions between excited atoms and birefringent dielectric surfaces. Eur. Phys. J. D **15**, 113–126 (2001)
46. Sandoghdar, V., Sukenik, C.I., Hinds, E.A., Haroche, S.: Direct measurement of the van der Waals interaction between an atom and its images in a micron-sized cavity. Phys. Rev. Lett. **68**(23), 3432–3435 (1992)
47. Anderson, A., Haroche, S., Hinds, E.A., Jhe, W., Meschede, D.: Measuring the van der Waals forces between a Rydberg atom and a metallic surface. Phys. Rev. A **37**(9), 3594–3597 (1988)

48. Failache, H., Saltiel, S., Fichet, M., Bloch, D., Ducloy, M.: Resonant coupling in the van der Waals interaction between an excited alkali atom and a dielectric surface: An experimental study via stepwise selective reflection spectroscopy. Eur. Phys. J. D **23**, 237 (2003)

49. Harber, D.M., Obrecht, J.M., McGuirk, J.M., Cornell, E.A.: Measurement of the Casimir-Polder force through center-of-mass oscillations of a Bose-Einstein condensate. Phys. Rev. A **72**(3), 033,610–033,615 (2005)

50. Obrecht, J.M., Wild, R.J., Antezza, M., Pitaevskii, L.P., Stringari, S., Cornell, E.A.: Measurement of the temperature dependence of the Casimir-Polder force. Phys. Rev. Lett. **98**(6), 063,201–063,204 (2007)

51. Derevianko, A., Obreshkov, B., Dzuba, V.A.: Mapping out atom-wall interaction with atomic clocks. Phys. Rev. Lett. **103**(13), 133,201–133,204 (2009)

52. Trenary, M.: Reflection absorption infared spectroscopy and the structure of molecular adsorbates on metal surfaces. Annu. Rev. Phys. Chem. **51**(1), 381–403 (2000)

53. Rusina, G.G., Eremeev, S.V., Echenique, P.M., Benedek, G., Borisova, S.D., Chulkov, E.V.: Vibrations of alkali metal overlayers on metal surfaces. J. Phys. Condens. Matter **20**(22), 224007 (2008)

54. Neubrech, F., Weber, D., Enders, D., Nagao, T., Pucci, A.: Antenna sensing of surface phonon polaritons. J. Phys. Chem. **114**, 7299 (2010)

55. Marani, R., Cognet, L., Savalli, V., Westbrook, N., Westbrook, C.I, Aspect, A.: Using atomic interference to probe atom-surface interactions. Phys. Rev. A **61**(5), 053–402 (2000)

56. Lepoutre, S., Jelassi, H., Lonij, V.P.A., Trenec, G., Buchner, M., Cronin, A.D., Vigue, J.: Observation of atom wave phase shifts induced by van der Waals atom-surface interactions. EPL **88**, 200002 (2009)

57. Shimizu, F., Fujita, J.I.: Giant quantum reflection of neon atoms from a ridged silicon surface. J. Phys. Soc. Jpn. **71**(1), 5–8 (2002)

58. Oberst, H., Kouznetsov, D., Shimizu, K., Fujita, J.I., Shimizu, F.: Fresnel diffraction mirror for an atomic wave. Phys. Rev. Lett. **94**(1), 013–203 (2005)

59. Stern, O.: Beugung von molekularstrahlen am gitter einer krystallspaltflche. Naturwissenschaften **17**, 391 (1929)

60. Estermann, L., Stern, O.: Beugung von molekularstrahlen. Z. Phys. **61**, 95 (1930)

61. Benedek, G., Valbusa U.: Dynamics of Gas-Surface Interaction.Vol. 21. Springer Series in Chemical Physics, Berlin (1982)

62. Scoles, G.: Atomic and Molecular Beam Methods. Oxford University Press, New York, vol. I and II. (1988)

63. Girard, C., Humbert, J.: Quadrupolar and non-local effects in the physisorption of rare-gas atoms on metal surfaces. Chem. Phys. **97**, 87–94 (1985)

64. Vidali, G., Cole, M.W., Klein, J.R.: Shape of physical adsorption potentials. Phys. Rev. B **28**(6), 3064–3073 (1983)

65. DeKieviet, M., Dubbers, D., Klein, M., Pieles, U., Skrzipczyk, M.: Surface science using molecular beam spin echo. Surf. Sci. **377**(379), 1112–1117 (1997)

66. Stöferle, T., Warring, U., Lach, G., Jentschura, U.D., DeKieviet, M.F.M.: Quantum and classical reflection of He from single crystal gold. Manuscript in preparation (2011)

Chapter 13
Fluctuations, Dissipation
and the Dynamical Casimir Effect

Diego A. R. Dalvit, Paulo A. Maia Neto and Francisco Diego Mazzitelli

Abstract Vacuum fluctuations provide a fundamental source of dissipation for systems coupled to quantum fields by radiation pressure. In the dynamical Casimir effect, accelerating neutral bodies in free space give rise to the emission of real photons while experiencing a damping force which plays the role of a radiation reaction force. Analog models where non-stationary conditions for the electromagnetic field simulate the presence of moving plates are currently under experimental investigation. A dissipative force might also appear in the case of uniform relative motion between two bodies, thus leading to a new kind of friction mechanism without mechanical contact. In this paper, we review recent advances on the dynamical Casimir and non-contact friction effects, highlighting their common physical origin.

13.1 Introduction

The Casimir force discussed in this volume represents the average radiation pressure force upon one of the interacting bodies. When the zero-temperature limit

D. A. R. Dalvit (✉)
Theoretical Division MS B213, Los Alamos National Laboratory, Los Alamos,
NM 87545, USA
e-mail: dalvit@lanl.gov

P. A. M. Neto
Instituto de Física UFRJ, Caixa Postal 68528, Rio de Janeiro, RJ 21941-972, Brazil
e-mail: pamn@if.ufrj.br

F. D. Mazzitelli
Centro Atómico Bariloche, Comision Nacional de Energía Atómica, R8402AGP,
Bariloche, Argentina
and
Departamento de Física, FCEyN, Universidad de Buenos Aires, Ciudad Universitaria,
1428, Buenos Aires, Argentina
e-mail: fmazzi@df.uba.ar

D. Dalvit et al. (eds.), *Casimir Physics*, Lecture Notes in Physics 834,
DOI: 10.1007/978-3-642-20288-9_13, © Springer-Verlag Berlin Heidelberg 2011

is considered, the average is taken over the vacuum field state. Although the average electric and magnetic fields vanish, the Casimir force is finite because radiation pressure is quadratic in the field strength operators. In this sense, the Casimir force derives from the fluctuating fields associated with the field zero-point energy (or more precisely from their modification by the interacting bodies).

As any quantum observable, the radiation pressure itself fluctuates [1, 2]. For a single body at rest in empty space, the average vacuum radiation pressure vanishes (for the ground-state field cannot be an energy source), and all that is left is a fluctuating force driving a quantum Brownian motion [3]. The resulting dynamics is characterized by diffusion in phase space, thus leading to decoherence of the body center-of-mass [4, 5].

Besides diffusion, the radiation pressure coupling also leads to dissipation, with the corresponding coefficients connected by the fluctuation-dissipation theorem [6]. As in the classical Brownian motion, the fluctuating force on the body at rest is closely related to a dissipative force exerted when the body is set in motion. Since the vacuum state is Lorentz invariant, the Casimir dissipative force vanishes in the case of uniform motion of a single body in empty space, as expected from the principle of relativity. For a non-relativistic "mirror" in one spatial dimension (1D), the Casimir dissipative force is proportional to the second-order derivative of the velocity [7], like the radiation reaction force in classical electrodynamics. Casimir dissipation is in fact connected to the emission of photon pairs by the accelerated (electrically neutral) mirror, an effect known as the dynamical Casimir effect (DCE). The power dissipated in the motion of the mirror is indeed equal to the total radiated power in DCE as expected from energy conservation.

The creation of photons in a 1D cavity with one moving mirror was first analyzed by Moore [8], and explicit results were later derived in Ref. [9]. Relativistic results for the dissipative Casimir force upon a single mirror in 1D and the connection with DCE were derived in a seminal paper by Fulling and Davies [10]. At this earlier stage, the main motivation was the analogy with the Hawking radiation associated with black-hole evaporation [11, 12].

The interplay between Casimir dissipation and fluctuations was investigated only much later [3, 13, 14], in connection with a major issue in quantum optics: the fundamental quantum limits of position measurement (this was motivated by the quest for interferometric detection of gravitational waves) [15–17]. Linear response theory [18] provides a valuable tool for computing the Casimir dissipative force on a moving body from the fluctuations of the force on the body at rest, which is in general much simpler to calculate. This method was employed to compute the dissipative force on a moving, perfectly reflecting sphere [19] and on a plane surface experiencing a time-dependent perturbation [20] (the latter was also computed by taking the full time-dependent boundary conditions (b.c.) [21, 22]).

In all these configurations, Casimir dissipation turns out to be very small when realistic physical parameters are taken into account. The predicted orders of magnitude are more promising when considering a closely related effect: quantum non-contact friction in the shear relative motion between two parallel surfaces [23, 24]. In contrast with the radiation reaction dissipative effect discussed so far, quantum friction takes place even for a uniform relative motion. On the other hand, quantum friction requires the material media to have finite response times (dispersion). From Kramers–Kronig relations [25], the material media must also be dissipative, and the resulting friction depends on the imaginary part of the dielectric constant ε.

Whereas the direct measurement of the Casimir radiation reaction dissipative force seems to be beyond hope, the corresponding photon emission effect might be within reach in the near future. The properties of the radiated photons have been analyzed in great detail over recent years. For a single moving plane mirror, the frequency spectrum was computed in 1D [26] as well as in 3D [27] in the non-relativistic approximation. A variety of 3D geometries were considered, including deforming mirrors [28], parallel plates [29], cylindrical waveguides [30], and spherical cavities containing either scalar [31, 32] or electromagnetic fields [33, 34].

Closed rectangular [35–37] or cylindrical [38] microwave cavities with one moving wall are by far the best candidates for a possible experimental implementation, with the mechanical oscillation frequency Ω tuned into parametric resonance with cavity field modes. Because of the parametric amplification effect, it is necessary to go beyond the perturbative approximation in order to compute the intracavity photon number even in the non-relativistic regime [39–41].

As the microwave field builds up inside the cavity, cavity losses (due to transmission, dissipation or diffraction at the rough cavity walls) become increasingly important. Finite transmission at the mirrors of a 1D cavity was taken into account within the scattering approach developed in Refs. [26, 42, 43]. Master equations for the reduced density operator of the cavity field in lossy 3D cavities were derived in Refs. [44, 45]. Predictions for the total photon number produced at very long times obtained from the different models are in disagreement, so that a reliable estimation of the DCE magnitude under realistic conditions is still an open theoretical problem.

It is nevertheless clear that measuring DCE photons is a highly non-trivial challenge (see for instance the proposal [46] based on superradiance amplification). For this reason, in recent years the focus has been re-oriented towards analog models of DCE. Although dynamical Casimir photons are in principle emitted even in the case of a global 'center-of-mass' oscillation of a cavity, the orders of magnitude are clearly more favorable when some cavity length is modulated. In this case, one might modulate the optical cavity length by changing the intracavity refractive index (or more generally material optical constants) instead of changing the physical cavity length. For instance, the conductivity of a semiconductor slab can be rapidly changed with the help of a short optical pulse, simulating the motion of a metallic mirror [47–50] and thereby producing photons exactly as in the DCE [51, 52]. An experiment along these lines is currently under way [53] (see Ref. [54] for an update). Alternatively, one might select a setup for operation of an optical parametric oscillator such that it becomes formally equivalent to DCE with a modulation frequency in the optical domain [43].

Alongside the examples in quantum and nonlinear optics, one can also devise additional analogues of DCE in the field of circuit quantum electrodynamics

[55, 56]. For instance, a co-planar waveguide with a superconducting quantum interference device (SQUID) at its end is formally equivalent to a 1D model for a single mirror [57, 58]. When a time-dependent magnetic flux is applied to the SQUID, it simulates the motion of the mirror. More generally, Bose–Einstein condensates also provide interesting analogues for DCE [59] and Casimir-like dissipation [60], with electromagnetic vacuum fluctuations replaced by zero-point fluctuations of the condensate.

Reviews on fluctuations and Casimir dissipation on one hand and on DCE on the other hand can be found in Refs. [61, 62] and Refs. [63–65], respectively. This review paper is organized as follows. In Sect. 13.2, we discuss the interplay between fluctuations, dissipation and the photon creation effect for a single mirror in free space. Section 13.3 presents a short introduction to non-contact quantum friction. In Sect. 13.4, photon creation in resonant cavities with either moving walls or time-dependent material properties is presented in detail. Section 13.5 briefly discusses experimental proposals, and Sect. 13.6 contains some final remarks.

13.2 Dissipative Effects of the Quantum Vacuum

13.2.1 1D Models

We start with the simplest theoretical model: a non-relativistic point-like 'mirror' coupled to a massless scalar field $\phi(x, t)$ in 1D. We assume Dirichlet boundary conditions, at the instantaneous mirror position $q(t)$:

$$\phi(q(t), t) = 0. \tag{13.1}$$

In the non-relativistic approximation, we expect the vacuum radiation pressure force $f(t)$ to be proportional to some derivative of the mirror's velocity. As a quantum effect, the force must also be proportional to $\hbar$, and then dimensional analysis yields

$$f(t) \propto \frac{\hbar q^{(3)}(t)}{c^2}, \quad \text{(1D)} \tag{13.2}$$

where $q^{(n)}(t) \equiv d^n q(t)/dt^n$. Note that (13.2) is consistent with the Lorentz invariance of the vacuum field state, which excludes friction-like forces proportional to $q^{(1)}(t)$ for a single moving mirror (but not in the case of relative motion between two mirrors discussed in Sect. 13.3).

In order to compute the dimensionless prefactor in (13.2), we solve the b.c. (13.1) to first order in $q(t)$ as in Ref. [7], with the mirror's motion treated as a small perturbation. However, instead of analyzing in the time domain, we switch to the frequency domain, which allows us to understand more clearly the region of validity of the theoretical model leading to (13.2). We write the Fourier transform of the field as a perturbative expansion:

$$\Phi(x, \omega) = \Phi_0(x, \omega) + \delta\Phi(x, \omega), \tag{13.3}$$

where the unperturbed field $\Phi_0(x, \omega)$ corresponds to a static mirror at $x = 0$: $\Phi_0(0, \omega) \equiv 0$. The boundary condition for $\delta\Phi(x, \omega)$ is derived from (13.1) by taking a Taylor expansion around $x = 0$ to first order in $Q(\Omega)$ (Fourier transform of $q(t)$):

$$\delta\Phi(0, \omega_o) = -\int_{-\infty}^{\infty} \frac{d\omega_i}{2\pi} Q(\omega_o - \omega_i)\partial_x\Phi_0(0, \omega_i). \tag{13.4}$$

Equation (13.4) already contains the frequency modulation effect at the origin of Casimir dissipation: the motion of the mirror (frequency Ω) generates an output amplitude at the sideband frequency $\omega_o = \omega_i + \Omega$ proportional to $Q(\Omega)$ from a given input field frequency ω_i.

In order to find the Casimir dissipative force, we take the Fourier transform of the appropriate component $T_{11} = \frac{1}{2}[\frac{1}{c^2}(\partial_t\phi)^2 + (\partial_x\phi)^2]$ of the energy-momentum tensor and then replace the total field $\Phi(x, \omega)$ containing the solution $\delta\Phi(x, \omega)$ of Eq. (13.4). After averaging over the vacuum state, we obtain the resulting force

$$F(\Omega) = \chi(\Omega)Q(\Omega), \tag{13.5}$$

$$\chi(\Omega) = 2i\frac{\hbar}{c^2}\int_{-\infty}^{\infty} \frac{d\omega_i}{2\pi}(\Omega + \omega_i)|\omega_i|. \tag{13.6}$$

After regularization, it is simple to show that the contribution $\int_{-\infty}^{-\Omega} d\omega_i(\ldots)$ cancels the contribution $\int_0^{\infty} d\omega_i(\ldots)$ in (13.6). Thus, only field frequencies in the interval $-\Omega \leq \omega_i \leq 0$ contribute to the dynamical radiation pressure force, yielding

$$F(\Omega) = i\frac{\hbar\Omega^3}{6\pi c^2}Q(\Omega), \tag{13.7}$$

in agreement with (13.2) with a positive prefactor ($\frac{1}{6\pi}$) as expected for a dissipative force. This result was first obtained in Ref. [7] within the perturbative approach and coincides with the non-relativistic limit of the exact result derived much earlier in Ref. [10]. It may also be obtained as a limiting case of the result for a partially transmitting mirror, which was derived either from the perturbative approach outlined here [14] or by developing the appropriate Schwinger-Keldysh effective action within a functional approach to the dissipative Casimir effect [66]. When considering a moving dielectric half-space, one obtains as expected one-half of the r.h.s. of (13.7) in the limit of an infinite refractive index [67]. The final result for $F(\Omega)$ is exactly the same as in (13.7) when we replace the Dirichlet b.c. (13.1) by the Neumann b.c. at the instantaneously co-moving Lorentz frame (primed quantities refer to the co-moving frame) $\partial_{x'}\phi|_{x'=q(t')} = 0$ [68]. Dirichlet and Neumann b.c. also yield the same force in the more general relativistic regime [69]. On the other hand, for the Robin b.c.

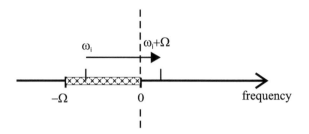

Fig. 13.1 For a given mechanical modulation frequency Ω (assumed to be positive in this diagram), the field modes contributing to the dissipative dynamical Casimir effect lie in the interval $[-\Omega, 0]$, which corresponds to negative input field frequencies ω_i, yielding positive sideband frequencies

$$\partial_{x'}\phi|_{x'=q(t')} = \frac{1}{\beta}\phi|_{x'=q(t')} \tag{13.8}$$

(β is a constant parameter), the force susceptibility $\chi(\Omega)$ displays a non-monotonic dependence on Ω, and is nearly suppressed at $\Omega \sim 2.5c/\beta$ [70].

We could have anticipated that high-frequency modes with $\omega_i \gg \Omega$ would not contribute because they "see" the mirror's motion at frequency Ω as a quasi-static perturbation, and indeed the dissipative Casimir force originates from low-frequency modes for which the motion is non-adiabatic. But there is a more illuminating interpretation that explains why the contribution comes precisely from the frequency interval $[-\Omega, 0]$. In Fig. 13.1, we show that this interval corresponds to the field modes leading to frequency sidebands $\omega_o = \omega_i + \Omega$ (see Eq. (13.4)) *across* the border between positive and negative frequencies (for $\Omega < 0$ the corresponding interval is $[0, -\Omega]$ and the analysis is essentially the same). In other words, for these specific modes the motional frequency modulation leads to mixing between positive and negative frequencies. Bearing in mind the correspondence between positive (negative) frequencies and annihilation (creation) operators, this mixing translates into a Bogoliubov transformation coupling output annihilation operators to input creation ones, and viceversa [71, 72] (examples will be presented in Sect. 13.4). The important conclusion is that sideband generation for these modes corresponds to photon creation (and also annihilation in the case of a general initial field state), whereas outside the interval $[-\Omega, 0]$, where no mixing occurs, the sideband effect corresponds to photon inelastic scattering with neither creation nor annihilation.

From this discussion, we can also surmise the important property that the dynamical Casimir photons have frequencies bounded by the mechanical frequency Ω, as long as the perturbative non-relativistic approximation holds [26]. Moreover, the Casimir photons are emitted in pairs, with photon frequencies satisfying $\omega_1 + \omega_2 = \Omega$. Hence the frequency spectrum is symmetrical around $\Omega/2$, where it has a maximum for the Dirichlet case [26] but not generally for the Robin b.c. [73].

In short, the derivation of (13.2) from (13.4) highlights the direct connection between the dissipative dynamical force and the dynamical Casimir photon emission effect. The dissipative force thus plays the role of a Casimir radiation reaction force, damping the mirror's mechanical energy as Casimir photons are emitted. In fact, the expression given by (13.2) has the same form as the radiation reaction force in classical electrodynamics, apart from a dimensionless pre-factor inversely proportional to the fine-structure constant $e^2/(\hbar c)$.

The discussion on the field modes actually contributing in (13.4) also allows us to address the domain of validity of the various assumptions employed in the derivation presented above. We have assumed in (13.1) that the field vanishes at the instantaneous mirror's position, no matter how fast the mirror and field oscillate. However, the electric currents and charge density inside a real metallic mirror respond to field and position changes over a finite time scale, so that we expect our oversimplified model to be physically meaningful only at low frequencies (more general results and discussions are presented in Refs. [14, 74]). Since the relevant field frequencies are bounded by the mechanical frequencies ($|\omega_i| \leq |\Omega|$), the model is consistent as long as typical mechanical frequencies are much smaller than the frequency scales characterizing the metallic response—typically the plasma frequency of metals.

A second point to be clarified is the connection between the perturbative linear approximation employed above and the non-relativistivistic approximation. When deriving (13.4) from (13.1), we have taken the long-wavelength approximation to expand the field around $x = 0$. Let us assume, to simplify the discussion, that the mirror oscillates with frequency Ω and amplitude q_0. The non-relativistic regime then translates into $\Omega q_0/c \ll 1$ and all relevant field modes correspond to long wavelengths $2\pi c/\omega_i \gg q_0$ since they satisfy the inequality $|\omega_i| \leq \Omega$. More generally, the long-wavelength approximation follows from the non-relativistic condition provided that there is an inertial reference frame for which the motion is spatially bounded.

13.2.2 Casimir-Driven Decoherence

As in classical Brownian motion, the dissipative effect is closely related to the fluctuations provided by the reservoir, which in our case is the quantum vacuum field. This connection provides yet another tool for computing the dissipative Casimir force: by using linear response theory, one can derive (13.7) from the correlation function of the force on a static plate [14]. (This method has been employed for different geometries in 3D [13, 19, 20, 30]). More interestingly, if we take the mirror's position as a full dynamical observable rather than a prescribed function of time, vacuum radiation pressure plays the role of a Langevin fluctuating force [3, 66], leading to diffusion in phase space, which adds to the associated average dissipative force (13.7).

Let us first analyze the mechanical effect of the dissipative Casimir force in the context of classical dynamics. We consider a point-like mirror of mass M in a harmonic potential well of frequency Ω. Taking the Casimir dissipative force given by (13.7) into account, and neglecting for the time being any associated stochastic force, we write the mirror's equation of motion as

$$\frac{d^2q(t)}{dt^2} = -\Omega^2 q(t) + \frac{\hbar}{6\pi Mc^2}\frac{d^3q(t)}{dt^3}. \tag{13.9}$$

We assume that the oscillator's zero-point energy $\hbar\Omega/2$ is much smaller than the rest mass energy Mc^2, and then find oscillatory solutions of (13.9) which are damped at the rate

$$\Gamma = \frac{1}{12\pi}\frac{\hbar\Omega}{Mc^2}\Omega \ll \Omega. \tag{13.10}$$

This result provides a good illustration of how weak the Casimir dissipation is for a single mirror in a vacuum. On the other hand, the associated diffusion in phase space is more relevant, particularly in the context of the full quantum theory discussed below, since it provides an efficient decoherence mechanism for non-classical quantum states.

The quantum description of the mirror's dynamics can be developed from the Hamiltonian for the radiation pressure coupling with a dispersive semi-transparent mirror (transparency frequency ω_c) [75]. One derives a master equation for the reduced density operator ρ of the mirror, which can also be cast in the form of a Fokker–Planck equation for the Wigner function $W(x, p, t)$ representing the mirror quantum state [4, 5]:

$$\partial_t W = -(1 - \Delta M/M)\frac{p}{M}\partial_x W + M\Omega^2 x\partial_p W + 2\Gamma\partial_x(xW) + D_1\frac{\partial^2 W}{\partial x^2} - D_2\frac{\partial^2 W}{\partial x\partial p}. \tag{13.11}$$

The time-dependent coefficients ΔM (mass correction), Γ (damping coefficient), D_1 and D_2 (diffusion coefficients) are written in terms of the correlation function of the field linear momentum operator. The perfectly reflecting limit corresponding to (13.1) is obtained when $\Omega \ll \omega_c$, in line with our previous discussion since $1/\omega_c$ represents the characteristic time scale for the material medium response. In this limit, for $t \gtrsim 1/\omega_c$, $\Gamma(t)$ rapidly approaches the expected constant value as given by (13.10), whereas $D_1(t)$ approaches the asymptotic value

$$D_1 = \hbar\Gamma/(M\Omega) \tag{13.12}$$

for $t \gtrsim 2\pi/\Omega$. This connection between diffusion and damping plays the role of a fluctuation-dissipation theorem for the vacuum state (zero temperature).

Under the time evolution described by the Fokker–Planck equation (13.11), an initially pure state gradually evolves into a statistical mixture. The physical reason behind this decoherence effect is the buildup of entanglement between the mirror and the field due to the radiation pressure coupling and the associated dynamical Casimir photon creation [4, 5]. As an example of initial state, we consider the "Schrödinger's cat" superposition of coherent states $|\psi\rangle = (|\alpha\rangle + |-\alpha\rangle)/\sqrt{2}$ with the amplitude $\alpha = iP_0/\sqrt{2M\hbar\omega_0}$ along the imaginary axis ($\pm P_0$ are the average momenta corresponding to each state's component). The corresponding Wigner function $W(x,p)$ contains an oscillating term proportional to $\cos(2P_0 x/\hbar)$ which represents the coherence of the superposition. Clearly, the diffusion term proportional to D_1 in (13.11) will wash out these oscillations along the position axis, thus transforming the cat state into a mixture of the two coherent states. From (13.11) and (13.12), the corresponding decoherence time scale is

$$t_d = \frac{\hbar^2}{2P_0^2 D_1} = 4\left(\frac{\delta p}{2P_0}\right)^2 \Gamma^{-1}, \tag{13.13}$$

where $\delta p = \sqrt{M\hbar\Omega/2}$ is the momentum uncertainty of the coherent state (we have assumed that $\Omega t_d \gg 1$). Since $2P_0$ measures the distance between the two components in phase space, (13.13) shows that decoherence is stronger when the state components are further apart, corresponding to thinner interference fringes in phase space. Decoherence from entanglement driven by dynamical Casimir photon creation is thus very effective for macroscopic superpositions in spite of the smallness of the corresponding damping coefficient Γ. As the radiation pressure control of microresonators improves all the way to the quantum level [76, 77], Casimir driven decoherence might eventually become of experimental relevance.

13.2.3 3D Models

The orders of magnitude can be reliably assessed only by considering the real-world three-dimensional space. We start with the simplest geometry in 3D: a plane mirror parallel to the xy plane, of area A and moving along the z-axis. For an infinite plane, we expect the dissipative Casimir force to be proportional to A, so that we have to modify (13.2) to include a squared length. For a scalar field satisfying a Dirichlet b.c. analogous to (13.1), the derivation is very similar to the one outlined above [7]:

$$f(t) = -\frac{\hbar A q^{(5)}(t)}{360\pi^2 c^4}. \quad (3D,\ scalar) \tag{13.14}$$

For the electromagnetic field, the model of perfect reflectivity provides an accurate description of metallic mirrors at low frequencies. For a mirror moving along the z-axis, the electric and magnetic fields in the instantaneously co-moving Lorentz frame S' satisfy

$$\hat{\mathbf{Z}} \times \mathbf{E}'|_{\text{mirror}} = 0, \quad \hat{\mathbf{Z}} \cdot \mathbf{B}'|_{\text{mirror}} = \mathbf{0}. \tag{13.15}$$

It is useful to decompose the fields into transverse electric (TE) and transverse magnetic (TM) polarizations (where 'transverse' means perpendicular to the incidence plane defined by $\hat{\mathbf{Z}}$ and the propagation direction). For the TE component, we define the vector potential in the usual way: $\mathbf{E}^{(\mathrm{TE})} = -\partial_t \mathbf{A}^{(\mathrm{TE})}$, $\mathbf{B}^{(\mathrm{TE})} = \nabla \times \mathbf{A}^{(\mathrm{TE})}$ under the Coulomb gauge $\nabla \cdot \mathbf{A}^{(\mathrm{TE})} = \mathbf{0}$. Since $\mathbf{A}^{(\mathrm{TE})} \cdot \hat{\mathbf{Z}} = 0$, $\mathbf{A}^{(\mathrm{TE})}$ is invariant under the Lorentz boost from the co-moving frame to the laboratory frame. The resulting b.c. is then similar to (13.1):

$$\mathbf{A}^{(\mathrm{TE})}(x, y, q(t), t) = \mathbf{0}. \tag{13.16}$$

On the other hand, $\mathbf{A}^{(\mathrm{TM})}$ has a component along the z-axis, so that the Coulomb gauge is no longer invariant under the Lorentz boost, resulting in complicated b.c. also involving the scalar potential. It is then convenient to define a new vector potential $\mathscr{A}^{(\mathrm{TM})}$ as

$$\mathbf{E}^{(\mathrm{TM})} = \nabla \times \mathscr{A}^{(\mathrm{TM})}, \quad \mathbf{B}^{(\mathrm{TM})} = \partial_t \mathscr{A}^{(\mathrm{TM})}$$

under the gauge $\nabla \cdot \mathscr{A}^{(\mathrm{TM})} = 0$. Like $\mathbf{A}^{(\mathrm{TE})}$, $\mathscr{A}^{(\mathrm{TM})}$ is also invariant under Lorentz boosts along the z-axis. From (13.15), one derives that $\mathscr{A}^{(\mathrm{TM})}$ satisfies a Neumann b.c. at the co-moving frame, yielding

$$\left[\partial_z + \dot{q}(t)\partial_t + \mathcal{O}(\dot{q}^2)\right]\mathscr{A}^{(\mathrm{TM})}(x, y, q(t), t) = \mathbf{0}. \tag{13.17}$$

The condition of perfect reflectivity then results in two independent problems: a Dirichlet b.c. for TE modes, and a Neumann b.c. in the instantaneously co-moving frame for TM modes. The TM contribution turns out to be 11 times larger than the TE one, which coincides with (13.14). The resulting dissipative force is then [78]

$$f(t) = -\frac{\hbar A q^{(5)}(t)}{30\pi^2 c^4}. \quad \text{(3D, electromagnetic)} \tag{13.18}$$

As in the 1D case, the dissipative Casimir force plays the role of a radiation reaction force, associated with the emission of photon pairs with wave-vectors satisfying the conditions $|\mathbf{k}_1| + |\mathbf{k}_2| = \Omega/c$ and $\mathbf{k}_{1\parallel} = -\mathbf{k}_{2\parallel}$ from translational symmetry parallel to the plane of the mirror. The angular distribution of emitted photons displays an interesting correlation with polarization: TE photons are preferentially emitted near the normal direction, whereas TM ones are preferentially emitted at larger angles, near a grazing direction if the frequency is smaller than $\Omega/2$ [27].

Results beyond the model of perfect reflection were obtained in Ref. [79] for a dielectric half-space (see also Ref. [80] for the angular and frequency spectra of emitted photons). In this case, there is also photon emission (and the associated dissipative radiation reaction force) if the dielectric mirror moves sideways, or if a dielectric sphere rotates around a diameter [81]. We will come back to this type of arrangement when discussing non-contact quantum friction in the next section.

To conclude this section, we compute the total photon production rate for a perfectly reflecting oscillating mirror directly from (13.18). By energy conservation, the total radiated energy is the negative of the work done on the mirror by the dissipative Casimir force:

$$E = - \int\limits_{-\infty}^{\infty} f(t) q^{(1)}(t) dt. \tag{13.19}$$

We evaluate the integral in (13.19) using the result (13.18) for an oscillatory motion of frequency Ω and amplitude q_0 exponentially damped over a time scale $T \gg 1/\Omega$: $E = \hbar T A q_0^2 \Omega^6 / (120 \pi^2 c^4)$. Since the spectrum is symmetrical with respect to the frequency $\Omega/2$, we can derive the number of photons N from the radiated energy using the relation $E = N \hbar \Omega / 2$. The total photon production rate is then given by

$$\frac{N}{T} = \frac{1}{15} \frac{A}{\lambda_0^2} \left(\frac{v_{max}}{c} \right)^2 \Omega \tag{13.20}$$

with $v_{max} \equiv \Omega q_0$ and $\lambda_0 \equiv 2\pi c/\Omega$ representing the typical scale of the relevant wavelengths. With $v_{max}/c \sim 10^{-7}$, $\Omega/(2\pi) \sim 10$ GHz and $A \sim \lambda_0^2 \sim 10$ cm^2, we find $N/T \sim 10^{-5}$ photons/s or approximately one photon pair every two days!

The dynamical Casimir effect is clearly very small for a single oscillating mirror. Adding a second parallel plane mirror, the photon production rate is enhanced by a factor $(\Omega L/c)^{-2} \sim 10^6$ for separation distances L in the sub-micrometer range [29]. But at such short distances, finite conductivity of the metallic plates, not considered so far, is likely to reduce the photon production rate. In this type of arrangement, a much larger effect is obtained by considering the shear motion of one plate relative to the other (instead of a relative motion along the normal direction). As discussed in the next section, because of finite conductivity a large friction force is predicted at short distances, which results from the creation of pairs of excitations inside the metallic medium [24, 82]. As for the emission of photon pairs, the orders of magnitude are more promising when considering a closed cavity with moving walls, to be discussed in Sect. 13.4.

13.3 Quantum Friction

There is an intimate connection between the dynamical Casimir effect and the possibility that electrically neutral bodies in relative motion may experience non-contact friction due to quantum vacuum fluctuations, the so-called "quantum friction". As we have discussed in the previous section, dielectric bodies in accelerated motion radiate Casimir photons. Shear motion of two bodies, even at constant relative speed, can also radiate energy. Just as in the case of a single accelerated mirror in a vacuum, shear motion cannot be removed by a change of reference frame. A frictional force between two perfectly smooth parallel planes shearing against each other with a relative velocity **v** results from the exchange of photons between the two surfaces. These photons carry the information of the motion of one surface to the other one, and as a result linear momentum is exchanged between the plates, leading to friction.

In order to illustrate the physics of quantum friction we will follow here an approach due to Pendry [24] who considered the simplest case of zero temperature and the non-retarded (van der Waals) limit. The nice feature of this approach is that it manifestly connects to the intuitive picture of motion-induced (virtual) photons as mediators of momentum exchange between the shear surfaces. A dielectric surface, although electrically neutral, experiences quantum charge fluctuations, and these have corresponding images on the opposing dielectric surface. Since the surfaces are in relative parallel motion, the image lags behind the fluctuating charge distribution that created it, and this results in a frictional van der Waals force. Note that for ideal perfect metals, the image charges arrange themselves instantaneously (do not lag behind), and therefore no quantum friction is expected in this case.

We model each of the dielectric surfaces as a continuum of oscillators with Hamiltonian

$$\hat{H}_\alpha = \sum_{\mathbf{k}j} \hbar\omega_{\alpha;\mathbf{k}j}(\hat{a}^\dagger_{\alpha;\mathbf{k}j}\hat{a}_{\alpha;\mathbf{k}j} + 1/2),\qquad(13.21)$$

where $\hat{a}^\dagger_{\alpha;\mathbf{k}j}$ and $\hat{a}_{\alpha;\mathbf{k}j}$ are creation and annihilation bosonic operators associated with the upper ($\alpha = u$) or lower ($\alpha = l$) plate. Each mode on each surface is defined by $\mathbf{k}$, which is a wave-vector parallel to the planar surface, and by j, which denotes degrees of freedom perpendicular to the surface. Following Pendry we restrict ourselves to the non-retarded limit (very long wavelengths for the EM modes). In this limit the EM field is mainly electrostatic, only the static TM polarization matters for a dielectric surface (since the static TE field is essentially a magnetic field that does not interact with the non-magnetic surface), and the intensity decays exponentially from the surfaces (evanescent fields). The coupling between the oscillator modes belonging to different surfaces is mediated by the EM field, and it is assumed to be a position–position interaction of the form

$$\hat{H}_{\mathrm{int}}(t) = \sum_{\mathbf{k}jj'} C_{\mathbf{k}jj'}(d)\hat{x}_{u;\mathbf{k}j} \otimes \hat{x}_{l;-\mathbf{k}j'}\, e^{-ik_x vt}.\qquad(13.22)$$

In the non-relativistic limit the effect of the surfaces shearing with speed v along the x direction is contained in the last exponential factor. This type of Hamiltonian follows from the effective electrostatic interaction between the fluctuating charges in the dielectrics and its expansion to lowest order in the displacement of each oscillator from its equilibrium position (equivalently, it also follows from the non-retarded and static limit of the dipole–dipole interactions between fluctuating dipoles in each surface). The coupling factors $C_{\mathbf{k}jj'}(d)$ can be obtained by analyzing how each oscillator dissipates energy into the vacuum gap. This is done in Ref. [24] in two ways, by invoking a scattering type of approach relating the fields at the interphases with reflection amplitudes, and by considering how the fluctuating charge distributions in each surface dissipate energy. The result is $C_{kjj'}(d) = (\beta_{kj}\beta_{kj'}/2k\varepsilon_0)e^{-|k|d}$, where

$$\beta_{kj}^2 = \left(\frac{dN}{d\omega}\right)^{-1} \frac{4k\omega\varepsilon_0}{\pi} \text{Im}\left[\frac{\varepsilon(\omega) - 1}{\varepsilon(\omega) + 1}\right]. \tag{13.23}$$

Note the exponential decay due to the evanescent nature of the EM field. In this equation $dN/d\omega$ is the density of oscillator modes at frequency ω and $\varepsilon(\omega)$ is the complex dielectric permittivity of the plates (assumed to be identical). Although the interaction Hamiltonian does not depend explicitly on the quantized EM field (because this derivation is semiclassical), one can infer the quantum processes of creation and absorption that take place by expanding the product $\hat{x}_{u;kj} \otimes \hat{x}_{l;-kj'}$:

$$\hat{x}_{u;kj} \otimes \hat{x}_{l;-kj'} = -\frac{1}{2}[\hat{a}_{u,kj}^\dagger - \hat{a}_{u,kj}] \otimes [\hat{a}_{l,-kj'}^\dagger - \hat{a}_{l,-kj'}]. \tag{13.24}$$

Imagine the system of the two dielectrics is initially in the ground state at zero temperature, $|\psi(t = 0)\rangle = |\psi_g\rangle_u \otimes |\psi_g\rangle_l$, where $|\psi_g\rangle_\alpha = \prod_{kj} |\psi_{g,kj}\rangle_\alpha$ is the product of the harmonic oscillators' ground states for surface α. Eq. (13.24) implies that two motion-induced virtual photons created from an EM vacuum produce one excitation in each surface, i.e., there is a non-zero probability of transition to states $|1;kj\rangle_u \otimes |1;-kj'\rangle_l$. The transition probability can be computed using time-dependent perturbation theory for the perturbation $H_{int}(t)$. To first order, the transition probability from the ground state into each of these two-excitation states is

$$P_{kjj'}(t) = \frac{\beta_{kj}^2\beta_{kj'}^2}{4k^2\varepsilon_0^2} \frac{e^{-2d|k|}}{4\omega_{u,kj}\omega_{l,-kj}} \frac{4\sin^2[(\omega_{u,kj} + \omega_{l,-kj'} - k_x v)t/2]}{(\omega_{u,kj} + \omega_{l,-kj'} - k_x v)^2}. \tag{13.25}$$

In the limit of large times ($t \to \infty$) we use that $\sin^2(\Omega t/2)/(\Omega/2)^2 \approx \pi t\delta(\Omega)$ (here $\delta(\Omega)$ is Dirac's delta function), and therefore the transition probability grows linearly in time. We can find the frictional force equating the frictional work $F_x v$ with the rate of change in time of the energy of the excitations, namely

$$F_x v = \frac{dU}{dt} = \sum_{kjj'} \hbar(\omega_{u,kj} + \omega_{l,-kj'})\frac{dP_{kjj'}}{dt}, \tag{13.26}$$

and the r.h.s. is time-independent since the transition probabilities grow linearly in time. Using the expression (13.23) for β_{kj}^2 and the transition probabilities (13.25) at large times, and writing the sums over the dielectric degree of freedom j as $\sum_j = \int_0^\infty d\omega dN/d\omega$ (and similarly for j'), one finally obtains the following expression for the frictional force

$$F_x = \frac{\hbar}{\pi} \int \frac{d^2\mathbf{k}}{(2\pi)^2} k_x e^{-2|k|d} \int_0^{k_x v} d\omega \, \text{Im}\left[\frac{\varepsilon(\omega) - 1}{\varepsilon(\omega) + 1}\right] \text{Im}\left[\frac{\varepsilon(k_x v - \omega) - 1}{\varepsilon(k_x v - \omega) + 1}\right]. \tag{13.27}$$

In the literature there are other more rigorous approaches to calculate quantum friction that go beyond the non-retarded quasi-static limit considered above, and that can take into account effects of relativistic motion as well as finite temperature. One of these approaches [83] follows the spirit of the Lifshitz–Rytov theory [84], considering the fluctuating electromagnetic (EM) field as a *classical* field whose stochastic fluctuations satisfy the fluctuation-dissipation relation that relates the field fluctuations with the absorptive part of the dielectric response of the plates. The EM field is a solution to Maxwell's equations with classical fluctuating current densities on the plates as source fields, and it satisfies the usual EM b.c. imposed on the comoving reference frames on each plate. The relation between the EM fields in different frames is obtained via Lorentz transformations. An alternative full quantum-mechanical approach considers the *quantum* EM field in interaction with (quantized) noise polarizations and noise currents within the plates [85]. As before, the fields in each reference frame are related by Lorentz transformations. In this approach the quantum expectation value of the noise currents is given by the (quantum) fluctuation-dissipation relation.

Quantum friction can also happen for neutral atoms moving close to surfaces. The theoretical methods to compute the frictional force in these cases are similar to the surface–surface quantum friction, and we refer the reader to some of the relevant works [86–88]. See also the Chap. 11 by Intravaia *et al.* in this volume for additional discussions of quantum friction in the atom-surface context.

13.4 Resonant Photon Creation in Time Dependent Cavities

As mentioned in the Introduction, photon creation can be enhanced in closed cavities: if the external time dependence involves a frequency that is twice the frequency of a mode of the electromagnetic field in the (unperturbed) cavity, parametric amplification produces a large number of photons. As we will see, under certain circumstances (ideal three dimensional cavities with non equidistant frequencies in the spectrum) the number of photons in the resonant mode may grow exponentially. Parametric amplification can take place by changing the length of the cavity with a moving surface, but also by changing its effective length through time dependent electromagnetic properties of the cavity.

In order to simplify the notation, in this section we will use the natural units $\hbar = c = 1$.

13.4.1 Dynamical Casimir Effect in 1D Cavities

As in Sect. 13.2, we start with a massless real scalar field in a 1D cavity with one mirror fixed at $x = 0$ and the other performing an oscillatory motion

$$L(t) = L_0[1 + \varepsilon \sin(\Omega t)], \tag{13.28}$$

where Ω is the external frequency and $\varepsilon \ll 1$. As we will be mainly concerned with situations where $\Omega L_0 = O(1)$, the maximum velocity of the mirror will be of order ε, and therefore small values of ε correspond to a non-relativistic motion of the mirror. We shall assume that the oscillations begin at $t = 0$, end at $t = T$, and that $L(t) = L_0$ for $t < 0$ and $t > T$. The scalar field $\phi(x,t)$ satisfies the wave equation and Dirichlet b.c. $\phi(x = 0, t) = \phi(L(t), t) = 0$. When the mirror is at rest, the eigenfrequencies are multiples of the fundamental frequency π/L_0. Therefore, in order to analyze resonant situations we will assume that $\Omega = q\pi/L_0$, $q = 1, 2, 3, \ldots$.

Inside the cavity we can write

$$\phi(x,t) = \sum_{k=1}^{\infty} \left[a_k \psi_k(x,t) + a_k^\dagger \psi_k^*(x,t) \right], \tag{13.29}$$

where the mode functions $\psi_k(x,t)$ are positive frequency modes for $t < 0$, and a_k and $a_k^\dagger$ are time-independent bosonic annihilation and creation operators, respectively. The field equation is automatically verified by writing the modes in terms of Moore's function $R(t)$ [8] as

$$\psi_k(x,t) = \frac{i}{\sqrt{4\pi k}} \left(e^{-ik\pi R(t+x)} - e^{-ik\pi R(t-x)} \right). \tag{13.30}$$

The Dirichlet boundary condition is satisfied provided that $R(t)$ satisfies Moore's equation

$$R(t + L(t)) - R(t - L(t)) = 2. \tag{13.31}$$

These simple expressions for the modes of the field are due to conformal invariance, a symmetry for massless fields in one spatial dimension.

The solution to the problem involves finding a solution $R(t)$ in terms of the prescribed motion $L(t)$. For $t < 0$ the positive frequency modes are given by $R(t) = t/L_0$ for $-L_0 \le t \le L_0$, which is indeed a solution to Eq. (13.31) for $t < 0$. For $t > 0$, Eq. (13.31) can be solved, for example, using a perturbative expansion in ε similar to the one employed in Sect. 13.2 for a single mirror. However, as the external frequency is tuned with the unperturbed modes of the cavity, in general there will be resonant effects, which produce secular terms proportional to $\varepsilon^m (\Omega t)^n$ with $m \le n$. Thus this expansion is valid only for short times $\varepsilon \Omega t \ll 1$. It is possible to obtain a non-perturbative solution of Eq. (13.31) using Renormalization Group (RG) techniques [39]. The RG-improved solution automatically adds the most secular terms, $(\varepsilon \Omega t)^n$, to all orders in ε, and is valid for longer times $\varepsilon^2 \Omega t \ll 1$. The RG-improved solution is [39, 89]

$$R(t) = \frac{t}{L_0} - \frac{2}{\pi q} \text{Im} \ln \left[1 + \xi + (1 - \xi) e^{\frac{iq\pi t}{L_0}} \right], \tag{13.32}$$

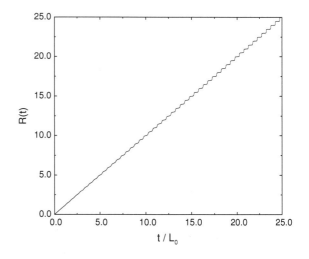

where $\xi = \exp[(-1)^{q+1}\pi q \varepsilon t/L_0]$. As shown in Fig. 13.2, the function $R(t)$ develops a staircase shape for long times [39, 90]. Within regions of t between odd multiples of L_0 there appear q jumps, located at values of t satisfying $\cos(q\pi t/L_0) = \mp 1$. the upper sign corresponding to even values of q and the lower one to odd values of q.

The vacuum expectation value of the energy density of the field is given by [10] $\langle T_{00}(x,t)\rangle = -f(t+x) - f(t-x)$, where

$$f = \frac{1}{24\pi}\left[\frac{R'''}{R'} - \frac{3}{2}\left(\frac{R''}{R'}\right)^2 + \frac{\pi^2}{2}(R')^2\right]. \tag{13.33}$$

For $q = 1$ ("semi-resonant" case) no exponential amplification of the energy density is obtained, whereas for $q \geq 2$ ("resonant" cases) the energy density grows exponentially in the form of q traveling wave packets which become narrower and higher as time increases (see Fig. 13.3). Note that, as the energy density involves the derivatives of the function $R(t)$, there is one peak for each jump of $R(t)$.

The number of created particles can be computed from the solution given by Eq. (13.32). Photons are created resonantly in all modes with $n = q + 2j$, with j a non-negative integer. This is due to the fact that the spectrum of a one dimensional cavity is *equidistant*: although the external frequency resonates with a particular eigenmode of the cavity, intermode coupling produces resonant creation in the other modes. At long times, the number of photons in each mode grows linearly in time, while the total number of photons grows quadratically and the total energy inside the cavity grows exponentially [35]. These different behaviors are due to the fact that the number of excited modes, i.e. the number of modes that reach a growth linear in time, increases exponentially.

The production of massless particles in one dimensional cavities has been analyzed numerically in Ref. [91]. As expected, the numerical evaluations are in

Fig. 13.3 Energy density profile between plates for fixed time $t/L_0 = 20.4$ for the $q = 4$ case. The amplitude coefficient is $\varepsilon = 0.01$

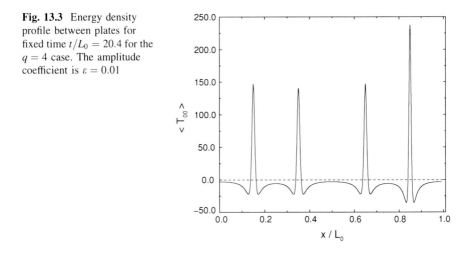

perfect agreement with the analytic results described above in the case of small amplitudes $\varepsilon \leq 0.01$.

The force on the moving mirror can be computed as the discontinuity of $\langle T_{11} \rangle$ at $x = L(t)$. The force produced by the outside field is one-half the expression derived in Sect. 13.2. This is much smaller than the intracavity contribution and will be neglected. Therefore

$$\langle F \rangle \approx \langle T_{11}(L(t), t) \rangle = \langle T_{00}(L(t), t) \rangle, \tag{13.34}$$

where the energy momentum tensor is evaluated *inside* the cavity. This expression reproduces the usual attractive result when the mirror is at rest ($t < 0$)

$$\langle F \rangle = -\frac{\pi}{24L_0^2}, \tag{13.35}$$

that is, the static Casimir effect in 1D. However, at long times it becomes an exponentially increasing pressure due to the presence of real photons in the cavity [35].

All this treatment can be extended to the case of Neumann b.c. $n^\mu \partial_\mu \phi|_{\text{mirror}} = 0$, where n^μ is a unit two-vector perpendicular to the trajectory of the mirror. The modes of the field can be written in terms of Moore's function $R(t)$ as

$$\phi_k(x, t) = \frac{1}{\sqrt{4\pi k}} \left(e^{-ik\pi R(t+x)} + e^{-ik\pi R(t-x)} \right). \tag{13.36}$$

Note the change of sign between Eq. (13.30) for Dirichlet modes, and Eq. (13.36) for Neumann modes. The spectrum of motion-induced photons is the same for both Dirichlet and Neumann b.c. [68], but not for the mixed configuration with one Dirichlet mirror and one Neumann mirror [92]. The one dimensional DCE has also been investigated for cavities with Robin b.c. [93].

13.4.2 Photon Creation in 3D Cavities

The one dimensional DCE with Dirichlet b.c. described in the previous section is not only of academic interest: it describes photon creation for the TEM modes in a 3D cylindrical cavity with a non-simply connected section [38]. However, in order to analyze TE and TM modes in general 3D cavities, a new approach is needed, since conformal invariance is no longer useful in 3D.

We shall first describe in some detail the simpler case of a scalar field in a rectangular cavity satisfying Dirichlet b.c. [36], and then comment on the extension to the case of the electromagnetic field in cylindrical cavities with an arbitrary section.

13.4.2.1 Scalar Field

We consider a rectangular cavity formed by perfectly reflecting walls with dimensions L_x, L_y, and L_z. The wall placed at $z = L_z$ is at rest for $t < 0$ and begins to move following a given trajectory, $L_z(t)$, at $t = 0$. We assume this trajectory as prescribed for the problem (not a dynamical variable) and that it works as a time-dependent boundary condition for the field. The field $\phi(\mathbf{x}, t)$ satisfies the wave equation and the b.c. $\phi|_{\text{walls}} = 0$ for all times. The Fourier expansion of the field for an arbitrary moment of time, in terms of creation and annihilation operators, can be written as

$$\phi(\mathbf{x}, t) = \sum_{\mathbf{n}} \hat{a}_{\mathbf{n}}^{\text{in}} u_{\mathbf{n}}(\mathbf{x}, t) + \text{H.c.}, \tag{13.37}$$

where the mode functions $u_{\mathbf{n}}(\mathbf{x}, t)$ form a complete orthonormal set of solutions of the wave equation with vanishing b.c..

When $t \leq 0$ (static cavity) each field mode is determined by three positive integers n_x, n_y and n_z, namely

$$u_{\mathbf{n}}(\mathbf{x}, t < 0) = \frac{1}{\sqrt{2\omega_{\mathbf{n}}}} \sqrt{\frac{2}{L_x}} \sin\left(\frac{n_x \pi}{L_x} x\right) \sqrt{\frac{2}{L_y}} \sin\left(\frac{n_y \pi}{L_y} y\right)$$

$$\times \sqrt{\frac{2}{L_z}} \sin\left(\frac{n_z \pi}{L_z} z\right) e^{-i\omega_{\mathbf{k}} t}, \tag{13.38}$$

with $\omega_{\mathbf{n}} = \pi \sqrt{\left(\frac{n_x}{L_x}\right)^2 + \left(\frac{n_y}{L_y}\right)^2 + \left(\frac{n_z}{L_z}\right)^2}$.

When $t > 0$ the boundary condition on the moving wall becomes $\phi(x, y, z = L_z(t), t) = 0$. In order to satisfy it we expand the mode functions in Eq. (13.37) with respect to an *instantaneous basis* [94]

$$u_{\mathbf{n}}(\mathbf{x}, t > 0) = \sum_{\mathbf{m}} Q_{\mathbf{m}}^{(\mathbf{n})}(t) \sqrt{\frac{2}{L_x}} \sin\left(\frac{m_x \pi}{L_x} x\right) \sqrt{\frac{2}{L_y}} \sin\left(\frac{m_y \pi}{L_y} y\right)$$

$$\times \sqrt{\frac{2}{L_z}} \sin\left(\frac{m_z \pi}{L_z(t)} z\right) = \sum_{\mathbf{m}} Q_{\mathbf{m}}^{(\mathbf{n})}(t) \varphi_{\mathbf{m}}(\mathbf{x}, L_z(t)), \tag{13.39}$$

with the initial conditions

$$Q_{\mathbf{m}}^{(\mathbf{n})}(0) = \frac{1}{\sqrt{2\omega_{\mathbf{n}}}} \delta_{\mathbf{m},\mathbf{n}} , \quad \dot{Q}_{\mathbf{m}}^{(\mathbf{n})}(0) = -i \frac{\sqrt{\omega_{\mathbf{n}}}}{2} \delta_{\mathbf{m},\mathbf{n}}. \tag{13.40}$$

In this way we ensure that, as long as $L_z(t)$ and $\dot{L}_z(t)$ are continuous at $t = 0$, each field mode and its time derivative are also continuous functions. The expansion in Eq. (13.39) for the field modes must be a solution of the wave equation. Taking into account that the $\varphi_{\mathbf{k}}$'s form a complete and orthonormal set and that they depend on t only through $L_z(t)$, we obtain a set of (exact) coupled equations for $Q_{\mathbf{m}}^{(\mathbf{n})}(t)$ [36]:

$$\ddot{Q}_{\mathbf{m}}^{(\mathbf{n})} + \omega_{\mathbf{m}}^2(t) Q_{\mathbf{m}}^{(\mathbf{n})} = 2\lambda(t) \sum_{\mathbf{j}} g_{\mathbf{mj}} \dot{Q}_{\mathbf{j}}^{(\mathbf{n})} + \dot{\lambda}(t) \sum_{\mathbf{j}} g_{\mathbf{mj}} Q_{\mathbf{j}}^{(\mathbf{n})}$$

$$+ \lambda^2(t) \sum_{\mathbf{j},\mathbf{l}} g_{\mathbf{lm}} g_{\mathbf{lj}} Q_{\mathbf{j}}^{(\mathbf{n})}, \tag{13.41}$$

where

$$\omega_{\mathbf{m}}(t) = \pi \sqrt{\left(\frac{m_x}{L_x}\right)^2 + \left(\frac{m_y}{L_y}\right)^2 + \left(\frac{m_z}{L_z(t)}\right)^2} ; \quad \lambda(t) = \frac{\dot{L}_z(t)}{L_z(t)}. \tag{13.42}$$

The coefficients $g_{\mathbf{mj}}$ are defined by

$$g_{\mathbf{mj}} = -g_{\mathbf{jm}} = L_z(t) \int_0^{L_z(t)} dz \frac{\partial \varphi_{\mathbf{m}}}{\partial L_z} \varphi_{\mathbf{j}}. \tag{13.43}$$

The annihilation and creation operators $\hat{a}_{\mathbf{m}}^{\text{in}}$ and $\hat{a}_{\mathbf{m}}^{\dagger \text{in}}$ correspond to the particle notion in the 'in' region ($t < 0$). If the wall stops for $t > t_{\text{final}}$, we can define a new set of operators, $\hat{a}_{\mathbf{m}}^{\text{out}}$ and $\hat{a}_{\mathbf{m}}^{\dagger \text{out}}$, associated with the particle notion in the 'out' region ($t > t_{\text{final}}$). These two sets of operators are connected by means of the Bogoliubov transformation

$$\hat{a}_{\mathbf{m}}^{\text{out}} = \sum_{\mathbf{n}} (\hat{a}_{\mathbf{n}}^{\text{in}} \alpha_{\mathbf{nm}} + \hat{a}_{\mathbf{n}}^{\dagger \text{in}} \beta_{\mathbf{nm}}^*). \tag{13.44}$$

The coefficients $\alpha_{\mathbf{nm}}$ and $\beta_{\mathbf{nm}}$ can be obtained as follows. When the wall returns to its initial position the right hand side in Eq. (13.41) vanishes and the solution is

$$Q_{\mathbf{m}}^{(\mathbf{n})}(t > t_{\text{final}}) = A_{\mathbf{m}}^{(\mathbf{n})} e^{i\omega_{\mathbf{m}}t} + B_{\mathbf{m}}^{(\mathbf{n})} e^{-i\omega_{\mathbf{m}}t}, \tag{13.45}$$

with $A_{\mathbf{m}}^{(\mathbf{n})}$ and $B_{\mathbf{m}}^{(\mathbf{n})}$ being some constant coefficients to be determined by the continuity conditions at $t = t_{\text{final}}$. Inserting Eq. (13.45) into Eqs.(13.37) and (13.39) we obtain an expansion of ϕ in terms of $\hat{a}_{\mathbf{m}}^{\text{in}}$ and $\hat{a}_{\mathbf{m}}^{\dagger\text{in}}$ for $t > t_{\text{final}}$. Comparing this with the equivalent expansion in terms of $\hat{a}_{\mathbf{m}}^{\text{out}}$ and $\hat{a}_{\mathbf{m}}^{\dagger\text{out}}$ it is easy to see that

$$\alpha_{\mathbf{n}\mathbf{m}} = \sqrt{2\omega_{\mathbf{m}}} B_{\mathbf{m}}^{(\mathbf{n})}, \quad \beta_{\mathbf{n}\mathbf{m}} = \sqrt{2\omega_{\mathbf{m}}} A_{\mathbf{m}}^{(\mathbf{n})}. \tag{13.46}$$

The amount of photons created in the mode $\mathbf{m}$ is the average value of the number operator $\hat{a}_{\mathbf{m}}^{\dagger\text{out}} \hat{a}_{\mathbf{m}}^{\text{out}}$ with respect to the initial vacuum state (defined through $\hat{a}_{\mathbf{m}}^{\text{in}} |0_{\text{in}}\rangle = 0$). With the help of Eqs. (13.44) and (13.46) we find

$$\langle \mathcal{N}_{\mathbf{m}} \rangle = \langle 0_{\text{in}} | \hat{a}_{\mathbf{m}}^{\dagger\text{out}} \hat{a}_{\mathbf{m}}^{\text{out}} | 0_{\text{in}} \rangle = \sum_{\mathbf{n}} 2\omega_{\mathbf{m}} |A_{\mathbf{m}}^{(\mathbf{n})}|^2. \tag{13.47}$$

In the approach described so far we worked at the level of the dynamical equation for the quantum scalar field. Alternatively, one can analyze the problem using the *effective Hamiltonian method* developed in Ref. [95]. The idea is the following. Assume that a massless scalar field is confined within a time dependent volume and satisfies Dirichlet b.c.. At the classical level, the field can be expanded in terms of a basis of functions $f_\alpha(\mathbf{x}, t)$ that fulfill the b.c. at each time, that is

$$\phi(\mathbf{x}, t) = \sum_\alpha q_\alpha(t) f_\alpha(\mathbf{x}, t). \tag{13.48}$$

For the rectangular cavities considered in this section these functions can be chosen to be $\varphi_{\mathbf{m}}(\mathbf{x}, L_z(t))$. Inserting this expansion into the Klein–Gordon Lagrangian, one ends up with a Lagrangian for the generalized coordinates $q_\alpha(t)$, which is a quadratic function of $q_\alpha(t)$ and $\dot{q}_\alpha(t)$, i.e. it describes a set of coupled harmonic oscillators with time dependent frequencies and couplings. This system can be quantized following the usual procedure, and the final results for the number of created photons are equivalent to those obtained in Eq. (13.47).

13.4.2.2 Parametric Amplification in 3D

As in Sect. 13.4.1, we are interested in resonant situations where the number of photons created inside the cavity could be enhanced for some specific external frequencies. So we study the trajectory given in Eq. (13.28). To first order in ε, the equations for the modes Eq. (13.41) take the form

$$\ddot{Q}_{\mathbf{m}}^{(\mathbf{n})} + \omega_{\mathbf{m}}^2 Q_{\mathbf{m}}^{(\mathbf{n})} = 2\varepsilon \left(\frac{\pi m_z}{L_z}\right)^2 \sin(\Omega t) Q_{\mathbf{m}}^{(\mathbf{n})} - \varepsilon\Omega^2 \sin(\Omega t) \sum_{\mathbf{j}} g_{\mathbf{m}\mathbf{j}} Q_{\mathbf{j}}^{(\mathbf{n})}$$
$$+ 2\varepsilon\Omega \cos(\Omega t) \sum_{\mathbf{j}} g_{\mathbf{m}\mathbf{j}} \dot{Q}_{\mathbf{j}}^{(\mathbf{n})} + O(\varepsilon^2). \tag{13.49}$$

It is known that a naive perturbative solution of these equations in powers of the displacement ε breaks down after a short amount of time, of order $(\varepsilon\Omega)^{-1}$. As in the 1D case discussed in the previous section, this happens for those particular values of the external frequency Ω such that there is a resonant coupling with the eigenfrequencies of the static cavity. In this situation, to find a solution valid for longer times (of order $\varepsilon^{-2}\Omega^{-1}$) we proceed as follows. We assume that the solution of Eq. (13.49) is of the form

$$Q_{\mathbf{m}}^{(\mathbf{n})}(t) = A_{\mathbf{m}}^{(\mathbf{n})}(t)e^{i\omega_{\mathbf{m}}t} + B_{\mathbf{m}}^{(\mathbf{n})}(t)e^{-i\omega_{\mathbf{m}}t}, \tag{13.50}$$

where the functions $A_{\mathbf{m}}^{(\mathbf{n})}$ and $B_{\mathbf{m}}^{(\mathbf{n})}$ are slowly varying. In order to obtain differential equations for them, we insert this ansatz into Eq. (13.49) and neglect second-order derivatives of $A_{\mathbf{m}}^{(\mathbf{n})}$ and $B_{\mathbf{m}}^{(\mathbf{n})}$. After multiplying the equation by $e^{\pm i\omega_{\mathbf{m}}t}$ we average over the fast oscillations. The resulting equations are

$$\frac{1}{\varepsilon}\frac{dA_{\mathbf{m}}^{(\mathbf{n})}}{dt} = -\frac{\pi^2 m_z^2}{2\omega_{\mathbf{m}}L_z^2}B_{\mathbf{m}}^{(\mathbf{n})}\delta(2\omega_{\mathbf{m}} - \Omega)$$

$$+ \sum_{\mathbf{j}}(-\omega_{\mathbf{j}} + \frac{\Omega}{2})\delta(-\omega_{\mathbf{m}} - \omega_{\mathbf{j}} + \Omega)\frac{\Omega}{2\omega_{\mathbf{m}}}g_{\mathbf{kj}}B_{\mathbf{j}}^{(\mathbf{n})}$$

$$+ \sum_{\mathbf{j}}\left[(\omega_{\mathbf{j}} + \frac{\Omega}{2})\delta(\omega_{\mathbf{m}} - \omega_{\mathbf{j}} - \Omega) + (\omega_{\mathbf{j}} - \frac{\Omega}{2})\delta(\omega_{\mathbf{m}} - \omega_{\mathbf{j}} + \Omega)\right]$$

$$\times \frac{\Omega}{2\omega_{\mathbf{m}}}g_{\mathbf{mj}}A_{\mathbf{j}}^{(\mathbf{n})}, \tag{13.51}$$

and

$$\frac{1}{\varepsilon}\frac{dB_{\mathbf{m}}^{(\mathbf{n})}}{dt} = -\frac{\pi^2 m_z^2}{2\omega_{\mathbf{m}}L_z^2}A_{\mathbf{m}}^{(\mathbf{n})}\delta(2\omega_{\mathbf{m}} - \Omega)$$

$$+ \sum_{\mathbf{j}}(-\omega_{\mathbf{j}} + \frac{\Omega}{2})\delta(-\omega_{\mathbf{m}} - \omega_{\mathbf{j}} + \Omega)\frac{\Omega}{2\omega_{\mathbf{m}}}g_{\mathbf{mj}}A_{\mathbf{j}}^{(\mathbf{n})}$$

$$+ \sum_{\mathbf{j}}\left[(\omega_{\mathbf{j}} + \frac{\Omega}{2})\delta(\omega_{\mathbf{m}} - \omega_{\mathbf{j}} - \Omega) + (\omega_{\mathbf{j}} - \frac{\Omega}{2})\delta(\omega_{\mathbf{m}} - \omega_{\mathbf{j}} + \Omega)\right]$$

$$\times \frac{\Omega}{2\omega_{\mathbf{m}}}g_{\mathbf{mj}}B_{\mathbf{j}}^{(\mathbf{n})}, \tag{13.52}$$

where we used the notation $\delta(\omega)$ for the Kronecker δ-function $\delta_{\omega 0}$.

The method used to derive these equations is equivalent to the "multiple scale analysis" [96] and to the slowly varying envelope approximation [97]. The equations are non-trivial (i.e., lead to resonant behavior) if $\Omega = 2\omega_{\mathbf{m}}$ (resonant condition). Moreover, there is intermode coupling between modes $\mathbf{j}$ and $\mathbf{m}$ if any of the conditions $|\omega_{\mathbf{m}} \pm \omega_{\mathbf{j}}| = \Omega$ is satisfied.

We derived the equations for three dimensional cavities. It is easy to obtain the corresponding ones for one dimensional cavities. The conditions for resonance and intermode coupling are the same. The main difference is that for one dimensional cavities the spectrum is equidistant. Therefore an infinite set of modes may be coupled. For example, when the external frequency is $\Omega = 2\omega_1$, the mode m is coupled with the modes $m \pm 2$. This has been extensively studied in the literature [35, 39, 98, 99].

In what follows we will be concerned with cavities with non-equidistant spectrum. Eqs.(13.51) and (13.52) present different kinds of solutions depending both on the mirror's frequency and the spectrum of the static cavity. In the simplest 'parametric resonance case' the frequency of the mirror is twice the frequency of some unperturbed mode, say $\Omega = 2\omega_m$. In order to find $A_m^{(n)}$ and $B_m^{(n)}$ from Eqs. (13.51) and (13.52) we have to analyze whether the coupling conditions $|\omega_m \pm \omega_j| = \Omega$ can be satisfied or not. If we set $\Omega = 2\omega_m$, the resonant mode $\mathbf{m}$ will be coupled to some other mode $\mathbf{j}$ only if $\omega_j - \omega_m = \Omega = 2\omega_m$. Clearly, the latter relation will be satisfied depending on the spectrum of the particular cavity under consideration.

Let us assume that this condition is not fullfilled. In this case, the equations for $A_m^{(n)}$ and $B_m^{(n)}$ can be easily solved and give

$$\langle \mathcal{N}_m \rangle = \sinh^2 \left[\frac{1}{\Omega} \left(\frac{m_z \pi}{L_z} \right)^2 \varepsilon t_f \right]. \tag{13.53}$$

In this uncoupled resonance case the average number of created photons in the mode $\mathbf{m}$ increases exponentially in time. Another way of looking at this particular situation is to note that, neglecting the intermode couplings, the amplitude of the resonant mode satisfies the equation of an harmonic oscillator with time dependent frequency. For the particular trajectory given in Eq. (13.28), the dynamics of the mode is governed by Eq. (13.49) with $g_{mj} = 0$, that is

$$\ddot{Q}_m^{(n)} + \left[\omega_m^2 - 2\varepsilon \left(\frac{\pi m_z}{L_z} \right)^2 \sin(\Omega t) \right] Q_m^{(n)} = 0, \tag{13.54}$$

which is the well known Mathieu equation [96]. The solutions to this equation have an exponentially growing amplitude when $\Omega = 2\omega_m$.

There are simple situations in which there is intermode coupling. For instance for a cubic cavity of size L, the fundamental mode $(1,1,1)$ is coupled to the mode $(5,1,1)$ when the external frequency is $\Omega = 2\omega_{111}$. In this case the number of photons in each mode grows with a lower rate than that of the uncoupled case [36]

$$\langle \mathcal{N}_{111} \rangle \simeq \langle \mathcal{N}_{511} \rangle \simeq e^{0.9\varepsilon t_f / L}. \tag{13.55}$$

We will describe some additional examples of intermode coupling in the next subsection, in the context of a full electromagnetic model.

It is worth stressing that the creation of scalar particles in 3D cavities has been studied numerically in Ref. [100]. At long times, the numerical results coincide with the analytical predictions derived from Eqs. (13.51) and (13.52), both in the presence and absence of intermode coupling.

13.4.2.3 The Electromagnetic Case

The previous results have been generalized to the case of the electromagnetic field inside a cylindrical cavity, with an arbitrary transversal section [38]. Let us assume that the axis of the cavity is along the z-direction, and that the caps are located at $z = 0$ and $z = L_z(t)$. All the surfaces are perfect conductors.

When studying the electromagnetic field inside these cavities it is convenient to express the physical degrees of freedom in terms of the vector potentials $\mathbf{A}^{(TE)}$ and $\mathscr{A}^{(TM)}$ introduced in Sect. 13.2.3. These vectors can be written in terms of the so called "scalar Hertz potentials" as $\mathbf{A}^{(TE)} = \hat{\mathbf{Z}} \times \nabla\phi^{TE}$ and $\mathscr{A}^{(TM)} = \hat{\mathbf{Z}} \times \nabla\phi^{TM}$. For perfect reflectors the b.c. do not mix TE and TM polarizations, and therefore the electromagnetic field inside the cavity can be described in terms of these two *independent* scalar Hertz potentials: no crossed terms appear in Maxwell's Lagrangian or Hamiltonian.

The scalar Hertz potentials satisfy the Klein–Gordon equation. The b.c. of both potentials on the static walls of the cavity are

$$\phi^{TE}\big|_{z=0} = 0; \qquad \frac{\partial\phi^{TE}}{\partial n}\bigg|_{trans} = 0, \tag{13.56}$$

$$\frac{\partial\phi^{TM}}{\partial z}\bigg|_{z=0} = 0; \qquad \phi^{TM}\big|_{trans} = 0, \tag{13.57}$$

where $\partial/\partial n$ denotes the normal derivative on the transverse boundaries. On the other hand, the b.c. on the moving mirror has been already discussed in Sect. 13.2 (see Eqs. (13.16) and (13.17)). In terms of the Hertz potentials they read as

$$\phi^{TE}\big|_{z=L_z(t)} = 0; \qquad (\partial_z + \dot{L}_z\,\partial_t)\phi^{TM}\big|_{L_z(t)} = 0. \tag{13.58}$$

The energy of the electromagnetic field

$$H = \frac{1}{8\pi}\int d^3x(\mathbf{E}^2 + \mathbf{B}^2) = H^{TE} + H^{TM} \tag{13.59}$$

can be written in terms of the scalar potentials as

$$H^{(P)} = \frac{1}{8\pi}\int d^3x\Big[\dot{\phi}^{(P)}(-\nabla_\perp^2)\dot{\phi}^{(P)} + \phi^{(P)\prime}(-\nabla_\perp^2)\phi^{(P)\prime} + \nabla_\perp^2\phi^{(P)}\nabla_\perp^2\phi^{(P)}\Big], \tag{13.60}$$

where dots and primes denote derivatives with respect to time and z respectively. The supraindex P corresponds to TE and TM and $\nabla_\perp$ denotes the gradient on the xy plane.

The quantization procedure has been described in detail in previous papers [37, 38, 101]. At any given time both scalar Hertz potentials can be expanded in terms of an instantaneous basis

$$\phi^{(P)}(\mathbf{x}, t) = \sum_n a_n^{IN} C_n^{(P)} u_n^{(P)}(\mathbf{x}, t) + \text{c.c.}, \tag{13.61}$$

where a_n^{IN} are bosonic operators that annihilate the IN vacuum state for $t < 0$, and $C_n^{(P)}$ are normalization constants that must be appropriately included to obtain the usual form of the electromagnetic Hamiltonian (13.60) in terms of annihilation and creation operators.

For TE modes, the mode functions are similar to those of the scalar field satisfying Dirichlet b.c. described in the previous section

$$u_n^{TE} = \sum_m Q_{m,TE}^{(n)}(t) \sqrt{2/L_z(t)} \sin\left(\frac{m_z \pi z}{L_z(t)}\right) v_{m_\perp}(\mathbf{x}_\perp). \tag{13.62}$$

For TM modes. the choice of the instantaneous basis is less trivial and has been derived in detail in Ref. [37]

$$u_n^{TM} = \sum_m [Q_{m,TM}^{(n)}(t) + \dot{Q}_{m,TM}^{(n)}(t) g(z, t)] \sqrt{2/L_z(t)} \cos\left(\frac{m_z \pi z}{L_z(t)}\right) r_{m_\perp}(\mathbf{x}_\perp). \tag{13.63}$$

Here the index $\mathbf{m} \neq 0$ is a vector of non-negative integers. The function $g(z, t) = \dot{L}_z(t) L_z(t) \xi(z/L_z(t))$ (where $\xi(z)$ is a solution to the conditions $\xi(0) = \xi(1) = \partial_z \xi(0) = 0$, and $\partial_z \xi(1) = -1$) appears when expanding the TM modes in an instantaneous basis and taking the small ε limit. There are many solutions for $\xi(z)$, but all of them can be shown to lead to the same physical results [37]. The mode functions $v_{m_\perp}(\mathbf{x}_\perp)$ and $r_{m_\perp}(\mathbf{x}_\perp)$, are described below for different types of cavities.

The mode functions $Q_{m,TE/TM}^{(n)}$ satisfy second order, mode-coupled linear differential equations similar to Eq. (13.49) [37]. As before, for the "parametric resonant case" ($\Omega = 2\omega_n$ for some n) there is parametric amplification. Moreover, for some particular geometries and sizes of the cavities, different modes $\mathbf{n}$ and $\mathbf{m}$ can be coupled, provided either of the resonant coupling conditions $\Omega = |\omega_n \pm \omega_m|$ are met. When intermode coupling occurs it affects the rate of photon creation, typically resulting in a reduction of that rate.

The number of motion-induced photons with a given wavevector $\mathbf{n}$ and polarization TE or TM can be calculated in terms of the Bogoliubov coefficients. When the resonant coupling conditions are not met, the different modes will not be coupled during the dynamics. As in the scalar case, the system can be described by a Mathieu equation (13.54) for a single mode. As a consequence, the number of motion-induced photons in that given mode will grow exponentially. The growth rate is different for TE and TM modes [37]

$$\langle \mathcal{N}_{\mathbf{n},\mathrm{TE}}(t) \rangle = \sinh^2(\lambda_{\mathbf{n},\mathrm{TE}}\varepsilon t); \quad \langle \mathcal{N}_{\mathbf{n},\mathrm{TM}}(t) \rangle = \sinh^2(\lambda_{\mathbf{n},\mathrm{TM}}\varepsilon t), \tag{13.64}$$

where $\lambda_{n,\mathrm{TE}} = n_z^2/2\omega_n$ and $\lambda_{n,\mathrm{TM}} = (2\omega_n^2 - n_z^2)/2\omega_{\mathbf{n}}$. When both polarizations are present, the rate of growth for TM photons is larger than for TE photons, *i.e.*, $\lambda_{\mathbf{n},\mathrm{TM}} > \lambda_{\mathbf{n},\mathrm{TE}}$. As in the case of the scalar field, these equations are valid for $\varepsilon^2 \Omega t \ll 1$.

We describe some specific examples:

Rectangular section. For a waveguide of length $L_z(t)$ and transversal rectangular shape (lengths L_x, L_y), the TE mode function is

$$v_{n_x,n_y}(\mathbf{x}_\perp) = \frac{2}{\sqrt{L_x L_y}} \cos\left(\frac{n_x \pi x}{L_x}\right) \cos\left(\frac{n_y \pi y}{L_y}\right), \tag{13.65}$$

with n_x and n_y non-negative integers that cannot be simultaneously zero. The spectrum is

$$\omega_{n_x,n_y,n_z} = \sqrt{(n_x \pi/L_x)^2 + (n_y \pi/L_y)^2 + (n_z \pi/L_z)^2}, \tag{13.66}$$

with $n_z \geq 1$. The TM mode function is

$$r_{m_x,m_y}(\mathbf{x}_\perp) = \frac{2}{\sqrt{L_x L_y}} \sin\left(\frac{m_x \pi x}{L_x}\right) \sin\left(\frac{m_y \pi y}{L_y}\right), \tag{13.67}$$

where m_x, m_y are positive integers. The spectrum is given by ω_{m_x,m_y,m_z}, with $m_z \geq 0$.

Let us analyze the particular case of a cubic cavity of size L under the parametric resonant condition $\Omega = 2\omega_{\mathbf{k}}$. The fundamental TE mode is doubly degenerate $((1,0,1)$ and $(0,1,1))$ and uncoupled to other modes. The number of photons in these TE modes grows as $\exp(\pi \varepsilon t \sqrt{2}L)$. The fundamental TM mode $(1,1,0)$ has the same energy as the fundamental TE mode, and it is coupled to the TM mode $(1,1,4)$. Motion-induced TM photons are produced exponentially as $\exp(4.4\varepsilon t/L)$, much faster than TE photons.

Circular section. For a waveguide with a transversal circular shape of radius R, the TE mode function is

$$v_{nm}(\mathbf{x}_\perp) = \frac{1}{\sqrt{\pi}} \frac{1}{R J_n(y_{nm})\sqrt{1 - n^2/y_{nm}^2}} J_n\left(y_{nm}\frac{\rho}{R}\right) e^{in\phi}, \tag{13.68}$$

where J_n denotes the Bessel function of nth order, and y_{nm} is the mth positive root of the equation $J_n'(y) = 0$. The eigenfrequencies are given by

$$\omega_{n,m,n_z} = \sqrt{\left(\frac{y_{nm}}{R}\right)^2 + \left(\frac{n_z \pi}{L_z}\right)^2}, \tag{13.69}$$

where $n_z \geq 1$. The TM mode function is

$$r_{nm}(\mathbf{x}_\perp) = \frac{1}{\sqrt{\pi}} \frac{1}{R J_{n+1}(x_{nm})} J_n\left(x_{nm}\frac{\rho}{R}\right) e^{in\phi}, \tag{13.70}$$

where x_{nm} is the mth root of the equation $J_n(x) = 0$. The spectrum is given by Eq.(13.69) with y_{nm} replaced by x_{nm} and $n_z \geq 0$. Denoting the modes by (n, m, n_z), the lowest TE mode is $(1,1,1)$ and has a frequency $\omega_{111} = (1.841/R)$ $\sqrt{1 + 2.912(R/L_z)^2}$. This mode is uncoupled to any other modes, and the number of photons in this mode grows exponentially in time as $\exp(\pi\varepsilon t/\sqrt{1 + 0.343(L_z/R)^2}L_z)$ when parametrically excited. The lowest TM mode $(0,1,0)$ is also uncoupled and has a frequency $\omega_{010} = 2.405/R$. The parametric growth is $\exp(4.81\varepsilon t/R)$. For L_z large enough ($L_z > 2.03R$), the resonance frequency ω_{111} of the lowest TE mode is smaller than that for the lowest TM mode. Then the $(1,1,1)$ TE mode is the fundamental oscillation of the cavity.

13.4.3 Time Dependent Electromagnetic Properties

From a theoretical point of view, it is possible to create photons from the vacuum not only for a cavity with a moving mirror, but also when the electromagnetic properties of the walls and/or the media inside the cavity change with time. Given the difficulties in a possible experimental verification of the DCE for moving mirrors, the consideration of time dependent properties is not only of academic interest, but it is also relevant for the analysis of the experimental proposals discussed in Sect. 13.5

A setup that has attracted both theoretical and experimental attention is the possibility of using short laser pulses in order to produce periodic variations of the conductivity of a semiconductor layer placed inside a microwave cavity. The fast changes in the conductivity induce a periodic variation in the effective length of the cavity, and therefore the creation of photon pairs [47–50]. This setup has been analyzed at the theoretical level [51, 102–104], and there is an ongoing experiment aimed at the detection of the motion induced radiation [53] (see Sect. 13.5)

For the sake of clarity we discuss in detail the model of a massless scalar field within a rectangular cavity with perfect conducting walls with dimensions L_x, L_y, and L_z described in Ref. [51]. At the midpoint of the cavity ($x = L_x/2$) there is a plasma sheet. We model the conductivity properties of such material by a delta-potential with a time dependent strength $V(t)$. This is a time dependent generalization of the model introduced in Ref. [75]. The strength of the potential is given by

$$V(t) = 4\pi \frac{e^2 n(t)}{m^*}, \tag{13.71}$$

where e is the electron charge, m^* the electron's effective mass in the conduction band and $n(t)$ the surface density of carriers. We assume that the irradiation of the plasma sheet produces changes in this quantity. The ideal limit of perfect conductivity corresponds to $V \to \infty$, and $V \to 0$ to a 'transparent' material. The strength of the potential varies between a minimum value, V_0, and a maximum $V_{\max}$. The Lagrangian of the scalar field within the cavity is given by

$$\mathcal{L} = \frac{1}{2}\partial_\mu\phi\partial^\mu\phi - \frac{V(t)}{2}\delta(x - L_x/2)\phi^2, \tag{13.72}$$

where $\delta(x)$ is the one-dimensional Dirac delta function. The use of an infinitely thin film is justified as long as the width of the slab is much smaller than the wavelengths of the relevant electromagnetic modes in the cavity. The corresponding Lagrange equation reads,

$$(\nabla^2 - \partial_t^2)\phi = V(t)\delta(x - L_x/2)\phi. \tag{13.73}$$

We divide the cavity into two regions: region I $(0 \leq x \leq L_x/2)$ and region II $(L_x/2 \leq x \leq L_x)$. Perfect conductivity at the edges of the cavity imposes Dirichlet b.c. for the field. The presence of the plasma sheet introduces a discontinuity in the x-spatial derivative, while the field itself remains continuous,

$$\phi_1(x = L_x/2, t) = \phi_{11}(x = L_x/2, t),$$
$$\partial_x\phi_1(x = L_x/2, t) - \partial_x\phi_{11}(x = L_x/2, t) = -V(t)\phi(x = L_x/2, t). \tag{13.74}$$

We will consider a set of solutions that satisfies automatically all b.c..

$$\psi_{\mathbf{m}}(\mathbf{x}, t) = \sqrt{\frac{2}{L_x}}\sin(k_{m_x}(t)x)\sqrt{\frac{2}{L_y}}\sin\left(\frac{\pi m_y y}{L_y}\right)\sqrt{\frac{2}{L_z}}\sin\left(\frac{\pi m_z z}{L_z}\right), \tag{13.75}$$

where m_y, m_z are positive integers. The function $\psi_{\mathbf{m}}$ depends on t through $k_{m_x}(t)$, which is the m_x-th positive solution to the following transcendental equation

$$2k_{m_x}\tan^{-1}\left(\frac{k_{m_x}L_x}{2}\right) = -V(t). \tag{13.76}$$

To simplify the notation, in what follows we will write k_m instead of k_{m_x}. Note that, when $V(t) \to \infty$, the solutions to this equation become the usual ones for perfect reflectors, $k_m = mL_x/2$, with m a positive integer.

Let us define

$$\Psi_{\mathbf{m}}(\mathbf{x}, t) = \begin{cases} \psi_{\mathbf{m}}(x, y, z, t) & 0 \leq x \leq L_x/2 \\ -\psi_{\mathbf{m}}(x - L_x, y, z, t) & L_x/2 \leq x \leq L_x \end{cases} \tag{13.77}$$

These functions satisfy the b.c. and the orthogonality relations

$$(\Psi_{\mathbf{m}}, \Psi_{\mathbf{n}}) = [1 - \sin(k_m(t)L_x)/k_m(t)L_x]\delta_{\mathbf{m},\mathbf{n}}.$$

There is a second set of modes with a node on the cavity midpoint. As these solutions do not "see" the slab, they will be irrelevant in what follows.

For $t \leq 0$ the slab is not irradiated, consequently V is independent of time and has the value V_0. The modes of the quantum scalar field that satisfy the Klein–Gordon Eq. (13.73) are

$$u_{\mathbf{m}}(\mathbf{x}, t) = \frac{e^{-i\bar{\omega}_{\mathbf{m}}t}}{\sqrt{2\bar{\omega}_{\mathbf{m}}}} \Psi_{\mathbf{m}}(\mathbf{x}, 0), \tag{13.78}$$

where $\bar{\omega}_{\mathbf{m}}^2 = (k_m^0)^2 + \left(\frac{\pi m_y}{L_y}\right)^2 + \left(\frac{\pi m_z}{L_z}\right)^2$ and k_m^0 is the m-th solution to Eq. (13.76) for $V = V_0$. At $t = 0$ the potential starts to change in time and the set of numbers $\{k_m\}$ acquires a time dependence through Eq. (13.76).

Using Eq. (13.78) we expand the field operator ϕ as

$$\phi(\mathbf{x}, t) = \sum_{\mathbf{m}} \left[b_{\mathbf{m}} u_{\mathbf{m}}(\mathbf{x}, t) + b_{\mathbf{m}}^\dagger u_{\mathbf{m}}^*(\mathbf{x}, t) \right], \tag{13.79}$$

where $b_{\mathbf{m}}$ are annihilation operators. Notice that in the above equation we omitted the modes with a node at $x = L_x/2$ because their dynamics is not affected by the presence of the slab.

For $t \geq 0$ we write the expansion of the field mode u_s as

$$u_s(\mathbf{x}, t > 0) = \sum_{\mathbf{m}} P_{\mathbf{m}}^{(s)}(t) \Psi_{\mathbf{m}}(\mathbf{x}, t). \tag{13.80}$$

Assume a time dependent conductivity given by

$$V(t) = V_0 + (V_{\max} - V_0) f(t), \tag{13.81}$$

where $f(t)$ is a periodic and non-negative function, $f(t) = f(t + T) \geq 0$, that vanishes at $t = 0$ and attains its maximum at $f(\tau_e) = 1$. In each period, $f(t)$ describes the excitation and relaxation of the plasma sheet produced by the laser pulse. Typically, the characteristic time of excitation τ_e is the smallest time scale and satisfies $\tau_e \ll T$. Under certain constraints, large changes in V induce only small variations in k through the transcendental relation between k and V (see Eq. (13.76)). In this case, a perturbative treatment is valid and a linearization of such a relation is appropriate. Accordingly we write

$$k_n(t) = k_n^0 (1 + \varepsilon_n f(t)), \tag{13.82}$$

where

$$\varepsilon_n = \frac{V_{\max} - V_0}{L_x (k_n^0)^2 + V_0 \left(1 + \frac{V_0 L_x}{4}\right)}. \tag{13.83}$$

The restriction for the validity of the perturbative treatment is $V_0 L_x \gg V_{\max}/V_0 > 1$. These conditions are satisfied for realistic values of L_x, V_0, and $V_{\max}$.

Replacing Eq. (13.80) into $(\nabla^2 - \partial_t^2)u_s = 0$ we find a set of coupled differential equations for the amplitudes $P_{\mathbf{m}}^{(s)}(t)$. The dynamics is described by a set of coupled harmonic oscillators with periodically varying frequencies and couplings, as already discussed in this section. It is of the same form as the equations that describe the modes of a scalar field in a three dimensional cavity with an oscillating boundary. For the same reasons as before, a naive perturbative solution of previous equations in powers of ε_n breaks down after a short amount of time when the external frequency is tuned with some of the eigenfrequencies of the cavity. Assuming that $f(t)$ is a sum of harmonic functions of frequencies $\Omega_j = j2\pi/T$, the resonance condition is $\Omega_j = 2\bar{\omega}_{\mathbf{n}}$ for some j and $\mathbf{n}$. If there is no intermode coupling, a nonperturbative solution gives an exponential number of created photons in that particular mode

$$\langle \mathcal{N}_{\mathbf{n}}(t) \rangle = \langle b_{\mathbf{n}}^\dagger b_{\mathbf{n}} \rangle = \sum_s 2\bar{\omega}_{\mathbf{n}} |A_{\mathbf{n}}^{(s)}(t)|^2 \approx \sinh^2\left(\frac{(k_n^0)^2 f_j}{\Omega_j}\varepsilon_n t\right), \qquad (13.84)$$

where f_j is the amplitude of the oscillations of $f(t)$ with frequency Ω_j.

A full electromagnetic calculation has been presented in Ref. [104]. It was shown there that the scalar model presented here describes the TE electromagnetic modes inside the cavity. The treatment of TM modes involves an independent scalar field, with a potential proportional to $\delta'(x - L_x/2)$. Moreover, the model has also been generalized to the case of arbitrary positions of the plasma sheet within the cavity [104]. The number of created TE photons depends strongly on the position of the layer, and the maximum number is attained when it is located at the midpoint of the cavity. On the other hand, for TM modes this dependence is rather weak.

In the treatment above no dissipation effects are considered (the delta-potential is real). Similar calculations for lossless dielectric slabs with time-dependent and real permittivities [103] also neglect dissipation. However, it has been pointed out that dissipative effects may be relevant in the evaluation of created photons [102]. In general, one expects the electromagnetic energy to be dissipated in the cavity walls, in the plasma sheet, and/or in dielectric slabs contained in the cavity. In resonant situations without dissipation, we have seen that the dynamics of the relevant electromagnetic mode is described by a harmonic oscillator with time dependent frequency. A phenomenological way of taking into account dissipative effects is to replace this equation with that of a damped oscillator [105] . Of course this model cannot be consistently quantized unless one includes a noise term, otherwise the usual commutation relations are violated. Using the *quantum noise operator approach* [106] one can estimate the rate of photon creation in this model and, provided the dissipation is not too large, the number of photons still grows exponentially, although at a smaller rate. However, it has been recently argued [43] that these results should be valid only in the short time limit, while in the long time limit the system should reach a stationary state with a constant number of photons inside the cavity. As the calculations in [43] involve 1D cavities, this point deserves further investigation.

13.5 Experimental Perspectives

Since the first theoretical predictions about motion induced radiation, it was clear that the experimental observation of this effect was not an easy task. As mentioned at the end of Sect. 13.2, the photon creation produced by a single accelerated mirror is extremely small in realistic situations (see Eq. (13.20)).

The most promising situation seems to be the photon creation by parametric amplification described in Sect. 13.4. However simple numerical estimations show that, even in the most favorable cases, it is difficult to observe the DCE in the laboratory. In all the 3D examples discussed in Sect. 13.4, the number of created photons grows exponentially in time as

$$\langle \mathcal{N} \rangle = \sinh^2(\eta \omega \varepsilon t), \tag{13.85}$$

where ω is the frequency of the resonant mode and η is a number of order 1 related to the geometry of the cavity. Here ε denotes the relative amplitude of the oscillations in the moving mirror case, or the relative amplitude of the oscillations of the relevant component of the wavevector in the case of time dependent conductivity (see Eq. (13.82)). This equation is valid as long as $\varepsilon^2 \omega t \ll 1$ and neglects any dissipative effects. As the electromagnetic cavity has a finite Q-factor, a rough estimation of the maximum number of created photons $\langle \mathcal{N}_{max} \rangle$ is obtained by setting $t_{max} = Q/\omega$ in the above equation. As mentioned at the end of Sect. 13.4 , there is no agreement in the literature about this estimation. Calculations based on the use of a master equation [44] give an exponential growth with a rate diminished by a factor $\Gamma = 1 - 1/(2Q\varepsilon)$ (see also Ref. [107]). On the other hand, it was shown that, in the case of 1D cavities, the total number of photons inside the cavity should reach a constant value proportional to the finesse of the cavity at long times [43]. It was argued in the same work that the exponential growth in the presence of dissipation would be valid only at short times. In any case, it is clear that a *necessary* condition to have an observable number of photons is that $2Q\varepsilon > 1$.

Assuming a cavity of length $L_0 \simeq 1$ cm, the oscillation frequency of the mirror should be in the GHz range in order to meet the parametric resonance condition. A plausible possibility for reaching such high mechanical oscillation frequencies is to consider surface vibrations, instead of a global motion of the mirror [35]. In this context, the maximum attainable values of the relative amplitude would be around $\varepsilon \simeq 10^{-8}$, and therefore the quality factor of the cavity should be greater than 10^8 in order to have a non-negligible number of photons. Microwave superconducting cavities with Q-factors as high as 10^{12} have been built [108]. However, the Q-factor would be severely limited by the presence of an oscillating wall. Therefore, it is an extraordinary challenge to produce extremely fast oscillations while keeping the extremely high Q-factors needed in the DCE. Moreover, the oscillations should be tuned with high precision to parametric resonance with a cavity mode.

13.5.1 High Frequency Resonators and Photon Detection via Superradiance

A concrete setup for producing and detecting motion induced photons has been proposed in Ref. [46]. A Film Bulk Acoustic Resonator (FBAR) is a device that consists of a piezoelectric film sandwiched between two electrodes. An aluminum nitride FBAR of thickness corresponding to a half of the acoustic wavelength can be made to vibrate up to a frequency of 3 GHz, with an amplitude of $\varepsilon = 10^{-8}$. The expected maximum power of motion-induced photons produced by such a FBAR depends of course on the Q-factor of the cavity. It can be estimated to be

$$P_{max} = \langle \mathcal{N}_{max} \rangle \hbar \omega / t_{max}. \tag{13.86}$$

Assuming that $Q\varepsilon = O(1)$, this gives $P_{max} \simeq 10^{-22} W$, which is too small for direct detection.

However, this low power could be detected using ultracold atoms. Let us consider a cavity filled with an ensemble of population-inverted atoms in a hyperfine state whose transition frequency is equal to the resonance frequency of the cavity. Then the Casimir photons can trigger a stimulated emission of the atoms, and therefore they can be indirectly detected by this form of *superradiance*. Reference [46] contains a description of the experimental setup that could be used to observe the DCE, and a discussion about the rejection of signals not produced by the Casimir photons. In particular, stimulated amplification could also be triggered by the spontaneous decay of one of the atoms (superfluorescence). In order to discriminate between both effects it could be necessary to attain larger values of $Q\varepsilon$.

13.5.2 Time Dependent Conductivity Induced by Ultra-Short Laser Pulses

In order to avoid the experimental complications associated with the high frequency motion of the mirror, it is possible to produce effective changes in the length of the cavity by inducing abrupt changes in the reflectivity of a slab contained in the cavity, as already mentioned in Sect. 13.4.3. This can be done by illuminating a semiconducting slab with ultra short laser pulses [47–50]. An experiment based on this idea is currently being carried out by the group of Padova [109].

In this case, a numerical estimation of the maximum number of photons created in the cavity looks, at first sight, much more promising than in the case of moving mirrors. Using a slab of a thickness around 1 mm, it is possible to reach values of ε as large as 10^{-4}, and therefore the constraints on the Q-factor of the cavity are considerably milder. Moreover, it is experimentally possible to generate trains of thousands of laser pulses with a repetition frequency on the order of a GHz.

Fig. 13.4 Superconducting
cavity of the experimental
setup of MIR experiment at
Padova to measure the
dynamical Casimir effect
(Courtesy of Giuseppe
Ruoso)

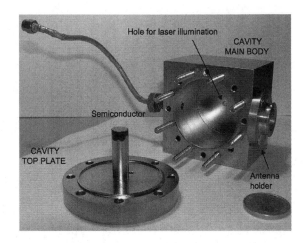

In Padova's setup (see Fig. 13.4), a high $Q \approx 10^6$ superconducting cavity contains a GaAs semiconducting slab. The laser pulses are tuned at 4.70 GHz, twice the frequency of the fundamental TE mode of the cavity. However, it has been pointed out [102] that dissipative effects may play an important role in this kind of experiment. Indeed, the changes in the conductivity of the slab are due to the creation of electron-hole pairs by the laser pulses, and during this process the dielectric permittivity acquires an imaginary part. The associated dissipation prevents photon creation unless severe constraints on the properties of the semiconductor are fulfilled: it must have a very short recombination time (tenths of ps), and a high mobility (around $1m^2(Vs)^{-1}$). A slab with these characteristics has been constructed by irradiating a GaAs sample with fast neutrons, in order to reduce the recombination time of the original sample (about 1 ns) while keeping constant the value of the mobility [109]. Photons are detected using a loop antenna inside the cavity. The minimum number of photons that can be detected is around 100, below the expected signal of Casimir photons [54, 109].

A related setup is illumination of a *superconductor* instead of a semiconductor surface. The advantage in this case is that dissipative effects are less important, because the variation of the imaginary part of the permittivity is much smaller for superconductors than for semiconductors in the microwave region [110]. Moreover, since the abrupt changes in the conductivity are due to local heating of the surface (and not to the creation of electron-hole pairs as in the semiconductor), the intensity of the laser can be considerably smaller, reducing unwanted effects of energy accumulation inside the cavity.

13.5.3 Optical Parametric Oscillators

Standard nonlinear optics can be interpreted, in some cases, as a time-dependent modulation of the refractive index. Reference [43] considered an optical parametric oscillator (OPO) with a pump laser beam of frequency Ω and amplitude E_{pump} interacting with a very thin $\chi^{(2)}$ nonlinear crystal slab placed on the interior side of the cavity mirror. For a type-I arrangement, the total polarization component along a suitable crystal symmetry axis may be written in terms of the intracavity electric field component along the same direction as [43]

$$P(t) = \varepsilon_0 \left(\chi^{(1)} + \frac{1}{2} \chi^{(2)} E_{\text{pump}} \sin(\Omega t - \theta) \right) E(t), \tag{13.87}$$

where ε_0 is the vacuum permittivity, $\chi^{(1)}$ and $\chi^{(2)}$ are the relevant components of the linear and second-order nonlinear susceptibility tensors, and θ is a phase that depends on the pump beam phase and the position of the crystal.

The total susceptibility as given by Eq. (13.87) (including the nonlinear second-order term) corresponds to an effective refractive index oscillating at the pump beam frequency Ω, thus leading to a modulation of the optical cavity length. This is formally equivalent to modulating the physical cavity length by bouncing the mirror with frequency Ω. But in the OPO the pump beam frequency is in the optical range, as are the generated photons with frequencies satisfying $\omega + \omega' = \Omega$. For ω and ω' corresponding to cavity modes, parametric amplification is enhanced and the resulting photon flux is typically several orders of magnitude higher than in the case of mechanical motion [43].

13.5.4 Superconducting Coplanar Waveguides

Another possibility to induce fast variations of the b.c. on the electromagnetic field is to consider a coplanar waveguide terminated by a SQUID [57, 58]. A time-dependent magnetic flux can be applied to control the effective inductance of the SQUID, which in turn produces a time-dependent Robin boundary condition for the phase field (time integral of the electric field), equivalent to that of a transmission line with a variable length. This setup simulates a moving Robin mirror in 1D with an effective velocity that might be close to the speed of light. As a consequence, the first-order non-relativistic results [26, 73], based on the perturbative approach outlined in Sect. 13.2, must be modified by the inclusion of higher-order frequency sidebands [57].

As in the previous examples, it is crucial to check if the flux of Casimir photons can be discriminated from other sources of photons, like the classical thermal contribution. The analysis presented in Ref. [57] shows that this is the case, for realistic values of the parameters, at temperatures below 70 mK.

13.6 Final Remarks

We have reviewed some theoretical and experimental advances in the analysis of moving bodies or time dependent boundary conditions coupled to the vacuum fluctuations of the electromagnetic field.

Accelerated neutral bodies produce the emission of real photons, while experiencing a radiation reaction force. When the dynamics of the bodies is treated quantum mechanically, the interaction with the vacuum fluctuations not only causes this dissipative force, but also an appreciable amount of decoherence, which is a consequence of the entanglement between the mirrors and the electromagnetic field. This is a particular example of quantum Brownian motion, where the Brownian particle (mirror) loses coherence while being subjected to a damping force due to its coupling to the environment (the quantum field).

When two neutral bodies are in relative motion, we expect velocity-dependent forces between them. There is a particularly interesting situation in which two parallel, non-perfectly conducting slabs are in relative parallel motion with constant velocity. In this case, there is a *vacuum friction* between the slabs even in the absence of real photons. The effect can be understood in terms of the interaction of image charges, or as the interchange of virtual photons between the surfaces. There are similar friction forces for neutral atoms moving near surfaces.

The rate of photon creation produced by a single accelerated body in free space is deceptively small in realistic situations. However, in closed cavities a much larger number of photons may be produced by parametric amplification. Indeed in an ideal 3D cavity one expects an exponential growth in the number of photons when its size varies periodically at an appropriate resonant frequency, making detection of photon creation not an impossible task. The calculation of photon creation in the presence of ideal conductors have been performed in 1D and 3D using different analytical approximations. The results are consistent and have been confirmed by fully numerical calculations. However, the case of moving mirrors with finite conductivity (i.e electromagnetic cavities with a finite Q-factor) is not a completely settled issue. In 1D cavities, at short times the growth of the total number of created photons is still exponential (with a different rate), while at large times the total number of photons should reach saturation. This problem has not yet been solved for 3D cavities. The difficulties in evaluating the DCE for mirrors with finite conductivity recalls a similar situation in the static Casimir effect, where the evaluation of the Casimir force depends strongly on the theoretical model used to describe the conductivity of the bodies, and there are interesting correlations between finite conductivity, temperature and geometry. These correlations may have relevant counterparts in the dynamical problem. In any case, although difficult, the direct experimental detection of the motion induced radiation is not out of reach, as long as one can keep a very high Q-factor in a cavity with moving walls. In particular, there is a specific proposal that involves nanoresonators in a high Q-cavity filled with a gas of cold atoms to detect a small number of photons through superradiance.

An exponentially large number of photons can also be produced when some electromagnetic property of the cavity varies periodically with time. Of particular importance is the case in which the conductivity of a semiconductor or superconductor slab placed inside an electromagnetic cavity is modulated using short laser pulses. Theoretical estimates show that this setup could be implemented, with milder requirements on the Q-factor of the cavity. Once again, there is no comprehensive theoretical model that takes into account the (dissipative) response of the slab to the laser pulses, and its relevance for the photon creation process. However, this is a promising alternative and there is an ongoing experiment at Padova based on this setup.

There are other possibilities to produce fast variations of the boundary conditions on the electromagnetic field, that involve optical parametric oscillators or superconducting waveguides. The theoretical analyses suggest that it should be easier to detect the photons created in these setups than in the case of moving mirrors.

In summary, there is a plethora of interesting effects related to the electromagnetic vacuum fluctuations in the presence of moving bodies and/or other time dependent external conditions. The eventual experimental confirmation of some of these effects will certainly produce an increasing activity on this subject in the near future, as was the case for the static Casimir effect following the first realization of precise experiments in that area since 1997.

Acknowledgements The work of DARD was funded by DARPA/MTO's Casimir Effect Enhancement program under DOE/NNSA Contract DE-AC52-06NA25396. PAMN thanks CNPq and CNE/FAPERJ for financial support and the Universidad de Buenos Aires for its hospitality during his stay at Buenos Aires. FDM thanks the Universidad de Buenos Aires, CONICET and ANPCyT for financial support. We are grateful to Giuseppe Ruoso for providing a picture of MIR experiment at Padova.

References

1. Barton, G.: On the fluctuations of the Casimir force. J. Phys. A: Math. Gen. **24**, 991–1005 (1991)
2. Barton, G.: On the fluctuations of the Casimir force. 2: the stress-correlation function. J. Phys. A: Math. Gen. **24**, 5533–5551 (1991)
3. Jaekel, M.-T., Reynaud, S.: Quantum fluctuations of position of a mirror in vacuum. J. Phys. (Paris) I **3**, 1–20 (1993)
4. Dalvit, D.A.R., Maia Neto, P.A.: Decoherence via the Dynamical Casimir Effect. Phys. Rev. Lett. **84**, 798–801 (2000)
5. Maia Neto, P.A., Dalvit, D.A.R.: Radiation pressure as a source of decoherence. Phys. Rev. A. **62**, 042103 (2000)
6. Callen, H.B., Welton, T.A.: Irreversibility and generalized noise. Phys. Rev. **83**, 34–40 (1951)
7. Ford, L.H., Vilenkin, A.: Quantum radiation by moving mirrors. Phys. Rev. D **25**, 2569–2575 (1982)
8. Moore, G.T.: Quantum theory of electromagnetic field in a variable-length one-dimensional cavity. J. Math. Phys. **11**, 2679 (1970)
9. Castagnino, M., Ferraro, R.: The radiation from moving mirrors: The creation and absorption of particles. Ann. Phys. (NY) **154**, 1–23 (1984)
10. Fulling, S.A., Davies, P.C.W.: Radiation from a moving mirror in two dimensional space-time-conformal anomaly. Proc. R. Soc. A **348**, 393–414 (1976)
11. Hawking, S.W.: Black-hole explosions. Nature (London) **248**, 30–31 (1974)
12. Hawking, S.W.: Particle creation by black-holes. Commun. Math. Phys. **43**, 199–220 (1975)

13. Braginsky, V.B., Khalili, F.Ya.: Friction and fluctuations produced by the quantum ground-state. Phys. Lett **161**, 197–201 (1991)
14. Jaekel, M.-T., Reynaud, S.: Fluctuations and dissipation for a mirror in vacuum. Quantum Opt. **4**, 39–53 (1992)
15. Braginsky, V.B., Vorontsov, Y.u.I.: Quantum-mechanical limitations in macroscopic experiments and modern experimental techniques. Usp. Fiz. Nauk. **114**, 41–53 (1974)
16. Caves, C.: Defense of the standard quantum limit for free-mass position. Phys. Rev. Lett. **54**, 2465–2468 (1985)
17. Jaekel, M.-.T., Reynaud, S.: Quantum limits in interferometric measurements. Europhys. Lett. **13**, 301–306 (1990)
18. Kubo, R.: Fluctuation-dissipation theorem. Rep. Prog. Phys. **29**, 255–284 (1966)
19. Maia Neto, P.A., Reynaud, S.: Dissipative force on a sphere moving in vacuum. Phys. Rev. A **47**, 1639–1646 (1993)
20. Barton, G.: New aspects of the Casimir effect: fluctuations and radiative reaction. In: Berman, P.Cavity Quantum Electrodyamics, Supplement: Advances in Atomic, Molecular and Optical Physics. Academic Press, New York (1993)
21. Maia Neto, P.A., Machado, L.A.S.: Radiation Reaction Force for a Mirror in Vacuum. Braz. J. Phys. **25**, 324–334 (1995)
22. Golestanian, R,.Kardar, M.: Mechanical Response of Vacuum. Phys. Rev. Lett. **78**, 3421–3425 (1997); Phys. Rev. A **58**, 1713–1722 (1998)
23. Volotikin, A.I., Persson, B.N.J.: Near-field radiative heat transfer and noncontact friction. Rev. Mod. Phys. **79**, 1291–1329 (2007)
24. Pendry, J.B.: Shearing the vacuum-quantum friction. J. Phys.:Condens. Matter **9**, 10301–10320 (1997)
25. Nussenzveig, H.M.: Causality and dispersion relations. Academic Press, New York (1972)
26. Lambrecht, A., Jaekel, M.-T., Reynaud, S.: Motion induced radiation from a vibrating cavity. Phys. Rev. Lett. **77**, 615–618 (1996)
27. Maia Neto, P.A., Machado, L.A.S.: Quantum radiation generated by a moving mirror in free space. Phys. Rev. A. **54**, 3420–3427 (1996)
28. Montazeri, M., Miri, M.: Radiation from a dynamically deforming mirror immersed in the electromagnetic vacuum. Phys. Rev. A. **77**, 053815 (2008)
29. Mundarain, D.F., Maia Neto, P.A.: Quantum radiation in a plane cavity with moving mirrors. Phys. Rev. A. **57**, 1379–1390 (1998)
30. Maia Neto, P.A.: The dynamical Casimir effect with cylindrical waveguides. J. Opt. B: Quantum Semiclass. Opt. **7**, S86–S88 (2005)
31. Pascoal, F., Celeri, L.C., Mizrahi, S.S., Moussa, M.H.Y.: Dynamical Casimir effect for a massless scalar field between two concentric spherical shells. Phys. Rev. A. **78**, 032521 (2008)
32. Pascoal, F., Celeri, L.C., Mizrahi, S.S., Moussa, M.H.Y., Farina, C.: Dynamical Casimir effect for a massless scalar field between two concentric spherical shells with mixed boundary conditions. Phys. Rev. A. **80**, 012503 (2009)
33. Eberlein, C.: Theory of quantum radiation observed as sonoluminescence. Phys. Rev. A. **53**, 2772–2787 (1996)
34. Mazzitelli, F.D., Millán, X.O.: Photon creation in a spherical oscillating cavity. Phys Rev. A. **73**, 063829 (2006)
35. Dodonov, V.V., Klimov, A.B.: Generation and detection of photons in a cavity with a resonantly oscillating boundary. Phys. Rev. A. **53**, 2664–2682 (1996)
36. Crocce, M., Dalvit, D.A.R., Mazzitelli, F.D.: Resonant photon creation in a three dimensional oscillating cavity. Phys. Rev. A. **64**, 013808 (2001)
37. Crocce, M., Dalvit, D.A.R., Mazzitelli, F.D.: Quantum electromagnetic field in a three dimensional oscillating cavity. Phys. Rev. A. **66**, 033811 (2002)
38. Crocce, M., Dalvit, D.A.R., Lombardo, F., Mazzitelli, F.D.: Hertz potentials approach to the dynamical Casimir effect in cylindrical cavities of arbitrary section. J. Opt. B: Quantum Semiclass. Opt. **7**, S32–S39 (2005)

39. Dalvit, D.A.R., Mazzitelli, F.D.: Renormalization-group approach to the dynamical Casimir effect. Phys. Rev. A. **57**, 2113–2119 (1998)
40. Lambrecht, A., Jaekel, M.-T., Reynaud, S.: Frequency up-converted radiation from a cavity moving in vacuum. Eur.Phys.J. D. **3**, 95–104 (1998)
41. Dalvit, D.A.R., Mazzitelli, F.D.: Creation of photons in an oscillating cavity with two moving mirrors. Phys. Rev. A. **59**, 3049–3059 (1999)
42. Jaekel, M.-T., Reynaud, S.: Motional Casimir force. J. Phys. I. **2**, 149–165 (1992)
43. Dezael, F.X., Lambrecht, A.: Analogue Casimir radiation using an optical parametric oscillator. Europhys. Lett. **89**, 14001 (2010)
44. Dodonov, V.V.: Dynamical Casimir effect in a nondegenerate cavity with losses and detuning. Phys. Rev. A. **58**, 4147–4152 (1998)
45. Schaller, G., Schützhold, R., Plunien, G., Soff, G.: Dynamical Casimir effect in a leaky cavity at finite temperature. Phys. Rev. A. **66**, 023812 (2002)
46. Kim, W.-J., Brownell, J.H., Onofrio, R.: Detectability of dissipative motion in quantum vacuum via superradiance. Phys. Rev. Lett. **96**, 200402 (2006)
47. Yablonovitch, E.: Accelerating reference frame for electromagnetic waves in a rapidly growing plasma: Unruh-Davies-Fulling-DeWitt radiation and the nonadiabatic Casimir effect. Phys Rev. Lett. **62**, 1742–1745 (1989)
48. Yablonovitch, E., Heritage, J.P., Aspnes, D.E., Yafet, Y.: Virtual photoconductivity. Phys. Rev. Lett. **63**, 976–979 (1989)
49. Lozovik, Y.E., Tsvetus, V.G., Vinogradov, E.A.: Femtosecond parametric excitation of electromagnetic field in a cavity. JETP Lett. **61**, 723–729 (1995)
50. Lozovik, Y.E., Tsvetus, V.G., Vinogradov, E.A.: Parametric excitation of vacuum by use of femtosecond laser pulses. Phys. Scr. **52**, 184–190 (1995)
51. Crocce, M., Dalvit, D.A.R., Lombardo, F., Mazzitelli F., D.: Model for resonant photon creation in a cavity with time dependent conductivity. Phys. Rev. A. **70**, 033811 (2004)
52. Mendonça, J.T., Guerreiro, A.: Phys. Rev. A. **80**, 043603 (2005)
53. Braggio, C., Bressi, G., Carugno, G., Del Noce, C., Galeazzi, G., Lombardi, A., Palmieri, A., Ruoso, G., Zanello, D.: A novel experimental approach for the detection of the dynamic Casimir effect. Europhys. Lett. **70**, 754–760 (2005)
54. Braggio, C., Bressi, G., Carugno, G., Della Valle, F., Galeazzi, G., Ruoso, G.: Characterization of a low noise microwave receiver for the detection of vacuum photons. Nucl. Instrum. Methods Phys. Res. A **603**, 451–455 (2009)
55. Takashima, K., Hatakenaka, N., Kurihara, S., Zeilinger, A.: Nonstationary boundary effect for a quantum flux in superconducting nanocircuits. J. Phys. A. **41**, 164036 (2008)
56. Castellanos-Beltran, M.A., Irwin, K.D., Hilton, G.C., Vale, L.R., Lehnert, K.W.: Amplification and squeezing of quantum noise with a tunable Josephson metamaterial. Nat. Phys. **4**, 928–931 (2008)
57. Johansson, J.R., Johansson, G., Wilson, C.M., Nori, F.: Dynamical Casimir effect in a superconducting coplanar waveguide. Phys. Rev. Lett. **103**, 147003 (2009)
58. Wilson, C. M., Duty, T., Sandberg, M., Persson, F., Shumeiko, V., Delsing, P.: Photon generation in an electromagnetic cavity with a time-dependent boundary, arXiv:1006.2540
59. Carusotto, I., Balbinot, R., Fabbri, A., Recati, A.: Density correlations and analog dynamical Casimir emission of Bogoliubov phonons in modulated atomic Bose–Einstein condensates. Eur. Phys. J. D. **56**, 391–404 (2010)
60. Roberts, D., Pomeau, Y.: Casimir-like force arising from quantum fluctuations in a slow-moving dilute Bose–Einstein condensate. Phys. Rev. Lett. **95**, 145303 (2005)
61. Jaekel, M.-T., Reynaud, S.: Movement and fluctuations of the vacuum. Rep. Prog. Phys. **60**, 863–887 (1997)
62. Kardar, M., Golestanian, R.: The "friction" of vacuum, and other fluctuation-induced forces. Rev. Mod. Phys. **71**, 1233–1245 (1999)
63. Dodonov, V.V.: Nonstationary Casimir effect and analytical solutions for quantum fields in cavities with moving boundaries. Adv. Chem. Phys. **119**, 309–394 (2001)

64. Dodonov, V.V.: Dynamical Casimir effect: Some theoretical aspects. J. Phys.: Conf. Ser. **161**, 012027 (2009)
65. Dodonov, V. V.: Current status of the dynamical Casimir effect. arXiv:1004.3301 (2010)
66. Fosco, C.D, Lombardo, F.C., Mazzitelli, F.D.: Quantum dissipative effects in moving mirrors: a functional approach. Phys. Rev. D. **76**, 085007 (2007)
67. Barton, G., Eberlein, C.: On quantum radiation from a moving body with finite refractive-index. Ann. Phys. (New York) **227**, 222–274 (1993)
68. Alves, D.T., Farina, C., Maia Neto, P.A.: Dynamical Casimir effect with Dirichlet and Neumann boundary conditions. J. Phys. A: Math. Gen. **36**, 11333–11342 (2003)
69. Alves, D.T., Granhen, E.R., Lima, M.G.: Quantum radiation force on a moving mirror with Dirichlet and Neumann boundary conditions for a vacuum, finite temperature, and a coherent state. Phys. Rev. D. **77**, 125001 (2008)
70. Mintz, B., Farina, C., Maia Neto, P.A., Rodrigues, R.: Casimir forces for moving boundaries with Robin conditions. J. Phys. A: Math. Gen. **39**, 6559–6565 (2006)
71. Dodonov, V.V., Klimov, A.B., Man'ko, V.I.: Generation of squeezed states in a resonator with a moving wall. Phys. Lett. A. **149**, 225–228 (1990)
72. Dodonov, V.V., Klimov, A.B.: Long-time asymptotics of a quantized electromagnetic-field in a resonator with oscillating boundary. Phys. Lett. A. **167**, 309–313 (1992)
73. Mintz, B., Farina, C., Maia Neto, P.A., Rodrigues, R.: Particle creation by a moving boundary with a Robin boundary condition. J. Phys. A: Math. Gen. **39**, 11325–11333 (2006)
74. Jaekel, M.-T., Reynaud, S.: Causality, stability and passivity for a mirror in vacuum. Phys. Lett. A. **167**, 227–232 (1992)
75. Barton, G., Calogeracos, A.: On the quantum electrodynamics of a dispersive mirror. 1: Mass shifts, radiation, and radiative reaction. Ann. Phys. (New York) **238**, 227–267 (1995)
76. Verlot, P., Tavernarakis, A., Briant, T., Cohadon, P.-.F., Heidmann, A.: Scheme to probe optomechanical correlations between two optical beams down to the quantum level. Phys. Rev. Lett. **102**, 103601 (2009)
77. Schliesser, A., Arcizet, O., Riviere, R., Anetsberger, G., Kippenberg, T.J.: Resolved-sideband cooling and position measurement of a micromechanical oscillator close to the Heisenberg uncertainty limit. Nature Phys. **5**, 509–514 (2009)
78. Maia Neto, P.A.: Vacuum radiation pressure on moving mirrors. J. Phys. A: Math. Gen. **27**, 2167–2180 (1994)
79. Barton, G., North, C.A.: Peculiarities of Quantum Radiation in Three Dimensions from Moving Mirrors with High Refractive Index. Ann. Phys. (New York) **252**, 72–114 (1996)
80. Gütig, R., Eberlein, C.: Quantum radiation from moving dielectrics in two, three and more spatial dimensions. J. Phys. A: Math. Gen. **31**, 6819–6838 (1998)
81. Barton, G.: The quantum radiation from mirrors moving sideways. Ann. Phys. (New York) **245**, 361–388 (1996)
82. Pendry, J.B.: Quantum friction—fact or fiction?. New J. Phys. **12**, 033028 (2010)
83. Volotikin, A.I., Persson, B.N.J.: Theory of friction: the contribution from a fluctuating electromagnetic field. J. Phys.:Condens. Matter. **11**, 345–359 (1999)
84. Lifshitz, E.M.: The theory of molecular attractive forces between solids. Sov. Phys. JETP. **2**, 73–83 (1956)
85. Buhmann, S.Y, Welsch, D.-G.: Dispersion forces in macroscopic quantum electrodynamics. Progr. Quantum Electron. **31**, 51–130 (2007)
86. Dedkov, G.V., Kyasov, A.A.: Electromagnetic and fluctuation-electromagnetic forces of interaction of moving particles and nanoprobes with surfaces: a non-relativistic consideration. Phys. Solid State. **44**, 1809–1832 (2002)
87. Hu, B.L., Roura, A., Shresta, S.: Vacuum fluctuations and moving atom/detectors: from the Casimir-Polder to the Unruh-Davies-DeWitt-Fulling effect. J. Opt. B: Quantum Semiclass. Opt. **6**, S698–S705 (2004)
88. Scheel, S., Buhmann, S.Y.: Casimir-Polder forces on moving atoms. Phys. Rev. A. **80**, 042902 (2009)

89. Dodonov, V.V., Klimov, A.B., Nikonov, D.E.: Quantum phenomena in resonators with moving walls. J. Math. Phys. **34**, 2742 (1993)

90. Petrov, N.P.: The dynamical Casimir effect in a periodically changing domain: a dynamical systems approach. J. Opt B: Quant. Semiclass. Optics. **7**, S89–S99 (2005)

91. Ruser, M.: Vibrating cavities: a numerical approach. J. Opt. B: Quant. Semiclass. Optics. **7**, S100–S115 (2005)

92. Alves, D.T., Farina, C., Granhen, E.R.: Dynamical Casimir effect in a resonant cavity with mixed boundary conditions. Phys. Rev. A. **73**, 063818 (2006)

93. Farina, C., Azevedo, D., Pascoal, F.: Dynamical Casimir effect with Robin boundary conditions in a three dimensional open cavity. In: Milton, K.A., Bordag, M. (eds.) Proceedings of QFEXT09, p. 334. World Scientific, Singapore (2010). arXiv:1001.2530

94. Law, C.K.: Resonance response of the quantum vacuum to an oscillating boundary. Phys. Rev. Lett. **73**, 1931–1934 (1994)

95. Schützhold, R., Plunien, G., Soff, G.: Trembling cavities in the canonical approach. Phys.Rev. A. **57**, 2311–2318 (1998)

96. Bender, C.M., Orszag, S.A.: Advanced Mathematical Methods for Scientists and Engineers. McGraw Hill, New York (1978)

97. Boyd, R.: Nonlinear Optics. 3rd edn. Academic Press, Burlington USA (2008)

98. Ji, J.-Y., Soh, K.-S., Cai, R.-G., Kim, S.P.: Electromagnetic fields in a three-dimensional cavity and in a waveguide with oscillating walls. J. Phys. A. **31**, L457–L462 (1998)

99. Dodonov, V.V.: Resonance excitation and cooling of electromagnetic modes in a cavity with an oscillating wall. Phys. Lett. A. **213**, 219–225 (1996)

100. Ruser, M.: Numerical investigation of photon creation in a three-dimensional resonantly vibrating cavity: Transverse electric modes. Phys. Rev. A. **73**, 043811 (2006)

101. Hacyan, S., Jauregui, R., Soto, F., Villarreal, C.: Spectrum of electromagnetic fluctuations in the Casimir effect. J. Phys. A: Math. Gen. **23**, 2401 (1990)

102. Dodonov, V.V., Dodonov, A.V.: The nonstationary Casimir effect in a cavity with periodical time-dependent conductivity of a semiconductor mirror. J. Phys. A: Math. Gen. **39**, 6271–6281 (2006)

103. Uhlmann, M., Plunien, G., Schützhold, R., Soff, G.: Resonant cavity photon creation via the dynamical Casimir effect. Phys.Rev. Lett. **93**, 193601 (2004)

104. Naylor, W., Matsuki, S., Nishimura, T., Kido, Y.: Dynamical Casimir effect for TE and TM modes in a resonant cavity bisected by a plasma sheet. Phys. Rev. A. **80**, 043835 (2009)

105. Dodonov, V.V.: Photon distribution in the dynamical Casimir effect with an account of dissipation. Phys. Rev. A. **80**, 023814 (2009)

106. Lax, M.: Quantum noise. IV. Quantum theory of noise sources. Phys. Rev. **145**, 110–129 (1966)

107. Mendonça, J.T., Brodin, G., Marklund, M.: Vacuum effects in a vibrating cavity: time refraction, dynamical Casimir effect, and effective Unruh acceleration. Phys. Lett. A. **372**, 5621–5624 (2008)

108. Arbet-Engels, V., Benvenuti, C., Calatroni, S., Darriulat, P., Peck, M.A., Valente, A.M., Van't Hof, C.A.: Superconducting niobium cavities, a case for the film technology. Nucl. Instrum. Methods Phys. Res. A. **463**, 1–8 (2001)

109. Agnesi, A., Braggio, C., Bressi, G., Carugno, G., Galeazzi, G., Pirzio, F., Reali, G., Ruoso, G., Zanello, D.: MIR status report: an experiment for the measurement of the dynamical Casimir effect. J. Phys. A: Math. Gen. **41**, 164024 (2008)

110. Segev, E., Abdo, B., Shtempluck, O., Buks, E., Yurke, B.: Prospects of employing superconducting stripline resonators for studying the dynamical Casimir effect experimentally. Phys. Lett. A. **370**, 202–206 (2007)